Remote Sensing for Target Object Detection and Identification

Remote Sensing for Target Object Detection and Identification

Special Issue Editors

Gemine Vivone
Paolo Addesso
Amanda Ziemann

MDPI • Basel • Beijing • Wuhan • Barcelona • Belgrade • Manchester • Tokyo • Cluj • Tianjin

Special Issue Editors

Gemine Vivone
IMAA, CNR
Italy

Paolo Addesso
DIEM, University of Salerno
Italy

Amanda Ziemann
Los Alamos National Laboratory
USA

Editorial Office
MDPI
St. Alban-Anlage 66
4052 Basel, Switzerland

This is a reprint of articles from the Special Issue published online in the open access journal *Remote Sensing* (ISSN 2072-4292) (available at: https://www.mdpi.com/journal/remotesensing/special_issues/target_detection).

For citation purposes, cite each article independently as indicated on the article page online and as indicated below:

LastName, A.A.; LastName, B.B.; LastName, C.C. Article Title. *Journal Name* **Year**, *Article Number*, Page Range.

ISBN 978-3-03928-332-3 (Pbk)
ISBN 978-3-03928-333-0 (PDF)

Contents

About the Special Issue Editors

Gemine Vivone received a B.S. (summa cum laude), M.S. (summa cum laude), and Ph.D. (highest rank) degrees in information engineering from the University of Salerno, Salerno, Italy, in 2008, 2011, and 2014, respectively. He is a researcher at the National Research Council (Italy). His main research interests focus on statistical signal processing, detection of remotely sensed images, data fusion, and tracking algorithms. Dr. Vivone serves as a referee for several remote sensing and image processing journals. Dr. Vivone was the Lead Guest Associate Editor of a Special Stream for *IEEE Geoscience and Remote Sensing Letters,* he was the Guest Associate Editor of a Special Issue for MDPI *Remote Sensing* and a Co-Editor of a Special Issue for *International Journal of Image and Data Fusion.* Dr. Vivone is currently an Associate Editor for *IEEE Geoscience and Remote Sensing Letters* (GRSL) and he is an Editorial Board Member for MDPI *Remote Sensing.* Dr. Vivone received the Symposium Best Paper Award at the IEEE International Geoscience and Remote Sensing Symposium (IGARSS) in 2015 and the Best Reviewer Award of the IEEE Transactions on Geoscience and Remote Sensing in 2017.

Paolo Addesso (assistant professor) received a Laurea degree (cum laude) in electronic engineering and a Ph.D. degree in information engineering from the University of Salerno, Fisciano, Italy, in 2000 and 2005, respectively. He is currently an assistant professor with the University of Salerno. His research interests include remote sensing, detection and estimation theory, signal processing, and gravitational waves data analysis.

Amanda Ziemann (scientist) received B.S. and M.S. degrees in applied mathematics from the Rochester Institute of Technology (RIT) in 2010 and 2011, respectively, as well as a Ph.D. in imaging science from RIT in 2015. She was an Agnew National Security Postdoctoral Fellow at Los Alamos National Laboratory (LANL), a United States Department of Energy national laboratory, and is currently a staff scientist in the Space Data Science and Systems Group at LANL. She is a referee for several international journals. Her research interests include remote sensing, spectral imaging, signal detection, and data fusion.

Editorial

Editorial for Special Issue "Remote Sensing for Target Object Detection and Identification"

Gemine Vivone [1,2,*] , **Paolo Addesso** [2] **and Amanda Ziemann** [3]

1   IMAA, CNR, 85050 Tito, Italy
2   DIEM, University of Salerno, 84084 Fisciano, Italy; paddesso@unisa.it
3   Los Alamos National Laboratory, Los Alamos, NM 87544, USA; ziemann@lanl.gov
*   Correspondence: gvivone@unisa.it

Received: 27 December 2019; Accepted: 2 January 2020; Published: 6 January 2020

Abstract: This special issue gathers fourteen papers focused on the application of a variety of target object detection and identification techniques for remotely-sensed data. These data are acquired by different types of sensors (both passive and active) and are located on various platforms, ranging from satellites to unmanned aerial vehicles. This editorial provides an overview of the contributed papers, briefly presenting the technologies and algorithms employed as well as the related applications.

Keywords: target detection; target identification; SAR; visible; infrared; hyperspectral

---

Target object detection and identification is among the primary uses for a remote sensing system. It is of paramount importance in several fields, including environmental and urban monitoring, hazard and disaster management, and defense and military applications. In recent years, these analyses have made use of the tremendous amount of data acquired by sensors mounted on satellite, airborne, and unmanned aerial vehicle (UAV) platforms.

The papers included in this special issue exploit different remote sensing phenomenologies for target object detection and identification; this includes synthetic aperture radar (SAR) imaging, which uses active sensors operating in the microwave domain, and multispectral and hyperspectral imaging, which uses passive sensors that typically capture visible and/or infrared radiation. The selection of one particular technology depends on both the specific application and the desired signal detection technique. As such, these aspects will be highlighted when summarizing the aforementioned papers.

Data acquired by SAR sensors are used in two papers [1,2], both focusing on environmental and hazard monitoring. In particular, Biondi et al. [1] present a robust procedure to evaluate water flow elevation by using SAR data (e.g., using COSMO-SkyMed images). By tracking the double-bounce reflections from a bridge crossing a river over time, it is possible to estimate the distance between the river surface and the bridge and, consequently, the water flow level. The paper by Liu et al. [2] is focused on the assessment of flood hazard for power grids using SAR data, where the aim is to assess the safety of the transmission line towers. This is performed by identifying indicators such as the shortest distance from a tower to a flood, the proportion of flood in a search area, and the difference in elevation between the tower base and the flood level.

Another group of papers [3–9] proposes object detection and recognition approaches that use images (or videos) acquired in the visible and near-infrared (VNIR) wavelength range, making use of the high (or very high) spatial resolution and high spectral content. Indeed, the latter are key features in order to identify shapes, thus enabling more reliable object detection and recognition. The use of convolutional neural networks (CNNs) in particular is becoming more widespread, as demonstrated in the work of Zhang et al. [3]. In this paper, a region-based object detection is performed, relying on the so-called feature pyramid network (FPN) which combines high and low resolution features without any additional memory consumption. Alternatively, Ma et al. [4] employ CNNs to perform a stable and robust multi-model decision fusion, which jointly uses contextual features and object spatial structure information. Another interesting application of CNNs is described in Zhang et al. [5], in which vehicle detection for traffic monitoring systems is performed using satellite video data. In contrast, Li et al. [6] focus their work on the design of a parallel hardware architecture, based on multiple neural processing units (NPUs), for performing a power-efficient object detection by using CNNs. Liu et al. [7] explore alternative frameworks to CNNs with the aim of avoiding time-consuming training phases. Specifically, in [7], the authors exploit an unsupervised saliency detection method aimed at the identification of oil tanks when the images are affected by various disturbance factors, such as different colors and shadows (caused by changes in view angles and illumination conditions). The problem of vehicle detection is also addressed by Cao et al. in [8], where the authors present a new object matching framework based on affine-function transformations by using images acquired by UAVs (i.e., the DJI Phantom 4 Pro). Finally, Yang et al. [9] perform anomaly detection by using high spectral resolution hyperspectral data from visible to infrared wavelengths. The anomalies are caused by rare and sparse small objects whose spectra are significantly different from the background. In order to deal with the high dimensionality of the problem and to reduce the computational burden, an approach based on low-rank representations is presented.

The third group of papers [10–14] focuses on object detection using infrared sensors. Zhang et al. [10] propose a method based on a low rank sparse decomposition that uses a non-convex optimization with an Lp-norm constraint in order to identify small targets in sequences of infrared images. In this paper, an efficient solver based on the alternating direction method of multipliers (ADMM) is presented. The detection of small targets by using infrared radiation is also the main topic of the contribution by Zhang et al. [11], in which a low-rank-based method with a regularization term based on the nuclear norm is proposed. This approach is able to properly solve the tensor robust principal component analysis (TRPCA) problem which models the separation of targets from the background. Again, ADMM is employed to provide a computationally efficient solver. Sun et al. [12] address the infrared small target detection problem using a noise model based on a non-independent identically distributed mixture of Gaussians, which is able to deal with real and complex scenarios in which the noise can change in different frames of a sequence of infrared images. The final target identification paper is focused on a flux density-based algorithm, which is able to identify the different infrared gradient vector fields between target and noise. Li et al. [13] present a thermal infrared (TIR) target tracking algorithm based on semantic features. Specifically, a mask sparse representation is used to distinguish the reliable pixels (for target tracking) from the unreliable ones in each TIR frame. The last step uses this model to improve a particle filter-based approach for TIR target tracking. The final paper is authored by Niu et al. [14], and in this paper, the authors present a study about the observability of an Earth entry orbital test vehicle (OTV) via ground-based infrared sensors. The physical foundation of this work relies on the high-temperature flow field that originates during the entry phase of an OTV. A suitable physical model is developed in order to simulate the infrared signature of the Earth entry OTV, which is useful in computing the so-called maximum detecting range, and is more broadly useful in designing remote ground-based detection systems.

Acknowledgments: We are grateful for having the opportunity to lead this special issue. We would like to thank the journal editorial team and reviewers for conducting the review process, and all of the authors for their submissions.

Conflicts of Interest: The authors declare no conflict of interest.

References

1. Biondi, F.; Tarpanelli, A.; Addabbo, P.; Clemente, C.; Orlando, D. Pixel Tracking to Estimate Rivers Water Flow Elevation Using Cosmo-SkyMed Synthetic Aperture Radar Data. *Remote Sens.* **2019**, *11*, 2574. [CrossRef]
2. Liu, L.; Du, R.; Liu, W. Flood Distance Algorithms and Fault Hidden Danger Recognition for Transmission Line Towers Based on SAR Images. *Remote Sens.* **2019**, *11*, 1642. [CrossRef]

3. Zhang, X.; Zhu, K.; Chen, G.; Tan, X.; Zhang, L.; Dai, F.; Liao, P.; Gong, Y. Geospatial Object Detection on High Resolution Remote Sensing Imagery Based on Double Multi-Scale Feature Pyramid Network. *Remote Sens.* **2019**, *11*, 755. [CrossRef]

4. Ma, W.; Guo, Q.; Wu, Y.; Zhao, W.; Zhang, X.; Jiao, L. A Novel Multi-Model Decision Fusion Network for Object Detection in Remote Sensing Images. *Remote Sens.* **2019**, *11*, 737. [CrossRef]

5. Zhang, J.; Jia, X.; Hu, J. Local Region Proposing for Frame-Based Vehicle Detection in Satellite Videos. *Remote Sens.* **2019**, *11*, 2372. [CrossRef]

6. Li, L.; Zhang, S.; Wu, J. Efficient Object Detection Framework and Hardware Architecture for Remote Sensing Images. *Remote Sens.* **2019**, *11*, 2376. [CrossRef]

7. Liu, Z.; Zhao, D.; Shi, Z.; Jiang, Z. Unsupervised Saliency Model with Color Markov Chain for Oil Tank Detection. *Remote Sens.* **2019**, *11*, 1089. [CrossRef]

8. Cao, S.; Yu, Y.; Guan, H.; Peng, D.; Yan, W. Affine-Function Transformation-Based Object Matching for Vehicle Detection from Unmanned Aerial Vehicle Imagery. *Remote Sens.* **2019**, *11*, 1708. [CrossRef]

9. Yang, Y.; Zhang, J.; Song, S.; Liu, D. Hyperspectral Anomaly Detection via Dictionary Construction-Based Low-Rank Representation and Adaptive Weighting. *Remote Sens.* **2019**, *11*, 192. [CrossRef]

10. Zhang, T.; Wu, H.; Liu, Y.; Peng, L.; Yang, C.; Peng, Z. Infrared Small Target Detection Based on Non-Convex Optimization with Lp-Norm Constraint. *Remote Sens.* **2019**, *11*, 559. [CrossRef]

11. Zhang, L.; Peng, Z. Infrared Small Target Detection Based on Partial Sum of the Tensor Nuclear Norm. *Remote Sens.* **2019**, *11*, 382. [CrossRef]

12. Sun, Y.; Yang, J.; Li, M.; An, W. Infrared Small-Faint Target Detection Using Non-i.i.d. Mixture of Gaussians and Flux Density. *Remote Sens.* **2019**, *11*, 2831. [CrossRef]

13. Li, M.; Peng, L.; Chen, Y.; Huang, S.; Qin, F.; Peng, Z. Mask Sparse Representation Based on Semantic Features for Thermal Infrared Target Tracking. *Remote Sens.* **2019**, *11*, 1967. [CrossRef]

14. Niu, Q.; Meng, X.; He, Z.; Dong, S. Infrared Optical Observability of an Earth Entry Orbital Test Vehicle Using Ground-Based Remote Sensors. *Remote Sens.* **2019**, *11*, 2404. [CrossRef]

Article

Pixel Tracking to Estimate Rivers Water Flow Elevation Using Cosmo-SkyMed Synthetic Aperture Radar Data

Filippo Biondi [1,*,†], **Angelica Tarpanelli** [2], **Pia Addabbo** [3], **Carmine Clemente** [4] **and Danilo Orlando** [5]

[1] Electromagnetic Laboratory, Engineering Faculty, Università degli Studi dell'Aquila, Piazzale E. Pontieri, 67100 Monteluco di Roio, L'Aquila AQ, Italy
[2] Consiglio Nazionale delle Ricerche (CNR), Istituto di ricerca per la Protezione Idrogeologica (IRPI), via Madonna Alta, 126, 06128 Perugia, Italy; angelica.tarpanelli@irpi.cnr.it
[3] Science and Technology for Transportations Faculty, Università degli Studi "Giustino Fortunato", viale Raffale Delcogliano, 12, 82100 Benevento, Italy; p.addabbo@unifortunato.eu
[4] Center for Signal and Image Processing, Department of Electronic and Electrical Engineering, University of Strathclyde, Glasgow G1 1XW, UK; carmine.clemente@strath.ac.uk
[5] Engineering Faculty, Università degli Studi "Niccolò Cusano", Via Don Carlo Gnocchi, 3, 00166 Roma, Italy; danilo.orlando@unicusano.it
* Correspondence: filippo.biondi@marina.difesa.it; Tel.: +39-335-833-4216
† Current address: Via Luca Benincasa 21/B 06073 Corciano Perugia, Italy.

Received: 21 September 2019; Accepted: 30 October 2019; Published: 2 November 2019

Abstract: The lack of availability of historical and reliable river water level information is an issue that can be overcome through the exploitation of modern satellite remote sensing systems. This research has the objective of contributing in solving the information-gap problem of river flow monitoring through a synthetic aperture radar (SAR) signal processing technique that has the capability to perform water flow elevation estimation. This paper proposes the application of a new method for the design of a robust procedure to track over the time double-bounce reflections from bridges crossing rivers to measure the gap space existing between the river surface and bridges. Specifically, the difference in position between the single and double bounce is suitably measured over the time. Simulated and satellite temporal series of SAR data from COSMO-SkyMed data are compared to the ground measurements recorded for three gauges sites over the Po and Tiber Rivers, Italy. The obtained performance indices confirm the effectiveness of the method in the estimation of water level also in narrow or ungauged rivers.

Keywords: synthetic aperture radar (SAR); rivers water-flow elevation estimation; pixel-tracking; phase unwrapping

---

1. Introduction

The flow of water in rivers and streams is of great interest because it represents the easiest access to water, a fundamental natural resource for human beings, animals and cultivations. The ability to quantify the flow of water in terms of river discharge and flow volume depends on the monitoring of water surface height of water bodies [1]. In-situ gauges have been installed along rivers to measure the water height and to describe its variation in the space and time. Unfortunately, the ground network is not uniformly distributed worldwide (many rivers in developing countries are still unmonitored) and since the 1980s, we have also been assisting the decline of gauge stations in the developed countries [2]. For this reason, the advanced capability of satellite sensors to monitor inland water and the direct access to their data motivated scientists to integrate and reinforce the traditional monitoring of surface water with this new source of information.

River discharge estimation from satellite remote sensing of river hydraulic variables has been investigated in recent decades [3–6]. Nadir altimetry has been largely used for measuring the river surface height from space [7,8]. From the first studies carried out with Geosat [7] to the more recent analyses with Jason-2 and SARAL [8,9], the improved capability of the altimeters allowed to monitor even narrow rivers, for example, the Po river (~300 m wide) or the Garonne River (~200 m wide). Despite these encouraging results, the use of radar altimetry for narrow rivers is still limited because of uncertainty in the evaluation of the water surface elevation due to the local topography that contaminates the returned radar signal. With the synthetic aperture radar (SAR) technology applied to altimeters, reliable measurements of water level are obtained for rivers of 200 m width [10] and according to the requirements of the next SWOT mission, rivers of 100 m width will be accurately observed by the new Karin sensors [11]. However, the nominal orbit of the satellites often does not guarantee the global coverage of all narrow rivers. The designed inter-track diamond distances (when the ground tracks of the low Earth orbits satellites, traveling on ascending and descending directions, are depicted, geometric figures in the shape of diamonds are obtained) and the revisit time represent obstacles to the monitoring of water courses. To overcome these issues, low cost satellite constellations are investigated in order to provide global coverage and finer temporal resolution. Moreover, the use of high resolution measurements is fundamental to ensure the level information of small water bodies. In the literature, some examples show the use of SAR images to derive information about the water level. For example, Reference [12] showed the use of SAR images from ENVISAT and RADARSAT to indirectly estimate water level of the Severn River (UK) and the Red River (US). Other research has investigated the use of Along Track Interferometry from SAR (ATI-SAR) to obtain water level estimation. However, in order to obtain two interferometric InSAR images with a short time delay from a moving platform, it is necessary to install two antennas separated by the corresponding spatial baseline oriented along the flight direction. Accordingly, the technique is called ATI, which is different if compared to the cross-track interferometry (XTI) used for topographic reconstruction. ATI can be suitably exploited to estimate the surface velocity of water masses and classical dual-sensor ATI geometric configuration was first proposed in Reference [13], in which the authors describe a new method to measure sea surface currents. The experimental results refer to an airborne implementation of the technique, tested over the San Francisco Bay near the time of maximum tidal flow and leading to a map of the east-west component of the water current. This study also underlines that only the line-of-sight (LOS) component of the targets velocities is measured by ATI. In Reference [14], the authors investigated for the first time the application of ATI for the estimation of ocean currents. Data have been acquired by the shuttle radar topography mission (SRTM), which used an auxiliary antenna yielding a baseline aligned with the azimuth direction of 7 m. Unfortunately, the ATI configuration requires the use of multiple receivers and it is clear that, in order to obtain high sensitivity in estimating low-rate velocities, it is necessary to design relatively long baseline ATI geometries. However, there are no satellite systems configured in this way and therefore it is necessary to identify other solutions.

Additionally, a common issue for techniques based on interferometric SAR is the dependency of the results on the quality of the image pair registration. Interferogram formation requires images to be co-registered with an accuracy finer than a few tenths of a resolution cell to avoid significant loss of phase coherence. For InSAR products co-registration, a 2-D polynomial of low order is usually chosen as a warp function, and the polynomial parameters are estimated through least squares fit from the shifts measured on image windows [15]. A direct consequence of accurate co-registration is the unlocking of the use of pixel tracking techniques. The use of pixel-tracking has proved to be effective for the precise space offset measurements of pixels located on co-registered sets of SAR images observing the same scenes by long temporal series. To this end, it is possible to process a couple of along-track interferometry single-look complex (SLC) SAR images observing the same scene. Specifically, the images must be formed in a short time interval, varying from some milliseconds to a few seconds. The phase differences between all range-azimuth resolution cells composing the two images are proportional to the Doppler shift of the backscattered signal. This technique has shown good potential for applications such as monitoring glacier movements, volcanic activities and co-seismic tears in the solid earth resulting from severe earthquakes, addressing some of the limitations of conventional differential InSAR (DInSAR) techniques, particularly their sparse coverage and the impact of highly vegetated areas [16]. Similar techniques have also been applied in low resolution SAR imagery, measuring large Earth deformations [17,18] and recently it was found that pixel tracking is also suitable for micro-Doppler estimation of maritime targets [19,20].

In this paper, we introduce a pixel tracking technique based on the localized spectral analysis with the objective to track in time the double-bounce scattering effect in order to measure the height of rivers. To achieve this objective, the proposed technique exploits the measurement of the distance between bridges or other man-made objects on the embankments. A necessary condition is that the structures are perpendicular to the slant-range direction. The height estimation is obtained by measuring the cross-slant-range distance between the edge of the structure and the echo of second bounce reflected by the river surface. This physical phenomenon is tracked over time in order to observe the trend of variation of the water level. The applicability of this method is supported by the fact that the double bounce echoes can be easily detected thanks to the shading effect generated by the bridge infrastructure. This echo is detected in the range cells immediately contiguous to those containing the bridge. The time tracking of the double-bounce echo shift is possible because in the instants of radar observation the river surface is operating as a mirror at variable distance. This distance is proportional to the height variation of the river's water surface. The echo space-time shift is due to the particular geometric SAR configuration where the images are observed in slant coordinates. In order to proceed with the estimation of the hydrometric levels it is therefore essential to develop a reliable tracking algorithm. Unfortunately, the variation in pixels of the slant-range coordinates is very small, because of the short distance existing between the bridge and the water surface. In order to obtain robust measurements, it is necessary to design an efficient pixel dilation stage, to be performed before the tracking algorithm [21]. The absolute shifts are derived after phase unwrapping [22]. In this paper the proposed method is assessed by estimating water surface elevation for narrow-medium rivers (50–300 m) using temporal series of COSMO-SkyMed (CSK) data. To the best of the author's knowledge, this is the first attempt to use SAR images directly to estimate river water level.

The remainder of the paper is organized as follows—details of the signal processing techniques are described in the following section. In Section 3 results of a set of simulated data are presented whereas in Section 4 the experimental results using real data (CSK) are reported. Section 5 provides a discussion about the experimental results with the help of performance indices while Section 6 concludes the paper.

2. Rivers Water Flow Elevation Retrieval

In this section, the proposed methodology for the retrieval of rivers water flow elevation is described in depth. We will discuss the geometry of observation in Section 2.1 and the workflow of the algorithm in Section 2.2.

2.1. The Observation Geometry

The geometry of the proposed electromagnetic measurement system is depicted in Figure 1. Precisely, Figure 1a shows the scheme of an existing river on the SAR representation plane, in the range-azimuth reference system. The water of the river is represented flowing along the range direction and from left to right (this is represented by the red arrow visible on the right side of the figure). A bridge is present in the middle of the figure and crossing the azimuth dimension. Figure 1b represents what is shown in Figure 1a but in the height-range coordinates. It is clear that the Ref point, which is the focused single-bounce backscattered echo, is constituted by the front-edge of the bridge that is visible on the Single Look Complex (SLC) image. Its position is constant during the entire time series. The positions of points P1, P2 and P3 on the SLC projection screen, represents the double-bounce backscattered echoes of the same target. The distance of the two-way propagation system depends on the heights of the water levels L1, L2 and L3, which are variable with respect to the time.

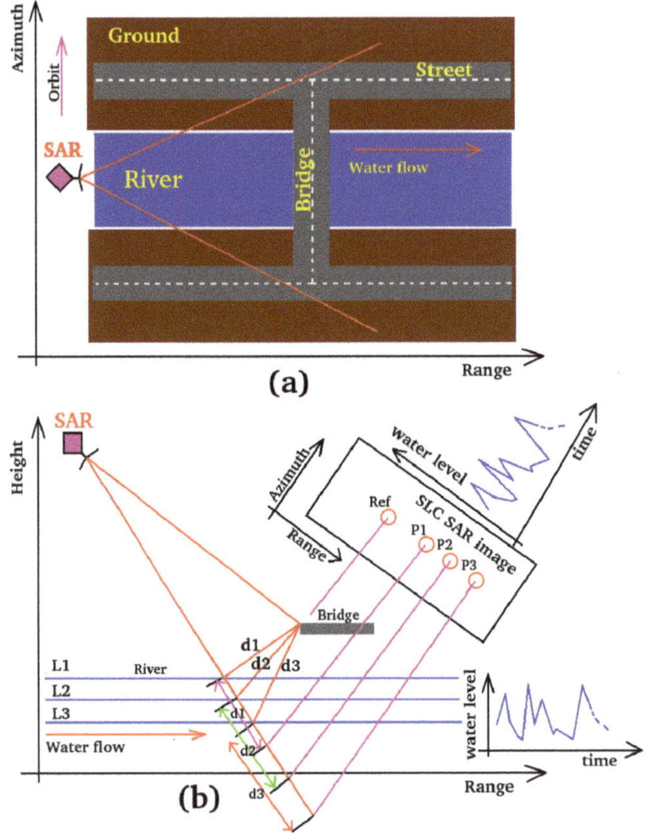

Figure 1. Observation geometry for the river water flow elevation estimation: (**a**) Range-azimuth representation (**b**); Range-height representation.

2.2. The Processing Scheme

The workflow of the proposed estimation technique is depicted in Figure 2 and comprises 5 main processing blocks. The main rationale of the proposed procedure is based on the pixel tracking. Sub-pixel offset tracking (SPOT) is a relevant technique to measure large-scale ground displacements in both range and azimuth directions. The technique is complementary to differential interferometric SAR and persistent scatterers interferometry when the radar phase information is unstable [23,24]. In this paper, we apply the SPOT in an alternative way, as explained in the following.

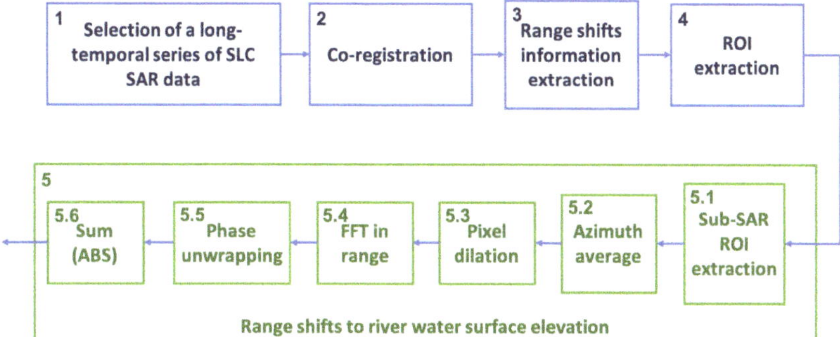

Figure 2. Flow chart of the main procedure to extract water level information from a temporal series of synthetic aperture radar (SAR) data.

The starting point of this algorithm is a long temporal series of InSAR data but, instead of investigating the deformations of the ground, we investigated the movements of the double-bounce scattering effect of man made structures localized on the water surface. Although the variations in the water levels of rivers vary over time much faster with respect to the movements of the ground, we also apply the SPOT technique to trace their hydrometric levels. To this end, it is necessary to design a specific mathematical model that is described in detail below. Indeed, as shown in Figure 2, the starting point (block 1) is the selection of a long temporal series of SAR data, which is based on the desired temporal observation period. This series is, then, processed using images pairs in order to track the hydrometric levels variation over the time. Specifically, the magnitude at the output of the two-dimensional matched filter of the receiver chain can be expressed as the following matrix (for simplicity, we assume that target position in the range-azimuth plane is $(0,0)$) [25] [Chapter 4], for each interferometric complex pair, indexed by i and belonging to a time series of length G:

$$r_{c,D}^{i}(n_c, n_D) = A \frac{\sin(n_c \delta_{R_c}/B_c)}{(n_c \delta_{R_c}/B_c)} \times \frac{\sin(n_D \delta_{R_D}/B_D)}{(n_D \delta_{R_D}/B_D)}, i = 1, \ldots, G$$

$$n_c = -N_c/2, \ldots, 0, \ldots, N_c/2,$$

$$n_D = -N_D/2, \ldots, 0, \ldots, N_D/2, \tag{1}$$

$$N_c, N_D \in \mathbb{N} \text{ (even)},$$

where:

- A is the backscattering coefficient;
- N_c is the number of pixels of the image along the range;
- N_D is the number of pixels of the image along the azimuth;
- δ_{R_c} is the chirp resolution;
- δ_{R_D} is the Doppler resolution;
- n_c is the chirp wavenumber;
- n_D is the Doppler wavenumber;
- B_c is the bandwidth of the transmitted chirp signal;
- B_D is the synthesized Doppler band.

Such data will necessarily have to be co-registered (the coregistration process consists of perfectly aligning the pixels of any slave image to the corresponding pixels of the master image. The alignment process is very precise and can also be accurately performed at the sub-pixel level). The co-registration stage is performed as the second stage of the processing chain. After co-registration, the stage 3 exploits the range shifts for an initial coarse estimation of the double-bounce shifts and error correction. To give a sketch of these steps, let consider the offset components of the sub-pixel normalized cross-correlation, that according to References [15,26] are described by the complex parameter $D^i_{tot_{(c,D)}}$ referred to as total displacement, which is given by:

$$D^i_{tot_{(c,D)}} = D^i_{displ_{(c,D)}} + D^i_{topo_{(c,D)}}$$
$$+ D^i_{orbit_{(c,D)}} + D^i_{control_{(c,D)}} + D^i_{atmosphere_{(c,D)}} + D^i_{noise_{(c,D)}}, i = 1, \ldots, G, \tag{2}$$

where:

- $D^i_{displ_{(c,D)}}$ is the offset component of the signal position presented in (1), generated by the variation of the river water level and detected as a sub-pixel misalignment existing between the first SAR image (master) and the i-th slave SAR image;
- $D^i_{topo_{(c,D)}}$ is the offset component generated by the earth displacement when located on highly sloped terrain;
- $D^i_{orbit_{(c,D)}}$ is the offset caused by residual errors of the satellite orbits;
- $D^i_{control_{(c,D)}}$ is the offset component generated by general attitude and control errors of the flying satellite trajectory;
- $D^i_{atmosphere_{(c,D)}}$ and $D^i_{noise_{(c,D)}}$ are the contributions accounting for change in the atmospheric and ionospheric dielectric constant and for decorrelation phenomena (spatial, temporal, thermal, etc.), respectively.

Note that the above equation accounts for the general case where displacement exists in both range and azimuth dimensions. In the operating scenario considered here, the displacement generated by the double-bounce scattering component of the bridge, when perturbed by the temporal variations of the river water levels, is significantly greater with respect to any other sporadic displacements due to physical phenomena. As depicted in Figure 1, this phenomenon is maximum when the longitudinal axis of the bridge is observed perpendicular to the range direction. With the above remarks in mind, we can assume that the contribution in the displacement $D^i_{tot_{(c,D)}}$ is greater along the range dimension. Thus, neglecting the azimuth component, we can consider $D^i_{tot_{(c,D)}} = D^i_{tot_c}$.

Figure 3 is a schematic representation of the parameters estimated by the coregistration procedure. The square number one is a focused pixel of the master image and the square number two is the same pixel but located on the slave image. The parameters $D^i_{tot_c}$ and θ^i are the distance between the master and slave pixel centers and the angle respect to the horizontal axis respectively. In the present case, since the shift of the double-bounce scattering occurs in range, the parameter $\theta^i = 0$ so $D^i_{tot_D} = 0$. Moreover, the atmospheric time-variation during the very short acquisition time interval has little influence on the temporal component of the last displacement parameters because of its low accuracy. All errors are compensated for, choosing only high energy and stable points and subtracting the initial offsets in order to retrieve the shifts contributions only generated by the target displacement.

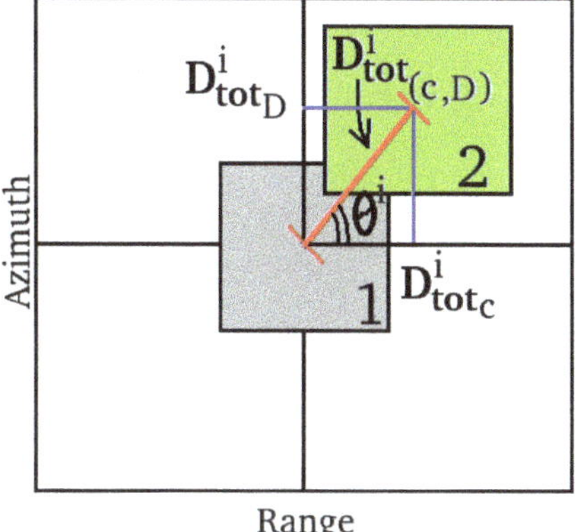

Figure 3. Schematic representation of isolated pixels with a certain shift due to space displacement.

Once all the image pairs are co-registered and $D^i_{tot_c}$ is computed, we proceed with the estimation of the water level on a specific region of interest (ROI). The ROI extraction is performed by the block number 4 and the last computational block (number 5) performs a precise range shifts estimation and water level estimation as detailed in the sub-blocks (5.1–5.6). In particular, the data-input ROI are fed to the block 5.1 that extracts the sub-ROI, containing only the bridge. Processing stage number 5.2 performs the azimuth average of the sub-ROI and computational block 5.3 is designed to perform the pixel dilation by oversampling [21]. Computational block 5.4 performs the one-dimensional DFT (in the range direction) of the sub-ROI average. To this end, we compute

$$DFT(r^i_{c,D}(n_c - D^i_{tot_c}, n_D)) = R^i_{c,D}(k_c, k_D) \exp\left(-j2\pi \frac{k^i_c}{N^i_c} D^i_{tot_c}\right), \tag{3}$$

where

$$R^i_{c,D}(k_c, k_D) = DFT\left(r^i_{c,D}(n_c, n_D)\right)$$
$$= \sum_{n_c=0}^{\tilde{N}_c-1} \sum_{n_D=0}^{\tilde{N}_D-1} \left(\frac{A \sin(n_c \delta_{R_c}/B_c) \sin(n_D \delta_{R_D}/B_D)}{(n_c \delta_{R_c}/B_c)(n_D \delta_{R_D}/B_D)} \exp\left(-j2\pi \frac{k_c}{N_c} n_c\right) \exp\left(-j2\pi \frac{k_D}{N_D} n_D\right) \right), \tag{4}$$

with $\tilde{N}_c \times \tilde{N}_D$ the dimension of the sub-ROI under test and we have clearly exploited the shift theorem of the DFT.

The last two processing stages (blocks number 5.5 and 5.6) convert the phase variations from the output of the previous computational stage into effective water levels of the rivers. To this aim, let us observe that the term $\exp\left(-j2\pi \frac{k_c}{N_c}(D^i_{tot_c})\right)$ represents a rotational vector where the frequency is proportional to the temporal-space shift amount $D^i_{tot_c}$. It follows that the higher the shift of the double-bounce pixel, the higher the oscillation frequency of the phasor. This frequency increasing will generate more turns of the phasor around the circle angle. All these turns, when unwrapped, represent a distance proportional to the height of the river water. As a phase unwrapping algorithm, we select the one-dimensional case of that developed in Reference [22]. Once the absolute phase value is estimated, it is suitably scaled through a constant factor obtained by comparing absolute phase measurement with the true height of the river. This operation is equivalent to an algorithm calibration procedure. Specifically, the river height at the i-th time instant is evaluated as follows (note that $\angle\{R^i_{c,D}(k_c, k_D)\}$ is known)

$$H^i_{water} = A_c\,Unwrapp\left(\angle\exp\left(-j2\pi\frac{k}{N}(D^i_{tot_c})\right)\right),\ i = 1,\ldots,G, \tag{5}$$

where the term A_c is the calibration parameter and the function $Unwrapp(.)$ performs the phase unwrapping algorithm described in Reference [22]. As for the the constant A_c, it is obtained as

$$A_c = \frac{H^1_{water}}{Unwrapp\left(\angle\exp\left(-j2\pi\frac{k^1_c}{N^1_c}(D^1_{tot_c})\right)\right)}. \tag{6}$$

where H^1_{water} is a measurement coming from fixed ground stations.

Finally, the computational stage number 5.5 performs the phase unwrapping of the frequency variation exponential term reported in (4) and (5) and the last stage number 5.6 estimates the integral below the unwrapped function.

3. Test on Simulated Data

The simulated data are generated in order to emulate as closely as possible the most suitable observation and geometry characteristics. The bridge is designed with a longitudinal axis perfectly perpendicular to the range direction, a bridge width of about 30 pixels and a maximum variation of double-bounce scattering of about 1.5 pixels. The radar is designed with the same features as the COSMO-SkyMed sensor.

In this specific case, simulated data consist of a sub-ROI reflectivity range profile for which the energy is shown in Figure 4a. From inspection of the figure, it is possible to notice the beginning and the end of the bridge edges. These scattering events, denoted by the numbers 1 and 2, are generated by the direct reflection events and they are directly projected onto the slant-range line. The energy peak generated by the double-bounce scattering mechanism is the one indicated by the number 3 and, thanks to the layover effect, is located behind the main scatterers. The position in time of peaks 1 and 2 remain stable over time because the bridge does not move while peak 3 changes its position due to the variation of the height level of the river's water surface. According to the DFT property (6), a time-variation position of the peak number 3 is corresponding to a frequency variation of its DFT. The variation of the phasor oscillation frequency will increase its absolute phase value (unwrapped phase) [21]. This distance, appropriately scaled and then calibrated at the first observation performed by the ground station (terrain gauge), is being used as an indication of the river's height. The input signal is represented by the sinusoidal sub-pixel variation indicated by arrow number 3. In Figure 4b), the unwrapped phases related to each interferometric pair are reported (the lines with different colors are corresponding to the different observations). The reconstructed level variation is depicted in Figure 4c. From Figure 4b we see that the output response of the function $Unwrapp(.)$ is linear so the parameter A_c of Equation (5) represents the angular coefficient of the linear phases. The ripple in Figure 4c is present due to an oversampling factor of 10 has been used. The higher the oversampling factor, the less visible this noise will be. In the experimental data using the satellite sensor, a very high oversampling factor of 64 will be chosen.

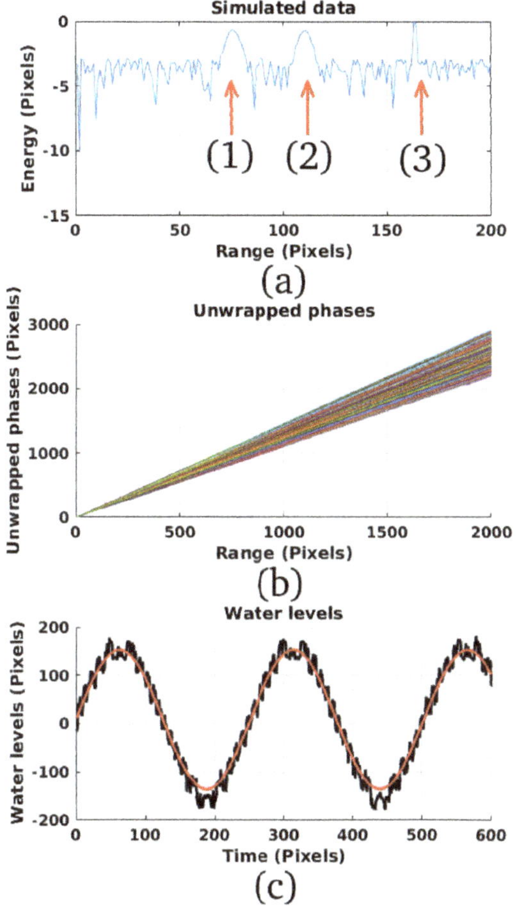

Figure 4. Results of simulated data. (**a**): reflectivity profile. (**b**): Unwrapped phases. (**c**): water-levels estimated results (pixels).

4. Test on Cosmo-SkyMed Data

The performance of the procedure has been evaluated processing a long temporal series of CSK data using three different case studies which are described in detail in Section 4.1. The results estimated from satellite data are validated by comparison with the in-situ observations in Section 4.2.

4.1. Case Studies and In-Situ Observations

The processed data belong to the persistent Earth observation mission called MAPITALY procured by the Italian Space Agency (ASI). This mission performs the interferometric observation of the whole Italian territory with a revisiting time of about 10 days. For the analysis, three datasets of CSK images were considered based on three study areas. The results estimated from satellite data were validated by comparison with the ground observations of water level recorded in a consistent period with respect to the satellite images. In particular, three gauged stations along two Italian rivers were considered—Pontelagoscuro along the Po River, Ponte Nuovo and Ripetta along the Tiber River.

The first case study observes the Po River located on the Northern Italy at the Pontelagoscuro station. Figure 5a shows the georeferenced representation of the SAR long temporal series. Inside the SAR acquisition footprints there is a small red box with a yellow marker inside, geolocated on the following coordinates: datum WGS-84 (EPSG): 4240, 44° 53′ 16.66″ N 11° 36′ 29.42″ E. This box is tagged in yellow with the number 1. This ROI-footprint is represented in detail in Figure 5b where the bridge is observed by an optical image and contoured by the same red box. The ground water levels are registered by the Agenzia Interregionale del Fiume Po.

The second data-set is measured over the Tiber river located in Central Italy. Figure 6a represents some footprints of the SAR observations. Inside the SAR acquisition footprints there is a small red box with a green marker inside, geolocated on the following coordinates—datum WGS-84 (EPSG): 4240, 43° 00′ 37.84″ N 12° 25′ 44.89″ E. This box is tagged in yellow with the number 1. This ROI-footprint is represented in detail in Figure 6b where the bridge is observed by an optical image and contoured by the same red box. The observed water levels are registered by the Servizio Idrografico of the Umbria Region.

The last case study is composed of satellite observations concentrated on the city of Rome. The optical representation of the environment where the river water level has been estimated is reported in Figure 7a. The picture also reports the footprints of the long temporal series of the interferometric SAR observations. The measurements are focused on the Cavour bridge where the optical representation of the infrastructure is reported in Figure 7b. The data are registered by the Agenzia Regionale Protezione Civile of the Lazio Region.

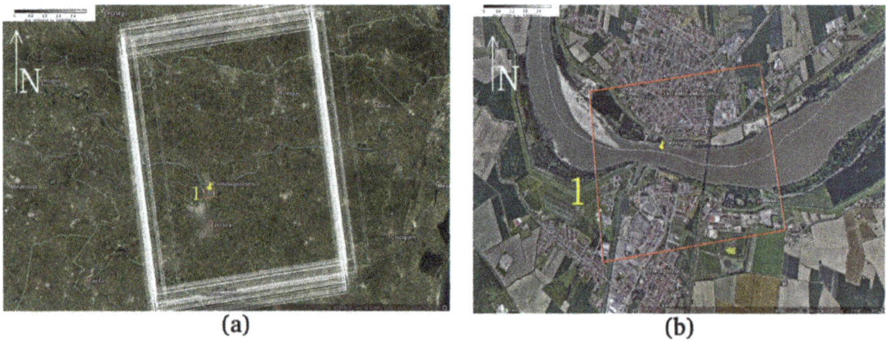

Figure 5. Case study 1 optical representations. (**a**): georeferenced footprints of the SAR long temporal series. (**b**): region of interest (ROI) footprint.

Figure 6. Case study 2 optical representations. (**a**): georeferenced footprints of the SAR long temporal series. (**b**): ROI footprint.

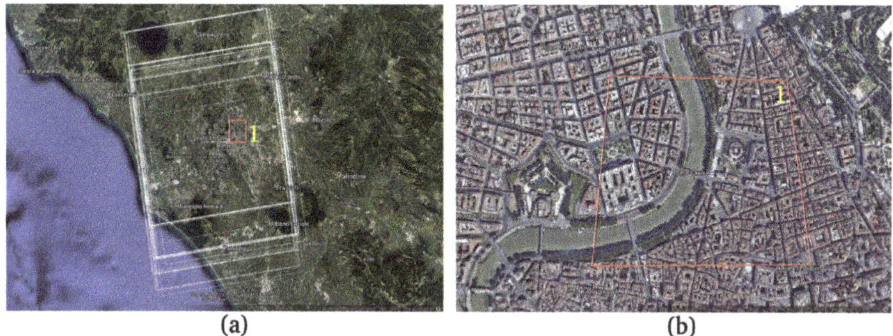

(a)   (b)

Figure 7. Case study 3 optical representations. (**a**): georeferenced footprints of the SAR long temporal series. (**b**): ROI footprint.

The number of processed images for study areas and the period of analysis is specified in Table 1. The number of images varies from 27 images for the study area at Rome and 106 for Pontelagoscuro station. The period of images is different, ranging from 2 to 9 years. For all the experiments, we set the cross-correlation window size to 128 × 128 pixels and the oversampling factor to 64 in both the range and azimuth directions. This value was found to be an optimal pixel dilation level. All the coregistration parameters are reported in Table 2.

Table 1. Satellite datasets used in the analysis and corresponding river gauged stations

STATION	RIVER	COORDINATES (WGS-84)	Time of Obs.	Images Number	River Width [m]
Pontelagoscuro	Po	44° 53′ 18.84″ N, 11° 36′ 28.89″ E	May 2009–Aug. 2018	106	340
Ponte Nuovo	Tiber	43° 00′ 37.11″ N, 12° 25′ 45.15″ E	Mar. 2011–Apr. 2017	76	60
Ripetta	Tiber	41° 54′ 17.59″ N 12° 28′ 27.84″ E	Sept. 2009–Oct. 2016	37	100

Table 2. Coregistration parameters

Parameter	Value
Initial shifts	Coarse cross-correlation
Number of points	4000
Correlation threshold	0.8
Oversampling factor	200
search pixel window	48x48 pixel
Points skimming (minimum points)	30
Use of DEM	Yes
Doppler Centr. Est. Strategy	Polynomials

4.2. Experimental Results

In Figure 8a, the SLC-ROI is reported, referring to the small patch number 1 depicted in Figure 5a,b. This ROI contains the electromagnetic representation of the bridge, used to calculate the water levels of the Po river. The detailed sub-ROI SLC image is shown in Figure 8b. This sub-ROI consists of the data input of the computational block number 5 depicted by Figure 2. Figure 9a is the long temporal series time average sub-ROI particular. Figure 9b gives the description of the scattering events related to the bridge. The purple line number 1 represents the beginning of the bridge structure. Purple line 2 shows the position of the end of the bridge, this scattering event is considered to be a stationary landmark in time. The blue line 3 represents the lowest deviation of the double-bounce scattering event with respect to the reference point, which means that the river is in flood. Line 4 represents the slightest deviation of the double-bounce scattering event, if the pixel is represented it means that the river is dry. The variance of the double-bounce layovered backscattered echoes is strictly located inside the spatial gap contained by the blue layers 3 and 4. The temporal trend of this layover scattering line is a function of the river water level. The trend of the estimated water levels is depicted in Figure 10a. The blue line represents the water-levels measured by the CSK satellite system and the red function represents the ground truths given by the ground observation station. The unwrapped phases of the FT result are shown in Figure 10b.

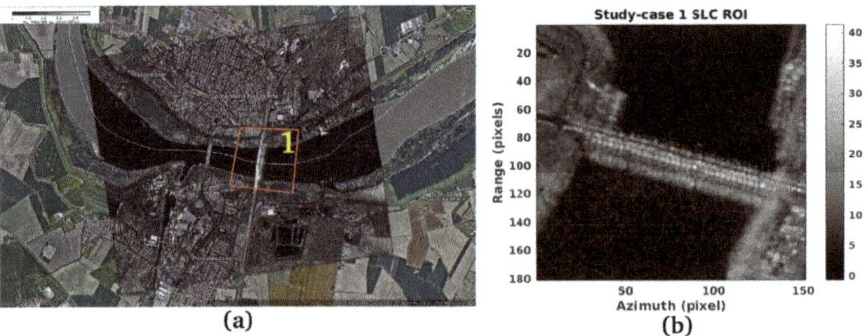

Figure 8. Case study 1 magnitude SAR images. (**a**): Geolocated image with ROI footprint representation. (**b**) Particular of the ROI magnitude in the slant coordinates.

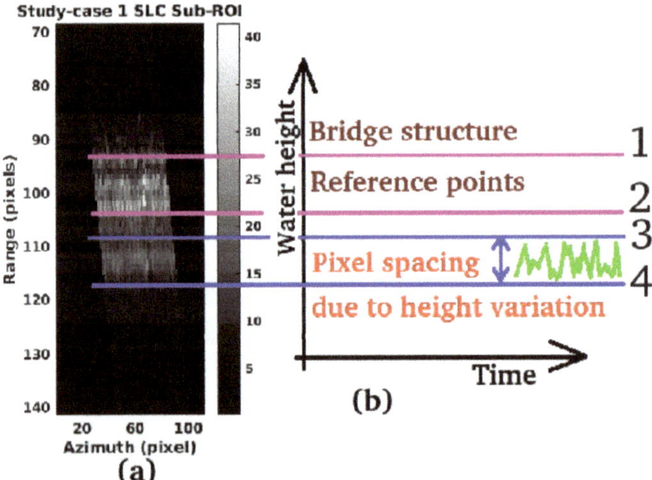

Figure 9. Case study 1 Sub-ROI parameters. (**a**): Sub-ROI SAR image representation in the slant coordinates. (**b**): Scattering parameters on the reference points and pixel spacing trend.

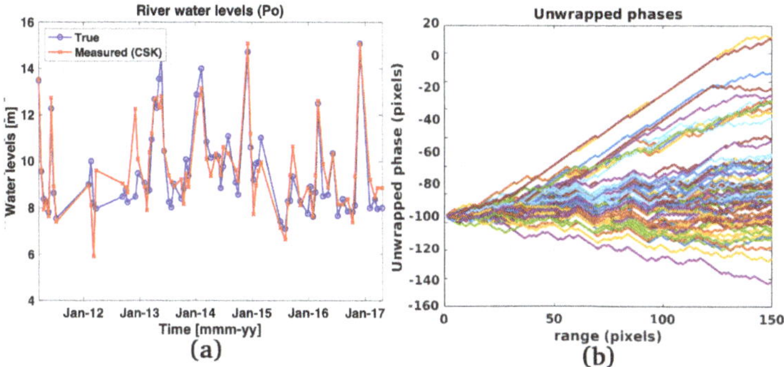

Figure 10. Case study 1 experimental measurements. (**a**): Water levels observed by the ground measurement station (blue line) and by the satellite (red line). (**b**): Unwrapped phases.

Case study 2 has the objective of studying the trend of the Tiber heights in the part of the river that crosses the Umbrian region located in the Italian Central Apennines. This part of the Tiber is narrower with respect to the Po and the radar observations are a bit noisier. Considering this phenomenon, the estimation of the heights of the rivers on this section is more difficult. In Figure 11a the SLC-ROI is reported, referring to the small patch number 1 depicted in Figure 7a,b. This ROI contains the electromagnetic representation of the bridge, used to calculate the water levels of the Po river. The detailed sub-ROI SLC image is shown in Figure 11b. This sub-ROI consists of the data input of the computational block number 5 depicted by Figure 2. Figure 12a is the long temporal series time average sub-ROI particular. Figure 12b gives the description of the scattering events related to the bridge. The purple line number 1 represents the beginning of the bridge structure. Purple line 2 shows the position of the end of the bridge, this scattering event is considered to be a stationary landmark in time. The blue line 3 represents the lowest deviation of the double-bounce scattering event with respect to the reference point, which means that the river is in flood. Line 4 represents the slightest deviation of the double-bounce scattering event, if the pixel is represented it means that the river is dry. The variance of the double-bounce layovered backscattered echoes is strictly located inside the spatial gap contained by the blue layers 3 and 4. The temporal trend of this layover scattering line is a function of the river water level. The trend of the estimated water levels is depicted in Figure 13a. The blue line represents the water-levels measured by the CSK satellite system and the red function represents the ground truths given by the ground observation station. The unwrapped phases of the FT result are shown in Figure 13b.

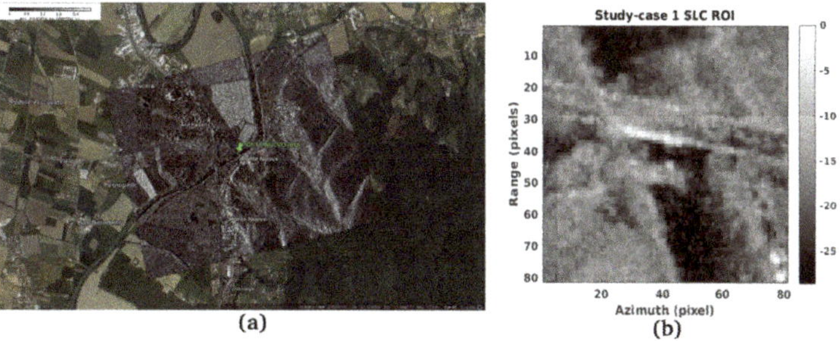

Figure 11. Case study 2 magnitude SAR images. (**a**): Geolocated image with ROI footprint representation. (**b**) Particular of the ROI magnitude in the slant coordinates.

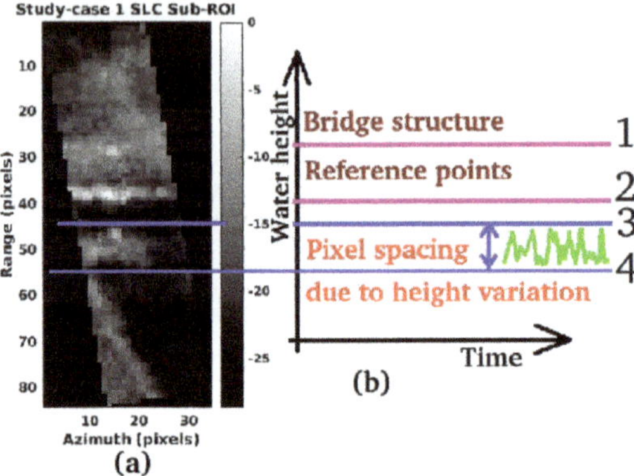

Figure 12. Case study 2 Sub-ROI parameters. (**a**): Sub-ROI SAR image representation in the slant coordinates. (**b**): Scattering parameters on the reference points and pixel spacing trend.

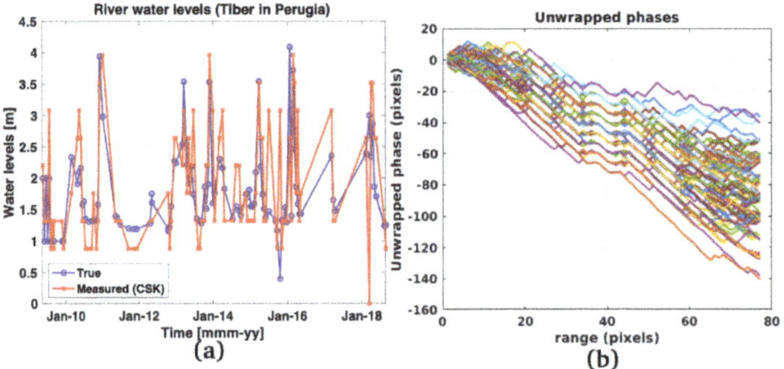

Figure 13. Case study 2 experimental measurements. (**a**): Water levels observed by the ground measurement station (blue line) and by the satellite (red line). (**b**): Unwrapped phases.

In the final case, it was planned to quantify the water height levels of the Tiber River as it flows through the city center of Rome. Also for this case study the experimental measurements were compared with the ground-based measuring facilities. Figure 14a shows the georeferenced SAR extended image where the ROI is visible inside the red box tagged by the yellow number 1. The geolocated SAR ROI is represented in detail in Figure 14b where the bridge is observed and contoured by the same red box. Finally, the optical representation of the *Cavour* bridge is shown in Figure 14c. The double-bounce electromagnetic scattering effects were generated by the *Cavour* bridge (the yellow arrow which is geolocated on the following coordinates: datum WGS-84 (EPSG): 4240, 41° 51′ 36.10″ N 12° 28′ 38.00″ E). The temporal trend of the river water levels is depicted in Figure 15a. In Figure 15b, the errors corresponding to the estimated values in Figure 15a are reported. In addition to the errors represented by a red line, the errors of the estimated values averaged over 5 and 10 samples are shown with a blue and a black line, respectively. As expected, it could be convenient, in the case when the data are very noisy, to exploit the smoothing effect due to averaging. Moreover, from the non-averaged errors of Figure 15b, it can be seen that some measurements are wrong but such uncompliant samples are very few; in fact, observing the trend of the average errors on 5 and 10 samples a significant drop in the error can be seen, which remains well below one meter.

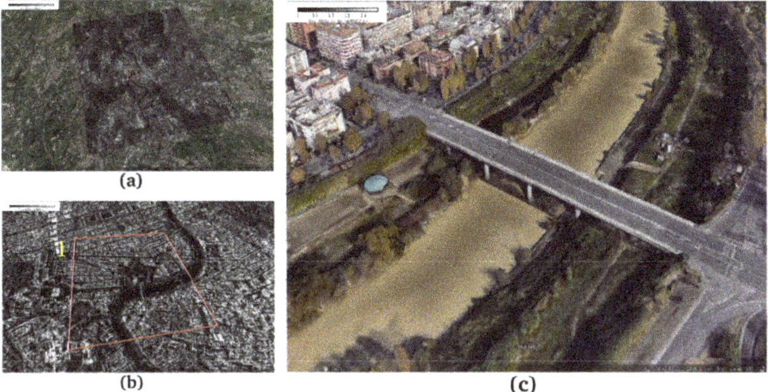

Figure 14. (**a**): Case study 3 ROI. (**b**): ROI footprint. (**c**): Cavour bridge optical image.

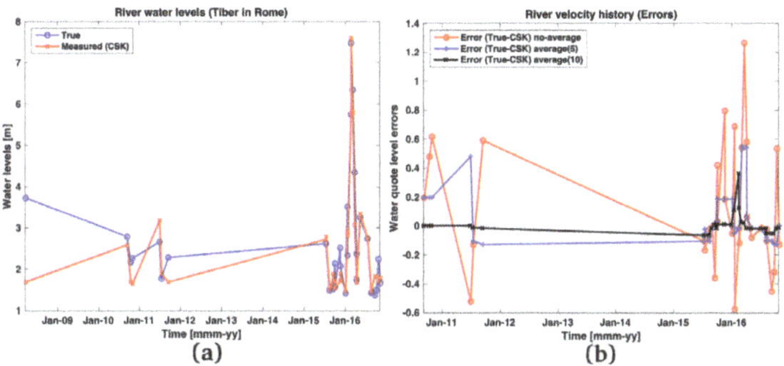

Figure 15. Case study 3 results. (**a**): Ground-station data (red) and satellite observations (blue), versus pixels. (**b**): Water level errors versus time. Blue: CSK versus true. Red: CSK versus true (5 observations average). Black: CSK versus true (10 observations average).

5. Discussion and Performance Assessment

This section provides comments on the experimental results observed in the three case studies. The experimental results show that the implemented algorithm is quite robust, although sometimes it fails to provide a reliable estimate. This is because there is a temporal misalignment from the actual SAR observation that occurs in Italy either early in the morning or late in the evening compared to the measurement of the instrument located in the immediate vicinity of the bridges. This time misalignment is also found to be many hours.

In Figure 16, the scatterplot is reported for each case study, representing the comparison between the water levels estimated by satellite and those observed by the in-situ station. For the Po and the Tiber in Rome (Figure 16 left and right) the water levels overlap the bisector line, as also indicated by the linear regression (red line). In the case of Tiber at Ponte Nuovo (Figure 16 center) the simulated water levels underestimate the ground-based observations up to 2 m, whereas they overestimate the higher water levels. Therefore, the red line does not lie upon the bisector line as the other two cases.

The worst performance are obtained for the Ponte Nuovo case study (as shown at the center of Figure 16). It is worth noticing that this case study has to be selected to provide an example of a difficult scene to analyze. This difficulty can be attributed to the width of the river that in this region is quite narrow (60 m as shown in Table 1) and is greatly contaminated by the surrounding vegetation. These characteristics can affect the satellite measurements and represent a good testbed for the proposed algorithm which, despite everything, would provide still reliable results as shown in the figure. A possible solution to improve the performance of the algorithm could be the installation of corner reflectors in the scene to mitigate the effect of the noise introduced in the radar measurements.

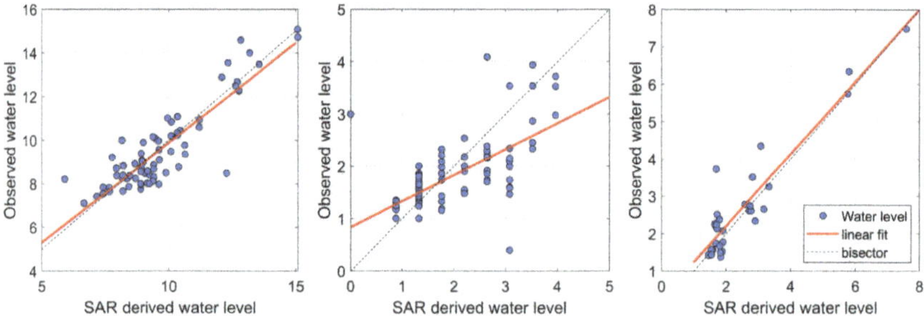

Figure 16. Scatterplot of the water levels estimated by satellite and observed by in-situ stations for Po at Pontelagoscuro (**left**), Tiber at Ponte Nuovo (**center**) and Tiber at Ripetta (**right**).

In order to quantify the performances of the analysis, four indicators are calculated:

- the Pearsone correlation coefficient (R),
- the Nash-Sutcliffe efficiency (NS) [27],
- the root-mean square error (RMSE), expressed in [m],
- the related root-mean square error (RRMSE), defined as the ratio between the RMSE and the mean of the observed water levels.

Table 3 shows the performances for the three case studies comparing the water levels estimated by the procedure and those observed by in-situ stations. As deduced by the scatterplots, best results are obtained for case study one and three, with coefficient of correlation greater than 0.88 and NS greater than 0.77.

Lower performances are obtained for case two with NS smaller than zero and RRMSE of 39%. However, if we calculate the performance considering the linear regression as shown in Figure 16 (center), the performances improve (NS = 0.43; RMSE = 0.51; RRMSE = 0.29). This means that the procedure can fail in terms of absolute values but can be a support to evaluate a variability of the water level if no other measurements are available, as for example in ungauged basins.

In the first case study, the value of the RMSE is quite high (0.91 m) comparing to the other two cases but the RRMSE is the lowest. This is due to the fact that the RMSE is an absolute figure of merit that is not related to a specific value as, on the contrary, the RRMSE. In fact, for the case 1, the water levels range from 6 to 15 m, whereas a narrower range characterizes the other two cases.

Table 3. Performance indicators of the simulated water levels versus the ground-based observations.

STATION	R	NS	RMSE [m]	RRMSE
Pontelagoscuro	0.88	0.77	0.91	0.10
Ponte Nuovo	0.65	−0.03	0.68	0.39
Ripetta	0.93	0.85	0.55	0.21

6. Conclusions

In this paper, an innovative procedure for estimating the water flow elevation of rivers is proposed. Today, a shared and worldwide database containing historical and reliable data concerning the water surface elevation of rivers has not been completed. Many areas are still unmonitored and due to the large importance of fresh water, an evaluation of the river system is fundamental. The main scope of this research is contributing to solving this information-gap problem by designing a SAR signal processing technique having the capability to perform water flow level estimation. The problem of measuring such data is usually solved by designing an ATI SAR geometry, which is constituted by two radars spatially distanced by a baseline extended in the azimuth direction. In the case of space-borne missions the performing of ATI can be an unusual and difficult task. For several single-antenna spaceborne SAR satellite systems, the refocusing of ATI observations from one raw data is a problem because of the not-oversampled nature of the received electromagnetic bursts. This phenomenon makes raw data very similar to a white random process and appearing interlaced Doppler bands completely disjoint. After the range-Doppler focusing process, this problem causes decorrelation when observing the ATI phase of distributed targets. Spaceborne LOS level measurements could be taken into consideration only for small and very coherent targets and in any case were located within the same radar resolution cell. This paper proposed the application of a robust technique for tracking the double-bounce reflections of some principal bridges crossing the rivers and to measure the gap space existing between the river surface and the bridges. The developed algorithm tracked over time the double-bounce scattering event position reflected on the river surface with respect to the single-bounce and direct backscattered echoes form the principal structures of the bridge. River water-flow data were indirectly retrieved by converting the time-domain water surface variation in velocities. The experiments were evaluated by processing simulated and a long temporal series of COSMO-SkyMed data.

Author Contributions: Conceptualization, F.B., A.T., P.A., C.C. and D.O.; Data curation, F.B. and A.T.; Formal analysis, F.B., A.T., P.A., C.C. and D.O.; Funding acquisition, F.B.; Investigation, F.B. A.T., P.A., C.C. and D.O.; Methodology, F.B., A.T., P.A., C.C. and D.O.; Resources, F.B., A.T., P.A. and C.C.; Software, F.B.; Supervision, F.B. and D.O.; Validation, F.B. and D.O.; Visualization, F.B.; Writing original draft, F.B. and D.O.; Writing review & editing, F.B., A.T., P.A., C.C. and D.O.

Funding: This research received no external funding.

Conflicts of Interest: The authors declare no conflict of interest.

References

1.   McGraw-Hill Book Company. *Open Channel Hydraulics, Ven Te Chow, 1959: Open Channel Hydraulics*; Open Channel Hydraulics, Kogakusha: Tokyo, Japan, 1959.

2.   Hannah, D.M.; Demuth, S.; van Lanen, H.A.J.; Looser, U.; Prudhomme, C.; Rees, G.; Stahl, K.; Tallaksen, L.M. Large-scale river flow archives: Importance, current status and future needs. *Hydrol. Process.* **2011**, *25*, 1191–1200. [CrossRef]

3.   Smith, L.C. Satellite remote sensing of river inundation area, stage, and discharge: A review. *Hydrol. Process.* **1997**, *11*, 1427–1439.:10<1427::AID-HYP473>3.0.CO;2-S. [CrossRef]

4.   Bjerklie, D.M.; Dingman, S.L.; Vorosmarty, C.J.; Bolster, C.H.; Congalton, R.G. Evaluating the potential for measuring river discharge from space. *J. Hydrol.* **2003**, *278*, 17–38. [CrossRef]

5.   Bjerklie, D.M.; Moller, D.; Smith, L.C.; Dingman, S.L. Estimating discharge in rivers using remotely sensed hydraulic information. *J. Hydrol.* **2005**, *309*, 191–209. [CrossRef]

6.   Bjerklie, D.M. Estimating the bankfull velocity and discharge for rivers using remotely sensed river morphology information. *J. Hydrol.* **2007**, *341*, 144–155. [CrossRef]

7.   Koblinsky, C.J.; Clarke, R.T.; Brenner, A.C.; Frey, H. Measurement of River Level Variations with Satellite Altimetry. *Water Resour. Res.* **1993**, *29*, 1839–1848. [CrossRef]

8.   Biancamaria, S.; Frappart, F.; Leleu, A.S.; Marieu, V.; Blumstein, D.; Desjonquères, J.D.; Boy, F.; Sottolichio, A.; Valle-Levinson, A. Satellite radar altimetry water elevations performance over a 200 m wide river: Evaluation over the Garonne River. *Adv. Space Res.* **2016**, *59*, 128–146. [CrossRef]

9.   Normandin, C.; Frappart, F.; Diepkilé, A.T.; Marieu, V.; Mougin, E.; Blarel, F.; Lubac, B.; Nadine, B.; Abdramane, B. Evolution of the Performances of Radar Altimetry Missions from ERS-2 to Sentinel-3A over the Inner Niger Delta. *Remote Sens.* **2018**, *10*, 833. [CrossRef]

10. Schneider, R.; Tarpanelli, A.; Nielsen, K.; Madsen, H.; Bauer-Gottwein, P. Evaluation of multi-mode Cryosat-2 altimetry data over the Po River against in situ data and a hydrodynamic model. *Adv. Water Resour.* **2018**, *112*, 17–26. [CrossRef]

11. Fu, L.L.; Alsdorf, D.; Rodríguez, E.; Morrow, R.; Mognard, N.; Lambin, J.; Vaze, P.; Lafon, T. The SWOT (Surface Water and Ocean Topography) Mission: Spaceborne Radar Interferometry for Oceanographic and Hydrological Applications. In Proceedings of the OceanObs'09: Sustained Ocean Observations and Information for Society, Venice, Italy, 21–25 September 2009; Hall, J., Harrison, D.E., Stammer, D., Eds.; ESA Publication WPP-306: New York, NY, USA, 2009; Volume 2.

12. Matgen, P.; Schumann, G.; Henry, J.B.; Hoffmann, L.; Pfister, L. Integration of SAR-derived river inundation areas, high-precision topographic data and a river flow model toward near real-time management. *Int. J. Appl. Earth Obs. Geoinf.* **2007**, *9*, 247–263. [CrossRef]

13. Goldstein, R.M.; Richard, M.; Zebker, H.A. Interferometric radar measurement of ocean surface currents. *Nature* **1987**, *328*, 707–709. [CrossRef]

14. Romeiser, R.; Breit, H.; Eineder, M.; Runge, H.; Flament, P.; De Jong, K.; Vogelzang, J. Current measurements by SAR along-track interferometry from a Space Shuttle. *IEEE Trans. Geosci. Remote Sens.* **2005**, *43*, 2315–2324. [CrossRef]

15. Nitti, D.O.; Hanssen, R.F.; Refice, A.; Bovenga, F.; Nutricato, R. Impact of DEM-assisted coregistration on high-resolution SAR interferometry,. *IEEE Trans. Geosci. Remote Sens.* **2011**, *49*, 1127–1143. [CrossRef]

16. Michel, R.; Avouac, J.P.; Taboury, J. Measuring ground displacements from SAR amplitude images: Application to the Landers earthquake. *Geophys. Res. Lett.* **1999**, *26*, 875–878. [CrossRef]

17. Strozzi, T.; Luckman, A.; Murray, T.; Wegmuller, U.; Werner, C.L. Glacier motion estimation using SAR offset-tracking procedures. *IEEE Trans. Geosci. Remote Sens.* **2002**, *40*, 2384–2391. [CrossRef]

18. Casu, F.; Manconi, A. Four-dimensional surface evolution of active rifting from spaceborne SAR data. *Geosphere* **2016**, *12*, 697–705. [CrossRef]

19. Biondi, F. COSMO-SkyMed Staring Spotlight SAR Data for Micro-Motion and Inclination Angle Estimation of Ships by Pixel Tracking and Convex Optimization. *Remote Sens.* **2019**, *11*, 766.

20. Biondi, F.; Addabbo, P.; Clemente, C.; Orlando, D. Micro-Motion Estimation of Maritime Targets Using Pixel Tracking in Cosmo-Skymed Synthetic Aperture Radar Data: An Operative Assessment. *Remote Sens.* **2019**, *11*, 1637. [CrossRef]

21. Oetken, G.; Parks, T.W.; Schussler, H. New results in the design of digital interpolators. *IEEE Trans. Acoust. Speech Signal Process.* **1975**, *23*, 301–309. [CrossRef]

22. Goldstein, R.M.; Zebker, H.A.; Werner, C.L. Satellite radar interferometry: Two-dimensional phase unwrapping. *Radio Sci.* **1988**, *23*, 713–720. [CrossRef]

23. Wang, Z.; Perissin, D.; Lin, H. Subway tunnels identification through Cosmo-SkyMed PSInSAR analysis in Shanghai. In Proceedings of the 2011 IEEE International Geoscience and Remote Sensing Symposium, Vancouver, BC, Canada, 24–29 July 2011.

24. Ullo, S.L.; Addabbo, P.; Di Martire, D.; Sica, S.; Fiscante, N.; Cicala, L.; Angelino, C.V. Application of DInSAR Technique to High Coherence Sentinel-1 Images for Dam Monitoring and Result Validation Through In Situ Measurements. *IEEE J. Sel. Top. Appl. Earth Obs. Remote Sens.* **2019**, *12*, 875–890. [CrossRef]

25. Richards, M.A. *Fundamentals of Radar Signal Processing*; Tata McGraw-Hill Education: New York, NY, USA, 2005.

26. Biondi, F.; Clemente, C.; Orlando, D. An atmospheric phase screen estimation strategy based on multi-chromatic analysis for differential interferometric synthetic aperture radar. *IEEE Trans. Geosci. Remote Sens.* **2019**. [CrossRef]

27. Nash, J.; Sutcliffe, J. River flow forecasting through conceptual models part I — A discussion of principles. *J. Hydrol.* **1970**, *10*, 282–290, doi:10.1016/0022-1694(70)90255-6. [CrossRef]

 *remote sensing*

Technical Note

Flood Distance Algorithms and Fault Hidden Danger Recognition for Transmission Line Towers Based on SAR Images

Lianguang Liu [1,2], Rujun Du [1,*] and Wenlin Liu [2]

[1] State Key Laboratory of Alternate Electrical Power System with Renewable Energy Sources, North China Electric Power University, Beijing 102206, China
[2] Dezhou Tianhe Benan Electric Power Technology Co., Ltd., Dezhou 253000, China
* Correspondence: 1172201051@ncepu.edu.cn; Tel.: +86-18911800856

Received: 8 June 2019; Accepted: 8 July 2019; Published: 10 July 2019

Abstract: Synthetic Aperture Radar (SAR) has been extensively used in the monitoring of natural hazards such as floods and landslides. Predicting whether natural hazards will cause serious harm to important facilities on the ground is an important subject of study. In this study, the distance between the water body and the tower and the flood ratio in the search area and the elevation are defined as the evaluation indicators of the flood hazard of the tower, indicating whether flooding will threaten the safety of the transmission line tower. Herein, transmission tower flood identification algorithms based on the center distance of the tower and the grid distance of the tower are proposed. SAR satellite image data of the flood with a resolution of 10 m are selected to prove the feasibility and effectiveness of the proposed fault identification algorithm. The simulation results show that the SAR satellite image data with a resolution of 10 m can identify the distance accuracy of the transmission tower flood hazard by up to 7 m, which can be used to identify the flood fault of the transmission line tower.

Keywords: hazard prevention; flood hazard; hidden danger identification; tower failure

---

1. Introduction

As the scale of the power grid increases, the scope, and thus the workload, of power grid inspections continue to expand. At present, power inspections in most countries still rely on the manual recording of data, which has disadvantages such as high cost, dangerous working conditions, and absence of inspection. Over the past two decades, aerial inspections have been employed [1], which can greatly improve detection efficiency and precision. However, this method is restricted by factors such as flight safety, airline control, weathers changes, and refueling. Unmanned aerial vehicles (UAV) are not widely used because of the safety issues and a lack of durability [2–4]. For these reasons, the development of satellite technology provides a new and important means of fault detection and hazard prediction for transmission lines.

Floods account for 40% of the losses caused by natural hazards worldwide. Flood hazards cause damage to the power grid by submerging power equipment and thus causing short circuit damage, internal discharge damage, and moisture damaged. Furthermore, the foundation of the transmission line tower can be washed away by floods, which may cause the tower to collapse, thus causing power interruptions remd threatening the safe and stable operation of the power system [5–8]. In recent years, the once-in-a-century floods in the Oder River and Nice River basins flowing through Poland, Czech Republic, Austria, and Germany caused economic losses of 5.9 billion dollars. Two catastrophic floods in the central and northern United States caused economic losses of 2215 billion dollars. In China,

Jiangxi and Hebei suffered severe floods due to heavy rainfall, many substations and several lines were shut down, and more than 300,000 users suffered blackouts [9,10]. In June 2015, in Georgia, floods damaged Tbilisi's transmission lines, resulting in power outages for about 22,000 consumers in two districts of Tbilisi. In the summer of 2016, continuous heavy rainfall in the Hubei Province of China caused severe floods in Tianmen, resulting in the outage of several substations, resulting in a total of 30·10 kV line failures, involving 1158 stations and 115,276 users. In the Fujian Province of China, floods and geological hazards occurred an average of 3.3 times a year over the past 10 years, resulting in an average annual direct economic loss of 3.6 billion yuan [11–13]. If transmission line flood faults can be located quickly, the loss can be reduced. Therefore, an identification algorithm of transmission line tower flood faults is a subject that needs to be studied.

Synthetic Aperture Radar (SAR) has been widely used in hazard monitoring. This paper proposes the use of SAR images to identify flood faults on transmission lines. At present, on the basis of the SAR image water feature, the water body part of an image can be accurately extracted [14]; this can be done in different scenarios such as floods in forest areas or floods in cities [15,16]. The introduction of some improved algorithms also makes the extraction range of floods more accurate [17–32]. In [33,34], methods for searching for mountain fire faults of transmission line towers are proposed. By combining our research with the abovementioned methods, we propose an identification algorithm for flood faults in transmission line towers. This is of great significance for the operation and maintenance of transmission towers.

The objectives and novelty of the study are as follows:

(1) This is the first study based on SAR satellite imagery on hidden flood hazards related to transmission line towers. When hazards occur, SAR has the advantages of a quick response, accurate positioning, and a wide coverage, all of which help identify towers that may be infringed upon by floods and aid inspectors design targeted emergency repair schemes to minimize the economic losses caused by power outages;

(2) We propose two fast methods to calculate the shortest distance between tower and flood based on the center distance of the tower and the grid distance of the tower. Furthermore, we can find the nearest flood within a certain distance between tower and tower. The two algorithms fill in the methodological gaps of calculating the shortest distance from tower to flood;

(3) We propose that the shortest distance from tower to flood, the proportion of flood in a search area and the elevation difference between the tower base and the flood level should be taken as indicators to give a certain weight to evaluate the hazard degree of the tower. This evaluation method is a rapid evaluation made in emergency situations when floods occur. It can reflect the distribution of floods around towers and the hazard degree of towers.

2. Flood Recognition Algorithm Based on SAR Imaging

2.1. Image Preprocessing

The SAR radar echo signals are superimposed onto each other, which causes the radar image to produce granular spots. This phenomenon is the result of the influence of speckle noise, which is the main cause of SAR image noise. The existence of speckles has an impact on the interpretation and extraction of objects in the image, especially in terms of the extraction and recognition of the contours and edges of the target, and may even cause the disappearance of features. In order to extract and identify the feature information more accurately, it is necessary to weaken the fluctuation of the luminance value and the influence of the speckle noise through filtering.

A comparison of the filtering effects of various filtering methods [30] on SAR images shows Lee filtering to be an effective option.

2.2. Identification Model For Flood Areas

For the processing of SAR images, threshold segmentation can be used to extract water bodies. This principle is based on the low scatter value of the water body in the SAR image. It is achieved by setting a suitable threshold to mark the image; thus, the values less than the threshold portion become the water body, and the portions larger than the threshold portion become the background, forming a binary image. The advantage of this algorithm is that it is fast and the principle is simple; however, the determination of the threshold is difficult. Among several commonly used threshold segmentation methods, Otsu's optimal global threshold segmentation method has a low false alarm rate and high water extraction accuracy [31]. The algorithm is as follows:

An image histogram distribution can be expressed by

$$P_q = \frac{n_q}{n} \quad q = 0, 1, 2, \ldots, L-1,\tag{1}$$

where n is the total number of image pixels, n_q is the number of pixels with a grayscale of q, and L is the number of all possible gray levels in the image. Suppose the target area C_1 contains gray levels $[0, 1, 2, \ldots, k]$, C_2 contains gray levels $[k+1, \ldots, L-1]$, and the threshold is k, the largest inter-class variance $\sigma_B^2(k)$ is

$$\sigma_B^2(k) = P_1(k)\big[m_1(k) - m_G\big]^2 + P_2(k)\big[m_2(k) - m_G\big]^2\tag{2}$$

$P_1(k)$ and $P_2(k)$ are the percentage of pixels of C_1 and C_2 in the whole image, respectively, $m_1(k)$ and $m_2(k)$ are the average gray value of the pixels in the C_1 and C_2 regions, respectively, and m_G is the average gray value of the whole image. The average gray value of the gray level k can be obtained by the following formula:

$$m(k) = \sum_{i=0}^{k} i p_i.\tag{3}$$

Expand Equation (2) and substitute $P_2(k) = 1 - P_1(k)$ to get the following formula:

$$\sigma_B^2(k) = \frac{[m_G P_1(k) - m(k)]^2}{P_1(k)[1 - P_1(k)]}.\tag{4}$$

Determining the values of m and $P_1(k)$ can determine the between-class variance. Under the condition of maximum between-class variance, the threshold of the segmented image is easier to determine. Since k is an integer in the range $[0, \ldots, L-1]$, it is possible to find the k value at the maximum of the variance between classes by continuous loop calculation, where k is the optimal threshold. When the k value is not unique, the average of the plurality of k values is the optimal threshold. The ratio of between-class variance to the grayscale variance of the total image is a separability measure that divides the image into two categories:

$$\eta(k) = \frac{\sigma_B^2(k)}{\sigma_G^2(k)}.\tag{5}$$

The algorithm automatically calculates the segmentation threshold by finding the maximum between-class variance between the two types of features. Therefore, there is a good segmentation effect when there is a significant difference between the gray value of the object in the region of interest and the gray value of other features. That is to say, the gray value frequency distribution of the image has obvious "peak and valley" characteristics, and the larger the difference between the peak value and the valley value, the more obvious the segmentation effect.

The grayscale value at the original image point (x, y) is $I(x, y)$, and the target in the original image is extracted with the threshold k. Herein, a is the value of the background and b is the value of the target. The binary image $F(x, y)$ is generated as follows:

$$F(x, y) = \begin{cases} a, & I(x, y) \geq k \\ b, & \\ & I(x, y) < k. \end{cases} \tag{6}$$

The value of a is 1, and the value of b is 0. The original image completes the binary transformation to extract the water body.

2.3. Flood Failure Evaluation Index

The severity of flood damage to the tower is positively related to the distance and the elevation. The possibility of tower collapse caused by flood scouring near the water area is far greater than that caused by a flood far away from the water area. The elevation difference between the tower base and the flood level can reflect whether the tower is submerged. The distance and the elevation can be used as the main parameter for the evaluation of hidden dangers of flood failure. On the basis of the SAR image taken by satellite, the flood area is extracted through image processing. Combining this with the information of the transmission line tower account, a judgment on whether the tower has been flooded is made, and the hazard situation of the transmission line is preliminarily evaluated. If the tower is not flooded, the distance between the tower and the edge of the flood should then be determined to quickly judge the impact of the flood hazard on the transmission line.

An image is made up of pixels, and each point in the image can be converted into coordinates. According to the account information of the tower, visual interpretation, and coordinate picking, the pixel coordinate point set of the tower in the binary image extracted by the water body is determined as follows:

$$U = \{a_1(x_1, y_1), a_2(x_2, y_2), \ldots, a_n(x_n, y_n)\} \tag{7}$$

where U is the set of tower coordinates, a_i is the name of the tower, x_i is the abscissa of the pixel coordinates of the binary image, and y_i is the ordinate of the pixel coordinates of the binary image. $F(x_i, y_i)$ is the value of the point (x_i, y_i) in the binary image. If $F(x_i, y_i) = 0$, it shows that the tower is located in the water extraction part, the tower has been flooded, and the tower is most seriously affected. If $F(x_i, y_i) = 1$, the tower is located in the non-water part and so has not been flooded. However, there may still be potential flood hazards. The elevation layer and the distance between the tower and water body boundary should be calculated. A search for transmission line towers around the waters that may be endangered by floods should be conducted to determine the extent of the damage to the towers. The elevation model can be extracted from any public available elevation model such as SRTM or ASTER Global Digital Elevation Model. The elevation model is established to calculate the elevation difference between the flood level and the tower foundation. The specific content of the algorithm will not be discussed in this paper. The calculation of the distance between the towers and the water body boundary is described below.

3. Tower Flood Failure Distance Algorithm and Criterion

On the basis of the extracted water body binary image characteristics, a search algorithm for the center distance of the tower and a search algorithm for the tower base grid distance are proposed.

3.1. Tower Center Distance Search Algorithm

According to operating experience and flood control standards, floods beyond 3 km do not pose a threat to transmission lines. Therefore, it is only necessary to search for transmission line towers within 3 km of the water body. In addition, the safe distance can be increased or decreased as needed. A circular area of 3 km around the tower is used as the search area. If there is a water body in the search area, it can be judged that there is a flood hazard on the transmission line. On the basis of the SAR image and the tower information, the position information of the tower that may be affected and the latitude and longitude information of each point of the image are determined.

In Figure 1, we choose the image coordinates. The upper left corner is the coordinate origin, the horizontal axis is the x-axis, and the vertical axis is the y-axis. The two towers A_1 and A_2 have a circle with a radius of $R(R = 3km)$ centered on A_1 and A_2. L_1 and L_2 are water body boundaries, and between L_1 and L_2 are water bodies. Let the image resolution be M (m), and each pixel is an $M \times M$ rectangle. From this it can be obtained that $M \times r = 3000$, r is the radius in the image, and the unit is a pixel. The equation for indicating the boundary circle of the A_1 tower search area by coordinates is

$$(x - x_1)^2 + (y - y_1)^2 = (\frac{3000}{M})^2.$$ (8)

Points in the search area satisfy the following formula:

$$(x - x_1)^2 + (y - y_{1)})^2 < (\frac{3000}{M})^2.$$ (9)

The search can be conducted from the center to the outside. If the existence point (x_i, y_i) satisfies Equation (9) and $F(x_i, y_i) = 0$, then there is a water body in the search area of the tower. The nearest distance from the water body to the tower should then be calculated. The distance between the tower $A_1(x_1, y_1)$ and the water body is calculated according to the distance formula between two points:

$$d = \sqrt{(x_1 - x_i)^2 + (y_1 - y_i)^2}.$$ (10)

The calculation accuracy is $\sqrt{2}M/2$.

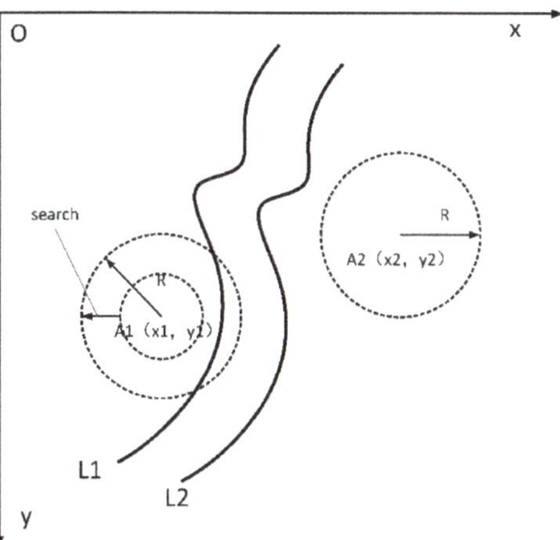

Figure 1. Tower center distance search method.

3.2. Tower Base Grid Distance Search Algorithm

The tower center distance search method can judge the tower flood failure. However, when the search area is wide and the number of towers is large, the calculation cost is high and so it takes a long time. The circular search area can be changed to the grid search area of the tower base. With the tower as the center, a square grid search area with a side length of 6 km is made. The four vertices of the grid search area are A, B, C, and D, as shown in Figure 2.

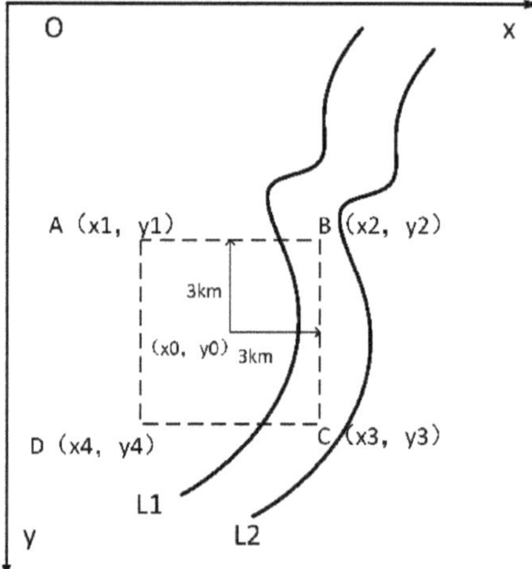

Figure 2. Tower base grid distance search method.

The coordinates of the tower are (x_0, y_0); the coordinates of the four endpoints are $A(x_1, y_1)$, $B(x_2, y_2)$, $C(x_3, y_3)$, and $D(x_4, y_4)$. The mathematical relationship between x_2 and x_0 is

$$M \cdot (x_2 - x_0) = 3000. \tag{11}$$

This is formulated as

$$x_2 = \frac{3000}{M} + x_0. \tag{12}$$

By the same logic,

$$\begin{cases} x_1 = x_4 = x_0 - \frac{3000}{M} \\ x_2 = x_3 = x_0 + \frac{3000}{M} \\ y_1 = y_2 = y_0 - \frac{3000}{M} \\ y_3 = y_4 = y_0 + \frac{3000}{M} \end{cases}. \tag{13}$$

It can be concluded that the points in the search area satisfy Equation (14). If there are points (x_i, y_i) in the binary image, which make $F(x_i, y_i) = 0$ and satisfy (14), then there is a potential flood hazard in the towers search area. According to Equation (10), the minimum distance between the tower and water body is calculated.

$$\begin{cases} x_0 - \frac{3000}{M} \leq x_i \leq x_0 + \frac{3000}{M} \\ y_0 - \frac{3000}{M} \leq y_i \leq y_0 + \frac{3000}{M} \end{cases} \tag{14}$$

Compared with the tower center distance search method, the tower base grid distance search method does not need to calculate the distance for each point, and the search process only involves a coordinate value comparison, which can improve the calculation speed of the algorithm. However, the grid search method searches for a square grid and searching near the vertex (a distance greater than 3 km) can cause false alarms. This type of false alarm can be avoided by distance quantitative comparison.

3.3. Flood Failure Criterion Based on Distance Algorithm, Flood Ratio and the Elevation

In the binary image, the function value of the water body is 0, and the function value of the non-aqueous body is 1. $\bar{F}$, the average value of the function in the search area, equals

$$\bar{F} = \frac{1}{n} \sum_{i=0}^{n} F(x_i, y_i). \tag{15}$$

If $\overline{F} = 1$, there is no water body in the search area, and there is no potential flood hazard for the transmission tower. If $\overline{F} < 1$, there are water bodies in the search area, and there are potential flood hazards for transmission line towers. The easiest way to get the F number is to count the number of ones within the window of 3000 × 3000 m and divide it by the number of pixels.

H is the flood ratio. The flood ratio can reflect the extent of flooding in the search area. The larger the H, the wider the flood range in the search area, and the greater the likelihood and severity of the flood damage to the tower.

The elevation Δh can reflect the difference in height between the tower base and the flood level. The smaller the Δh is, the closer the height of tower foundation and flood level is, and the greater the threat of flood disaster to tower is.

On the basis of the standard for construction, operation of the transmission line towers and overhaul experience accumulated over the years and the flood control standard, the flood hazard coefficient V of the tower is designed to measure the hazard situation. The coefficient V is mainly determined by the distance d between the tower and the flood, the flood ratio H in the search range of the tower and the elevation Δh as shown in the following equation:

$$V = C_1 \frac{d}{3000} + C_2(1 - H) + C_3 \frac{\Delta h}{1000} \tag{16}$$

The distance d from the tower to the flood is divided by the search radius (3 km), reflecting the distance between the flood in the search range and the tower. The closer the distance, the smaller the value, and the greater the degree of danger; C_1 is the weight of the distance. H is the proportion of floods. The larger the flood ratio, the larger the number. When the value is adjusted to $1 - H$, the larger the H, the smaller the value, and the weight coefficient is C_2. The elevation Δh between the tower base and the flood level is divided by 1000. If $\Delta h > 1000$ m, we think the tower is absolutely safe according to" Flood Control Standard of Transmission Line (GB50201-2014)". If $\Delta h < 1000$ m, the smaller $\Delta h/1000$ (a value less than 1) is, the closer the elevation difference between the tower base and flood level, the smaller the value, and the greater the degree of danger. C_3 is the weight of the elevation. Therefore, the closer the flood, the larger the flood ratio in the search area, the closer the elevation between the tower base and flood level, the smaller d/3000, $1 - H$ and $\Delta h/1000$ will be, and the smaller the V, the greater the degree of danger. In Formula (16), $C_1 = 0.3 C_2 = 0.2 C_3 = 0.5$. V is a number less than 1. The smaller the V, the greater the threat of flood to towers. By calculating the distance d and the flood ratio H and the elevation Δh in the search range of the tower, the tower flood hazard coefficient V is calculated. Thereafter, it is possible to judge the severity of the flood. On the basis of the statistical data of the State Grid, the Southern Power Grid and the Meteorological Bureau, the V coefficients of transmission lines and towers in some flood-stricken areas in China are calculated. According to the severity of actual hazards and the potential hazards of poles and towers, the hazard classification is divided according to the calculation of the coefficient and the actual situation. The damage degree of the tower is divided as follows:

1. When $d = 0$ m, the tower has been flooded and it is judged to be a super hazard;
2. When $V < 0.1$, the degree of hazard is judged to be in the A level;
3. When $0.1 \leq V < 0.4$, the degree of hazard is judged to be in the B level;
4. When $0.4 \leq V < 1$, the degree of hazard is judged to be in the C level.

After the hazard level is judged, it should be released quickly so that it can be utilized in time. The warning information released includes tower location, line name, voltage level, flood impact area, the location of flood relative to the tower, the distance between flood and tower, the elevation between the tower base and the flood level, etc. This is all possible as a result of the advantages of rapid satellite inspection and accurate positioning. According to the terrain, the soil, and the season, the values of C_1, C_2 and C_3 can be changed appropriately to adjust classification of the damage level caused by floods. According to various sets of data (the "Standard for flood control GB50201-2014" and "Research on hazard damage characteristics and hazard prevention technology of distribution network flood geological hazards", the "Application of Correction Technique of Initial Soil Water Storage Capacity Correction in Flood Forecasting", and the "Study on the Supply And Demand of Soil Erosion Control Service and Flood Control Service in Linfen Section of Fenhe River Basin"), the proportion of C_1 can be appropriately increased when the climate is humid and the soil is loose. This is because even if there is a certain distance from the flood, continuous erosion by the flood may cause the tower foundation to collapse, leading to tower collapse.

4. Case Analysis

To perform the analysis, the backscattering SAR image satellite image (resolution 10 m) in the flood-prone area is selected and Lee filter processing performed. After the processing, the water body is extracted using the threshold segmentation method, as shown in Figures 3–5.

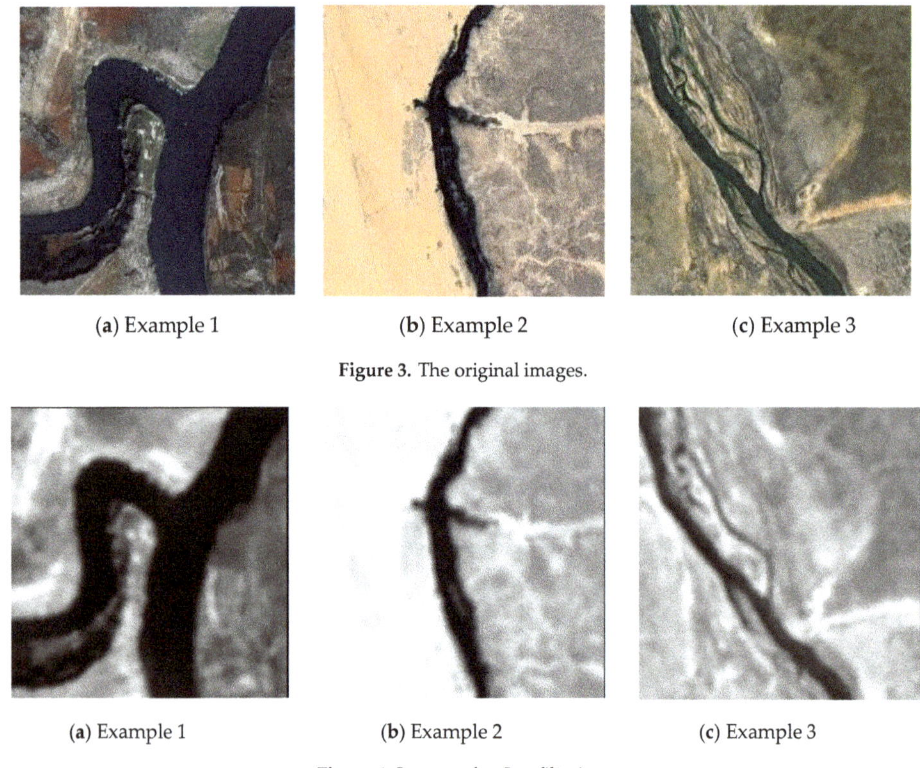

| (a) Example 1 | (b) Example 2 | (c) Example 3 |

Figure 3. The original images.

| (a) Example 1 | (b) Example 2 | (c) Example 3 |

Figure 4. Images after Lee filtering.

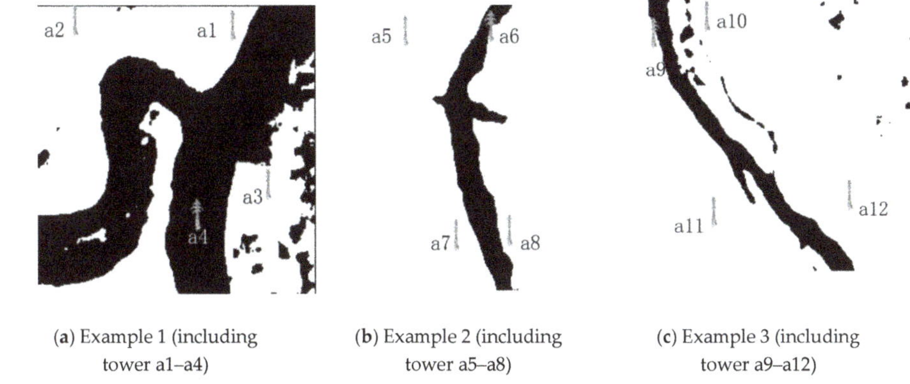

(**a**) Example 1 (including tower a1–a4) (**b**) Example 2 (including tower a5–a8) (**c**) Example 3 (including tower a9–a12)

Figure 5. The binary images of the water body after threshold segmentation.

A set of points was randomly selected around the water bodies of the three binary images of Figure 5, and each map selected four points, which are $a_1, a_2, \ldots, a_{12}$, as the positions of the towers. We calculated the distance from flood to tower, the elevation difference between flood level and tower foundation, and the flood ratio within the search range. The flood hazard coefficient was calculated. The results are shown in Table 1.

Comparing the calculation time of the two algorithms in Table 1, the tower grid distance algorithm is faster than the tower center distance algorithm. In order to make the test results more general, we randomly picked points in the flood area of Figure 3 as the tower coordinates. This was done to verify that the algorithm can accurately calculate the flood distance and make hidden danger evaluations no matter where the tower was located. Tower a4 was located in the water body, which was used to simulate the actual situation when a tower is submerged by flooding. The danger level S also reflects the fact that the tower was submerged by the flood, which proves the reliability of the calculation results. The results show that the combination of the distance and the elevation can accurately reflect the hidden flood hazards of transmission line towers. At the same time, the elevation reduces commission errors. The image resolution of the example is 10 m, and the calculation distance accuracy reaches 7 m. When using higher resolution images, the accuracy increases. In addition, combined with the water body around the tower, the hidden danger coefficient can effectively reflect the severity of the flood hazard caused by the tower.

Table 1. Simulation results.

Tower Number	Tower Coordinates	Coordinates of the Nearest Flood	The Alarm Distance (m)	Flood Ratio (%)	The Elevation Δh (m)	V Coefficient	Calculation Time of the Tower Center Distance Algorithm(s)	Calculation Time of Tower Base Grid Distance Algorithm(s)	The Affected Level
a1	(178,15)	(189,16)	110.45	11.4	350	0.2088	0.0212	0.0198	B
a2	(33,20)	(71,3)	416.29	10.5	240	0.1826	0.0288	0.0263	B
a3	(203,154)	(204,137)	170.29	12.3	270	0.1766	0.0224	0.0211	B
a4	(145,186)	(145,186)	0	45.5	0	0.0910	0.0162	0.0150	S
a5	(57,32)	(103,46)	480.83	3.8	110	0.1107	0.0299	0.0292	B
a6	(137,32)	(127,29)	104.40	20.1	170	0.1356	0.0188	0.0180	B
a7	(106,219)	(119,213)	143.17	14.4	130	0.1081	0.0231	0.0220	B
a8	(153,222)	(146,225)	76.16	22.5	200	0.1526	0.0185	0.0177	B
a9	(6,53)	(15,49)	98.49	30.1	80	0.1100	0.0171	0.0165	B
a10	(58,37)	(52,45)	100	18.7	50	0.0724	0.0191	0.0182	A
a11	(62,212)	(97,189)	418.80	4.1	160	0.1301	0.0330	0.0310	B
a12	(194,196)	(178,226)	340	8.3	130	0.1156	0.0263	0.0242	B

5. Discussion

The calculated results show that the accuracy of the two distance algorithms is the same. Because the two algorithms search for the water points around the towers comprehensively, and find the shortest distance from towers, the location of the water points is the same, but the search methods are different, resulting in different calculation speeds. In terms of algorithm speed, the tower-based grid search algorithm is faster. The reason for this is that compared with the tower center distance search method, the tower base grid distance search method does not need to calculate the distance of each point, and only involves the comparison of coordinate values in the search process, which shortens the calculation time. However, the search area divided by the grid search method is a square grid, and the water near the vertex (a distance greater than 3 km) may cause a false alarm. This kind of false alarm can be avoided by comparing the calculated distance with the safe distance quantitatively.

By comparing Figure 5a with Figure 3a, the non-water part is extracted from the binary image. The accuracy of the binary image extracted by water threshold segmentation has a certain influence on the calculation results of the shortest distance between tower and water body [33]. If the non-water part is extracted from the binary image, the water point calculated to be the shortest distance between the tower and the water point may be the extracted non-water point, resulting in errors in distance calculation and even false alarms [35,36]. If the actual water body is not extracted, the shortest distance calculated may not be the actual shortest distance (with other water body points being closer to the tower). At the same time, the calculation results of flood proportion in the tower search area also produce some errors. In order to make the evaluation index of flood hidden danger scenario and the calculation of the shortest flood distance more accurate, it is necessary to improve the accuracy of extracting binary maps from flood areas. In addition to the traditional Lee filtering and Otsu threshold segmentation, an improved method for water extraction is proposed in the literature [37–43]. In [25], the selection criterion of target blocks with water is proposed. Gauss distribution is used to fit the backscattering coefficient of ground objects. Combined with the improved Gamma model, the optimal threshold position is determined, the optimization criterion is constructed, the target threshold is solved by adaptive iteration, and the threshold accuracy is increased. In [37], the identified seed point is confirmed locally based on two parameters corresponding to intensities and percentage of occurrence of intensities around the seed. A densely populated range around the seed point is computed. From the seed point, regions are grown until the intensity value of that point is within the range to complete the task, with all flooded regions captured in the SAR image. Reference [38] proposes the Bayesian network, a system whereby remote sensing data (such as multi-temporal SAR intensity image and interferometric SAR coherent data) are combined with geomorphology and other ground information to coordinate the use of different information layers, which helps to more accurately detect flood-affected areas, and reduce false positives and omissions. Reference [39] proposes the use of interferometric data to distinguish zones where water receded from areas where it persisted for a longer time, and in one case, to measure changes in water level. In [40], water categories from Landsat images are extracted and water categories from TerraSAR-X images are subtracted; the remaining water represents the flooded area. According to the different scenarios of transmission line corridors, different threshold segmentation methods can be selected to achieve the optimal extraction of the water body.

According to the environment of transmission lines, considering the local temperature, humidity, soil, plants and other factors, the C_1, C_2 and C_3 values of flood hidden danger coefficient calculation formula are adjusted based on the analytic hierarchy process; in this way, the flood hidden danger coefficient can be widely and rationally used to evaluate the flood hidden danger of poles and towers, and whether poles and towers are threatened by floods can be reasonably judged [44–46]. At the same time, the safe distance from the tower to the flood can also be adjusted according to the environmental conditions of the transmission lines. For example, in areas with loose soil, the foundation of transmission line poles and towers may be affected to a certain extent even at greater distances from the flood, and so the impact of the flood is larger. How to set the most reasonable safe distance and the value of

C_1, C_2 and C_3 according to the specific environment and the service life of the tower, considering the influence of various factors comprehensively, is a problem worthy of further study [47–51].

The purpose of this study was to create a system in which the towers which suffered from flood hazards and transmission line towers that may have potential hazards are discovered quickly, to make a preliminary evaluation of the hazard situation, to provide a reference and basis for the emergency repair of transmission line towers, and to reduce the economic losses caused by power blackouts in flood hazard scenarios. This paper is a preliminary judgment of tower flood hazard scenarios based on SAR images in which only a limited number of indicators are selected; therefore, the most intuitive distance and the elevation in the satellite image is chosen as the main measure factor. If the tower is submerged, the design scheme should focus on dealing with it. If it is not submerged, it should be properly dealt with according to the distance, the elevation and the actual environment. The elevation is the determining factor signifying whether a tower is submerged. The distance can reflect whether there is a hidden danger. When the position of the tower is higher than the horizontal plane, the tower is not submerged, but it cannot be guaranteed that the tower is absolutely safe. With the erosion from water, the foundation soil of the tower may be loose, which may cause hidden dangers for the tower. Therefore, the distance cannot be ignored, and the closer the distance is, the more likelihood there is a potential problem. On the other hand, the transmission power could be located on the slope of a small hill 30 meters above the flood and less than 1 km from the flood front edge and there is no danger for the tower. The elevation will help the system to reduce false alarms. Equation (16), which is related to hazard discrimination, shows that the closer the distance and the smaller the elevation difference, the greater the hidden danger. The combination of distance and elevation makes the result more accurate and comprehensive. At the same time, the proportion of floods in the search area helps to assist in judging the hazard situation and potential flood hazards of each tower. Combined with the results of the example, we think that the proposed index and evaluation method can be used to judge the hidden danger degree of tower flooding.

In addition, the velocity and duration of the flood are also influencing factors. However, they are not easy to visualize in satellite imagery. Flood velocity, duration, water level, and other factors do affect the tower, but it is impossible to consider all factors as indicators because of the limitations of the length of the article.

This is the first study into transmission line tower faults based on satellite images. Our system involves choosing the SAR images to study the impact of flood hazards on transmission line towers, selecting the distance and the elevation index, maximizing the use of limited information, responding quickly when a hazard occurs, and providing sufficient references and guidance for line inspections. In summary, the authors believe that the choice of distance and the elevation as the main indicator is a suitable choice at this stage. In future studies, we will try to consider as many indicators as possible, including distance, water level, velocity, soil, humidity and so on, to make the tower flood risk assessment more accurate.

A future research direction may be the establishment of a high-resolution satellite image database of the transmission line corridor. Through the comparison of multi-stage satellite images, various kinds of geological hazards are found, and the use of higher resolution satellite images is conducive to improving the accuracy of the calculations. An intelligent algorithm is used to identify the location of the pole and tower, which makes the whole process more automated. By identifying and analyzing the environment of the line, the appropriate threshold segmentation algorithm can be selected intelligently.

6. Conclusions

On the basis of backscattering water SAR imaging, this paper proposes two kinds of tower flood failure algorithms. By combining these with the distance factor, the flood ratio and the elevation, the hidden danger coefficient can be calculated, which can effectively judge the flood failure of the transmission tower.

Through testing, the tower center distance algorithm and the grid distance algorithm can accurately calculate the nearest distance between the tower and flood. A 10 m resolution image can reach an accuracy of 7 m, with the grid distance algorithm being the faster of the two.

The binary images obtained using threshold segmentation influence the judgment. Extracting the non-water body part causes the flood hazard to be falsely reported. The water body information can be extracted by combining various threshold segmentation methods to improve the accuracy.

In future research, the tower coordinates can be intelligently identified without the need for manual selection. When searching the same area again, the tower position is determined directly according to the latitude and longitude coordinates, and the database is perfected to construct an expert system.

Author Contributions: Conceptualization, L.L.; Data curation, W.L.; Formal analysis, R.D.; Methodology, L.L.; Resources, W.L.; Supervision, L.L.; Validation, R.D.; Writing—original draft, R.D.; Writing—review and editing, R.D. and L.L.

Funding: This research received no external funding.

Acknowledgments: The authors would like to express gratitude to Junwei Han, who provided much help and instruction in data processing.

Conflicts of Interest: The authors declare no conflict of interest.

References

1. Yu, D.; Wu, Y.; Chen, F. Application of helicopter patrol technology in UHV AC transmission line. *Power Syst. Technol.* **2010**, *34*, 29–32.
2. Peng, X.; Liu, Z.; Mai, X.; Luo, Z.; Wang, K.; Xie, X. A transmission line inspection system based on remote sensing: System and Its Key Technologies. *Remote Sens. Inf.* **2015**, *30*, 51–57.
3. He, R.; Lu, C.; Yu, X.; Liu, X.; Zhang, X.; Mu, X.; Feng, G.; Wang, Y.; Wu, K. Development of a new type of unmanned transmission line inspection airship. In Proceedings of the International Conference on Measuring Technology & Mechatronics Automation IEEE Computer Society, Changsha, China, 10–11 February 2018; pp. 118–120.
4. Li, J.; Wang, L.; Shen, X. Unmanned aerial vehicle intelligent patrol-inspection system applied to transmission grid. In Proceedings of the 2018 2nd IEEE Conference on Energy Internet and Energy System Integration, Beijing, China, 20–22 October 2018; pp. 1–5.
5. Wang, Z. *The Comprehensive Assessment of the Power Loss and the Economic Impact of the Power Outages under the Flooding*; Hunan University: Changsha, China, 2012.
6. Banks, D.R. Telecomm disaster recovery planning for electric utilities. In Proceedings of the 2005 Rural Electric Power Conference, San Antonio, TX, USA, 8–10 May 2005; IEEE: Piscataway, NJ, USA, 2005.
7. Miao, X.; Chen, X. Natural disasters prevention of power communications system. In Proceedings of the 2010 International Conference on Power System Technology, Hangzhou, China, 24–28 October 2010; IEEE: Piscataway, NJ, USA, 2010.
8. Kwasinski, A.; Weaver, W.; Chapman, P.; Krein, P. Telecommunications power plant damage assessment caused by hurricane katrina—Site survey and follow-up results. In Proceedings of the 2006 International Telecommunications Energy Conference, Piscataway, RI, USA, 10 September 2006; IEEE: Piscataway, RI, USA, 2006.
9. Qiu, L.J. The power grid backbone in the flood–the sideline of flood prevention and protection of the State Grid Zhangzhou Power Supply Company in 2015. *Jiangxi Electr. Power.* **2015**, *39*, 16–17.
10. Liu, X. Seven days and seven nights—Hebei Lishui power supply company to resist the "7·21" catastrophic power grid repair documentary. *State Grid.* **2012**, *9*, 52–54.
11. Guo, Z. Analysis of the impact of flood disasters on the power grid. *China's Strate. Emerg. Ind.* **2018**, *8*, 75.
12. Wang, Y.; Wang, J. Research on hazard damage characteristics and hazard prevention technology of distribution network flood geological hazards. *Power Supply Consum.* **2016**, *33*, 12–18.
13. Hu, B. Look at the damage of flood to transmission and transformation lines and power supply equipment. *Electromech. Int. Mark.* **1999**, *7*, 22–23.

14. Bruno, C.; Canale, S.; Pirri, F. X-SAR SpotLigh images feature selection and water segmentation. In Proceedings of the 2012 IEEE International Conference on Imaging Systems and Techniques, Manchester, UK, 16–17 July 2012.

15. Juval, C.; Riihimäki, H.; Pulliainen, J.; Lemmetyinen, J.; Heilimo, J. Implications of boreal forest stand characteristics for X-band SAR flood mapping accuracy. *Remote Sens. Environ.* **2016**, *186*, 47–63.

16. Giustarini, L.; Hostache, R.; Matgen, P.; Bates, P.D.; Mason, D.C.; Schumann, G.J.-P. A change detection approach to flood mapping in urban areas using TerraSAR-X. *IEEE Trans. Geosci. Remote Sens.* **2013**, *51*, 2417–2430. [CrossRef]

17. Liu, J.; Xu, Z.; Chen, F.; Chen, F.; Zhang, L. Flood hazard mapping and assessment on the Angkor world heritage site, Cambodia. *Remote Sens.* **2019**, *11*, 98. [CrossRef]

18. Chaabani, C.; Chini, M.; Abdelfattah, R.; Hostache, R.; Chokmani, K. Flood mapping in a complex environment using Bistatic TanDEM-X/TerraSAR-X InSAR coherence. *Remote Sens.* **2018**, *10*, 1873. [CrossRef]

19. Martinis, S.; Twele, A.; Strobl, C.; Kersten, J.; Stein, E. A Multi-Scale flood monitoring system based on fully automatic MODIS and TerraSAR-X processing chains. *Remote Sens.* **2013**, *5*, 5598–5619. [CrossRef]

20. Boni, G.; Ferraris, L.; Pulvirenti, L.; Squicciarino, G.; Pierdicca, N.; Candela, L.; Pisani, A.R.; Zoffoli, S.; Onori, R.; Proietti, C.; et al. A prototype system for flood monitoring based on flood forecast combined with COSMO-SkyMed and Sentinel-1 data. *IEEE J. Sel. Top. Appl. Earth Obs. Remote Sens.* **2016**, *9*, 2794–2805. [CrossRef]

21. Guy, S.; Renaud, H.; Christian, P.; Lucien, H.; Patrick, M.; Florian, P.; Laurent, P. High-Resolution 3-D flood information from radar imagery for flood hazard management. *IEEE Trans. Geosci. Remote Sens.* **2007**, *45*, 1715–1725.

22. Pulvirenti, L.; Pierdicca, N.; Chini, M.; Guerriero, L. Monitoring flood evolution in vegetated areas using COSMO-SkyMed data: The Tuscany 2009 case study. *IEEE J. Sel. Top. Appl. Earth Obs. Remote Sens.* **2013**, *6*, 1807–1816. [CrossRef]

23. Amitrano, D.; Di Martino, G.; Iodice, A.; Riccio, D.; Ruello, G. Unsupervised rapid flood mapping using Sentinel-1 GRD SAR images. *IEEE Trans. Geosci. Remote Sens.* **2018**, *56*, 3290–3299. [CrossRef]

24. Deng, J.; Wang, K.; Deng, Y.; Huang, J. An effective way for automatically extracting water body information from SPOT-5 images. *J. Shanghai Jiaotong Univ. Agric. Sci.* **2005**, *2*, 198–201.

25. Chen, L.; Liu, Z.; Zhang, H. SAR image water extraction based on scattering characteristics. *Remote Sens. Technol. Appl.* **2014**, *29*, 963–969.

26. Zeng, L.; Li, L.; Wan, L. SAR-based fast flood mapping using Sentinel-1 imagery. *Geomat. World.* **2015**, *22*, 100–107.

27. Zhang, H.; Fang, W.; Xun, S.; Yang, Y. Flood identification method based on MODIS and GIS and information extraction of land use of submerged area. *J. Catastrophology.* **2010**, *25*, 22–26.

28. Wang, J.; Liu, T.; Yu, Z.; Hu, T.; Zhang, D.; Xun, D.; Wang, D. A research on town flood information rapid extraction based on COSMO-SkyMed and SPOT-5. *Remote Sens. Technol. Appl.* **2016**, *31*, 564–571.

29. Lang, F. *Research on Polarimetric SAR Imagery Filtering and Segmentation*; Wuhan University: Wuhan, China, 2014.

30. Chen, Z. *Flooded Area Classification by High-Resolution SAR Images*; Wuhan University: Wuhan, China, 2017.

31. An, C.; Niu, Z.; Li, Z.; Chen, Z. Otsu threshold comparison and SAR water segmentation result analysis. *J. Electron. Inf. Technol.* **2010**, *32*, 2215–2219. [CrossRef]

32. An, C.; Cheng, Z. SAR water segmentation based on Otsu and improved CV model. *Signal Process.* **2011**, *2*, 221–225.

33. Lu, J.; Liu, Y.; Wu, C.; Zhang, H.; Zhou, T. Study on satellite monitoring and alarm calculation algorithm of wild fire near transmission lines. *Proc. CSEE* **2015**, *35*, 5511–5519.

34. Lu, J.; Wu, C.; Yang, L.; Zhang, H.; Liu, Y.; Xu, X.J. Research and application of forest fire monitor and early-warning system for transmission line. *Power Syst. Prot. Control* **2014**, *42*, 89–95.

35. Cheng, G.; Han, J.; Lu, X. Remote sensing image scene classification: Benchmark and state of the art. *Proc. CSEE* **2017**, *105*, 1865–1883. [CrossRef]

36. Jiao, Y.; Wang, S.; Zhou, Y.; Wang, L. Uncertainty analysis of flood disaster assessment using radar imagery. In Proceedings of the 2007 IEEE International Geoscience & Remote Sensing Symposium 2007, Barcelona, Spain, 23–27 July 2007; pp. 4729–4732.

37. Arunangshu, C.; Debasish, C. Computerized seed and range selection method for flood extent extraction in SAR image using iterative region growing. *J. Indian Soc. Remote Sens.* **2019**, *47*, 563–571.
38. D'Addabbo, A.; Refice, A.; Pasquariello, G.; Lovergine, F.P.; Capolongo, D.; Manfreda, S. A bayesian network for flood detection combining SAR imagery and ancillary data. *IEEE Trans. Geosci. Remote Sens.* **2016**, *54*, 3612–3625. [CrossRef]
39. Pulvirenti, L.; Chini, M.; Pierdicca, N.; Boni, G. Use of SAR data for detecting floodwater in urban and agricultural areas: The Role of the Interferometric Coherence. *IEEE Trans. Geosci. Remote Sens.* **2016**, *54*, 1532–1544. [CrossRef]
40. Biswajeet, P.; Mahyat, S.T.; Mustafa, N.J. A new semiautomated detection mapping of flood extent from TerraSAR-X satellite image using rule-based classification and taguchi optimization techniques. *IEEE Trans. Geosci. Remote Sens.* **2016**, *54*, 4331–4342.
41. Matgen, P.; Hostache, R.; Schumann, G.; Pfister, L.; Hoffmann, L.; Savenije, H.H.G. Towards an automated SAR based flood monitoring system: Lessons learned from two case studies. *Phys. Chem. Earth Parts A/B/C* **2011**, *36*, 241–252. [CrossRef]
42. Pulvirenti, L.; Pierdicca, N.; Chini, M.; Guerriero, L. An algorithm for operational flood mapping from synthetic aperture radar (SAR) data based on the fuzzy logic. *Nat. Hazards Earth Syst. Sci.* **2011**, *11*, 529–540. [CrossRef]
43. Chini, M.; Hostache, R.; Giustarini, L.; Matgen, P. A hierarchical split-based approach for parametric thresholding of SAR images: Flood inundation as a test case. *IEEE Trans. Geosci. Remote Sens.* **2017**, *55*, 6975–6988. [CrossRef]
44. Hu, Z.; Li, X.; Sun, Y.; Gong, Z.; Wang, Y.; Zhu, L. Flood disaster response and decision-making support system based on remote sensing and GIS. In Proceedings of the 2007 IEEE International Geoscience & Remote Sensing Symposium, Barcelona, Spain, 23–28 July 2007; pp. 2435–2438.
45. Bayraktar, H.; Bayram, B. Fuzzy logic analysis of flood disaster monitoring and assessment of damage in SE Anatolia Turkey. In Proceedings of the 2009 International Conference on Recent Advances in Space Technologies, Istanbul, Turkey, 11–13 June 2009; pp. 13–17.
46. Chen, Y. The comprehensive ranking evaluation of flood disaster based on grey-cloud whitening-weight function. In Proceedings of the 2011 International Conference on Electronic & Mechanical Engineering & Information Technology IEEE, Harbin, China, 12–14 August 2011; pp. 1932–1934.
47. Zhang, Z.; Li, C. Analysis on decision-making model of plan evaluation based on grey relation projection and combination weight algorithm. *J. Syst. Eng. Electron.* **2018**, *4*, 789–796.
48. Krejčí, J.; Petri, D.; Fedrizzi, M. From measurement to decision with the analytic hierarchy process: Propagation of uncertainty to decision outcome. *IEEE Trans. Instrum. Meas. Year.* **2017**, *66*, 3228–3236. [CrossRef]
49. Kang, H.G.; Seong, P.H. A methodology for evaluating alarm-processing systems using informational entropy-based measure and the analytic hierarchy process. *IEEE Trans. Nucl. Sci.* **1999**, *46*, 2269–2280. [CrossRef]
50. Bauer-Marschallinger, B.; Paulik, C.; Hochstöger, S.; Mistelbauer, T.; Modanesi, S.; Ciabatta, L.; Massari, C.; Brocca, L.; Wagner, W. Soil moisture from fusion of scatterometer and SAR: Closing the Scale Gap with Temporal Filtering. *Remote Sens.* **2018**, *10*, 1030. [CrossRef]
51. Chai, X.; Zhang, T.; Shao, Y.; Gong, H.; Liu, L.; Xie, K. Modeling and mapping soil moisture of plateau pasture using RADARSAT-2 imagery. *Remote Sens.* **2015**, *7*, 1279–1299. [CrossRef]

Article

Geospatial Object Detection on High Resolution Remote Sensing Imagery Based on Double Multi-Scale Feature Pyramid Network

Xiaodong Zhang *, Kun Zhu, Guanzhou Chen, Xiaoliang Tan, Lifei Zhang, Fan Dai, Puyun Liao and Yuanfu Gong

State Key Laboratory of Information Engineering in Surveying Mapping and Remote Sensing, Wuhan University, Wuhan 430079, China; zkun@whu.edu.cn (K.Z.); cgz@whu.edu.cn (G.C.); xl_tan@whu.edu.cn (X.T.); lifeizhang@whu.edu.cn (L.Z.); daifan@whu.edu.cn (F.D.); LiaoPuyun@whu.edu.cn (P.L.); gongyuanfu@163.com (Y.G.)
* Correspondence: zxdlmars@whu.edu.cn; Tel.: +86-27-6877-8033

Received: 17 February 2019; Accepted: 23 March 2019; Published: 28 March 2019

Abstract: Object detection on very-high-resolution (VHR) remote sensing imagery has attracted a lot of attention in the field of image automatic interpretation. Region-based convolutional neural networks (CNNs) have been vastly promoted in this domain, which first generate candidate regions and then accurately classify and locate the objects existing in these regions. However, the overlarge images, the complex image backgrounds and the uneven size and quantity distribution of training samples make the detection tasks more challenging, especially for small and dense objects. To solve these problems, an effective region-based VHR remote sensing imagery object detection framework named Double Multi-scale Feature Pyramid Network (DM-FPN) was proposed in this paper, which utilizes inherent multi-scale pyramidal features and combines the strong-semantic, low-resolution features and the weak-semantic, high-resolution features simultaneously. DM-FPN consists of a multi-scale region proposal network and a multi-scale object detection network, these two modules share convolutional layers and can be trained end-to-end. We proposed several multi-scale training strategies to increase the diversity of training data and overcome the size restrictions of the input images. We also proposed multi-scale inference and adaptive categorical non-maximum suppression (ACNMS) strategies to promote detection performance, especially for small and dense objects. Extensive experiments and comprehensive evaluations on large-scale DOTA dataset demonstrate the effectiveness of the proposed framework, which achieves mean average precision (mAP) value of 0.7927 on validation dataset and the best mAP value of 0.793 on testing dataset.

Keywords: very-high-resolution (VHR) remote sensing imagery; object detection; multi-scale pyramidal features; multi-scale strategies

---

1. Introduction

Object detection on very-high-resolution (VHR) optical remote sensing imagery has attracted more and more attention. It not only needs to identify the category of the object, but also needs to give the precise location of the object [1]. The improvements of earth observation technology and diversity of remote sensing platforms have seen a sharp increase in the amount of remote sensing images, which promotes the research of object detection. However, the problems of the complex backgrounds, the overlarge images, the uneven size and quantity distribution of training samples, illumination and shadows make the detection tasks more challenging and meaningful [2–4].

The optical remote sensing image object detection has made great progress in recent years [5]. The existing detection methods can be divided into four main categories, namely, template matching-based methods, knowledge-based methods, object image analysis-based (OBIA-based) methods and machine learning-based methods [2]. The template matching-based methods [6–8] mainly contain rigid template matching and deformable template matching, which includes two steps, specifically, template generation and similarity measure. Geometric information and context information are the two most common knowledge for knowledge-based object detection algorithm [9–11]. The key of the algorithm is effectively transforming the implicit connotative information into established rules. OBIA-based image analysis [12] principally contains image segmentation and object classification. Notably, the appropriate segmentation parameters are the key factors, which will affect the effectiveness of the object detection. In order to more comprehensively and effectively characterize the object, machine learning-based methods [13,14] are applied. They first extract the features (e.g., histogram of oriented gradients (HOG) [15], bag of words (BoW) [16], Sparse representation (SR)-based features [17], etc.) of the object, then perform feature fusion and dimension reduction to concisely extract features. Finally, those features are fed into a classifier (e.g., Support vector machine (SVM) [18], AdaBoost [19], Conditional random field (CRF) [20], etc.) trained with a large amount of data for object detection. In conclusion, those methods rely on the hand-engineered features, however, they are difficult to efficiently process remote sensing images in the context of big data. In addition, the hand-engineered features can only detect specific targets, when applying them to other objects, the detection results are unsatisfactory [1].

In recent years, the deep learning algorithms emerging in the field of artificial intelligence (AI) are a new kind of computing model, which can extract advanced features from massive data and perform efficient information classification, interpretation and understanding. It has been successfully applied to the fields of machine translation, speech recognition, reinforcement learning, image classification, object detection and other fields [21–25]. Even in some applications, it has exceeded the human level [26]. Compared with the traditional object detection and localization methods, the deep learning-based methods have stronger generalization and features expression ability [2]. It learns effective representation of features by a large amount of data, and establishes relatively complex network structure, which fully exploits the association among data and builds powerful detectors and locators. Convolutional neural network (CNN) is a kind of deep learning model specially designed for two-dimensional structure images inspired by biological visual cognition (local receptive field) and it can learn the deep features of images layer by layer. The local receptive field of CNN can effectively capture the spatial relationship of the objects. The characteristics of weight sharing greatly reduces the training parameters of the network and the computational cost. Therefore, the CNN-based methods are being widely used when automatically interpreting images [2,27–30].

In the field of object detection, with the development of the large public natural image datasets (e.g., Pascal VOC [31], ImageNet [32]), and the significantly improved graphics processing units (GPUs), the CNN-based detection frameworks have achieved outstanding achievements [33]. The existing CNN-based detection methods can be roughly divided into two groups: the region-based methods and the region-free methods. The region-based methods first generate candidate regions and then accurately classify and locate the objects existing in these regions, and these methods have higher detection accuracy but slower speed. Conversely, the region-free methods directly regress the object coordinates and object categories in multiple positions of the image, and the whole detection process is one-stage. These region-free methods have faster detection speed but relatively poor accuracy [34]. Among numerous region-based methods, Region-based CNN (R-CNN) [35] is a pioneering work. It utilizes the selective search algorithm [36] to generate the region proposals, and then extracts features via CNN on these regions. The extracted features are fed into a trained SVM classifier, which classifies the category of the object. Finally, bounding box regression is used to correct the initial extracted coordinates and non-maximum uppression (NMS) is used to delete highly redundant bounding boxes to obtain accurate detection results. R-CNN [35] demands to perform feature extraction at each region proposal, so the process is time-consuming [37]. Besides, the forced image resizing process on the candidate regions before they are fed into the CNN also caused information loss. To solve the above problems, He et al. proposed Spatial Pyramid Pooling Network (SPP-Net) [38], which adds a spatial pyramid layer, namely, Region-of-Interest (RoI) pooling layer, on the top of the last convolutional layer. The RoI pooling layer divides the features and generates fixed-length outputs, therefore it can deal with the arbitrary-size input images. SPP-Net [38] performs one-time features extraction to obtain an entire-image feature map, and the region proposals share the entire-image feature map, which greatly speeds up the detection. On the basis of R-CNN, Fast-RCNN [39] adopts the multi-task loss function to carry out classification and regression simultaneously, which improves the detection, positioning accuracy and greatly improves the detection efficiency. However, using the selective search algorithm to generate region proposals is still very time-consuming because the algorithm implements on the central processing unit (CPU). In order to take advantage of the GPUs, Faster R-CNN [37], consisting of a region proposal network (RPN) and Fast R-CNN, was proposed. The two networks share convolution parameters, and they have been integrated into a unified network. Thus, the region-based object detection network achieves end-to-end operation. Feature pyramids play a crucial role in multi-scale object detection system, which combine resolution and semantic information over multiple scales. Feature pyramid network (FPN) [40] was proposed to simultaneously utilize low-resolution, semantically strong features and high-resolution, semantically weak features, it is superior to single-scale features for a region-based object detector and shows significant improvements in detecting small objects. In addition to the region-based object detection frameworks, there are many region-free object detection networks, including Over-Feat [41], you only look once (YOLO) [42] and single shot multi-box detector (SSD) [43], etc. These one-stage networks consider object detection as a regression problem, they do not generate region proposals and predict the class confidence and coordinates directly. They greatly improve the detection speed, although sacrificing some precision.

The CNN-based natural imagery object detection has made great progress, but high-precision and high-efficiency object detection for remote sensing images still has a long way to go. Different from natural images, remote sensing images usually show the following characteristics:

1. The perspective of view. Remote sensing images are usually obtained from a top-down view while natural images can be obtained from different perspectives, which greatly affects how objects are rendered on the images [1].
2. Overlarge image size. Remote sensing images are usually larger in size and range than natural images. Compared with natural image processing, remote sensing image processing is more time-consuming and memory-consuming.
3. Class imbalances. The imbalances mainly include category quantity and object size. Objects in natural scene images are generally uniformly distributed and not particularly numerous, but a single remote sensing image may contain one object or hundreds of objects and it may also simultaneously include large objects such as playgrounds and small objects like cars.
4. Additional influence factors. Compared with natural scene image, remote sensing image object detections are affected by illumination condition, image resolution, occlusion, shadow, background and border sharpness [33].

Therefore, constructing a robust and accurate object detection framework for remote sensing images is very challenging, but it is also of much significance. To overcome the size restrictions of the input images, the problem of small objects loss and retain the resolution of the objects, Chen et al. [1] put forward MultiBlock layer and MapBlock layer based on SSD [43]. The MultiBlock layer divides the input image into multiple blocks, the MapBlock layer maps the prediction results of each block to the original image. The network achieves a good effect on airplane detection. Considering the complex distribution of geospatial objects and the low efficiency for remote sensing imagery, Han et al. [33] proposed the P-R-Faster R-CNN, which achieves multi-class geospatial object detection by combining the robust properties of transfer mechanism and the sharable properties of Faster R-CNN. Guo et al. [3] proposed a unified multi-scale CNN for multi-scale geospatial object detection, which consists of a multi-scale object proposal network and a multi-scale object detection network. The network achieves the best precision on the Northwestern Polytechnical University very high spatial resolution-10 (NWPU VHR-10) [44] dataset. However, for small and dense objects detection on remote sensing images, they did not propose an effective solution, and did not make full use of the resolution and semantic information simultaneously, which may lead to unsatisfactory results in the case of more complex backgrounds, numerous data and overlarge image size [4,40]. Some frameworks [1,45–47] only have effects for certain types of objects. Besides, RoI pooling layer in these networks will cause misalignments between the inputs and their corresponding final feature maps, these misalignments affect the object detection accuracy, especially for small objects.

To solve the above problems, we presented an effective framework, namely, Double Multi-scale Feature Pyramid Network (DM-FPN), which makes full use of semantic and resolution features simultaneously. We also put forward some multi-scale training, inference and adaptive categorical non-maximum suppression (ACNMS) strategies. The main contributions of this paper are summarized as follows:

1. We have constructed an effective multi-scale geospatial object detection framework, which achieves good performance by simultaneously utilizing low-resolution, semantically strong features and high-resolution, semantically weak features. Accordingly, the RoI Align layer used in our framework can solve the misalignment caused by RoI pooling layer and it improves the object detection accuracy, especially for small objects.

2. We proposed several multi-scale training strategies, including the patch-based multi-scale training data and the multi-scale image sizes used during training. To overcome the size restrictions of the input images, we divided the image into blocks with a certain degree of overlap. The patch-based multi-scale training data strategy both enhance the resolution features of the small objects and integrally divide the large objects into a single patch for training. In order to increase the diversity of objects, we adopt multiple image sizes strategy for patches during training.

3. During the inference stage, we also proposed a multi-scale strategy to detect as many objects as possible. Besides, depending on the intensity of the object, we adopt the noval ACNMS strategy, which can effectively reduce redundancy among the highly overlapped objects and slightly overcome the uneven quantity distribution of training samples, enabling the framework preferably to detect both small and dense objects.

Experiment results evaluated on DOTA [48] dataset, a large-scale dataset for object detection in aerial images, indicating the effectiveness and superiority of the proposed framework. The rest of this paper is organized as follows. Section 2 introduces the related work involved in the paper. Section 3 elaborates the proposed framework in detail. Section 4 mainly includes the description of the datasets, evaluation criteria and experiment details. Section 5 implements ablation experiments and makes reliable analyses to the results. Section 6 discusses the proposed framework and analyzes its limitations. Finally, the conclusions are drawn in Section 7.

2. Related Works

In this section, we will first review some outstanding region-based object detection frameworks, they have achieved remarkable accomplishments on natural image object detection. Then we will introduce RoI Align layer, which can significantly improve the detection performance of small objects.

2.1. Region-Based Object Detection Networks

The region-based object detection networks are mainstream frameworks for high-precision object detection, including R-CNN, SPP-Net, Fast R-CNN and Faster R-CNN [35,37–39]. Their common process is to first generate numerous candidate areas by the region proposal algorithms [36,49,50]. Then, the networks employ CNN to extract abundant features from these candidate regions and infer the category and coordinates of objects on each region. Finally, a bounding box algorithm is utilized to get precise coordinates. Faster R-CNN integrates these steps to form a unified network and realizes end-to-end object detection. It consists of two modules, formally, RPN and Fast R-CNN, and the two tasks share convolutional features. Figure 1 shows the overall architecture of Faster R-CNN.

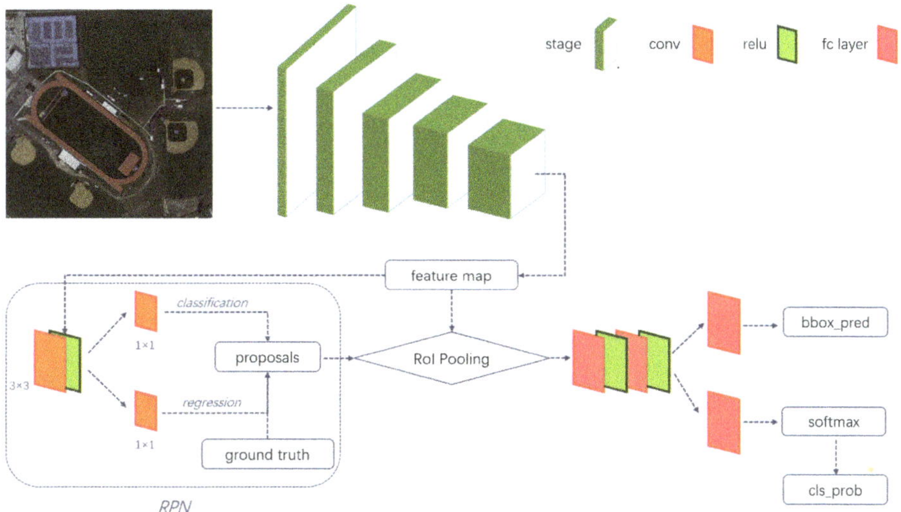

Figure 1. The architecture of Faster R-CNN. The "conv" represents convolutional layer, the "relu" represents activation function and the "fc layer" represents fully connected layer. The network outputs intermediate layers of the same size in the same "stage". The "bbox_pred" represents the position offset of the object and the "cls_prob" represents the probability of the category.

RPN is a kind of fully convolutional network [51], it deals with the arbitrary-size input image and outputs a set of region proposals with an objectness score. These candidate regions will be fed into the following Fast R-CNN for precise detection. The core scheme of RPN is "anchors", which simultaneously predicts multiple region proposals of diversiform scales and aspect ratios with a total number of k at each sliding window in the last shared convolutional layer. The features obtained from each sliding window will be imported into two sibling 1×1 convolutional layers, specifically, the box-classification layer (*cls*) and the box-regression layer (*reg*). The *cls* layer is used to identify a binary class label of being an object or not while the *reg* layer is used to correct the coordinates of the object. Therefore, the *cls* layer has 2k outputs while the *reg* layer has 4k outputs.

After RPN processing, we got a mass of candidate regions with class-agnostic and coordinate attributes. These regions will be fed into the subsequent Fast R-CNN for further category judgment and coordinate regression. Fast R-CNN adopts RoI pooling layer to extract fixed-length feature vectors from arbitrary-size candidate regions and these feature vectors are fed into categorical classification and regression layers to obtain the final detection results. The RPN and Fast R-CNN employ the approximate joint training scheme to share convolution. As such, an efficient and end-to-end object detection framework is constructed.

2.2. Feature Pyramid Network

Most region-based object detection frameworks only use the single-scale features for faster detection, such feature representations are very unfriendly to small objects. In Faster R-CNN, the backbone adopts Visual Geometry Group 16 weight layers (VGG16 [52]) and the last feature map reduces to 1/32 compared to the original image after 5 convolutional layers (with a pooling step of 2), some small objects like cars and ships will lose a large proportion of features after such operations. In the deep convolutional networks, the low-level layers have poor semantic but strong resolution while the high-level layers have rich semantic but scarce resolution [40]. Although some frameworks [43,53] adopt multi-scale feature maps that already computed from different layers, they abnegate low-level features and therefore lose the opportunity to take advantage of higher-resolution features. Combining strong resolution and semantic information will enhance the detection performance, especially for small objects. In a pioneering way, FPN leverages the in-network features obtained from the last layer of each stage in the convolutional networks (ConvNets). It combines coarse-resolution, semantically strong features with high-resolution, semantically weak features to construct a multi-scale pyramidal hierarchy network without additional memory consumption. We note that if the output feature maps have the same size, they are in the same stage. As shown in the Figure 2, the core mechanism of the FPN mainly includes bottom-up pathway, top-down pathway and lateral connections.

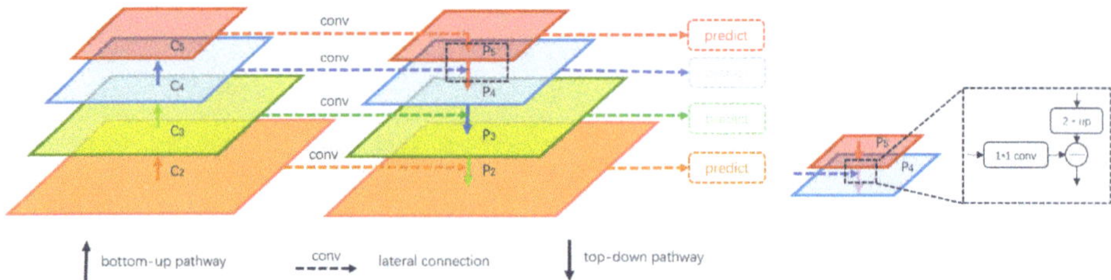

Figure 2. The core mechanism of the FPN mainly includes bottom-up pathway, top-down pathway and lateral connections.

- Bottom-up pathway. Actually, this operation is the forward propagation process of the network. During the operation, the last convolutional layer in each stage is extracted to establish a feature pyramid. Compared with other methods [54–56], this mechanism requires no additional memory footprint.
- Top-down pathway and lateral connections. The top-down pathway upsamples the feature map obtained from the bottom-up pathway to the same size as the semantically coarser, but spatially stronger feature maps. The lateral connections merge the same-size feature maps obtained from the bottom-up pathway and the top-down pathway respectively, which first undergoes a 1×1 convolutional layer to reduce channel dimensions. The mergence process is implemented by element-wise addition. Subsequently, a 3×3 convolution is executed on each merged feature map to eliminate the aliasing effect of upsampling.

2.3. ROI Align

ROI Align is a kind of regional feature aggregation method proposed in Mask R-CNN [57], which solves the problem of misalignment caused by RoI pooling during the two integer quantification operations. RoI pooling layer divides the region proposal on the last convolutional layer into a fixed-length (e.g., 7×7) feature map for subsequent classification and bounding box regression tasks. Since the coordinates of candidate regions are obtained by regression, generally speaking, they are floating-numbers. After rounding down, the data after the decimal point is abandoned. As shown in Figure 3a, there are two rounding operations during the pooling: the coordinates of candidate region are first quantified to integer, then the quantified RoI is divided into $k \times k$ bins on average, and each bin is quantified again thus introducing misalignments between the RoI and the final feature map. Such misalignments are harmful to objects detection task, especially for small objects.

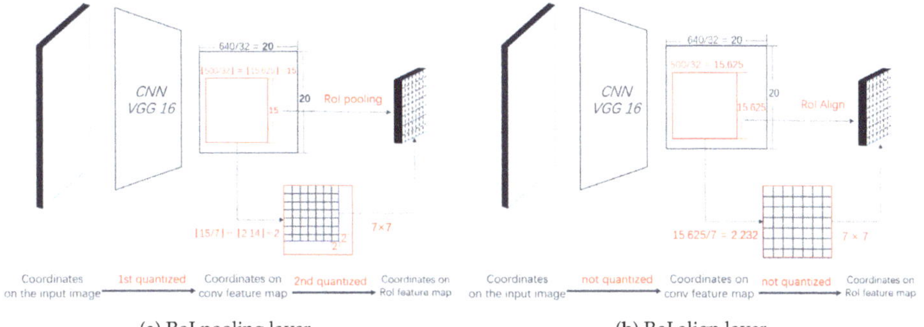

(**a**) RoI pooling layer.   (**b**) RoI align layer.

Figure 3. RoI align layer solves misalignments caused by RoI pooling layer.

RoI Align was proposed to solve the above deficiency of RoI Pooling, it abnegates all quantifications and utilizes bilinear interpolation to obtain the precise values. Formally, RoI Align retains the original floating-numbers instead of quantified integers. The alignment process is shown in Figure 3b. During the first quantification, the boundary coordinates of each candidate region are not round down to maintain floating-numbers. During the second quantification, each RoI is divided into $k \times k$ bins and this process is still not round down. Subsequently, four fixed sampled points are calculated by bilinear interpolation in each RoI bin, and the maximum or average pooling is performed to get align results. RoI Align solves the misalignments between the inputs and the extracted feature maps, which is significant for object detection on remote sensing images that contain numerous small objects.

3. Framework

In this section, we will elaborate the details of our proposed framework. In order to efficiently detect the objects on remote sensing images, we also propose some multi-scale training and inference strategies. Meanwhile, different ACNMS thresholds are selected according to the size and intensity of the category, which can improve the detector performance to some extent.

3.1. The Core Mechanism of the Proposed Network

3.1.1. The Overall Structure

The overall structure of the proposed framework named Double Multi-scale Feature Pyramid Network (DM-FPN) is shown in Figure 4.

The infrastructure of DM-FPN is based on Faster R-CNN [37] with FPN [40]. Formally, both the original region proposal network and the detection network were modified by FPN. DM-FPN combines coarse-resolution, semantically strong features with high-resolution, semantically weak features, and such operations have great advantages in detecting small objects. We adopt ResNet50 [58] as backbone of our framework. The convolution can be divided into 5 stages and the output of each stage's last residual block was selected as $\{C_2, C_3, C_4, C_5\}$, noting that they have strides of $\{4, 8, 16, 32\}$ pixels corresponding to the original image. We do not utilize the first stage because it is memory-consuming. This process is called the bottom-up pathway, which has been described in Section 2.2. The corresponding $\{P_2, P_3, P_4, P_5\}$ were obtained by top-down path, lateral connections and mergence. Actually, to eliminate the aliasing effect of upsampling, a 3×3 convolution is executed on each merged feature map to obtain the final feature maps $\{P_2, P_3, P_4, P_5\}$, which are shared by the region proposal network and the class-specific detection network.

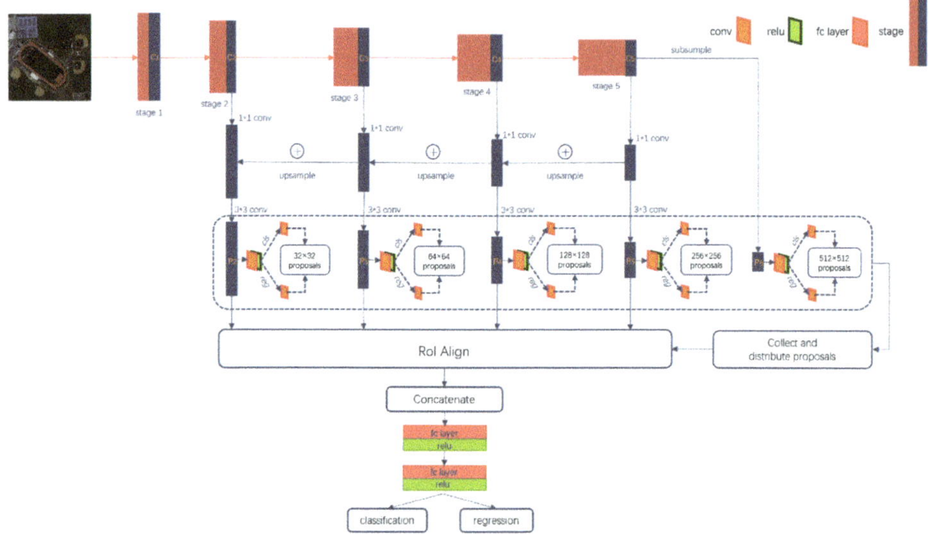

Figure 4. The overall structure of the proposed DM-FPN. It consists of a multi-scale region proposal network and a multi-scale object detection network. These two modules share convolutional layers.

3.1.2. Multi-Scale Region Proposal Network

The original RPN extracts region proposals on the last single-scale convolutional layer. In order to take advantage of the pyramid character of FPN, we need to extract candidate regions on multiple convolutional layers, namely, $\{P_2, P_3, P_4, P_5, P_6\}$, noting that P_6 is simply a stride 2 subsampling of P_5, which is only used in multi-scale region proposal network. The anchors own ranges of $\{32^2, 64^2, 128^2, 256^2, 512^2\}$ pixels on $\{P_2, P_3, P_4, P_5, P_6\}$ respectively. On each feature map, there are three aspect ratios, namely, $\{1{:}2, 1{:}1, 2{:}1\}$. As a result, there are a total of 15 anchors on these pyramidal feature maps. The selection of positive and negative samples is determined by the Intersection-over-Union (IoU) between the region proposal and ground-truth box. We note that IoU is defined as the ratio between the intersection and the union of two boxes. If an anchor has the highest IoU with a given ground-truth box or it has an IoU greater than 0.7 with any ground-truth box, then it will be assigned to the positive. Conversely, if an anchor has an IoU less than 0.3 for all ground-truth boxes, it's a negative sample. We abandon samples that are neither positive nor negative. In a mini-batch of 256, the ratio of positive to negative samples is 1:1. These rules apply to $\{P_2, P_3, P_4, P_5, P_6\}$ indistinguishably. Specially, the common ground-truth boxes are equally participated in the calculation with the pyramid anchors located on five-level feature maps. With these definitions, the loss function for an image is defined as:

$$L\left(\{p_i\}, \{t_i\}\right) = \frac{1}{N_{cls}} \sum_i (p_i, p_i^*) + \lambda \cdot \frac{1}{N_{reg}} \sum_i p_i^* L_{reg}\left(t_i, t_i^*\right) \tag{1}$$

where i represents the index of an anchor in a mini-batch while p_i is the predicted probability of anchor i being an object. If the anchor is positive, the ground-truth label p_i^* equals to 1, otherwise equals to 0. t_i is a vector that consists of four parameterized coordinates of the predicted bounding box, and t_i^* is that of the ground-truth box associated with a positive anchor. The classification loss L_{cls} is represented by the log loss, which identifies a binary class label of being an object or not. And the regression loss L_{reg} is constructed by the Smooth L1 loss. The above two loss functions are weighted by a balancing parameter λ. Usually, the *cls* term is normalized by the mini-batch size while the *reg* term is normalized by the number of anchors. In this paper, we specify that N_{cls} and N_{reg} are equal to 256 and 2000, respectively. We set λ is equals to 9 and thus both *cls* and *reg* terms are roughly equally weighted.

We note that we reserve the top 2000 region proposals based on their *cls* scores on {P_2, P_3, P_4, P_5, P_6} respectively, then we concatenate these candidate boxes and adopt Non-Maximum Suppression (NMS) with a fixed IoU threshold of 0.7 to retain the final 2000 RoIs, which will be fed into the subsequen class-specific detection network for exact object detection.

3.1.3. Multi-Scale Class-Specific Detection Network

Fast R-CNN [39] is a single-scale region-based object detection framework, which utilizes RoIs generated by RPN for object detection. Different from the previous networks that pooling RoI to single-scale feature map, we need to align RoIs from different scales to the multiple pyramidal feature maps. We assign an RoI of width w and height h (based on the input image) to the level P_k by:

$$k = \left\lfloor k_0 + \log_2(\sqrt{wh}/224) \right\rfloor \tag{2}$$

where 224 is the normative ImageNet pre-training size as FPN [40] does, and k_0 is the level that an RoI with a size of $w \times h = 224^2$ should be mapped into. Notably, we assigned k_0 equals to 4 as [40] does. These RoIs can be assigned to different levels according to their size. For example, if an anchor has a width of 188 and a height of 111, it should be mapped into the P_3 level. Subsequently, we adopt RoI align to extract 7×7 feature maps, which will be fed into two 1024-d fully-connected layers before the final classification and bounding box regression layers. Based on the above settings, both region proposal network and class-specific detection network can utilize multi-scale pyramidal features for object detection.

3.2. Multi-Scale Training Strategies

Multi-scale training strategies mainly include the patch-based multi-scale training data and the multi-scale image sizes used during training. Their descriptions are as follows:

1. Patch-based multi-scale training data. The size restrictions of the input images cause a lot of semantic information will lost in the deep convolutional layers, especially for small objects. Therefore, we slice remote sensing images into patches with a certain degree of overlap, and then send these image blocks into the network for training. At the same time, considering the uneven distribution of objects on the remote sensing image, which may include large objects such as playgrounds, and may also include small objects like cars, we enlarge and shrink remote sensing images by a factor of 2 and 0.5 respectively. The enlarged remote sensing images enhance the resolution features of the small objects while the shrunken remote sensing images integrally divide the large objects into a single patch for training.
2. Multi-scale image sizes used during training. In order to enhance the diversity of objects, we adopt multiple scales for patches during training. Each scale is the pixel size of a patch's shortest side and the network uniformly select a scale for each training sample at random.

3.3. Multi-Scale Inference Strategies

We scale images to detect as many objects as possible during inference, and the scaled images include enlarged and shrunken images, horizontally and vertically flipped images. Specifically, we first perform multi-scale process on each test image, then we slice it into patches with a certain degree of overlap according to its size and carry out detection on these image blocks. Finally, we apply ACNMS to these concatenate bounding boxes from each patch to get the final results.

3.4. Adaptive Categorical Non-Maximum Suppression (ACNMS)

NMS is a post-processing module in the object detection framework, which is mainly used to delete highly redundant bounding boxes. A single remote sensing image may contain one big object or hundreds small objects, thus there exists a class imbalance between different categories. In the previous multi-class object detection works [3,4,33], the NMS thresholds for different categories are the same, but we find that different NMS thresholds for different categories based on the category intensity (CI) can improve the accuracy of object detection to a certain extent. We define CI as:

$$CI = N_{IoC}/N_{img} \tag{3}$$

where N_{IoC} means the total number of instances for each category, N_{img} means the total number of images. If the CI of a category is greater than the given threshold, we set this category a larger NMS threshold than the generic NMS threshold. In general, NMS thresholds for denser objects are larger because they overlap each other more commonly.

4. Dataset and Experimental Settings

4.1. Dataset Description

We evaluated our proposed framework on DOTA [48] dataset, which contains 2806 aerial images with pre-divided 1411 training images, 458 validation images and 937 testing images. We note that the testing images have no labels, however, you can submit the test results in a fixed format to DOTA Evaluation Server (http://captain.whu.edu.cn/DOTAweb/evaluation.html). Those DOTA images are obtained from different sensors and platforms with crowdsourcing and the size ranges from 800×800 to 4000×4000 pixels. DOTA consists of 15 common categories, namely, plane, ship, storage tank, baseball diamond, tennis court, basketball court, ground track field, harbor, bridge, large vehicle, small vehicle, helicopter, roundabout, soccer ball field and swimming pool. The fully annotated DOTA dataset contains 188,282 instances, each of which is labeled by an oriented quadrilateral instead of an axis-aligned one, which is typically used for object annotation in natural scene images. Another common geospatial object detection dataset is NWPU VHR-10 [44], which contains 800 images in 10 categories with a total of 3651 instances. The average size of NWPU VHR-10 is 1000×1000 pixels. Compared with NWPU, DOTA is a larger annotated dataset for multi-class geospatial object detection, which has more complex backgrounds, larger image size and denser object distribution thus more reflective of the real-world applications [48]. Therefore, the evaluation on DOTA can better verify the effectiveness and robustness of our proposed network.

The benchmark of DOTA contains two detection tasks. Task 1 uses the initial oriented bounding boxes as ground truth. Task 2 uses the converted horizontal bounding boxes as ground truth. In this work, we only focus on the horizontal bounding box detection task with $(xmin, ymin, xmax, ymax)$ format, so we need to convert the labeled oriented bounding box into the minimum bounding rectangle for each image. Figure 5 shows some examples about the original annotations and their minimum bounding rectangles.

4.2. Evaluation Criteria

We adopted Precision-Recall Curve (PRC) and Average Precision (AP) as evaluation criteria in our experiments, which are widely used in the object detection works.

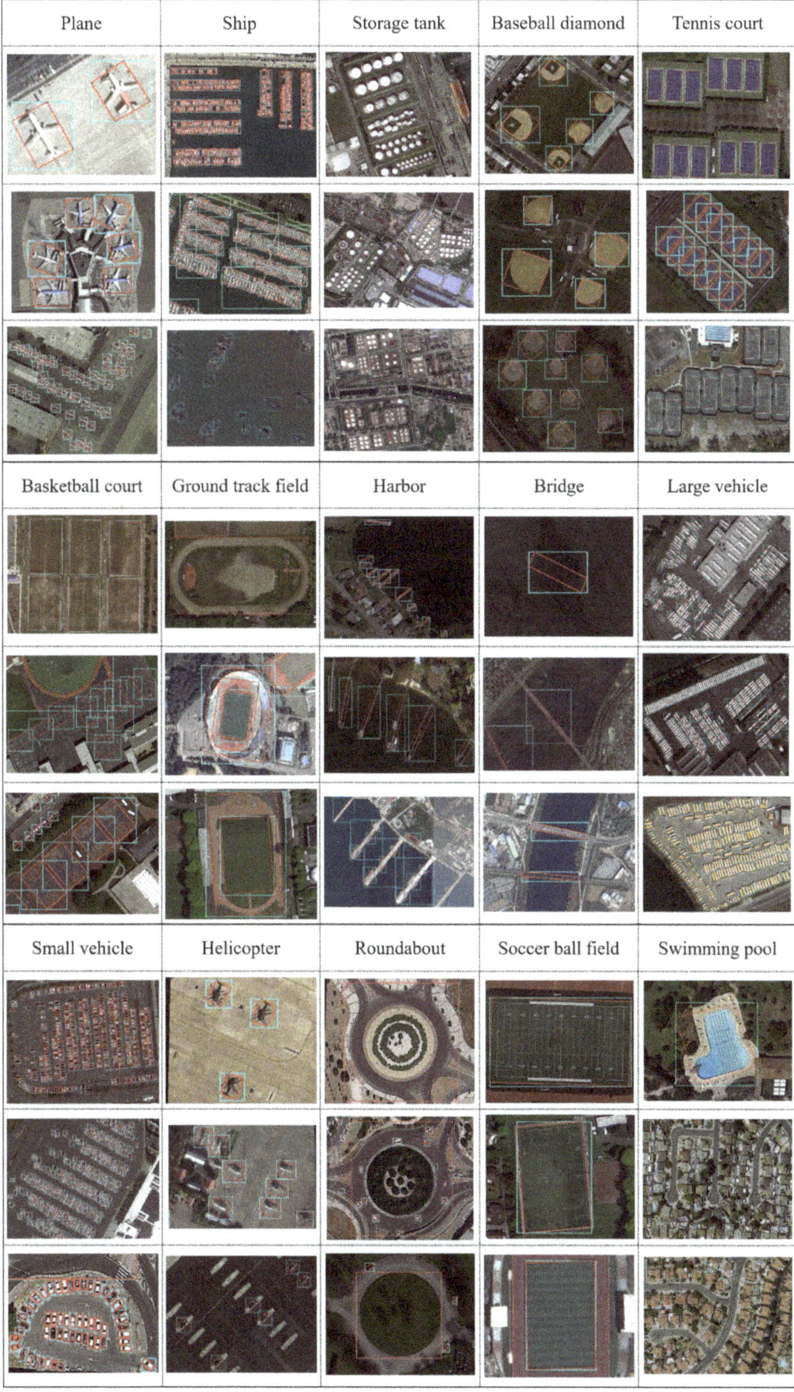

Figure 5. Examples of Annotated Images. The red quadrilaterals represent original annotations, the green rectangles represent minimum bounding rectangles.

4.2.1. Precision-Recall Curve

The precision metric is the ratio of the correct identification quantity to the total identification quantity while the recall metric is the proportion of the correct identification quantity to the total labeled quantity, which can be illustrated by the following two formulas:

$$precision = TP/(TP + FP) \tag{4}$$

$$recall = TP/(TP + FN) \tag{5}$$

we note that if the IoU value between the predicted bounding box and the ground truth is larger than 0.5, it will be considered as true positive (TP), otherwise, it will be considered as false positive (FP). In addition, false negative (FN) refers to the prediction boxes that overlap with ground truth but do not have the maximum overlap value. The precision-recall curve (PRC) describes the relationship between the precision metric and the recall metric, an object detector of a certain category is considered good if its prediction stays high as recall increases.

4.2.2. Average Precision

Average Precision (AP) is the averaged precision across all recall values between 0 and 1, namely, the area under the PRC. A higher AP indicates a better detector. Mean average precision(mAP) represents the average AP over all categories.

4.3. Baseline Methods

We compared the proposed framework with the classic region-based methods including Faster RCNN [37] and FPN [40] on DOTA validation dataset. For the testing dataset, we submitted the inference results to DOTA website because of lacking annotated labels, and we selected several current top-ranked results for comparison.

4.4. Implementation Details

We implemented our network on the open source Caffe2 (https://caffe2.ai/) framework and executed on a 64-bit Ubuntu 16.04 computer with 8GB memory GeForce GTX1070Ti GPU. We note the comparison models were implemented in their original environments without any additions.

4.4.1. Training

We first enlarged and shrunk the original images by a factor of 2 and 0.5 respectively, then we sliced the original and scaled images into patches of 1000×1000 pixels with an overlap of 500 pixels. All the original image patches, partial randomly selected enlarged and shrunken image patches were taken as our training samples with a total number of 31,396. These training samples will be fed into the network after data augmentation, which includes rotation and flip. We adopted three scales during training, they are 800×800, 900×900 and 1000×1000 pixels respectively. Each scale is the pixel size of a patch's shortest side and the network uniformly select a scale for each training sample at random. We adopted ResNet50 as our backbone, which was pre-trained on ImageNet dataset. We trained a total of 300k iterations with a learning rate of 0.0025 for the first 150k iterations, 0.00025 for the next 50k iterations, and 0.000025 for the remaining 100k iterations, which took us about 40 hours in total. The network was trained by stochastic gradient descent algorithm with a mini-batch of 2 images. Weight decay and momentum are 0.0001 and 0.9 respectively.

4.4.2. Inference

We implemented inference based on the image patches in order to detect as many objects as possible. To accelerate the inference, we sliced validation images into patches of 1000×1000 pixels with an overlap of 200 pixels. We performed detection on each diced image and then concatenated the predicted results from each patch. We set CI threshold to 10, and the ACNMS threshold is 0.38. Specifically, if the intensity of a category is greater than CI threshold, then its NMS threshold is 0.38, otherwise we set its NMS threshold to 0.3. Meanwhile, to verify the effectiveness of the multi-scale inference strategies, we also performed the same detections on the shrunken images, the horizontal rotation and vertical rotation images simultaneously. We did not perform detections on the enlarged images because of their vastly time-consuming.

5. Results and Analysis

5.1. Ablation Experiments

Ablation experiments were carried out to verify the effectiveness of the proposed multi-scale training, inference and ACNMS strategies. In the following subsection, we will gradually verify the relevant strategies. The multi-scale training and inference strategies can be expressed as Equation (6):

$$(p)\_based(x) + (s)\_scale \tag{6}$$

where p represents the patch sizes used for training, x represents the patch sources used for training and s represents the patch scales used for inference. For example, 800_based(4)+1_scale means that we resized the pre-divided patches into 800×800 pixels for training. These multi-scale training data include four data sources, specifically, the original images, the patches obtained from original images, enlarged and shrunken images. During inference, we performed detection on the patches only obtained from original images. The size of these patches is 1000×1000 pixels with an overlap of 200 pixels. Finally, we concatenated the bounding boxes from each patch and adopted ACNMS to get the final results. The detailed explanations are shown in Table 1.

Table 1. Details of multi-scale training and inference strategies.

Parameters	Connotation	Values	Details
p	Patch sizes used for training	0	Training with original images
		800	Training with patches of 800×800 pixels
		900	Training with patches of 900×900 pixels
		1000	Training with patches of 1000×1000 pixels
		$(800, 900, 1000)$	Training patches with a randomly selected size from $(800^2, 900^2, 1000^2)$ pixels
x	Patch sources used for training	0	Original images without slicing
		1	Patches from original images
		4	Original images, patches from original images, partial randomly selected enlarged and shrunken images simultaneously
s	Patch scales used for inference	0	Inference on the original images
		1	Inference on the patches from the original images
		4	Inference on the patches from original images, shrunken images, horizontal and vertical rotation images simultaneously

5.1.1. Patch-Based Training and Inference Strategies

In this section, we conducted two sets of ablation experiments to illustrate the superiority of patch-based training and inference strategies. We adopted (a), (b), (c), etc. to represent each method in Table 2. In each column, the bold number indicates the best detection result, and the other tables are the same. Table 2(a) carried out training using the original images without patches. For fair comparison, we resized the original images to 1000×1000 pixels and the inference was also performed on the original images. The training strategies of Table 2(b) were the same as Table 2(a), however, it performed inference on the patches obtained from the original images. Both training and inference of Table 2(c) were performed on the patches obtained from the original images.

Table 2. The AP values of ablation experiments for patch-based training and inference strategies.

Method	0_based(0)+0_scale (a)	0_based(0)+1_scale (b)	1000_based(1)+1_scale (c)
plane	0.7078	0.8015	**0.8986**
ship	0.6023	0.829	**0.8886**
storage tank	0.4213	0.5483	**0.7808**
baseball diamond	0.6478	0.4105	**0.8112**
tennis court	0.888	0.9064	**0.9078**
basketball court	0.4822	0.5279	**0.6671**
ground track field	0.4304	0.4104	**0.7225**
harbor	0.8391	0.7742	**0.8894**
bridge	0.2973	0.308	**0.6326**
large vehicle	0.675	0.7244	**0.764**
small vehicle	0.5571	0.6002	**0.679**
helicopter	0.3309	0.1027	**0.654**
roundabout	0.2943	0.3957	**0.722**
soccer ball field	0.4059	0.3982	**0.6588**
swimming pool	0.4472	0.5328	**0.6153**
mAP	0.5351	0.5513	**0.7528**

Comparing Table 2(a) and Table 2(b), we can observe that patch-based inference strategy has improved detection accuracy on most categories except baseball-diamond, ground-track-field, harbor, helicopter and soccer-ball-field. Through further experiments we found that the sizes of baseball-diamond, ground-track-field, harbor, and soccer-ball-field are so large that they often beyond the scope of a single patch, therefore, training with original images but prediction with patches are not conducive to these objects. However, the poor detection effect of helicopter is mainly caused by: (1) Quite a few samples, the sample number (630) of helicopter is far fewer than other categories; (2) Some helicopter samples are similar to airplane, and these two categories generally appear simultaneously. Nevertheless, the patch-based inference strategy is still slightly ascending.

With the patch-based training strategy, Table 2(c) shows the superiority compared to Table 2(b), it not only has an overwhelming advantage in mAP (0.5513 to 0.7528), but also increases the AP value of each category, which illustrates that the patch-based training strategy is targeted and more adequately understand the characteristics of the objects. Besides, the patch-based training strategy implicitly increases the sample number of each category, especially for the sample-scarce categories.

Computational efficiency is also an important indicator in evaluating a framework's performance, so we calculated the average running time for each strategy. The results are shown in Table 3.

Table 3. Average running time of patch-based training and inference strategies.

Method	0_based(0)+0_scale (a)	0_based(0)+1_scale (b)	1000_based(0)+1_scale (c)
Average running time per image (second)	0.3882	4.2434	3.8553

We note that the patch-based inference strategies (Table 3(b),(c)) consume more average running time than the original-image-based inference strategy (Table 3(a)), which is easy to understand because the patch-based inference strategy handles more images (patches). In addition, Table 3(c) takes less time than Table 3(b), which further demonstrates that the patch-based training strategy can more adequately extract the characteristics of the objects. The quantified PRCs over two ablation experiments are plotted in Figure 6.

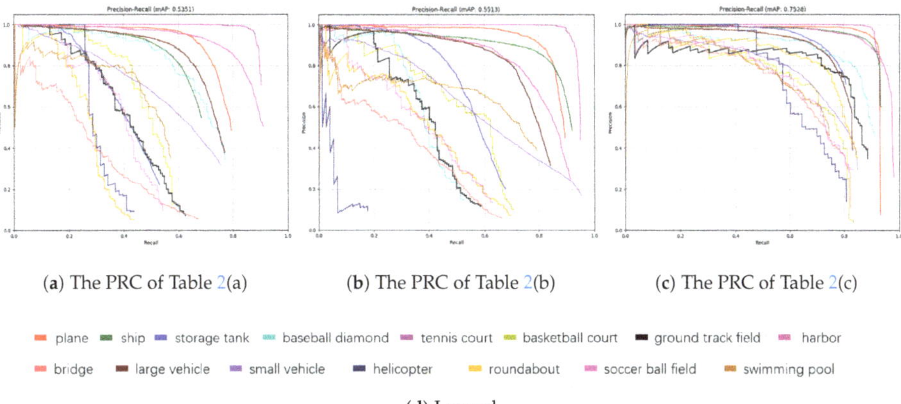

(**a**) The PRC of Table 2(a)   (**b**) The PRC of Table 2(b)   (**c**) The PRC of Table 2(c)

(**d**) Legend

Figure 6. The PRCs of training and inference strategies.

5.1.2. Multi-Scale Training Data and Multi-Scale Sizes Used during Training Strategies

Multi-scale training data consist of the original images, patches that based on the original images, the enlarged images and the shrunken images. Multi-scale sizes used during training refers to that an image or patch will be resized to a random scale from specified range before being fed into the framework and each scale is the pixel size of an image or patch's shortest side. We performed two relevant ablation experiments to verify the significance of multi-scale training data and multi-scale sizes used during training. The results are shown in Table 4.

Table 4. The AP values of ablation experiments for multi-scale strategies.

Method	1000_based(1)+1_scale (a)	800_based(4)+1_scale (b)	900_based(4)+1_scale (c)	1000_based(4)+1_scale (d)	(800,900,1000)_based(4)+1_scale (e)
plane	0.8986	0.899	0.9	0.8983	**0.9007**
ship	0.8886	0.8854	0.8856	0.891	**0.8919**
storage tank	0.7808	0.7781	0.7805	0.7794	**0.7817**
baseball diamond	0.8112	**0.8339**	0.8199	0.8172	0.8257
tennis court	0.9078	0.908	**0.9084**	0.908	0.908
basketball court	0.6671	0.6914	0.6976	**0.7275**	0.7061
ground track field	0.7225	0.7789	0.7681	**0.7966**	0.7683
harbor	0.8894	0.8832	0.8853	0.8894	**0.891**
bridge	0.6326	0.6232	0.6306	0.6362	**0.6444**
large vehicle	**0.764**	0.7504	0.752	0.7636	0.7599
small vehicle	0.679	0.6298	0.6416	0.7182	**0.7209**
helicopter	0.654	0.6815	0.7226	0.7222	**0.7385**
roundabout	0.722	0.7232	0.7173	0.7254	**0.7281**
soccer ball field	0.6588	0.6338	0.6724	0.673	**0.7122**
swimming pool	0.6153	0.7049	0.7215	0.672	**0.7253**
mAP	0.7528	0.7603	0.7669	0.7745	**0.7802**

The training data used in the Table 4(a) are only from the original images while the training data used in the remaining groups include the original images, the patches from the original images, the enlarged images and the shrunken images. Table 4(b)–(d) resize the training data to 800×800, 900×900, 1000×1000 pixels respectively. Table 4(e) utilizes multiple sizes including (800, 900, 1000) pixels, and the training data will be resized to a randomly selected size before being fed into the network. Apart from this, all experiment settings and inference strategies are identical.

Combining Table 4(a) and Table 4(d), we can find that multi-scale training data can really improve the accuracy (0.7528 to 0.7745), especially for large-size categories such as basketball-court (0.6671 to 0.7275), ground-track-field (0.7225 to 0.7966) and sample-scarce category such as helicopter (0.654 to 0.7222). The accuracy of Table 4(e) is higher than Table 4(b)–(d), which indicates that multi-scale training sizes are helpful in improving the accuracy. Comparisons between Table 4(b)–(d) illustrate that the larger the training image size, the higher the detection average accuracy.

Table 5 shows computational efficiency of multi-scale strategies. Similarly, the comparison between Table 4(a) and Table 4(d) illustrates that multi-scale training data improve the framework performance to a certain extent, so it performs better in terms of computational efficiency. The comparisons between the last four groups reveal that multi-scale sizes used during training not only improve the detection performance but also improve the computational efficiency.

Table 5. Average running time of multi-scale strategies.

Method	Average Running Time per Image (second)
1000_based(1)+1_scale (a)	3.8553
800_based(4)+1_scale (b)	4.103
900_based(4)+1_scale (c)	3.864
1000_based(4)+1_scale (d)	3.818
(800,900,1000)_based(4)+1_scale (e)	3.7654

The quantified PRCs over multi-scale training data and multi-scale sizes used during training are plotted in Figure 7.

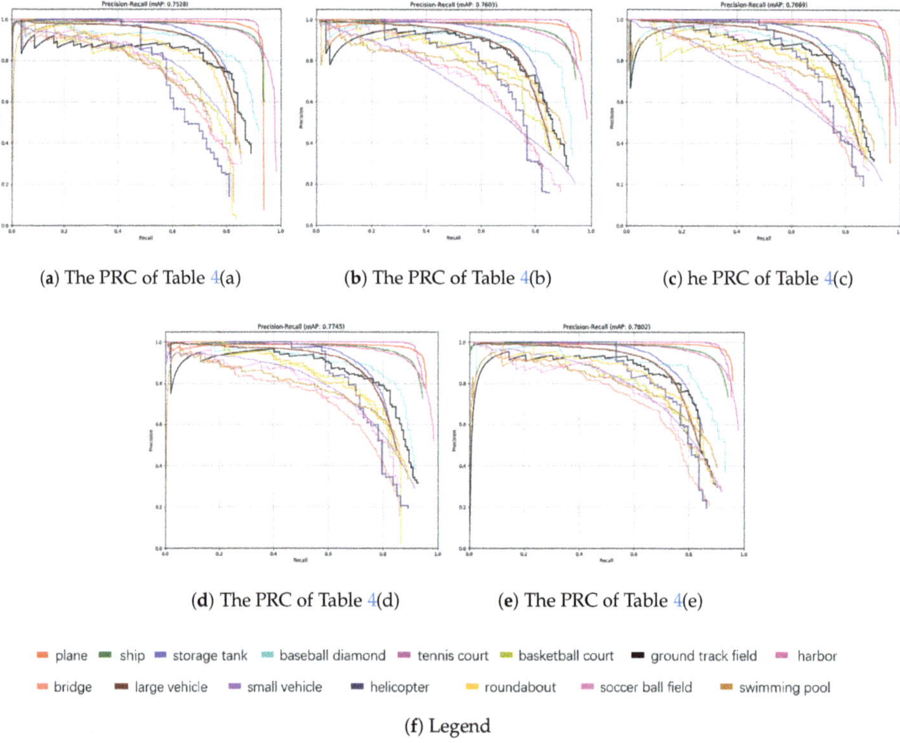

(**a**) The PRC of Table 4(a) (**b**) The PRC of Table 4(b) (**c**) he PRC of Table 4(c)

(**d**) The PRC of Table 4(d) (**e**) The PRC of Table 4(e)

(**f**) Legend

Figure 7. The PRCs of multi-scale strategies.

5.1.3. Multi-Scale Inference and ACNMS Strategies

We performed multi-scale inference on the original images, the shrunken images, the horizontal rotation and vertical rotation images simultaneously. For small and dense objects mainly including ship, large vehicle and small vehicle, we appropriately increase the NMS threshold according to their CI. The common NMS threshold is 0.3 while the ACNMS threshold is 0.38. The results are shown in Table 6.

Table 6. The AP values of ablation experiments for multi-scale inference and ACNMS strategies.

Method	(800,900,1000)_based(4)+1_scale (a)	(800,900,1000)_based(4)+1_scale[+] (b)	(800,900,1000)_based(4)+4_scales (c)	(800,900,1000)_based(4)+4_scales[+] (d)
plane	0.9007	0.9007	**0.901**	0.9004
ship	0.8919	0.8949	0.893	**0.895**
storage tank	0.7817	0.7817	0.8037	**0.8037**
baseball diamond	0.8257	0.8257	0.8265	**0.8294**
tennis court	0.908	0.908	**0.908**	0.9079
basketball court	0.7061	0.7061	0.7192	**0.7192**
ground track field	0.7683	0.7683	0.7985	**0.7985**
harbor	0.891	0.891	0.8924	**0.8924**
bridge	0.6444	0.6444	0.6652	**0.6653**
large vehicle	0.7599	0.78	0.7654	**0.8201**
small vehicle	**0.7209**	0.7208	0.7192	0.7183
helicopter	0.7385	0.7385	0.7447	**0.7447**
roundabout	0.7281	0.7281	0.7553	**0.7554**
soccer ball field	0.7122	0.7122	**0.7179**	0.7179
swimming pool	0.7253	**0.7253**	0.7197	0.7231
mAP	0.7802	0.7817	0.7887	**0.7927**

We note that the top right corner "+" in Table 6(b),(d) indicate that we utilized ACNMS strategy. The two comparisons between Table 6(a) and Table 6(c), Table 6(b) and Table 6(d) illustrate the effectiveness of multi-scale inference strategy, which has improved detection performance both in large and small objects such as storage tank, ground track field and roundabout. The two comparisons between Table 6(a) and Table 6(b), Table 6(c) and Table 6(d) illustrate the effectiveness of ACNMS strategy. We slightly improved the NMS threshold of ship, large vehicle and small vehicle because their CIs are far greater than other's. Specifically, the AP values of ship increase by 0.003 and 0.002 respectively in two comparison experiments, the AP values of large vehicle increase by 0.002 and 0.0055 respectively while the AP values of small vehicle remain unchanged. The relevant comparisons illustrate that increasing NMS threshold according to the category intensity does improve the detection accuracy.

Table 7 shows computational efficiency of multi-scale inference and ACNMS strategies. We note that the average running time of multi-scale inference is about three times longer than that of single-scale inference because the number of image (patch) processed by multi-scale inference is about three times more than that of single-scale inference. In addition, using ACNMS strategy does not increase additional average running time.

Table 7. Average running time of multi-scale inference and ACNMS strategies.

Method	Average Running Time per Image (second)
(800,900,1000)_based(4)+1_scale (a)	3.7654
(800,900,1000)_based(4)+1_scale[+] (b)	3.7237
(800,900,1000)_based(4)+4_scales (c)	12.5504
(800,900,1000)_based(4)+4_scales[+] (d)	12.7018

The quantified PRCs over multi-scale test and adaptive category NMS strategies are plotted in Figure 8.

5.2. Comparison with Other Methods

5.2.1. Comparison with Other Methods on DOTA Validation Dataset

We compared our framework with other region-based object detection networks mainly including Faster R-CNN [37] and FPN [40] on DOTA validation dataset. The selected networks had the same experimental settings as ours, however, they did not adopt our multi-scale training, inference and ACNMS strategies. Table 8 shows the comparison of different networks on DOTA validation dataset.

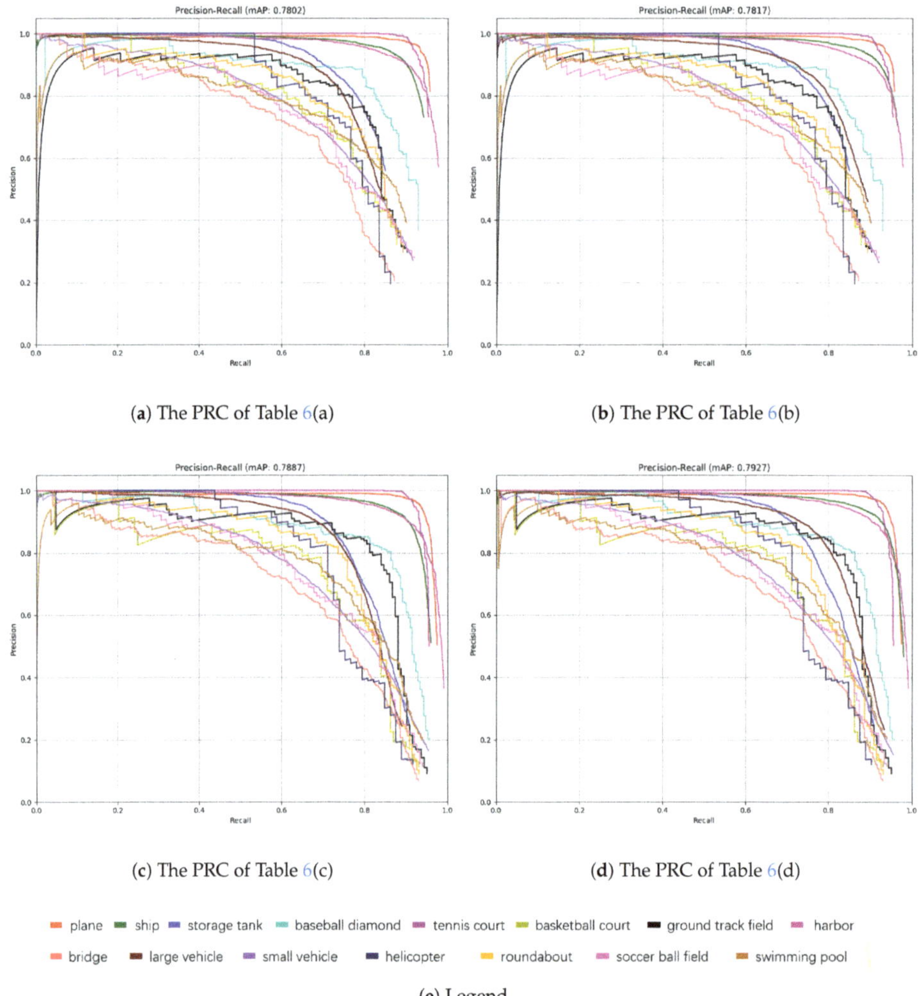

(**a**) The PRC of Table 6(a)

(**b**) The PRC of Table 6(b)

(**c**) The PRC of Table 6(c)

(**d**) The PRC of Table 6(d)

(**e**) Legend

Figure 8. The PRCs of multi-scale inference and ACNMS strategies.

We note that Faster R-CNN, FPN and Table 8(c) performed training and inference on the original images instead of patches. The proposed framework has an overwhelming advantage in mAP and AP values of each category. The mAP of Table 8(c) is 0.1712 higher than that of Faster R-CNN and 0.066 higher than that of FPN, which illustrate the superiority of the proposed network. The mAP of Table 8(d) is 0.4163 higher than that of Faster R-CNN, 0.3111 higher than that of FPN and 0.2451 higher than that of Table 8(c) , which illustrate the great superiority of the proposed network and the multi-scale training, inference and ACNMS strategies. The framework has great advantage in detecting small and dense objects such as ship, large vehicle, small vehicle and storage tank. The detection accuracy of sample-scarce objects such as helicopter and roundabout have also been greatly improved, which further confirms that the proposed framework has outstanding performance in detecting both small dense objects and large-scale objects.

Table 8. The AP values of ablation experiments with other frameworks on DOTA validation dataset.

Method	Faster R-CNN (a)	FPN (b)	0_based(0)+0_scale (c)	(800,900,1000)_based(4)+1_scale (d)
plane	0.4263	0.5404	0.7078	**0.9007**
ship	0.0909	0.3545	0.6023	**0.8919**
storage tank	0.1907	0.2656	0.4213	**0.7817**
baseball diamond	0.4852	0.6605	0.6478	**0.8257**
tennis court	0.8141	0.8179	0.888	**0.908**
basketball court	0.3612	0.4363	0.4822	**0.7061**
ground track field	0.385	0.464	0.4304	**0.7683**
harbor	0.5793	0.7114	0.8391	**0.891**
bridge	0.1972	0.377	0.2973	**0.6444**
large vehicle	0.4911	0.6115	0.675	**0.7599**
small vehicle	0.2852	0.4004	0.5571	**0.7209**
helicopter	0.3077	0.2727	0.3309	**0.7385**
roundabout	0.2312	0.3313	0.2943	**0.7281**
soccer ball field	0.3785	0.4072	0.4059	**0.7122**
swimming pool	0.2356	0.3862	0.4472	**0.7253**
mAP	0.3639	0.4691	0.5351	**0.7802**

The computational efficiency of different frameworks on DOTA validation dataset are shown in Table 9. There is no doubt that the first three groups consume less time than the last group because they performed training and inference on the original images instead of numerous patches. Besides, the proposed DM-FPN (Table 9(c)) can achieve higher object detection accuracy while maintain the same level of computational efficiency.

Table 9. Average running time of different frameworks on DOTA validation dataset.

Method	Faster R-CNN (a)	FPN (b)	0_based(0)+0_scale (c)	(800,900,1000)_based(4)+1_scale(d)
Average running time per image (second)	0.3268	0.2895	0.3882	3.7654

The quantified PRCs over different frameworks on DOTA validation dataset are plotted in Figure 9. We also visualized some detection results as shown in Figure 10.

5.2.2. Comparison with Other Frameworks on DOTA Testing Dataset

We submitted the inference results based on the testing dataset to DOTA Evaluation Server (http://captain.whu.edu.cn/DOTAweb/results.html) to verify the effectiveness of the proposed framework. Table 10 shows several current top rankings and our DM-FPN achieves the state-of-the-art performance (Our result is named of "CVEO" in Task 2, which achieves the best mAP of 0.793.). Specifically, DM-FPN achieves higher AP on 11 categories, especially in ship, small vehicle, large vehicle and swimming pool, which demonstrates that DM-FPN performs better on small and dense objects. In addition, some large-scale objects such as harbor and ground track field also achieve higher AP than the other frameworks, which further demonstrates that our proposed framework can achieve better results both in small dense objects and large-scale objects. The detection results on DOTA testing dataset are shown in Figure 11.

6. Discussion

We adopted DOTA dataset to train, verify and test the proposed DM-FPN, which achieved considerable results in the object detection of very-high-resolution optical remote sensing images with RGB three channels. DOTA is the largest dataset for object detection in aerial images, which contains numerous very-high-resolution remote sensing images and 15 common categories. The spatial resolution of the training dataset ranges [0.1, 5] meters, our framework achieves a better performance within this range. The differential spatial resolutions allow the detector to be more adaptive and robust for varieties of objects of the same category. In order to show the overall detection effect, we performed inferences on full images and the results are shown in Figure 12.

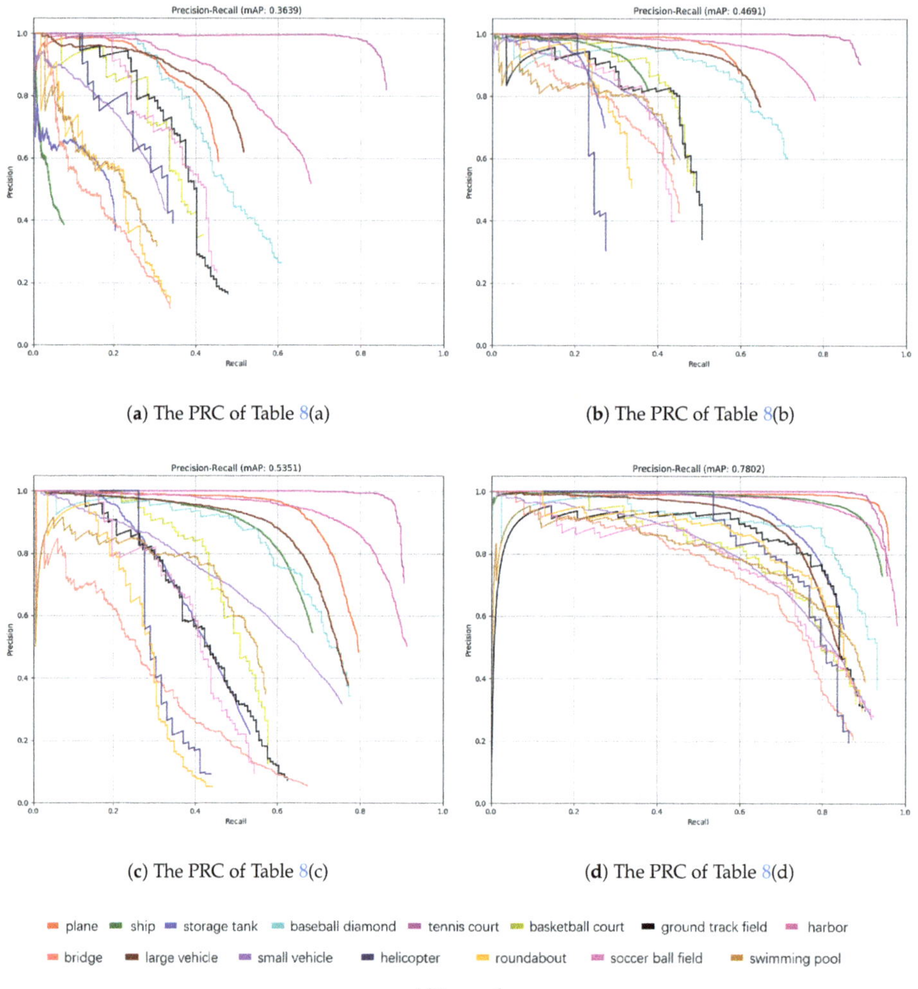

(**a**) The PRC of Table 8(a)

(**b**) The PRC of Table 8(b)

(**c**) The PRC of Table 8(c)

(**d**) The PRC of Table 8(d)

(**e**) Legend

Figure 9. The PRCs of different frameworks on DOTA validation dataset.

Table 10. The AP values of ablation experiments with other frameworks on DOTA testing dataset.

Method	changzhonghan	R2CNN_FPN_Tensorflow	FPN with Hobot-SNIPER	Improving Faster RCNN	Ours
plane	0.901	**0.902**	0.882	0.898	0.887
ship	0.851	0.781	0.839	0.851	**0.873**
storage tank	0.828	0.864	0.838	0.843	**0.871**
baseball diamond	0.819	0.819	0.797	0.824	**0.851**
tennis court	0.908	**0.909**	0.904	0.909	0.908
basketball court	0.836	0.824	0.803	0.797	**0.848**
ground track field	0.706	0.733	0.746	0.738	**0.789**
harbor	0.79	0.758	0.788	0.676	**0.833**
bridge	0.588	0.553	0.51	0.517	**0.621**
large vehicle	0.82	0.776	0.767	0.733	**0.833**
small vehicle	0.698	0.721	0.665	0.645	**0.782**
helicopter	**0.646**	0.638	0.601	0.499	0.64
roundabout	0.624	0.634	0.648	0.596	**0.693**
soccer ball field	0.584	0.645	0.627	0.549	**0.683**
swimming pool	**0.8**	0.782	0.753	0.737	0.782
mAP	0.759	0.754	0.738	0.73	**0.793**

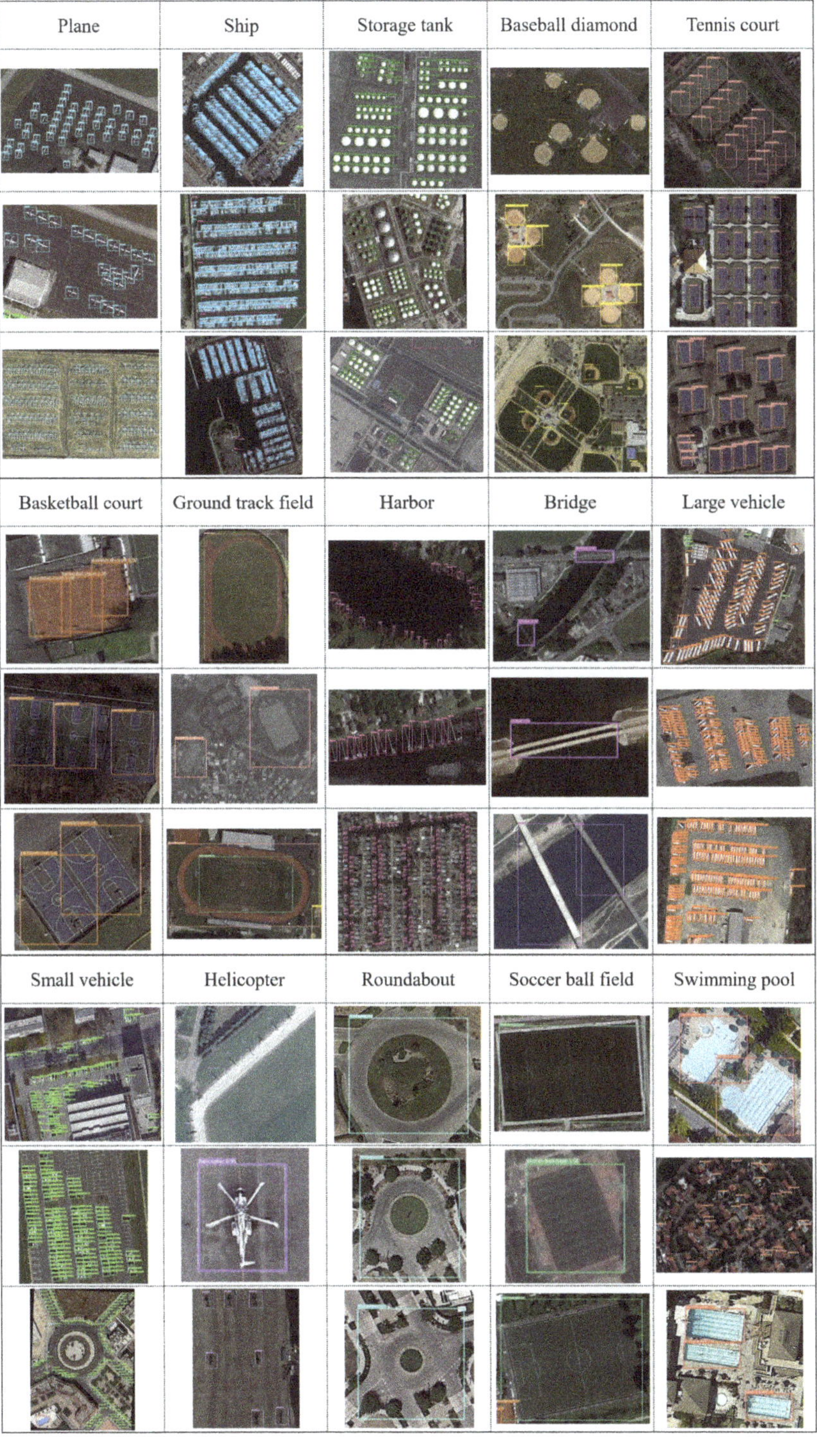

Figure 10. Detection results on DOTA validation dataset.

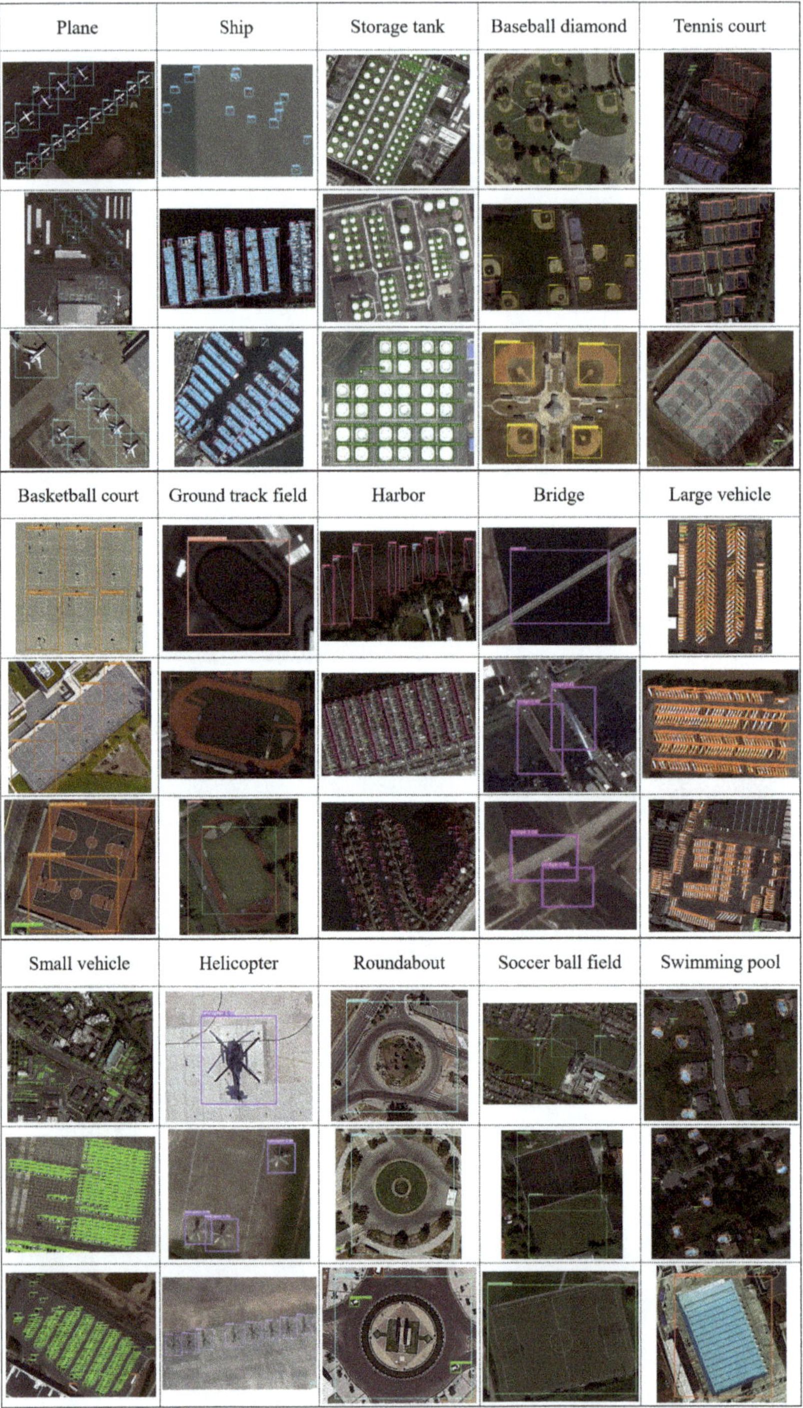

Figure 11. Detection results on DOTA testing dataset.

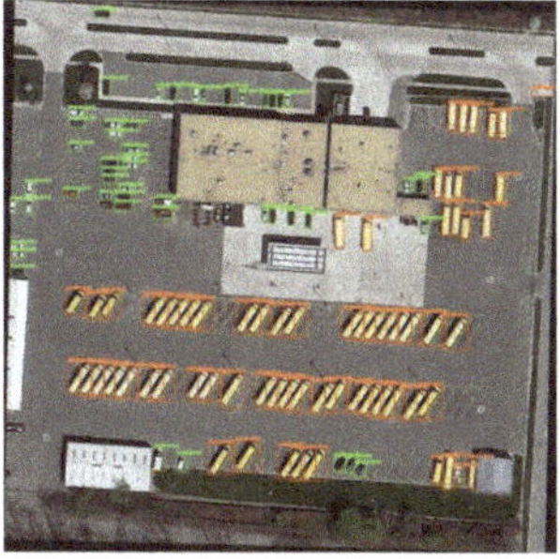

Large vehicle, small vehicle

Ship, harbor, small vehicle, swimming pool, roundabout

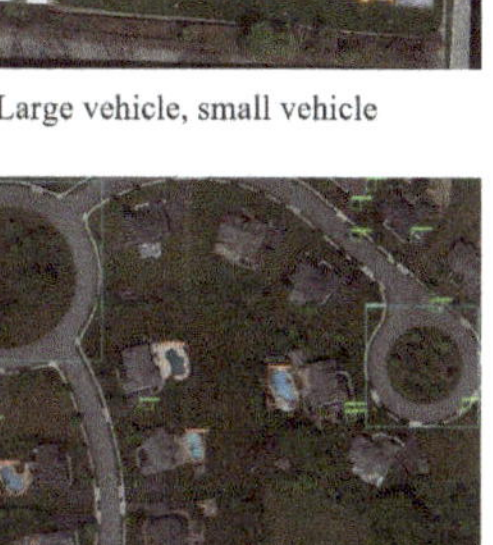

Swimming pool, roundabout, small vehicle

Small vehicle, storage tank, baseball diamond, tennis court, basketball court, swimming pool

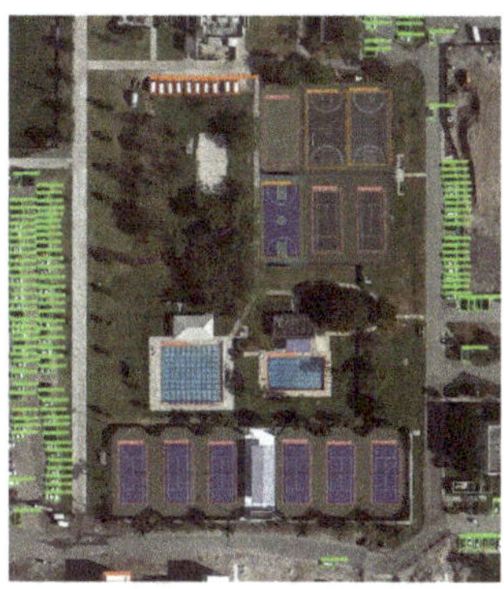

Large vehicle, small vehicle, tennis court, basketball court, swimming pool, soccer ball field

Harbor, ship

Ground track field, soccer ball field, basketball court, tennis court

Plane, small vehicle, large vehicle

Figure 12. Detection results on full images of DOTA.

The trained network performs better in detecting the existing 15 categories. However, the detection effects are not satisfactory in detecting the categories or scenes that did not appear in the training dataset, e.g., plane or helicopter over snow. It is also a common problem of all deep learning frameworks. If training samples are provided, the detection can still be performed hopefully.

7. Conclusions

In this paper, an effective region-based object detection framework named DM-FPN was proposed to solve small and dense object detection problem in VHR remote sensing imagery. DM-FPN makes full use of coarse-resolution, semantically strong features and high-resolution, semantically weak features simultaneously. We also proposed multi-scale training, inference and ACNMS strategies to solve the problem of the overlarge remote sensing images, the complex image backgrounds and the uneven size and quantity distribution of training samples.

Our framework was experimented on DOTA dataset. The internal ablation experiments (the same framework but different strategies) demonstrate the effectiveness of our proposed strategies while the external ablation experiments (different frameworks) demonstrate the effectiveness of our framework. In addition, we also submitted the inference results based on the testing dataset to DOTA Evaluation Server and DM-FPN achieves the state-of-the-art performance, especially in detecting small and dense objects.

In the future, we will improve our framework's performance in terms of detection speed and accuracy, thus constructing a faster and more accurate network for very-high-resolution remote sensing imagery object detection. At the same time, based on the work of this paper, we will expand our framework to the research of arbitrary-oriented bounding box object detection.

Author Contributions: X.Z. guided the algorithm design. K.Z. and G.C. designed the whole framework and experiments. K.Z. wrote the paper. G.C., X.T., L.Z. help organize the paper and performed the experimental analysis. F.D., P.L. help write python scripts of our framework. Y.G. contributed to the discussion of the design. K.Z. drafted the manuscript, which was revised by all authors. All authors read and approved the submitted manuscript.

Funding: This research was funded in part by LIESMARS Special Research Funding and the Fundamental Research Funds for the Central Universities.

Acknowledgments: The authors would like to thank Prof. Gui-Song Xia from State Key Laboratory for Information Engineering in Surveying, Mapping and Remote Sensing (LIESMARS), Wuhan University for providing the awesome remote sensing scene classification dataset DOTA. The authors would also like to thank the developers in the Caffe2 and Detectron developer communities for their open source deep learning frameworks.

References

1. Chen, Z.; Zhang, T.; Ouyang, C. End-to-end airplane detection using transfer learning in remote sensing images. *Remote Sens.* **2018**, *10*, 139. [CrossRef]
2. Cheng, G.; Han, J. A survey on object detection in optical remote sensing images. *ISPRS J. Photogramm. Remote Sens.* **2016**, *117*, 11–28. [CrossRef]
3. Guo, W.; Yang, W.; Zhang, H.; Hua, G. Geospatial object detection in high resolution satellite images based on multi-scale convolutional neural network. *Remote Sens.* **2018**, *10*, 131. [CrossRef]
4. Chen, S.; Zhan, R.; Zhang, J. Geospatial object detection in remote sensing imagery based on multiscale single-shot detector with activated semantics. *Remote Sens.* **2018**, *10*, 820. [CrossRef]
5. Lin, H.; Shi, Z.; Zou, Z. Maritime Semantic Labeling of Optical Remote Sensing Images with Multi-Scale Fully Convolutional Network. *Remote Sens.* **2017**, *9*, 480. [CrossRef]
6. Stankov, K. Detection of Buildings in Multispectral Very High Spatial Resolution Images Using the Percentage Occupancy Hit-or-Miss Transform. *IEEE J. Sel. Top. Appl. Earth Obs. Remote Sens.* **2014**, *7*, 4069–4080. [CrossRef]
7. Lin, Y.; He, H.; Yin, Z.; Chen, F. Rotation-Invariant Object Detection in Remote Sensing Images Based on Radial-Gradient Angle. *IEEE Geosci. Remote Sens. Lett.* **2014**, *12*, 746–750.

8.   Li, Y.; Zhang, Y.; Xin, H.; Hu, Z.; Ma, J. Large-Scale Remote Sensing Image Retrieval by Deep Hashing Neural Networks. *IEEE Trans. Geosci. Remote Sens.* **2018**, *56*, 950–965. [CrossRef]

9.   Baltsavias, E.P. Object extraction and revision by image analysis using existing geodata and knowledge: Current status and steps towards operational systems. *ISPRS J. Photogramm. Remote Sens.* **2004**, *58*, 129–151. doi:10.1016/j.ISPRSjprs.2003.09.002. [CrossRef]

10.   Leninisha, S.; Vani, K. Water flow based geometric active deformable model for road network. *ISPRS J. Photogramm. Remote Sens.* **2015**, *102*, 140–147. [CrossRef]

11.   Ok, A.O. Automated detection of buildings from single VHR multispectral images using shadow information and graph cuts. *ISPRS J. Photogramm. Remote Sens.* **2013**, *86*, 21–40. [CrossRef]

12.   Blaschke, T. Object based image analysis: A new paradigm in remote sensing? In Proceedings of the 2013 American Society for Photogrammetry and Remote Sensing Conference, Baltimore, MD, USA, 26–28 March 2013.

13.   Li, Y.; Wang, S.; Tian, Q.; Ding, X. Feature representation for statistical-learning-based object detection. *Pattern Recognit.* **2015**, *48*, 3542–3559. [CrossRef]

14.   Li, X.; Cheng, X.; Chen, W.; Gang, C.; Liu, S. Identification of Forested Landslides Using LiDar Data, Object-based Image Analysis, and Machine Learning Algorithms. *Remote Sens.* **2015**, *7*, 9705–9726. [CrossRef]

15.   Dalal, N.; Triggs, B. Histograms of oriented gradients for human detection. In Proceedings of the IEEE Conference on Computer Vision and Pattern Recognition, CVPR Workshops 2005, San Diego, CA, USA, 21–23 September 2005; Volume 1, pp. 886–893. [CrossRef]

16.   Fei-Fei, L.; Perona, P. A Bayesian hierarchical model for learning natural scene categories. In Proceedings of the IEEE Conference on Computer Vision and Pattern Recognition, CVPR Workshops 2005, San Diego, CA, USA, 21–23 September 2005; Volume 2, pp. 524–531. [CrossRef]

17.   Wright, J.; Yang, A.Y.; Ganesh, A.; Sastry, S.S.; Ma, Y. Robust Face Recognition via Sparse Representation. *IEEE Trans. Pattern Anal. Mach. Intell.* **2009**, *31*, 210–227. [CrossRef]

18.   Cortes, C.; Vapnik, V. Support-vector networks. *Mach. Learn.* **1995**, *20*, 273–297. [CrossRef]

19.   Freund, Y. Boosting a Weak Learning Algorithm by Majority. *Inf. Comput.* **1995**, *121*, 256–285. [CrossRef]

20.   Lafferty, J.; Mccallum, A.; Pereira, F.C.N.; Fper, F.P. Conditional random fields: Probabilistic models for segmenting and labeling sequence data. in Proceedings of the 18th International Conference on Machine Learning, Morgan Kaufmann, San Francisco, CA, USA, 28 June–1 July 2001; pp. 282–289.

21.   Lecun, Y.; Bengio, Y.; Hinton, G. Deep learning. *Nature* **2015**, *521*, 436–444. [CrossRef] [PubMed]

22.   Gu, J.; Wang, Z.; Kuen, J.; Ma, L.; Shahroudy, A.; Shuai, B.; Liu, T.; Wang, X.; Wang, G.; Cai, J.; Chen, T. Recent advances in convolutional neural networks. *Pattern Recognit.* **2018**, *77*, 354–377. [CrossRef]

23.   Schmidhuber, J. Deep Learning in Neural Networks: An Overview. *Neural Netw.* **2015**, *61*, 85–117. [CrossRef] [PubMed]

24.   Zhang, X.; Chen, G.; Wang, W.; Wang, Q.; Dai, F. Object-Based Land-Cover Supervised Classification for Very-High-Resolution UAV Images Using Stacked Denoising Autoencoders. *IEEE J. Sel. Top. Appl. Earth Obs. Remote Sens.* **2017**, *10*, 3373–3385. [CrossRef]

25.   Ma, L.; Li, M.; Ma, X.; Cheng, L.; Du, P.; Liu, Y. A review of supervised object-based land-cover image classification. *ISPRS J. Photogramm. Remote Sens.* **2017**, *130*, 277–293. doi:10.1016/J.Isprsjprs.2017.06.001. [CrossRef]

26.   He, K.; Zhang, X.; Ren, S.; Sun, J. Delving Deep into Rectifiers: Surpassing Human-Level Performance on ImageNet Classification. In Proceedings of the 2015 IEEE International Conference on Computer Vision Workshop, Santiago, Chile, 7–13 December 2015; pp. 1026–1034. [CrossRef]

27.   Fu, T.; Ma, L.; Li, M.; Johnson, B. Using convolutional neural network to identify irregular segmentation objects from very high-resolution remote sensing imagery. *J. Appl. Remote Sens.* **2018**, *12*, 1. [CrossRef]

28.   Cheng, G.; Zhou, P.; Han, J. Learning Rotation-Invariant Convolutional Neural Networks for Object Detection in VHR Optical Remote Sensing Images. *IEEE Trans. Geosci. Remote Sens.* **2016**, *54*, 7405–7415. [CrossRef]

29.   Li, K.; Cheng, G.; Bu, S.; You, X. Rotation-Insensitive and Context-Augmented Object Detection in Remote Sensing Images. *IEEE Trans. Geosci. Remote Sens.* **2018**, *56*, 2337–2348. [CrossRef]

30.   Cheng, G.; Zhou, P.; Han, J. RIFD-CNN: Rotation-Invariant and Fisher Discriminative Convolutional Neural Networks for Object Detection. In Proceedings of the 2016 IEEE CVPR, Las Vegas, NV, USA, 27–30 June 2016; pp. 2884–2893. [CrossRef]

31. Everingham, M.; Van Gool, L.; Williams, C.K.I.; Winn, J.; Zisserman, A. The Pascal Visual Object Classes (VOC) Challenge. *IJCV* **2010**, *88*, 303–338. [CrossRef]

32. Russakovsky, O.; Deng, J.; Su, H.; Krause, J.; Satheesh, S.; Ma, S.; Huang, Z.; Karpathy, A.; Khosla, A.; Bernstein, M.; et al. ImageNet Large Scale Visual Recognition Challenge. *IJCV* **2015**, *115*, 211–252. [CrossRef]

33. Han, X.; Zhong, Y.; Zhang, L. An Efficient and Robust Integrated Geospatial Object Detection Framework for High Spatial Resolution Remote Sensing Imagery. *Remote Sens.* **2017**, *9*, 666. [CrossRef]

34. Tao, K.; Sun, F.; Yao, A.; Liu, H.; Ming, L.; Chen, Y. RON: Reverse Connection with Objectness Prior Networks for Object Detection. In Proceedings of the IEEE Conference on Computer Vision and Pattern Recognition (CVPR), Honolulu, HI, USA, 21–26 July 2017.

35. Girshick, R.; Donahue, J.; Darrell, T.; Malik, J. Rich feature hierarchies for accurate object detection and semantic segmentation. In Proceedings of the IEEE conference on computer vision and pattern recognition, Columbus, OH, USA, 23–28 June 2014.

36. Uijlings, J.R.; Van De Sande, K.E.; Gevers, T.; Smeulders, A.W. Selective search for object recognition. *IJCV* **2013**, *104*, 154–171. [CrossRef]

37. Ren, S.; He, K.; Girshick, R.; Sun, J. Faster R-CNN: Towards Real-Time Object Detection with Region Proposal Networks. *IEEE Trans. Pattern Anal. Mach. Intell.* **2017**, *39*, 1137–1149. [CrossRef]

38. He, K.; Zhang, X.; Ren, S.; Sun, J. Spatial Pyramid Pooling in Deep Convolutional Networks for Visual Recognition. *IEEE Trans. Pattern Anal. Mach. Intell.* **2015**, *37*, 1904–1916. [CrossRef]

39. Girshick, R. Fast R-CNN. In Proceedings of the 2015 IEEE International Conference on Computer Vision, Santiago, Chile, 7–13 December 2015; pp. 1440–1448. [CrossRef]

40. Lin, T.Y.; Dollár, P.; Girshick, R.; He, K.; Hariharan, B.; Belongie, S. Feature pyramid networks for object detection. In Proceedings of the 30th IEEE Conference on Computer Vision and Pattern Recognition, Honolulu, HI, USA, 21–26 July 2016; pp. 936–944. [CrossRef]

41. Sermanet, P.; Eigen, D.; Zhang, X.; Mathieu, M.; Fergus, R.; Lecun, Y. Overfeat: Integrated recognition, localization and detection using convolutional networks. In Proceedings of the 2nd International Conference on Learning Representations (ICLR2014), Banff, AB, Canada, 14–16 April 2014.

42. Redmon, J.; Divvala, S.; Girshick, R.; Farhadi, A. You Only Look Once: Unified, Real-Time Object Detection. In Proceedings of the 2016 IEEE Conference on Computer Vision and Pattern Recognition, Las Vegas, NV, USA, 27–30 June 2016; pp. 779–788. [CrossRef]

43. Liu, W.; Anguelov, D.; Erhan, D.; Szegedy, C.; Reed, S.; Fu, C.Y.; Berg, A.C. SSD: Single Shot MultiBox Detector. In Proceedings of the 14th European Conference, Amsterdam, The Netherlands, 11–14 October 2016; pp. 21–37.

44. Cheng, G.; Han, J.; Zhou, P.; Guo, L. Multi-class geospatial object detection and geographic image classification based on collection of part detectors. *ISPRS J. Photogramm. Remote Sens.* **2014**, *98*, 119–132. [CrossRef]

45. Yang, X.; Sun, H.; Fu, K.; Yang, J.; Sun, X.; Yan, M.; Guo, Z. Automatic Ship Detection in Remote Sensing Images from Google Earth of Complex Scenes Based on Multiscale Rotation Dense Feature Pyramid Networks. *Remote Sens.* **2018**, *10*, 132. [CrossRef]

46. Xu, Y.; Zhu, M.; Li, S.; Feng, H.; Ma, S.; Che, J. End-to-End Airport Detection in Remote Sensing Images Combining Cascade Region Proposal Networks and Multi-Threshold Detection Networks. *Remote Sens.* **2018**, *10*, 1516. [CrossRef]

47. Cai, B.; Jiang, Z.; Zhang, H.; Zhao, D.; Yao, Y. Airport Detection Using End-to-End Convolutional Neural Network with Hard Example Mining. *Remote Sens.* **2017**, *9*, 1198. [CrossRef]

48. Xia, G.S.; Bai, X.; Ding, J.; Zhu, Z.; Belongie, S.; Luo, J.; Datcu, M.; Pelillo, M.; Zhang, L. DOTA: A Large-Scale Dataset for Object Detection in Aerial Images. In Proceedings of the 2018 IEEE/CVF Conference on Computer Vision and Pattern Recognition, Salt Lake City, Utah, USA, 18–22 June 2018.

49. Zitnick, C.L.; Dollár, P. Edge Boxes: Locating Object Proposals from Edges. In Proceedings of the 13th European Conference, Zurich, Switzerland, 6–12 September 2014.

50. Cheng, M.M.; Zhang, Z.; Lin, W.Y.; Torr, P.H.S. {BING}: Binarized Normed Gradients for Objectness Estimation at 300fps. In Proceedings of the 27th IEEE Conference on Computer Vision and Pattern Recognition, Columbus, OH, USA, 24–27 June 2014.

51. Long, J.; Shelhamer, E.; Darrell, T. Fully convolutional networks for semantic segmentation. *IEEE Trans. Pattern Anal. Mach. Intell.* **2014**, *39*, 640–651.

52. Simonyan, K.; Zisserman, A. Very Deep Convolutional Networks for Large-Scale Image Recognition. *arXiv* **2014**, arXiv:1409.1556.

53. Cai, Z.; Fan, Q.; Feris, R.; Vasconcelos, N. A Unified Multi-scale Deep Convolutional Neural Network for Fast Object Detection. In Proceedings of the 14th European Conference, Amsterdam, The Netherlands, 11–14 October 2016.

54. Honari, S.; Yosinski, J.; Vincent, P.; Pal, C. Recombinator Networks: Learning Coarse-to-Fine Feature Aggregation. In Proceedings of the 2016 IEEE Conference on Computer Vision and Pattern Recognition, Las Vegas, NV, USA, 27–30 June 2016.

55. Ghiasi, G.; C. Fowlkes, C. Laplacian Pyramid Reconstruction and Refinement for Semantic Segmentation. In Proceedings of the 14th European Conference, Amsterdam, The Netherlands, 11–14 October 2016; pp. 519–534. [CrossRef]

56. Pinheiro, P.O.; Lin, T.Y.; Collobert, R.; Dollár, P. Learning to Refine Object Segments. In Proceedings of the 14th European Conference, Amsterdam, The Netherlands, 11–14 October 2016

57. He, K.; Gkioxari, G.; Dollár, P.; Girshick, R. Mask R-CNN. *IEEE Trans. Pattern Anal. Mach. Intell.* **2017**, 1. [CrossRef]

58. He, K.; Zhang, X.; Ren, S.; Sun, J. Deep Residual Learning for Image Recognition. In Proceedings of the 2016 IEEE Conference on Computer Vision and Pattern Recognition, Las Vegas, NV, USA, 27–30 June 2016; pp. 770–778. [CrossRef]

Article

A Novel Multi-Model Decision Fusion Network for Object Detection in Remote Sensing Images

Wenping Ma [1], Qiongqiong Guo [1], Yue Wu [2,*], Wei Zhao [1], Xiangrong Zhang [1] and Licheng Jiao [1]

[1] Key Laboratory of Intelligent Perception and Image Understanding of Ministry of Education, International Research Center for Intelligent Perception and Computation, Joint International Research Laboratory of Intelligent Perception and Computation, School of Artificial Intelligence, Xidian University, Xi'an 710071, China; wpma@mail.xidian.edu.cn (W.M.); qqiongguo@126.com (Q.G.); weizhao_90@163.com (W.Z.); xrzhang@mail.xidian.edu.cn (X.Z.); lchjiao@mail.xidian.edu.cn (L.J.)

[2] School of Computer Science and Technology, Xidian University, Xi'an 710071, China

* Correspondence: ywu@xidian.edu.cn

Received: 28 January 2019; Accepted: 19 March 2019; Published: 27 March 2019

Abstract: Object detection in optical remote sensing images is still a challenging task because of the complexity of the images. The diversity and complexity of geospatial object appearance and the insufficient understanding of geospatial object spatial structure information are still the existing problems. In this paper, we propose a novel multi-model decision fusion framework which takes contextual information and multi-region features into account for addressing those problems. First, a contextual information fusion sub-network is designed to fuse both local contextual features and object-object relationship contextual features so as to deal with the problem of the diversity and complexity of geospatial object appearance. Second, a part-based multi-region fusion sub-network is constructed to merge multiple parts of an object for obtaining more spatial structure information about the object, which helps to handle the problem of the insufficient understanding of geospatial object spatial structure information. Finally, a decision fusion is made on all sub-networks to improve the stability and robustness of the model and achieve better detection performance. The experimental results on a publicly available ten class data set show that the proposed method is effective for geospatial object detection.

Keywords: convolutional neural networks (CNNs); object detection; remote sensing images; contextual information; part-based; multi-model

---

1. Introduction

Nowadays, optical remote sensing images with high spatial resolution are obtained conveniently due to the significant progress in remote sensing technology, which leads to a wide range of applications such as land planning, disaster control, urban monitoring, and traffic planning [1–4]. As one of the most fundamental and challenging tasks required for understanding remote sensing images, object detection has gained increasing attention in recent years. To deal with a variety of problems faced in optical remote sensing image object detection, numerous approaches have been proposed [5,6]. A deep review on object detection in optical remote sensing images can be found in [7].

As is known to all, a common method for object detection is to extract features. The quality of the extracted features is critical as it will directly affect the final result of object detection. Powerful feature representation can make an object more discriminative and its location more explicit, which makes the object easier to detect. On the contrary, insufficient ability to represent objects will result in inaccurate detection. Therefore, it is important for us to choose a method to extract features for object detection in remote sensing images. Currently, because of the advantage of directly generating more powerful feature representations from raw image pixels through neural networks, deep learning methods, especially CNN-based [4,8–25], are recognized as predominate techniques for extracting features in object detection. Therefore, we select a CNN-based approach to extract features for object detection in optical remote sensing images.

Object detection in remote sensing images becomes more complicated because of the diversity of illumination intensities, noise interference, and the influence of weather. At present, there are still a lot of problems to be solved, such as the diversity and complexity of geospatial object appearance, and the insufficient understanding of geospatial object spatial structure information.

In the field of optical remote sensing images, lots of object detection algorithms only pay attention to the features of objects themselves [16,17,26]. However, due to the diversity and complexity of geospatial object appearance, in many cases, relying solely on the characteristics of an object itself cannot effectively identify the object, and sometimes may even cause mis-detection between two objects which belong to two different classes but look very similar in appearance factor. For instance, recognizing a storage tank only through exploiting its features may be difficult as its appearance is just circular, and a bridge is often mistaken for part of the road (as shown in Figure 1). In this case, the application of auxiliary information can effectively help detect objects. Therefore, contextual information is a choice. Some existing works [18,20,27] take local contextual information into account and obtain good performance. For example, the work in [20] used features surrounding the regions of interest, thus alleviating false detection caused by object appearance ambiguity. Although those methods yield good results, there are still deficiencies. Also, relationships among objects play an important role in improving the performance of detection. Therefore, in addition to the use of local contextual information, the proposed method takes object-object relationship contextual information into consideration.

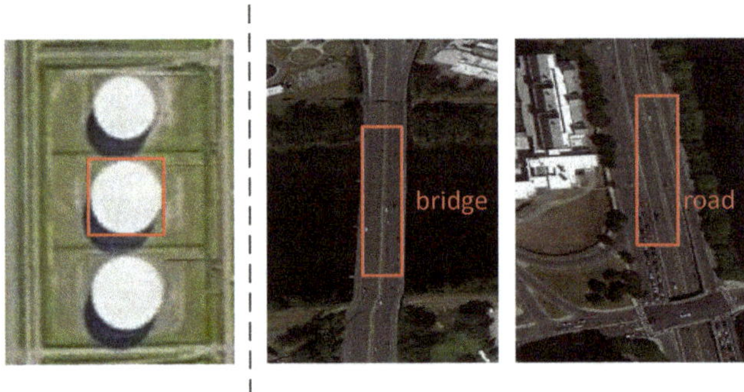

Figure 1. Examples difficult to detect. (**Left**) Only using the sample appearance features in the red rectangle, just a circle, is hard to identify the storage tank. (**Right**) The bridge and the road are easily confused.

The spatial structure of geospatial objects plays an important role in recognizing the objects. Optical remote sensing images with high spatial resolutions always contain abundant spatial structure information about objects. Therefore, investigating deeply the structural information about objects can result in good detection results. It is necessary to design an object detector to effectively alleviate the insufficient understanding of geospatial object spatial structure information. Each part of a geospatial object provides many local visual properties and much geometric information about the object. Paying attention to the various parts of an object can help us to understand more details about its spatial structure. There are lots of part-based models [28–32] concentrating on using the various parts of objects to improve detection performance. For example, Zhang et al. [28] proposed a generic discriminative part-based model (GDPBM), which divides a geospatial object with arbitrary orientation into several parts to achieve good performance for object detection in optical remote sensing images. Unlike the previous part-based approaches [28–32], which use traditional features such as histogram of oriented gradients (HOG) [33], the proposed method applies the CNN-based technique to extract high-level features for better feature representation. In addition, it is easier to obtain and process parts of objects in the proposed approach.

In this paper, we propose a novel multi-model decision fusion framework for object detection in remote sensing images. Aiming at the diversity and complexity of geospatial object appearance, we build a local contextual information and object-object relationship contextual information fusion sub-network. Focusing on the insufficient understanding of geospatial object spatial structure information, we construct a part-based multi-region feature fusion sub-network. Furthermore, unlike many methods just using single model, we make a decision fusion on several models for better stability and robustness. For the implementation of the multi-model decision fusion strategy, in addition to the above two sub-networks, we also fuse a baseline sub-network based on Faster R-CNN model.

In summary, the major contributions of this paper are presented as follows.

(1) We propose a local contextual information and object-object relationship contextual information fusion sub-network based on gated recurrent unit (GRU) to form discriminative feature representation, which can effectively recognize objects and reduce false detection between different types of objects with similar appearance. The object-object relationship contextual information is introduced for the first time in the field of remote sensing image object detection as far as we know.

(2) We propose a new part-based multi-region feature fusion sub-network to investigate more details of objects, which can diversify object features and enrich semantic information.

(3) We propose a multi-model decision fusion strategy to fuse the detection results of the three sub-networks, which can improve the stability and robustness of the model and obtain better algorithm performance.

The remainder of this paper is organized as follows. The second section gives a brief review of the related work on geospatial object detection, contextual information fusion, and the RoIAlign layer. In the third section, we introduce the proposed method in detail. The details of our experiments and results are presented in the fourth section. The last section concludes this paper with a discussion of the results.

2. Related Work

2.1. Geospatial Object Detection

In the past decades, the research on the field of remote sensing image object detection has made a breakthrough development. Many object detection algorithms have been proposed to address various problems [17,20,34]. For example, Cheng et al. [17] proposed a novel and effective approach to learn a rotation-invariant CNN (RICNN) model for addressing the problem of object rotation variations, which is achieved by introducing and learning a new rotation-invariant layer on the basis of the existing CNN frameworks. Han et al. [34] combined the weakly supervised learning (WSL) and high-level feature learning to tackle the problems of manual annotation and insufficiently powerful descriptors. Li et al. [20] put forward a novel region proposal network (RPN) including multiangle, multiscale, and multiaspect-ratio anchors to address the problem of geospatial object rotation variations, and also proposed a double-channel feature fusion network which can learn local and contextual properties to deal with the geospatial object appearance ambiguity issue.

Low-level features are often used for image analysis [35]. Employing the extracted low-level features of objects for object detection has been a very common method used by many scholars. Those low-level features contain scale-invariant feature transform (SIFT) [3,34,36], histogram of oriented gradients (HOG) [5,6,33], the bag-of-words (BoW) model [37–39], Saliency [40,41], etc. For example, Tuermer et al. [5] used the HOG feature and disparity maps to detect airborne vehicles in dense urban areas. Shi et al. [6] developed a circle frequency-HOG feature for ship detection by combining circle frequency features with HOG features. Han et al. [40] proposed to detect multiple-class geospatial objects through integrating visual saliency modeling and the discriminative learning of sparse coding. Although those low-level features show impressive success in some specific object detection tasks, they have certain limitations because they do not represent the high-level semantic information required for identifying objects, especially when visual recognition tasks become more challenging.

Currently, deep convolutional neural network (CNN) models are widely used in the field of visual recognition [42–44], such as object detection, owing to the powerful ability of CNN to capture both low-level and high-level features. The region-based convolutional neural network (R-CNN) [8] is considered as a milestone among CNN-based object detection approaches, and achieves superior performance. Subsequently, many advanced object detection algorithms in natural images, such as Fast R-CNN [9], Faster R-CNN [10], YOLO [11], SSD [12], Mask R-CNN [13], are proposed successively and yield unusually brilliant results. However, the aforementioned models can not be directly utilized for geospatial object detection, because the properties of remote sensing images and natural images are different and the direct application of those models to remote sensing images is not optimal. Researchers have done a lot of work in applying CNN-based models to detect geospatial objects in remote sensing images and achieved remarkable consequences [4,15–25,45]. For example, the work in [4] utilized a hyperregion proposal network (HRPN) and a cascade of boosted classifiers to detect vehicles in remote sensing images. Long et al. [16] proposed a new object localization framework based on convolutional neural networks to efficiently achieve the generalizability of the features used to describe geospatial objects, and obtained accurate object locations. Yang et al. [21] constructed a Markov random field (MRF)-fully convolutional network to detect airplanes.

2.2. Contextual Information Fusion

Contextual information is advantageous to various visual recognition tasks [18,20,27,46–53], such as object detection. For example, in order to promote object detection performance, the work in [48] developed a novel object detection model, attention to context convolution neural network (AC-CNN), through incorporating global and local contextual information into the region-based CNN detection framework. Bell et al. [49] presented the Inside-Outside Net (ION) to exploit information both inside and outside the regions of interest, which integrates the contextual information outside the regions of interest by using spatial recurrent neural networks. Furthermore, some recent works [50–52] proposed new architectures to investigate the contextual information about object-object relationships for better object detection performance. In the field of remote sensing images, the work in [20] fused local and contextual features to address the problem of object appearance ambiguity in object detection. Considering that the appearance is not enough to distinguish oil tanks from the complex background, Zhang et al. [27] applied trained CNN models to extract contextual features, which makes oil tanks easier to recognize. Xiao et al. [18] fused auxiliary features both within and surrounding the regions of interest to represent the complementary information of each region proposal for airport detection, effectively alleviating detection problems caused by the diversity of illumination intensities in remote sensing images. Motivated by those models, we believe that the local contextual information and the object-object relationship context are very useful for object detection in optical remote sensing images. It is necessary to remember features of the object itself before incorporating contextual information. The process of merging messages follows the memory characteristics of Gated Recurrent Units (GRU) [54]. Therefore, we use GRU to fuse the two types of features.

Next we introduce how the j-th hidden unit in a GRU cell works. First, the *reset* gate r_j is obtained by:

$$r_j = \sigma([\mathbf{W}_r\mathbf{x}]_j + [\mathbf{U}_r\mathbf{h}_{t-1}]_j) \tag{1}$$

where σ is the logistic sigmoid function, and $[.]_j$ indicates the j-th element of a vector. $\mathbf{x}$ is the input, while $\mathbf{h}_{t-1}$ denotes the previous hidden state. Both $\mathbf{W}_r$ and $\mathbf{U}_r$ are learnable weight matrices.

Similarly, the *update* gate z_j is calculated by:

$$z_j = \sigma([\mathbf{W}_z\mathbf{x}]_j + [\mathbf{U}_z\mathbf{h}_{t-1}]_j) \tag{2}$$

The actual activation of the proposed unit h_j is then calculated by:

$$h_j^t = z_j h_j^{t-1} + (1 - z_j)\widetilde{h}_j^t \tag{3}$$

where

$$\widetilde{h}_j^t = \phi([\mathbf{Wx}]_j + [\mathbf{U}(\mathbf{r} \odot \mathbf{h}_{t-1})]_j) \tag{4}$$

ϕ denotes *tanh* activate function, and $\odot$ indicates element-wise multiplication. $\mathbf{W}$ and $\mathbf{U}$ are weight matrices which are learned. As described in [54], the reset gate $\mathbf{r}$ effectively allows the hidden state to drop any information that is found to be irrelevant later in the future, which provides a more compact information representation. On the other side, the update gate $\mathbf{z}$ dominates how much information from the previous hidden state will carry over to the current hidden state. More details about GRU can be seen in Figure 2.

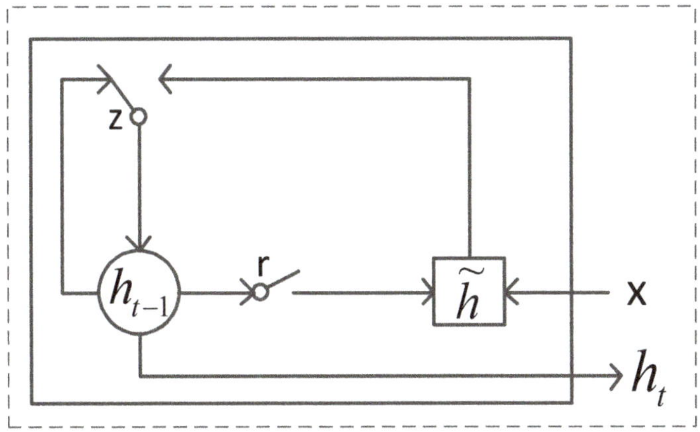

Figure 2. An illustration of a gated recurrent unit (GRU) [54]. The update gate z selects whether the hidden state h_t is to be updated with a new hidden state $\widetilde{h}$. The reset gate r decides whether the previous hidden state h_{t-1} is ignored.

2.3. The RoIAlign Layer

RoIAlign [13] is based on RoIPooling [10]. As we know, RoIPooling performs two quantizations, first quantizing a floating-number RoI to the discrete granularity of the feature map and then subdividing the quantized RoI into spatial bins which are themselves quantized. Unlike RoIPooling, RoIAlign avoids any quantization of the RoI boundaries or bins. In the execution of RoIAlign, bilinear interpolation [55] is exploited to calculate the exact values of the input features at four regularly sampled locations in each RoI bin. The result after bilinear interpolation is aggregated by average pooling.

3. Proposed Framework

The flowchart of the proposed object detection method is shown in Figure 3. The framework is based on the VGG16 model [56] and the popular detection frame Faster R-CNN [10]. First, given a remote sensing image, we employ the parts of VGG16 to extract object features and use the region proposal network (RPN) to generate region proposals. Unlike the work of Faster R-CNN using a RoI pooling layer to convert the features inside any valid region of interest into a small feature map with a fixed spatial extent, we apply the RoIAlign layer proposed in Mask R-CNN. There are misalignments between the RoIs and the extracted features in RoI pooling. RoIAlign can address the problem of misalignments introduced by quantizations, thus, enhancing the ability to detect small and intensive objects. Second, motivated by the work in [51] and for adapting to remote sensing images which contain complex backgrounds, we extract both local contextual information and object-object relationship contextual information, and fuse them by GRU. The fused feature is employed subsequently to obtain the classification and regression results of the contextual information fusion sub-network. Then, we divide the object in candidate regions generated by RPN into several parts and utilize the RoIAlign layer to pool each part. All parts are merged to gain better feature representations for detecting objects. After that, we perform classification and regression to obtain the consequences of the part-based multi-region sub-network. Finally, in the case of separately gaining results of the contextual information fusion sub-network, the part-based multi-region fusion sub-network, and the baseline sub-network, we execute a decision fusion on those results to acquire the bottom detection result, which we call multi-model decision fusion. Each component of the proposed framework is described as follows.

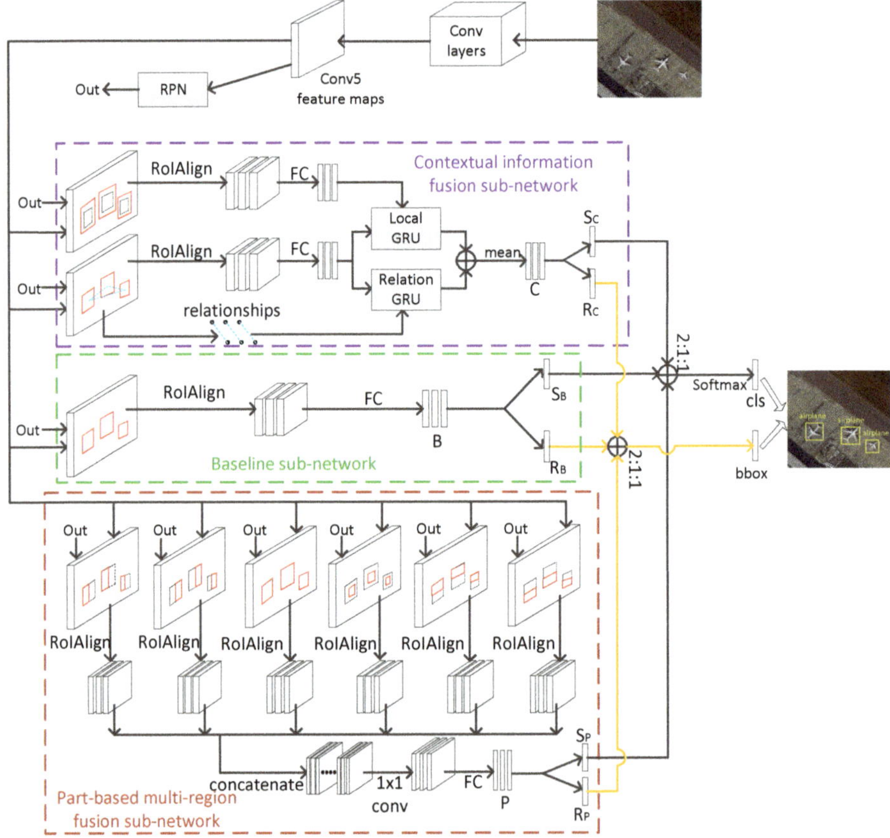

Figure 3. The proposed framework, which is made up of four parts. (1) A contextual information fusion sub-network; (2) a part-based multi-region fusion sub-network; (3) a baseline sub-network; (4) the last multi-model decision fusion part.

3.1. Local Contextual Information and Object-Object Relationship Contextual Information Fusion Sub-Network

Many works show the effectiveness of investigating features surrounding the regions of interest or relationships among objects [20,51]. Therefore, for object detection in remote sensing images, inspired by the work in [51], we construct our local contextual information and object-object relationship contextual information fusion sub-network. Different from [51] using global contextual information for the entire image, we employ local contextual features around objects. For some objects in remote sensing images, scenes far from them are more diverse, resulting in unstable contexts which are likely to be noise that affects the detection result. That is the reason we choose to exploit local contextual information for geospatial object detection. In addition, we replace RoI pooling with RoIAlign because of there existing a lot of dense and small objects in remote sensing images. The features to be fused in the sub-network consist of three parts: local contextual information, features in original candidate regions, and object-object relationship contextual information.

First, in conv5 layer, we extract the features from original proposal boxes and the 1.8× of original proposal boxes. The features in 1.8× of original proposal boxes are used as local contextual information. The RoIAlign layer and the fully connected layer act on the two types of features in succession. Second, we build relationships among objects [as illustrated in Figure 4]. The process is the same as [51]. There we set V to represent the collection of candidate boxes generated by RPN. The term v_i indicates the i-th candidate box. We calculate the relationship between v_i and v_j by:

$$e_{j \to i} = relu(W_p R^p_{j \to i}) * tanh(W_v[f^v_i, f^v_j]) \tag{5}$$

where $e_{j \to i}$ represents the influence of v_j on v_i and it is a scalar weight. W_p and W_v are weight matrices which are learned. The visual relationship vector is formed by concatenating visual feature f^v_i and f^v_j, indicated by $[f^v_i, f^v_j]$. The term $R^p_{j \to i}$ denotes the spatial position relationship. Visual feature f^v_i and f^v_j are results after *relu*, which are sparse. A lot of information will be lost if *relu* is used again. So *tanh* is applied to activate $W_v[f^v_i, f^v_j]$. $R^p_{j \to i}$ is obtained by:

$$
\begin{aligned}
R^p_{j \to i} = [&w_i, h_i, s_i, w_j, h_j, s_j, \frac{(x_i - x_j)}{w_j}, \frac{(y_i - y_j)}{h_j}, \\
&\frac{(x_i - x_j)^2}{w_j^2}, \frac{(y_i - y_j)^2}{h_j^2}, log(\frac{w_i}{w_j}), log(\frac{h_i}{h_j})]
\end{aligned}
\tag{6}
$$

where (x_i, y_i) means the center of RoI b_i. w_i and h_i are the width and height of b_i. s_i is the area of b_i. The final object-object relationship contextual information m_i is calculated by:

$$m_i = \max_{j \in V} pooling(e_{j \to i} * f^v_j) \tag{7}$$

It represents that we choose the box which has the greatest impact on v_i as the final relationship contextual message to be integrated. Then, we exploit GRUs to merge the three features gained in the previous operation, taking the processed features from original proposed boxes as the initial hidden states, both the relationship contexts and the processed features (local contextual information) which stem from 1.8× of original proposed boxes as inputs related to two GRUs. Afterwards, we average the outputs of the two GRUs and denote the final feature as C. Finally, we apply C to gain the class scores S_C and the predicted boxes R_C.

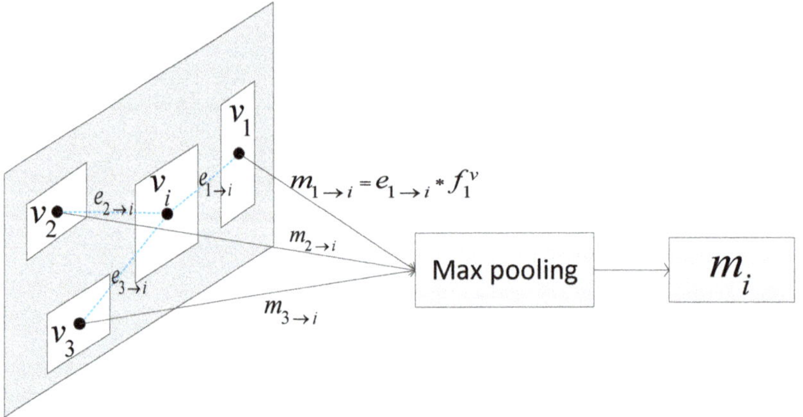

Figure 4. An illustration of building object-object relationship. The process is the same as [51]. For object v_i, the message $m_{1\rightarrow i}$ from object v_1 to object v_i is controlled by $e_{1\rightarrow i}$.

For large optical remote sensing images, it is necessary to use object-object relationship contextual information within meaningful limited regions in images instead of the entire images. That is because the effect of object-object relationship contextual information on the detection result is very little if the distance between two objects is too long. The images used in this paper are 400 pixels wide and 400 pixels high, just like limited regions cropped from large remote sensing images. Therefore we can obtain object-object relationship contextual information in the entire images.

3.2. Part-Based Multi-Region Fusion Sub-Network

For a specific object proposal, paying attention to each part of the object in it can help to obtain much useful spatial structure information about the object, so we can obtain more semantic information for better object detection performance. We use multiple parts of each object to acquire more local visual properties and geometric information, providing an enhanced feature representation.

The parts used include the original proposal box, the left-half part of the proposal box, the right-half part of the proposal box, the up-half part of the proposal box, the bottom-half part of the proposal box, and the inner part obtained by scaling the proposal box by a factor of 0.7 (see Figure 5). First, we gain those parts of each candidate region produced by RPN and perform the RoIAlign operation soon after. Second, we concatenate the pooled features along the channel axis. Then, a 1×1 convolution is implemented to reduce the dimension of the concatenated feature, which makes the feature adapt to the input shape of the fully connected layer. Later, the feature is fed into a fully connected layer to generate the final feature representation with more semantic information. We denote the final feature representation as P. Finally, we utilize P to gain the class scores S_P and the predicted boxes R_P.

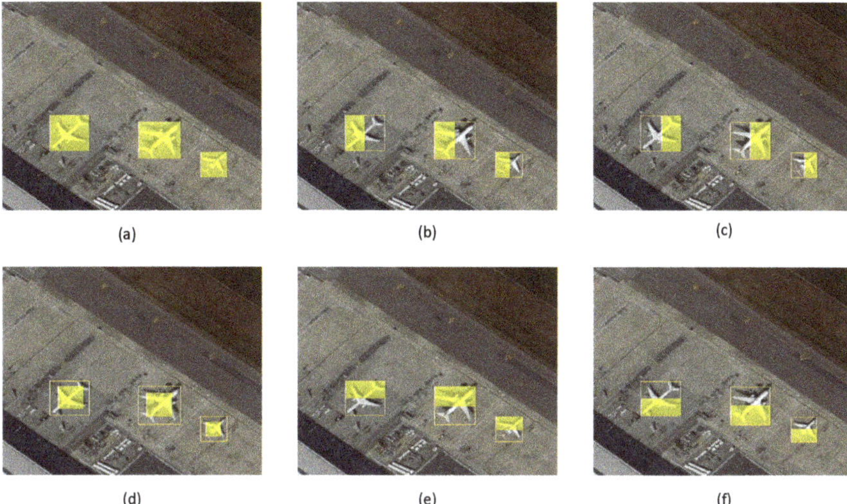

Figure 5. Illustration of object parts used in the proposed framework. (a) Original candidate boxes. (b) Left-half part of candidate boxes. (c) Right-half part of candidate boxes. (d) Inner part obtained by scaling candidate boxes by a factor of 0.7. (e) Up-half part of candidate boxes. (f) Bottom-half part of candidate boxes.

3.3. Multi-Model Decision Fusion Strategy

The multi-model decision fusion strategy, relying on several detection results, is more robust compared to the single model which may cause much false detection. In addition to exploiting the contextual information fusion sub-network and the part-based multi-region fusion sub-network, we also utilize a baseline sub-network that only uses the original proposal regions for object detection. In the baseline sub-network, we perform the RoIAlign operator as same as the two aforementioned sub-networks. Then we employ a fully connected layer to obtain the final feature denoted as B. Finally, we use B to gain the class scores S_B and the predicted boxes R_B.

After obtaining the three types of class scores S_C, S_P, S_B and predicted boxes R_C, R_P, R_B, we make a decision fusion on them. The decision fusion ratio of S_C, S_P, and S_B is 2:1:1, so do R_C, R_P, and R_B, which can provide better detection results in experiments. Then, we use a softmax layer to get the final class labels of all predicted boxes. The loss function employed in this paper is as same as that in Faster R-CNN [10].

4. Experiments and Results

In this part, we first introduce the data set and evaluation metrics used for the experiments. Then, we describe the implementation details and parameter settings of the proposed method. The results and some comparisons to other methods are discussed afterward. The models were trained on a computer with two Intel Xeon E5-2630 v4 CPUs and two NVIDIA GeForce GTX 1080 GPUs. The operating system and deep learning platform used were Ubuntu 16.04 and TensorFlow 1.3.0, respectively.

4.1. Data Set

We evaluate the performance of the proposed object detection method on a publicly available data set: NWPU VHR-10-v2 data set [20]. The data set stems from the positive image set of the original NWPU VHR-10 data set [31] and still contains ten classes of geospatial objects, including airplane, ship, storage tank, baseball diamond, tennis court, basketball court, ground track field, harbor, bridge, and vehicle. There are 1172 images (400 × 400 pixels) in the data set we use. The data set is challenging, because the objects are multi-category and multi-scale and the backgrounds are complex. In all experiments, the training data and test data we employ are the same as that in [20], 879 (75% of the data set) remote sensing images in the training data and 293 images in the test data.

4.2. Evaluation Metrics

Here, we evaluate the performance of object detection methods through two standard, universally agreed and widely used measures illustrated in [7], namely precision-recall curve (PRC) and average precision (AP).

4.2.1. Precision-Recall Curve (PRC)

The Precision metric measures the fraction of detections which are true positives, and the Recall metric weighs the fraction of positives which are correctly recognized. The number of true positives, the number of false positives, and the number of false negatives are denoted as TP, FP, and FN, respectively. Therefore, the Precision and Recall metrics can be obtained by:

$$Precision = \frac{TP}{(TP + FP)} \tag{8}$$

$$Recall = \frac{TP}{(TP + FN)} \tag{9}$$

The PRC metric is based on the overlapping area between the detection and the ground truth object. A detection is considered to be a true positive if the intersection over union (IoU) between the detection and the ground truth box exceeds a predetermined threshold; otherwise, the detection is marked as a false positive. What is more, if several detections overlap with a same ground truth bounding box, only one is regarded as the true positive, and others are labeled as false positives. The intersection over union IoU is formulated as:

$$IoU = \frac{area(detection \cap groundtruth)}{area(detection \cup groundtruth)} \tag{10}$$

4.2.2. Average Precision (AP)

The AP calculates the average value of Precision over the interval from Recall = 0 to Recall = 1, namely the area under the PRC. Therefore, the higher the AP value, the better the performance, and vice versa.

4.3. Implementation Details and Parameter Settings

The proposed model is based on the successful VGG16 network [56] that was pretrained on ImageNet [57]. To augment the training data, we flip all the training images horizontally. For training our model, we utilize the stochastic gradient descent with 0.9 momentum. The learning rate is initialized to 0.001 and we use it for 20 k iterations; then we continue training for 10k iterations with 0.0001. The last fully connected layers for classification and bounding box regression are randomly initialized with zero-mean Gaussian distributions with standard deviations of 0.001, simultaneously other fully connected layers and the 1×1 convolutional layer with standard deviations of 0.01. Biases are initialized to 0. For training RPN, each mini-batch arises from a single image which includes many positive and negative example anchors, and we randomly sample 128 anchors in an image to calculate the loss function of a mini-batch. The sampled positive and negative anchors have a ratio of up to 1:1. If there are fewer than 64 positive samples in an image, we pad the mini-batch with negative ones. The entire model is trained end-to-end. Furthermore, we consider a detection to be correct if the IoU between the predicted bounding box and the ground truth bounding box exceeds 0.5. Otherwise, the detection is considered as a false positive. In the implementation of the test, we employ Soft-NMS to reduce redundancy for better detection performance.

4.4. Evaluation of Local Contextual Information and Object-Object Relationship Contextual Information Fusion Sub-Network

To evaluate the efficiency of our local contextual information and object-object relationship contextual information fusion sub-network, we designed a basic set of experiments. First, we run the standard Faster R-CNN model as a benchmark experiment. Then, on the basis of the baseline sub-network, we incorporate the proposed sub-network which fuses both local contextual information and object-object relationship contextual information. In the experiments, we find that using the features extracted from the 1.8× of the original proposal boxes as local contextual features leads to better detection performance. In the field of remote sensing image object detection, some works [18,20,27] take local contextual information into account and therefore obtain good results. However, the object-object relationship contextual information has not been proven to be beneficial for detecting geospatial objects. To illustrate the usefulness of the object-object relationship contextual information, we implement an experiment in which we incorporate the sub-network only containing local contextual information into the baseline sub-network. The detailed experimental results are summarized in Table 1. As shown in Table 1, an improvement of 4.24 percent points in mean average precision (mAP) can be seen by adding the local contextual information and object-object relationship contextual information fusion sub-network compared to the Faster R-CNN baseline network. This validates that our local contextual information and object-object relationship contextual information fusion sub-network has a strong discriminating ability to represent features of geospatial objects, providing useful contextual cues for better detection performance. In addition, Table 1 shows the mAP improves from 92.42% (only using local contextual information) to 94.04% (using both local contextual information and object-object relationship contextual information), demonstrating that the object-object relationship contextual information plays an important role in achieving better detection performance for geospatial object detection. Furthermore, we execute an experiment to illustrate that local contextual information is more useful than global contextual information for the entire image in remote sensing image object detection. In the experiment, we replace local contextual information with global contextual information for the entire remote sensing image in the overall proposed framework. The results are shown in Table 1. As we can see, in terms of mAP over all ten object categories, applying local contextual information outperforms the use of global contextual information for the entire image by 2.4%. This demonstrates that the use of local contextual information is critical, leading to better detection results than using global contextual information for the entire remote sensing image.

4.5. Evaluation of Part-Based Multi-Region Fusion Network

To verify that the part-based multi-region fusion sub-network has a positive effect on geospatial object detection, we compared the overall proposed model (including the part-based multi-region fusion sub-network) with the previous variant where the framework only merges the baseline sub-network and the local contextual information and object-object relationship contextual information fusion sub-network. As can be seen from Table 1, incorporating the part-based multi-region fusion sub-network offers a further performance increase of 1.0 percent point. This demonstrates that fusing multiple parts of each geospatial object can investigate more spatial structural information about objects, which helps to diversify object features and enhance semantic information for forming powerful feature representation.

4.6. Evaluation of Multi-model Decision Fusion Strategy

In the proposed approach, we make a decision fusion on the results of three sub-networks, which include the local contextual information and object-object relationship contextual information fusion sub-network, the part-based multi-region fusion sub-network, and the baseline sub-network. To evaluate the effectiveness of the decision fusion ratio of 2:1:1 corresponding to those three sub-networks, we set 25 different ratios for contrast. These ratios consist of 1:1:1, 1:1:2, 1:1:3, 1:2:1, 1:2:2, 1:2:3, 1:3:1, 1:3:2, 1:3:3, 2:1:1, 2:1:2, 2:1:3, 2:2:1, 2:2:3, 2:3:1, 2:3:2, 2:3:3, 3:1:1, 3:1:2, 3:1:3, 3:2:1, 3:2:2, 3:2:3, 3:3:1, 3:3:2. The experimental results are illustrated in Table 2. As we can see, using the fusion ratio of 2:1:1 achieves the best result among all the experimental results, gaining a mAP value 95.04%. This indicates that the set fusion ratio of 2:1:1 is beneficial to the detection.

Table 1. Detection results of using sub-networks. C-Gl: Incorporate the Contextual Information Fusion Sub-network only containing global contextual information for the entire image. C-Lo: Incorporate the Contextual Information Fusion Sub-network only containing local contextual information. C-Re: Incorporate the Contextual Information Fusion Sub-network only containing object-object relationship contextual information. P: Incorporate the Part-based Multi-region Fusion Sub-network.

	C			P	mAP	Airplane	Ship	Storage Tank	Baseball Diamond	Tennis Court	Basketball Court	Ground Track Field	Harbor	Bridge	Vehicle
	C-Gl	C-Lo	C-Re												
Faster R-CNN (Baseline)					0.8980	1.0000	0.9225	0.9415	0.9521	0.9267	0.8429	1.0000	0.8788	0.6899	0.8254
ours		√			0.9242	1.0000	0.9106	0.9523	0.9593	0.9554	0.9116	1.0000	0.9235	0.7419	0.8873
ours		√	√		0.9404	0.9999	0.9184	0.9898	0.9757	0.9545	0.9484	0.9994	0.9497	0.7605	0.9072
ours		√	√	√	0.9504	0.9934	0.9227	0.9918	0.9668	0.9632	0.9756	1.0000	0.9740	0.8027	0.9136
ours	√		√	√	0.9264	0.9999	0.9139	0.9618	0.9630	0.9493	0.9424	1.0000	0.9172	0.7051	0.9115

Table 2. Comparison detection results of 25 different decision fusion ratios.

Fusion Ratio	mAP	Airplane	Ship	Storage Tank	Baseball Diamond	Tennis Court	Basketball Court	Ground Track Field	Harbor	Bridge	Vehicle
1:1:1	0.9386	1.0000	0.9303	0.9741	0.9740	0.9439	0.9506	1.0000	0.9689	0.7406	0.9029
1:1:2	0.9337	1.0000	0.9104	0.9616	0.9617	0.9421	0.9471	1.0000	0.9686	0.7421	0.9032
1:1:3	0.9345	1.0000	0.9061	0.9557	0.9748	0.9420	0.9459	1.0000	0.9715	0.7414	0.9079
1:2:1	0.9416	1.0000	0.9142	0.9921	0.9758	0.9557	0.9631	1.0000	0.9381	0.7603	0.9169
1:2:2	0.9320	1.0000	0.8993	0.9756	0.9528	0.9422	0.9506	1.0000	0.9601	0.7359	0.9033
1:2:3	0.9313	1.0000	0.9107	0.9696	0.9438	0.9414	0.9471	1.0000	0.9693	0.7285	0.9031
1:3:1	0.9403	1.0000	0.9131	0.9717	0.9774	0.9628	0.9500	1.0000	0.9607	0.7656	0.9012
1:3:2	0.9339	1.0000	0.9304	0.9762	0.9512	0.9412	0.9500	1.0000	0.9535	0.7345	0.9022
1:3:3	0.9330	1.0000	0.9315	0.9752	0.9583	0.9413	0.9462	1.0000	0.9576	0.7174	0.9028
2:1:1	**0.9504**	**0.9934**	**0.9227**	**0.9918**	**0.9668**	**0.9632**	**0.9756**	**1.0000**	**0.9740**	**0.8027**	**0.9136**
2:1:2	0.9391	1.0000	0.9204	0.9623	0.9743	0.9445	0.9495	1.0000	0.9705	0.7641	0.9053
2:1:3	0.9356	1.0000	0.9103	0.9564	0.9748	0.9440	0.9495	1.0000	0.9716	0.7476	0.9015
2:2:1	0.9379	0.9999	0.8866	0.9680	0.9661	0.9599	0.9512	1.0000	0.9598	0.7814	0.9059
2:2:3	0.9355	1.0000	0.9357	0.9710	0.9373	0.9436	0.9495	1.0000	0.9685	0.7465	0.9032
2:3:1	0.9352	1.0000	0.9136	0.9762	0.9621	0.9430	0.9512	1.0000	0.9502	0.7536	0.9025
2:3:2	0.9363	1.0000	0.9306	0.9762	0.9646	0.9426	0.9500	1.0000	0.9567	0.7398	0.9029
2:3:3	0.9337	1.0000	0.9281	0.9727	0.9436	0.9429	0.9506	1.0000	0.9598	0.7356	0.9032
3:1:1	0.9381	0.9934	0.9468	0.9768	0.9660	0.9798	0.9512	1.0000	0.9326	0.7674	0.8671
3:1:2	0.9405	1.0000	0.9325	0.9615	0.9741	0.9456	0.9495	1.0000	0.9705	0.7704	0.9014
3:1:3	0.9314	1.0000	0.8675	0.9605	0.9735	0.9447	0.9495	1.0000	0.9714	0.7459	0.9016
3:2:1	0.9383	1.0000	0.9142	0.9704	0.9734	0.9453	0.9500	1.0000	0.9662	0.7611	0.9023
3:2:2	0.9399	1.0000	0.9309	0.9699	0.9746	0.9447	0.9500	1.0000	0.9705	0.7554	0.9029
3:2:3	0.9405	1.0000	0.9344	0.9659	0.9740	0.9445	0.9495	1.0000	0.9707	0.7636	0.9029
3:3:1	0.9361	1.0000	0.9216	0.9762	0.9601	0.9445	0.9506	1.0000	0.9486	0.7565	0.9027
3:3:2	0.9371	1.0000	0.9298	0.9762	0.9676	0.9443	0.9500	1.0000	0.9591	0.7408	0.9033

4.7. Comparisons with Other Detection Methods

We compared the proposed approach with five state-of-the-art methods, including the collection of part detector (COPD) [31], a transferred CNN model from AlexNet [58], the rotation-invariant convolutional neural network (RICNN) [17], the rotation-insensitive and context-augmented object detector (RICAOD) [20], and Faster R-CNN [10]. In the implementation of the ten-class object detection task, the COPD is made up of 45 seed-based part detectors. Each part detector is a linear support vector machine (SVM) classifier and corresponds to a particular viewpoint of an object class, therefore the collection of them providing a solution for rotation-invariant detection of multi-class objects. Exploited as a common CNN feature extractor, the transferred CNN model has shown great success for PASCAL Visual Object Classes object detection. For dealing with the problem of object rotation variations, the RICNN is designed to introduce and learn a new rotation-invariant layer on the basis of the existing CNN architecture, AlexNet. The RICAOD utilizes multiangle anchors for rotation-invariant object detection and combines local and contextual features to address the problem of appearance ambiguity. The quantitative comparison results of the six different methods are shown in Table 3 and Figure 6, representing the AP values and PRCs, respectively. As can be observed in Table 3, in terms of mean AP over all ten object categories, the proposed approach outperforms the COPD method [31], the transferred CNN method [58], the RICNN method [17], the RICAOD method [20], and the Faster R-CNN method [10] by 40.15%, 35.43%, 21.93%, 7.92%, and 5.24%, respectively. In addition, we also obtain good detection accuracy in each category, especially airplane, storage tank, basketball, ground track field, and harbor, with very high AP values. Those fully demonstrate that the proposed method achieves much better performance compared to the existing state-of-the-art methods. Table 3 also shows the average running time of each image for the six different approaches. We can observe that the proposed method costs less computation time than other methods except Faster R-CNN.

For all results, it can be easily illustrated: due to the use of the contextual features containing local contextual features and object-object relationship contextual features, the proposed method obtains a discriminative feature representation ability to effectively recognize objects in spite of the diversity and complexity of object appearance, such as storage tank, bridge, and so on; the part-based multi-region fusion sub-network provides more spatial structural information about objects, so that more semantic information can be obtained to enhance the feature representation; the multi-model decision fusion strategy makes the algorithm more robust and provides better detection performance, because it acts like operating on three different single CNN-based models, each of which generates representative characteristics that describe the object.

Figure 7 shows a lot of geospatial object detection results. The green boxes denote true positives; the red boxes denote false positives; the yellow boxes indicate false negatives.

Table 3. Comparison detection results of six different methods.

	mAP	Airplane	Ship	Storage Tank	Baseball Diamond	Tennis Court	Basketball Court	Ground Track Field	Harbor	Bridge	Vehicle	Time per Image (Second)
COPD [31]	0.5489	0.6225	0.6937	0.6452	0.8213	0.3413	0.3525	0.8421	0.5631	0.1643	0.4428	1.16
Transferred CNN [58]	0.5961	0.6603	0.5713	0.8501	0.8093	0.3511	0.4552	0.7937	0.6257	0.4317	0.4127	5.09
RICNN [17]	0.7311	0.8871	0.7834	0.8633	0.8909	0.4233	0.5685	0.8772	0.6747	0.6231	0.7201	8.47
RICAOD [20]	0.8712	0.9970	0.9080	0.9061	0.9291	0.9029	0.8031	0.9081	0.8029	0.6853	0.8714	2.89
Faster R-CNN [10]	0.8980	**1.0000**	0.9225	0.9415	0.9521	0.9267	0.8429	1.0000	0.8788	0.6899	0.8254	**0.09**
ours	**0.9504**	0.9934	**0.9227**	**0.9918**	**0.9668**	**0.9632**	**0.9756**	**1.0000**	**0.9740**	**0.8027**	**0.9136**	0.75

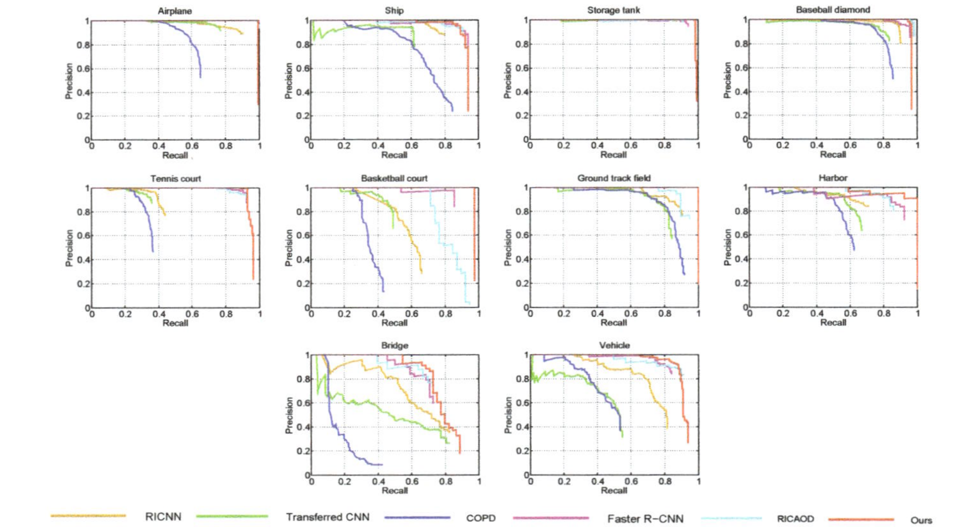

Figure 6. Precision-recall curves (PRCs) of the proposed method and other state-of-the-art methods for airplane, ship, storage tank, baseball diamond, tennis court, basketball court, ground track field, harbor, bridge, vehicle classes, respectively.

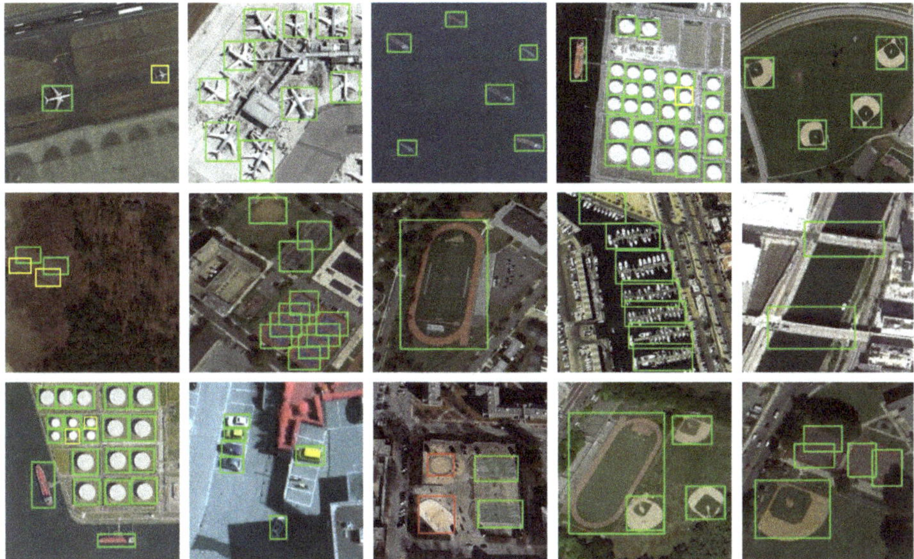

Figure 7. Some object detection results obtained by using the proposed method. The true positives, false positives, and false negatives are denoted by green, red, and yellow rectangles, respectively.

5. Conclusions

In this paper, we proposed a multi-model decision fusion framework for geospatial object detection. The framework combines a contextual information fusion sub-network, a part-based multi-region fusion sub-network, and a baseline sub-network to recognize and locate geospatial objects. The final detection results are obtained by way of making a decision fusion on the results of the three sub-networks. The proposed model presents a remarkable performance on the publicly available data set NWPU VHR-10-v2. All experiments show that: (1) local contextual information and object-object relationship contextual information are beneficial to effectively recognizing objects and alleviating the mis-detection between different types of objects with similar appearance; (2) the part-based multi-region fusion sub-network can provide more details of objects to alleviate the insufficient understanding of geospatial object spatial structure information; (3) the multi-model decision fusion strategy can lead to a more stable and robust model and achieve better algorithm performance; (4) the proposed framework can produce more accurate object detection results than other previous methods. In future work, for better detection performance, we will continue to improve the proposed framework. Many fine details of some small objects are lost due to the implementation of pooling, which can lead to the inability to identify the objects. Therefore, we will consider the use of features from lower convolutional layers. In addition, we will consider designing an operator to obtain more accurate localization of detected objects.

Author Contributions: Investigation, W.M., Q.G., Y.W. and W.Z.; Methodology, W.M. and Y.W.; Supervision, L.J.; Validation, X.Z.; Writing—original draft, W.M. and Q.G.; Writing—review and editing, Y.W. and W.Z.

Funding: The research was jointly supported by the National Natural Science Foundations of China (Nos. 61702392, 61671350, 61772400), and the China Postdoctoral Science Foundation (Nos. 2018T111022, 2017M623127).

Conflicts of Interest: The authors declare no conflict of interest.

References

1. Ahmad, K.; Pogorelov, K.; Riegler, M.; Conci, N.; Halvorsen, P. Social media and satellites. *Multimed. Tools Appl.* **2019**, *78*, 2837–2875. [CrossRef]
2. Ahmad, K.; Pogorelov, K.; Riegler, M.; Ostroukhova, O.; Halvorsen, P.; Conci, N.; Dahyot, R. Automatic detection of passable roads after floods in remote sensed and social media data. *arXiv* **2019**, *arXiv:1901.03298*.
3. Sirmacek, B.; Unsalan, C. Urban-Area and Building Detection Using SIFT Keypoints and Graph Theory. *IEEE Trans. Geosci. Remote Sens.* **2009**, *47*, 1156–1167. [CrossRef]
4. Tang, T.; Zhou, S.; Deng, Z.; Zou, H.; Lei, L. Vehicle Detection in Aerial Images Based on Region Convolutional Neural Networks and Hard Negative Example Mining. *Sensors* **2017**, *17*, 336. [CrossRef]
5. Tuermer, S.; Kurz, F.; Reinartz, P.; Stilla, U. Airborne Vehicle Detection in Dense Urban Areas Using HoG Features and Disparity Maps. *IEEE J. Sel. Top. Appl. Earth Obs. Remote Sens.* **2013**, *6*, 2327–2337. [CrossRef]
6. Shi, Z.; Yu, X.; Jiang, Z.; Li, B. Ship Detection in High-Resolution Optical Imagery Based on Anomaly Detector and Local Shape Feature. *IEEE Trans. Geosci. Remote Sens.* **2014**, *52*, 4511–4523.
7. Cheng, G.; Han, J. A Survey on Object Detection in Optical Remote Sensing Images. *ISPRS J. Photogramm. Remote Sens.* **2016**, *117*, 11–28. [CrossRef]
8. Girshick, R.; Donahue, J.; Darrell, T.; Malik, J. Rich Feature Hierarchies for Accurate Object Detection and Semantic Segmentation. In Proceedings of the IEEE Conference on Computer Vision and Pattern Recognition, Columbus, OH, USA, 23–28 June 2014; pp. 580–587.
9. Girshick, R. Fast R-CNN. In Proceedings of the IEEE International Conference on Computer Vision, Santiago, Chile, 11–18 December 2015; pp. 1440–1448.
10. Ren, S.; He, K.; Girshick, R.; Sun, J. Faster R-CNN: Towards Real-Time Object Detection with Region Proposal Networks. *IEEE Trans. Pattern Anal. Mach. Intell.* **2017**, *39*, 1137–1149. [CrossRef]
11. Redmon, J.; Divvala, S.; Girshick, R.; Farhadi, A. You Only Look Once: Unified, Real-Time Object Detection. In Proceedings of the IEEE Conference on Computer Vision and Pattern Recognition, Las Vegas, NV, USA, 26 June–1 July 2016; pp. 779–788.
12. Liu, W.; Anguelov, D.; Erhan, D.; Szegedy, C.; Reed, S.; Fu, C.Y.; Berg, A.C. SSD: Single Shot MultiBox Detector. In Proceedings of the European Conference on Computer Vision, Amsterdam, The Netherlands, 8–16 October 2016; pp. 21–37.
13. He, K.; Gkioxari, G.; Dollár, P.; Girshick, R. Mask R-CNN. In Proceedings of the IEEE International Conference on Computer Vision, Venice, Italy, 22–29 October 2017; pp. 2961–2969.
14. Gidaris, S.; Komodakis, N. Object Detection via a Multi-region and Semantic Segmentation-Aware CNN Model. In Proceedings of the IEEE International Conference on Computer Vision, Santiago, Chile, 11–18 December 2015; pp. 1134–1142.
15. Yang, X.; Sun, H.; Fu, K.; Yang, J.; Sun, X.; Yan, M.; Guo, Z. Automatic Ship Detection in Remote Sensing Images from Google Earth of Complex Scenes Based on Multiscale Rotation Dense Feature Pyramid Networks. *Remote Sens.* **2018**, *10*, 132. [CrossRef]
16. Long, Y.; Gong, Y.; Xiao, Z.; Liu, Q. Accurate Object Localization in Remote Sensing Images Based on Convolutional Neural Networks. *IEEE Trans. Geosci. Remote Sens.* **2017**, *55*, 2486–2498. [CrossRef]
17. Cheng, G.; Zhou, P.; Han, J. Learning Rotation-Invariant Convolutional Neural Networks for Object Detection in VHR Optical Remote Sensing Images. *IEEE Trans. Geosci. Remote Sens.* **2016**, *54*, 7405–7415. [CrossRef]
18. Xiao, Z.; Gong, Y.; Long, Y.; Li, D.; Wang, X.; Liu, H. Airport Detection Based on a Multiscale Fusion Feature for Optical Remote Sensing Images. *IEEE Geosci. Remote Sens. Lett.* **2017**, *14*, 1469–1473. [CrossRef]
19. Ren, Y.; Zhu, C.; Xiao, S. Deformable Faster R-CNN with Aggregating Multi-Layer Features for Partially Occluded Object Detection in Optical Remote Sensing Images. *Remote Sens.* **2018**, *10*, 1470. [CrossRef]
20. Li, K.; Cheng, G.; Bu, S.; You, X. Rotation-Insensitive and Context-Augmented Object Detection in Remote Sensing Images. *IEEE Trans. Geosci. Remote Sens.* **2018**, *56*, 2337–2348. [CrossRef]
21. Yang, Y.; Zhuang, Y.; Bi, F.; Shi, H.; Xie, Y. M-FCN: Effective Fully Convolutional Network-Based Airplane Detection Framework. *IEEE Geosci. Remote Sens. Lett.* **2017**, *14*, 1293–1297. [CrossRef]
22. Xu, Y.; Zhu, M.; Li, S.; Feng, H.; Ma, S.; Che, J. End-to-End Airport Detection in Remote Sensing Images Combining Cascade Region Proposal Networks and Multi-Threshold Detection Networks. *Remote Sens.* **2018**, *10*, 1516. [CrossRef]

23. Guo, W.; Yang, W.; Zhang, H.; Hua, G. Geospatial Object Detection in High Resolution Satellite Images Based on Multi-Scale Convolutional Neural Network. *Remote Sens.* **2018**, *10*, 131. [CrossRef]
24. Chen, S.; Zhan, R.; Zhang, J. Geospatial Object Detection in Remote Sensing Imagery Based on Multiscale Single-Shot Detector with Activated Semantics. *Remote Sens.* **2018**, *10*, 820. [CrossRef]
25. Liu, Y.; Zhang, Z.; Zhong, R.; Chen, D.; Ke, Y.; Peethambaran, J.; Chen, C.; Sun, L. Multilevel Building Detection Framework in Remote Sensing Images Based on Convolutional Neural Networks. *IEEE J. Sel. Top. Appl. Earth Obs. Remote Sens.* **2018**, *11*, 3688–3700. [CrossRef]
26. Wang, G.; Wang, X.; Fan, B.; Pan, C. Feature Extraction by Rotation-Invariant Matrix Representation for Object Detection in Aerial Image. *IEEE Geosci. Remote Sens. Lett.* **2017**, *14*, 851–855. [CrossRef]
27. Zhang, L.; Shi, Z.; Wu, J. A Hierarchical Oil Tank Detector With Deep Surrounding Features for High-Resolution Optical Satellite Imagery. *IEEE J. Sel. Top. Appl. Earth Obs. Remote Sens.* **2015**, *8*, 4895–4909. [CrossRef]
28. Zhang, W.; Sun, X.; Wang, H.; Fu, K. A generic discriminative part-based model for geospatial object detection in optical remote sensing images. *ISPRS J. Photogramm. Remote Sens.* **2015**, *99*, 30–44. [CrossRef]
29. Zhang, W.; Sun, X.; Fu, K.; Wang, C.; Wang, H. Object Detection in High-Resolution Remote Sensing Images Using Rotation Invariant Parts Based Model. *IEEE Geosci. Remote Sens. Lett.* **2013**, *11*, 74–78. [CrossRef]
30. Cheng, G.; Han, J.; Guo, L.; Qian, X.; Zhou, P.; Yao, X.; Hu, X. Object detection in remote sensing imagery using a discriminatively trained mixture model. *ISPRS J. Photogramm. Remote Sens.* **2013**, *85*, 32–43. [CrossRef]
31. Cheng, G.; Han, J.; Zhou, P.; Guo, L. Multi-class geospatial object detection and geographic image classification based on collection of part detectors. *ISPRS J. Photogramm. Remote Sens.* **2014**, *98*, 119–132. [CrossRef]
32. Cheng, G.; Han, J.; Zhou, P.; Guo, L. Scalable multi-class geospatial object detection in high-spatial-resolution remote sensing images. In Proceedings of the IEEE Geoscience and Remote Sensing Symposium, Quebec City, QC, Canada, 13–18 July 2014; pp. 2479–2482.
33. Dalal, N.; Triggs, B. Histograms of oriented gradients for human detection. In Proceedings of the IEEE Conference on Computer Vision and Pattern Recognition, San Diego, CA, USA, 20–26 June 2005; Volume 1, pp. 886–893.
34. Han, J.; Zhang, D.; Cheng, G.; Guo, L.; Ren, J. Object Detection in Optical Remote Sensing Images Based on Weakly Supervised Learning and High-Level Feature Learning. *IEEE Trans. Geosci. Remote Sens.* **2015**, *53*, 3325–3337. [CrossRef]
35. Wang, Q.; He, X.; Li, X. Locality and Structure Regularized Low Rank Representation for Hyperspectral Image Classification. *IEEE Trans. Geosci. Remote Sens.* **2019**, *57*, 911–923. [CrossRef]
36. Lowe, D.G. Distinctive Image Features from Scale-Invariant Keypoints. *Int. J. Comput. Vis.* **2004**, *60*, 91–110. [CrossRef]
37. Fei-Fei, L.; Perona, P. A Bayesian hierarchical model for learning natural scene categories. In Proceedings of the IEEE Conference on Computer Vision and Pattern Recognition, San Diego, CA, USA, 20–26 June 2005; Volume 2, pp. 524–531.
38. Zhang, D.; Han, J.; Cheng, G.; Liu, Z.; Bu, S.; Guo, L. Weakly Supervised Learning for Target Detection in Remote Sensing Images. *IEEE Geosci. Remote Sens. Lett.* **2014**, *12*, 701–705. [CrossRef]
39. Xu, S.; Fang, T.; Li, D.; Wang, S. Object Classification of Aerial Images with Bag-of-Visual Words. *IEEE Geosci. Remote Sens. Lett.* **2010**, *7*, 366–370.
40. Han, J.; Zhou, P.; Zhang, D.; Cheng, G.; Guo, L.; Liu, Z.; Bu, S.; Wu, J. Efficient, simultaneous detection of multi-class geospatial targets based on visual saliency modeling and discriminative learning of sparse coding. *ISPRS J. Photogramm. Remote Sens.* **2014**, *89*, 37–48. [CrossRef]
41. Li, Z.; Itti, L. Saliency and Gist Features for Target Detection in Satellite Images. *IEEE Trans. Image Process.* **2011**, *20*, 2017–2029. [PubMed]
42. Ma, W.; Zhang, J.; Wu, Y.; Jiao, L.; Zhu, H.; Zhao, W. A Novel Two-Step Registration Method for Remote Sensing Images Based on Deep and Local Features. *IEEE Trans. Geosci. Remote Sens.* **2019**, 1–10. [CrossRef]
43. Wang, Q.; Yuan, Z.; Du, Q.; Li, X. GETNET: A General End-to-End 2-D CNN Framework for Hyperspectral Image Change Detection. *IEEE Trans. Geosci. Remote Sens.* **2019**, *57*, 3–13. [CrossRef]
44. Wang, Q.; Liu, S.; Chanussot, J.; Li, X. Scene Classification With Recurrent Attention of VHR Remote Sensing Images. *IEEE Trans. Geosci. Remote Sens.* **2019**, *57*, 1155–1167. [CrossRef]

45. Nogueira, K.; Fadel, S.G.; Dourado, Í.C.; Werneck, R.D.O.; Muñoz, J.A.V.; Penatti, O.A.B.; Calumby, R.T.; Li, L.T.; dos Santos, J.A.; Torres, R.D.S. Exploiting ConvNet Diversity for Flooding Identification. *IEEE Geosci. Remote Sens. Lett.* **2018**, *15*, 1446–1450. [CrossRef]

46. Wiewiora, E.; Galleguillos, C.; Vedaldi, A.; Belongie, S.; Rabinovich, A. Objects in Context. In Proceedings of the IEEE International Conference on Computer Vision, Rio de Janeiro, Brazil, 14–20 October 2007; pp. 1–8.

47. Chen, Q.; Song, Z.; Dong, J.; Huang, Z.; Hua, Y.; Yan, S. Contextualizing Object Detection and Classification. *IEEE Trans. Pattern Anal. Mach. Intell.* **2015**, *37*, 13–27. [CrossRef]

48. Li, J.; Wei, Y.; Liang, X.; Dong, J.; Xu, T.; Feng, J.; Yan, S. Attentive Contexts for Object Detection. *IEEE Trans. Multimed.* **2017**, *19*, 944–954. [CrossRef]

49. Bell, S.; Zitnick, C.L.; Bala, K.; Girshick, R. Inside-Outside Net: Detecting Objects in Context with Skip Pooling and Recurrent Neural Networks. In Proceedings of the IEEE Conference on Computer Vision and Pattern Recognition, Las Vegas, NV, USA, 26 June–1 July 2016; pp. 2874–2883.

50. Chen, X.; Gupta, A. Spatial Memory for Context Reasoning in Object Detection. In Proceedings of the IEEE International Conference on Computer Vision, Venice, Italy, 22–29 October 2017; pp. 4086–4096.

51. Liu, Y.; Wang, R.; Shan, S.; Chen, X. Structure Inference Net: Object Detection Using Scene-Level Context and Instance-Level Relationships. In Proceedings of the IEEE Conference on Computer Vision and Pattern Recognition, Salt Lake City, UT, USA, 18–22 June 2018; pp. 6985–6994.

52. Hu, H.; Gu, J.; Zhang, Z.; Dai, J.; Wei, Y. Relation Networks for Object Detection. In Proceedings of the IEEE Conference on Computer Vision and Pattern Recognition, Salt Lake City, UT, USA, 18–22 June 2018; pp. 3588–3597.

53. Marcu, A.; Leordeanu, M. Dual Local-Global Contextual Pathways for Recognition in Aerial Imagery. *arXiv* **2016**, arXiv:1605.05462.

54. Cho, K.; van Merrienboer, B.; Gülçehre, Ç.; Bougares, F.; Schwenk, H.; Bengio, Y. Learning Phrase Representations using RNN Encoder-Decoder for Statistical Machine Translation. *arXiv* **2014**, arXiv:1406.1078.

55. Jaderberg, M.; Simonyan, K.; Zisserman, A.; Kavukcuoglu, K. Spatial Transformer Networks. In Proceedings of the Neural Information Processing Systems 2015, Montreal, QC, Canada, 7–12 December 2015; pp. 2017–2025.

56. Simonyan, K.; Zisserman, A. Very Deep Convolutional Networks for Large-Scale Image Recognition. *arXiv* **2014**, arXiv:1409.1556.

57. Deng, J.; Dong, W.; Socher, R.; Li, L.; Li, K.; Fei-Fei, L. ImageNet: A large-scale hierarchical image database. In Proceedings of the IEEE Conference on Computer Vision and Pattern Recognition, Miami, FL, USA, 20–25 June 2009; pp. 248–255.

58. Krizhevsky, A.; Sutskever, I.; Hinton, G.E. ImageNet Classification with Deep Convolutional Neural Networks. In Proceedings of the International Conference on Neural Information Processing Systems, Lake Tahoe, ND, USA, 3–8 December 2012; pp. 1097–1105.

*remote sensing*

Letter

Local Region Proposing for Frame-Based Vehicle Detection in Satellite Videos

Junpeng Zhang , **Xiuping Jia *** and **Jiankun Hu**

School of Engineering and Information Technology, University of New South Wales, Canberra 2612, Australia; junpeng.zhang@student.unsw.edu.au (J.Z.); J.Hu@adfa.edu.au (J.H.)
* Correspondence: x.jia@adfa.edu.au

Received: 31 August 2019; Accepted: 8 October 2019; Published: 12 October 2019

Abstract: Current new developments in remote sensing imagery enable satellites to capture videos from space. These satellite videos record the motion of vehicles over a vast territory, offering significant advantages in traffic monitoring systems over ground-based systems. However, detecting vehicles in satellite videos are challenged by the low spatial resolution and the low contrast in each video frame. The vehicles in these videos are small, and most of them are blurred into their background regions. While region proposals are often generated for efficient target detection, they have limited performance on satellite videos. To meet this challenge, we propose a Local Region Proposing approach (LRP) with three steps in this study. A video frame is segmented into semantic regions first and possible targets are then detected in these coarse scale regions. A discrete Histogram Mixture Model (HistMM) is proposed in the third step to narrow down the region proposals by quantifying their likelihoods towards the target category, where the training is conducted on positive samples only. Experiment results demonstrate that LRP generates region proposals with improved target recall rates. When a slim Fast-RCNN detector is applied, LRP achieves better detection performance over the state-of-the-art approaches tested.

Keywords: satellite videos; region proposals; convolutional neural networks; tiny and dim target detection; component mixture model

---

1. Introduction

As one of the most promising developments in remote sensing imagery, the satellite videos captured by Skybox and JL-1, have facilitated several emerging research and applications, including super resolution [1,2], video encoding [3,4] and target tracking [5,6]. They expand the earth observation capacity to rapid motion monitoring, such as vehicle and ship tracking [5,7,8]. To reveal these rapid motions, targets of interests need to be located throughout the satellite video first, and the extracted targets in each frame are then associated to construct the trajectories of targets of interest. Therefore, target detection in satellite videos is a fundamental and critical step for target tracking and motion pattern analysis.

Detecting objects of interest in a video can be achieved by the motion-based detectors, which search the changed pixels in a sequence of images by comparing with an estimated background model [9,10]. Various algorithms, such as Frame-Difference [5,11,12], Median Background [13], Gaussian Mixture Model (GMM) [14,15] and Visual Background Extractor (ViBe) [7,16,17], were developed for moving object detection. However, these approaches are prone to the inadequate background modelling and affected by the problem of parallax caused by the motion of the camera.

Alternatively, the image-based object detectors can extract objects of interest from a video frame by frame [18], whose performance is less affected by the parallax motion. By taking the advantage of the discriminative learning methods, these approaches employ a classifier to scan over possible locations of targets in an image by sliding window [19–21]. To reduce the number of the candidate locations to examine, region proposals, which refer a sparse set of potential target locations, are introduced to replace sliding windows over the entire image. For common computer vision tasks, generating region proposals are commonly guided by the object saliency, such as the edges [22–24], or based on superpixels [25–29] or segmentation masks [30,31]. In aerial videos, the coherent regions extracted by Maximally Stable Extremal Regions (MSER) [32,33] or Top-hat-Otsu [34] are also adopted for region proposal generation. Due to the weak contrast between targets and background in satellite videos, saliency-based approaches result in degraded region proposal performance —either generating too many region proposals or producing a low target recall rate. These approaches also lack the mechanisms for quantifying the region proposals' likelihood of being a target, and place the entire burden of handling a large number of region proposals in the target recognition stage. Convolutional Neural Networks were applied for searching region proposals in recent years. These approaches can provide the confidence score for each region proposal, and a significant portion of false alarms in the region proposals are removed before the recognition state [35–38]. However, they heavily rely on the training of a reliable region proposal network using a large amount of training samples.

To improve the region proposal performance to handle dim and small target detection in satellite video, we propose a Local Region Proposing (LRP) approach with three steps in this study. Our observation is that vehicles in satellite videos appear small and dim globally. Therefore we propose to perform segmentation at a coarse scale to form semantic region first. Possible locations of small targets in each semantic region are then extracted. To reduce the false alarm further and alleviate the computation burden on further target recognition stage, a discrete Histogram Mixture Model (HistMM) is proposed to quantify their likelihoods towards the target category. HistMM presents little difficulty in cooperating with most detectors, as it is estimated separately and only positive samples are required for estimating the model.

The remaining part of this paper is structured as follows. Section 2 presents the proposed local region proposal approach, after which the experimental results are presented in Section 3. We conclude this paper in Section 4 with remarks on the promising direction for future study.

2. Local Region Proposing

Figure 1 shows the Local Region Proposing approach (LRP) developed in this study is composed of three steps. First semantic regions are extracted by coarse-scale segmentation, then possible target locations are searched in each extracted region. The Histogram Mixture Model is developed for removing obvious false alarms from the region proposals.

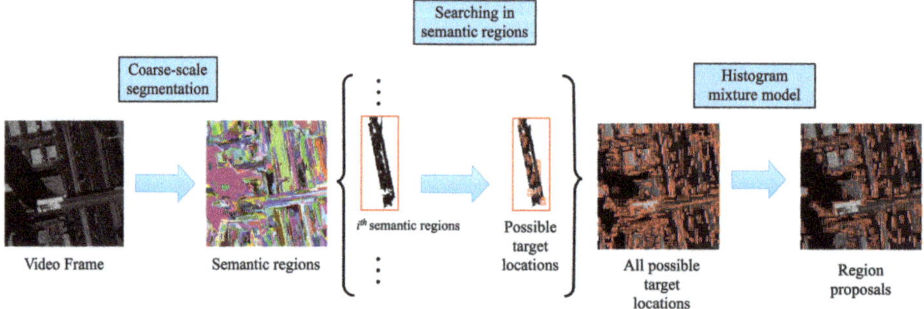

Figure 1. Overview of the proposed region proposal algorithm.

2.1. Semantic Region Extraction

Extracting semantic regions from a video frame can be by segmentation at a coarse scale, and the majority of pixels in each extracted region are more likely from a single land cover type.

The Felzenszwalb's graph-based segmentation approach [39] is a typical method for extracting the semantic regions.

By this graph-based segmentation approach, the scale of the generated superpixels can be controlled by a parameter k. Increasing k would lead to more coarse-scale superpixels, and these superpixels tend to present regions from different land cover types. The semantic regions are allowed to be larger than the target size on purpose. Decreasing k would generate fine-scale superpixels. However, it is often difficult to make superpixels to associate with small targets in satellite videos, due to the low spatial resolution and the low contrast of targets, for example, vehicles, to the background in satellite videos.

2.2. Searching Possible Locations in Semantic Regions

Unlike most dominating saliency object-based approaches, such as Selective Search [26,40], which merge superpixels to form region proposals, the proposed LRP searches region proposals inside semantic regions, where an adaptive threshold is introduced to accommodate the statistics of individual regions.

Note the set of extracted semantic regions as $\mathcal{R}$, for a semantic region that contains m pixels, the set of the pixels' coordinates is noted as $r = \{(x_0, y_0), (x_1, y_1), \ldots, (x_m, y_m)\} \in \mathcal{R}$. The intensity of a pixel at location (x, y) is referred to $I(x, y)$. The blobs with high local saliency are constructed by the pixels with intensities over a threshold thr_r, $I(x, y) > thr_r$, $(x, y) \in r$. The threshold thr_r is defined by

$$thr_r = \mu_r + f * \sigma_r,\qquad(1)$$

where μ_r and σ_r are the mean and standard deviation of pixel intensities in this local region r. The factor f is the expected saliency against the backgrounds. For each extracted blob, a corresponding boundary box is extracted as a possible location.

In the complex scenarios of satellite videos, this searching strategy may be affected by the presence of crowded vehicles and the blurred boundaries of vehicles, which results in merged proposals or incomplete proposals within an original boundary box extracted. We handle these cases by generating multiple proposals. The large boxes should be divided into sub regions to match the target size approximately and the small boxes should be expanded by half of the target size in each direction as a conservative treatment. Figure 2a shows an example where 4 region proposals are generated. To address those incomplete proposals, as shown in Figure 2b, the given bounding box is expanded in each directions.

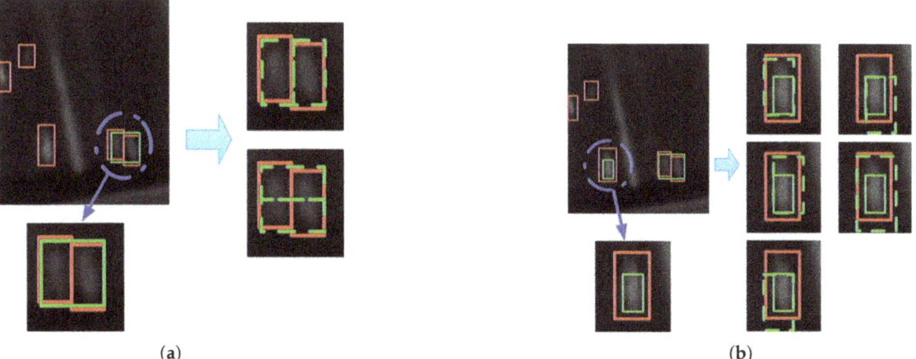

(a)   (b)

Figure 2. Generating multiple region proposals from a possible location. The red box refers to the groundtruth, green solid box refers to the extracted possible location, and green dash boxes refer to the generated region proposals. (**a**) and (**b**) illustrate two examples of generating region proposals by splitting and expanding original region proposals, respectively.

2.3. Histogram Mixture Model

2.3.1. Histogram Mixture Model for Removing Obvious False Alarms

The proposed Histogram Mixture Model (HistMM) measures the likelihoods of the generated region proposals towards their corresponding target category, so that obvious false alarms could be removed at an early stage. The HistMM is a mixture model built on a set of histograms, and training or estimating HistMM depends only on positive training samples.

Note the entire set of initial region proposals on a video frame as $\mathcal{X}_{rp} = \{x_0, x_1, \ldots, x_{n_{rp}}\}$, and n_{rp} is the number of initial region proposal on a given frame. For a region proposal $\forall x \in \mathcal{X}_{rp}$, it is marked as either target or background. We decide if x belongs to the target category (T) or the background category (B) by a Bayesian decision function,

$$R = \frac{p(T|x)}{p(B|x)} = \frac{p(x|T)p(T)}{p(x|B)p(B)}, \tag{2}$$

in which R measures the membership rate of x belonging to the target category versus belonging to the background category. $R \geq 1$ implies x is a target. The corresponding decision function for x that belongs to T can be simplified as

$$p(x|T) \geq c_t, \tag{3}$$

where c_t is a threshold.

The $p(x|T)$ refers to the likelihood of a region proposal x to the target category. We model it by a mixture model composed by a set of $n_{\mathcal{H}}$ histograms, $\mathcal{H} = \{h_1, h_2, \ldots, h_{n_{\mathcal{H}}}\}$. In this paper, we assume that each histogram contributes equally to the likelihood $p(x|T)$, therefore, the possibility of a proposal r that belongs to T is defined as,

$$p(x|T) = \frac{1}{n_{\mathcal{H}}} \sum_{i=1}^{n_{\mathcal{H}}} p(x|h_i). \tag{4}$$

The decision function in Equation (3) can be then interpreted as

$$p(x|T) = \frac{1}{n_{\mathcal{H}}} \sum_{i=1}^{n_{\mathcal{H}}} p(x|h_i) \geq c_t \Rightarrow \exists h \in \mathcal{H}, p(x|h) \geq c_t, \tag{5}$$

which means the likelihood to at least one histogram $\hat{h}_i$ in $\mathcal{H}$ is larger than c_t. On the contrary, a region proposals is a background when all likelihoods toward histograms in $\mathcal{H}$ are less than the threshold c_t, as

$$p(x|h) < c_t, \forall h \in \mathcal{H}. \tag{6}$$

For a given pair of a region proposal x and a histogram in $h \in \mathcal{H}$, we appropriate $p(x|h)$ by the Intersection of Histogram (IoH) between the histogram h and the histogram extracted from the region proposal x. For simplicity, we employ the Histogram of Color (HoC) for calculating $p(x|h)$, as

$$p(x|h) = IoH(h, HoC(x)) = \sum \min(h, HoC(x)), \tag{7}$$

which sums up the minimum values in all pairs of corresponding bins from h and $HoC(x)$. As shown in Figure 3, the $IoHs$ on $HoCs$ are distinct for distinguishing targets and backgrounds, although less information is provided due to the dim appearance of the vehicles.

Our HistMM removes obvious false alarms by the threshold c_t. A larger c_t tends to removal more possible false alarms, whereas it also risks abandoning some target instances. A smaller c_t may improve the coverage of targets in the region proposals, but the remaining number of proposals would be high. The detailed effects of different parameter settings are discussed in Section 3.2.

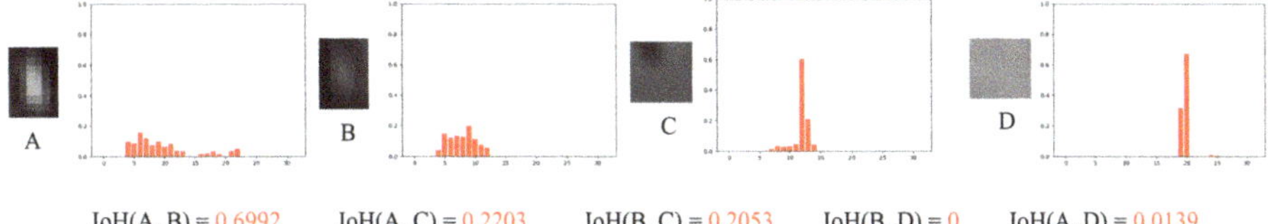

IoH(A, B) = 0.6992   IoH(A, C) = 0.2203   IoH(B, C) = 0.2053   IoH(B, D) = 0   IoH(A, D) = 0.0139

Figure 3. Histogram of Color can distinguish targets from backgrounds. Region proposal A and B are vehicles, whereas the region proposal C and D are obvious false alarms. For the four selected region proposals, their corresponding HoC are extracted, as shown in the right part of the figure. For A and B, the *IoH* is high, while both C and D have low *IoH* due to the extremely low similarities.

2.3.2. Estimating Histogram Mixture Model

For a set of n_{rp} possible region proposals $\mathcal{X}_{rp}$ on a video frame, we predict a region proposal $x \in \mathcal{X}_{rp}$ as a target or a background by Equation (6), as summarized in Algorithm 1. The complexity for predicting region proposals by HistMM grows linearly with the size of $\mathcal{X}_{rp}$, $\mathcal{O}(n_{\mathcal{H}} \times n_{rp})$. Therefore, our proposed HistMM is computationally feasible and scalable for the case with a large number of region proposals.

Algorithm 1 Removing Obvious False Alarms by Histogram Mixture Model (HistMM)

Input: $\mathcal{X}_{rp} = \{x_0, x_1, \ldots, x_{n_{rp}}\}$, $c_t > 0$, and $\mathcal{H} = \{h_1, h_2, \ldots, h_{n_{\mathcal{H}}}\}$
Output: $\mathcal{X}_{rp}$
1: **for** $x \in \mathcal{X}_{rp}$ **do**
2:     **if** $\forall h \in \mathcal{H}, p(x|h) \leq c_t$ **then**
3:         Remove x from $\mathcal{X}_{rp}$.
4:     **end if**
5: **end for**
6: **return** $\mathcal{X}_{rp}$

HistMM is estimated by a recursive learning algorithm on the positive samples of groundtruths [14,41]. Note the estimated set of histograms by $\hat{\mathcal{H}} = \{\hat{h}_1, \hat{h}_2, \ldots, \hat{h}_{n_{\mathcal{H}}}\}$, and all the positive samples in the groundtruths is denoted by $\mathcal{X}_{gt}$. For a groundtruth $x_{gt} \in \mathcal{X}_{gt}$, a histogram $\hat{h}_m$, $m \in \{1, \ldots, n_{\mathcal{H}}\}$, is updated by

$$\hat{\pi}_m \leftarrow \hat{\pi}_m + o_m(x_{gt})$$
$$\hat{h}_m \leftarrow \frac{\hat{h}_m \times \hat{\pi}_m + HoC(x_{gt}) \times o_m}{\hat{\pi}_m + o_m}, \tag{8}$$

where $\hat{\pi}_m$ counts the updates of estimated histogram $\hat{h}_m$, and, as $\hat{\pi}_m$ increases, the lower fraction of the new samples are taken into $\hat{h}_m$. $o_m(x_{gt})$ defines the x_{gt}'s ownership of an estimated histogram $\hat{h}_m$ as

$$o_m(x_{gt}) = \begin{cases} 1, & p(x_{gt}|\hat{h}_m) \geq c_t \quad \text{and} \quad m = \argmax_{i \in \{0,1,\ldots,n_{\mathcal{H}}-1\}} p(x_{gt}|\hat{h}_i) \\ 0, & \text{otherwise} \end{cases}, \tag{9}$$

by which $o_m(x_{gt}) = 1$ indicts that the new sample x_{gt} updates the histogram $\hat{h}_m$ by Equation (8). Otherwise, $o_m(x_{gt}) = 0$ means no nearby histogram component exists for this sample x_{gt}, and a new histogram component $\hat{h}_{n_{\mathcal{H}}}$ is added to $\hat{\mathcal{H}}$. $\hat{\pi}_{n_{\mathcal{H}}}$ is then initialized as 1 and the added histogram component $\hat{h}_{n_{\mathcal{H}}}$ is initialized by $HoC(x_{gt})$. This update procedure continues until it finishes iterating over the groundtruth set $\mathcal{X}_{gt}$, as summarized in Algorithm 2.

Algorithm 2 Training procedure of Histogram Mixture Model (HistMM)

Input: $\mathcal{X}_{gt} = \{x_1, \ldots, x_{n_{gt}}\}, c_t > 0$
Output: $\hat{\mathcal{H}}$

1: **for** $x \in \mathcal{X}_{gt}$ **do**

2:     **if** $\exists \hat{h} \in \hat{\mathcal{H}}, p(x|\hat{h}) \geq c_t$ **then**

3:         Find the updating histogram $\hat{h}_m$ and the ownership $o_m(x)$ by Equation (9).

4:         Update $\hat{h}_m$ by

$$\hat{\pi}_m \leftarrow \hat{\pi}_m + o_m(x)$$

$$\hat{h}_m \leftarrow \frac{\hat{h}_m \times \hat{\pi}_m + HoC(x) \times o_m}{\hat{\pi}_m + o_m}.$$

5:     **else**

6:         Initialize a new component by $HoC(x)$, and add it to $\mathcal{H}$.

7:     **end if**

8: **end for**

9: **return** $\hat{\mathcal{H}}$

---

3. Experimental Results

3.1. Datasets

Two satellite video datasets, SkySat-Las Vegas dataset and SkySat-Burj Khalifa dataset, were used for experimental evaluation of the proposed method for efficient region proposal. For both datasets, the satellite videos were collected by SkySat, which recorded 1800 frames with 30 frames per second. The spatial resolution of each frame in this video is 1.5 m and the frame size is 1920 × 1080 pixels.

The SkySat-Las Vegas dataset refers to the satellite video captured over Las Vegas, USA in March 2014. As illustrated in Figure 4a, two sub-regions were selected for training and one sub-region was selected for evaluation.

The SkySat-Burj Khalifa dataset refers to the satellite video, which is captured over Burj Khalifa, United Arab Emirates on April, 2014. This video is 60 seconds long, which counts up to 30 frames per second. As shown in Figure 4b, 3 sub-regions were selected from the original video, two of which were for training and the remaining one for evaluation.

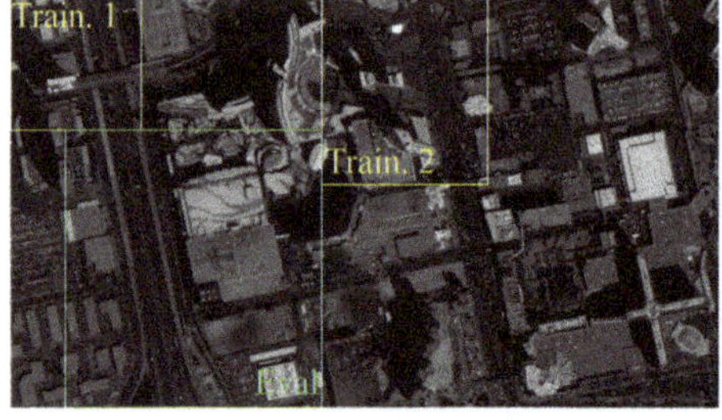

(a) SkySat-Las Vegas dataset   (b) SkySat-Burj Khalifa dataset

Figure 4. Two typical frames from the two satellite video datasets used. (The regions surrounded by the rectangle in yellow color are for training, while the regions in green color are for testing.)

For both datasets, vehicles on five frames from each datasets were annotated, and their corresponding boundary boxes were provided as labelled samples. As we can see in Table 1, the average target sizes are very small.

Table 1. Detailed information for the datasets.

Dataset	Region	Size	Average Vehicle Size
SkySat-Las Vegas	Train. 1	360×360	7.09×5.12
	Train. 2	580×1070	6.27×5.03
	Eval	720×700	7.54×6.00
SkySat-Burj Khalifa	Train. 1	300×400	6.52×5.11
	Train. 2	450×650	7.07×5.28
	Eval	500×670	6.97×5.80

3.2. Parameter Discussion

The LRP approach is mainly controlled by 3 parameters: the local region scale k, the threshold factor f and the threshold c_t in HistMM. The effect of each of them is discuss below. Their performance were evaluated in terms of the coverage of targets (recall), where a targets is recalled if there is at least 50% of IoU between any proposals and the ground-truth bounding box. These evaluations were conducted by the Leave-One-Out Cross Validation (LOOCV) strategy on training set of the SkySat-Las Vegas dataset.

- Semantic region Scale k controls size of the semantic regions generated. A larger k is preferred as it will generate a coarse segmentation as required. The semantic regions are allowed to be larger than the target size on purpose. As presented in Figure 5, reducing k gives fine-scale segmentation and leads to an increased number of region proposals with lower recall rate, while with increasing k, LRP generates fewer region proposals with improved recall rate.
- Threshold Factor f controls the segmentation threshold in each semantic region. Selecting a large f would result in fragmented region proposals and decrease recall scores. As illustrated in Figure 5, increasing f from 1.0 to 3.5, the recall scores experience a drop of over 40%.
- HistMM Threshold c_t is the Bayesian decision threshold in the HistMM for removing obvious false alarms as presented Section 2.3. The HistMM model with a smaller c_t tends to keep more obvious false alarms, which leads to unnecessarily more region proposals decreases. On the other hand, increasing c_t would filter out more obvious false alarms from the searched region proposals. As shown in Figure 6, when c_t increases to 0.5, the number of region proposals (N_{rp}) reduces significantly, while the recall scores holds nearly stable about 80%, which presents the most efficient case.

When c_t was set to 0.5 based on the cross validation on using the training data, the number of region proposals are reduced by over 60% by HistMM with almost no decrease in recall rate, las presented in Table 2 and Figure 7, which demonstrates the effectiveness of the proposed HistoMM model.

Table 2. Evaluation on the effectiveness of HistMM.

Dataset	Recall			N_{rp}		
	Before	After	Diff	Before	After	Diff
SkySat-Las Vegas	75.92%	75.10%	-0.82%	30,614	10,100	-67.01%
SkySat-Burj Khalifsa	77.31%	76.83%	-0.48%	17,017	6525	-61.66%

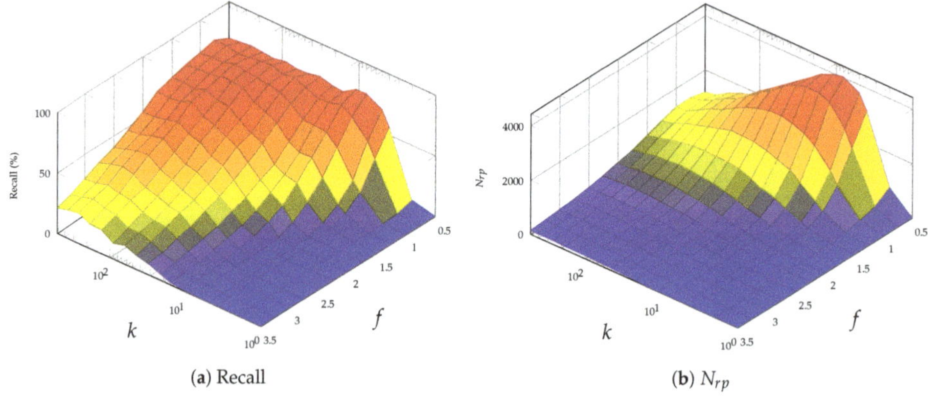

(**a**) Recall (**b**) N_{rp}

Figure 5. Region performance evaluation with different k and f.

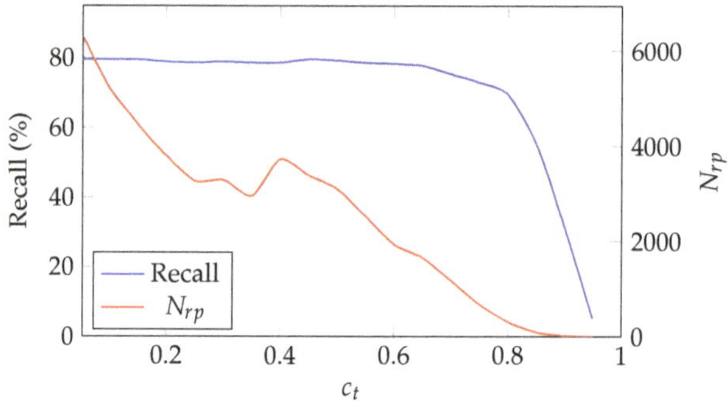

Figure 6. Region proposal performance by different c_t with $k = 81, f = 1.25$.

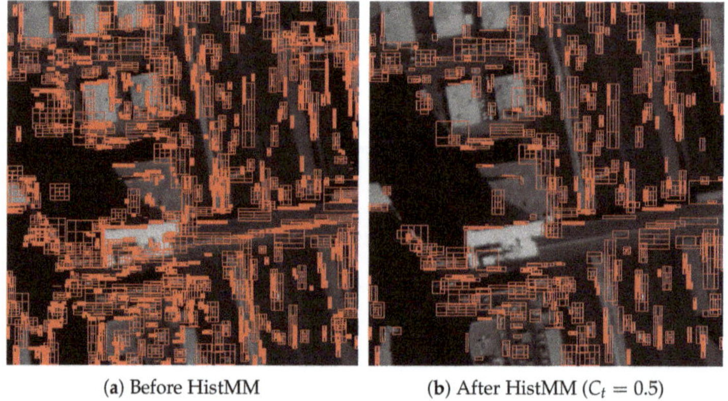

(**a**) Before HistMM (**b**) After HistMM ($C_t = 0.5$)

Figure 7. Visualization on region proposals before and after HistoMM.

3.3. Comparison of Region Proposal Approaches

The region proposal performance was compared with a set of existing region proposals approaches for both common object detection tasks as well as aerial object detection tasks. Inspired by the systematic region proposal evaluation research [42], the proposed region proposal scheme was evaluated against Superpixels (SP) [39,42], Selective Search (SS) [26] and Region Proposal Network

(RPN) [36]. SP generates a region proposal for each extracted superpixel, and SS merges neighboring superpixels as region proposals. For both SS and SP the extraordinarily tiny or large region proposals are considered impossible for vehicles in satellite videos and removed by post-processing. In addition to these well-known region proposals techniques, two approaches for aerial object detection are also included for comparison, which are Maximally Stable Extremal Regions (MSER) [33] or Top-hat-Otsu [34].

Qualitatively, the region proposals generated by our LRP are more concentrated on possible targets, while those saliency object-based approaches, SS and SP, produce more evenly distributed region proposals, as shown in Figure 8. A similar phenomenon is observed on the results by RPN, as both RPN and our LRP remove those obvious false alarms from the background.

Then quantitative performance evaluation on different approaches was conducted in terms of recall scores. Benefiting from the adopted searching strategy and the HistMM, LRP generates a reasonable number of region proposals with good coverage of the possible targets. As presented in Table 3 and Figure 9, our LRP achieves the highest recall$_{@0.5}$ scores on both evaluation datasets. In term of the number of the generated region proposals, it seems like our LRP generates more region proposals than SP, but it should be noted that more than one region proposals are generated by LRP for most possible targets, as shown in Figure 8. Although RPN generates more region proposals with better recall rates, it takes advantage of the finetune scheme from our Fast R-CNN model.

Table 3. Evaluation on region proposal performance.

Method	SkySat-Las Vegas			SkySat-Burj Khalifsa		
	N_{rp}	Recall	Time (s)	N_{rp}	Recall	Time (s)
SP	4092	37.95%	1.98	7922	51.38%	1.28
SS	18,222	20.00%	588.97	11,728	19.34%	264.00
MSER	15,347	37.73%	0.48	10,569	55.80%	0.34
Top-hat-Ostu	1329	2.01%	0.02	1280	29.28%	0.01
RPN (Finetuned from Fast-RCNN-LRP)	13,288	90.00%	0.72	7908	90.05%	0.48
lLRP	9874	80.00%	4.23	7424	79.56%	3.60

Besides, we also compare the detection performance by using a slim Fast-RCNN detector. This slim Fast-RCNN receives 128×128 video frame as input, and it includes two groups of convolutional layers and a branch of fully connected layers for classification, where the branch for boundary box regression are replaced with carefully selected anchor distribution. Each group of convolutional layers contains three layers with kernel in the same size of 3×3, and the number of output channels is 16 and 32 for the first and second convolutional layer group, respectively. After each convolutional layer, a non-linear transformation is conducted by a Rectifier Linear Unit (ReLU) [43,44], which is followed by a Batch Normalization (BN) layer [45]. The output size by Roi Pooling is 2×2, which is followed by two fully connected layers with 512 and 32 hidden neural units, respectively. A Faster R-CNN model is also included for comparison. Due to the limited number of training samples, directly training a Faster R-CNN model is challenging, therefore, this Faster R-CNN model is finetuned from our Fast R-CNN-LRP. The performance evaluation is based on the PASCAL VOC metrics, where we use Average Precision (AP) instead of Mean Average Precision (mAP), since only one target category is contained in both datasets.

Compared with detection results by SP and SS approaches, our approach recalls most of the targets with the highest AP scores, as presented in Table 4 and Figure 10. Compared with the state-of-the-art Faster-RCNN model, the developed LRP with Fast-RCNN model achieves slightly improved detection performance. As illustrated in Figure 11, fewer false alarms with higher detection scores are produced by the Fast R-CNN model using the proposed LRP approach.

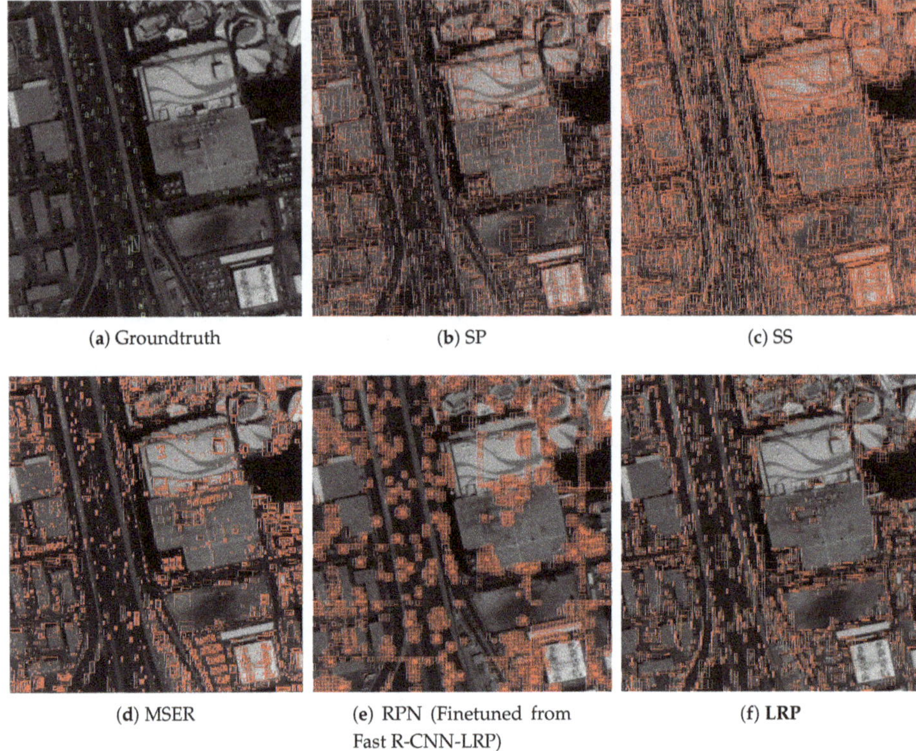

(a) Groundtruth **(b)** SP **(c)** SS

(d) MSER **(e)** RPN (Finetuned from Fast R-CNN-LRP) **(f) LRP**

Figure 8. Visualization on generated region proposals by different approaches on SkySat-Las Vegas Dataset.

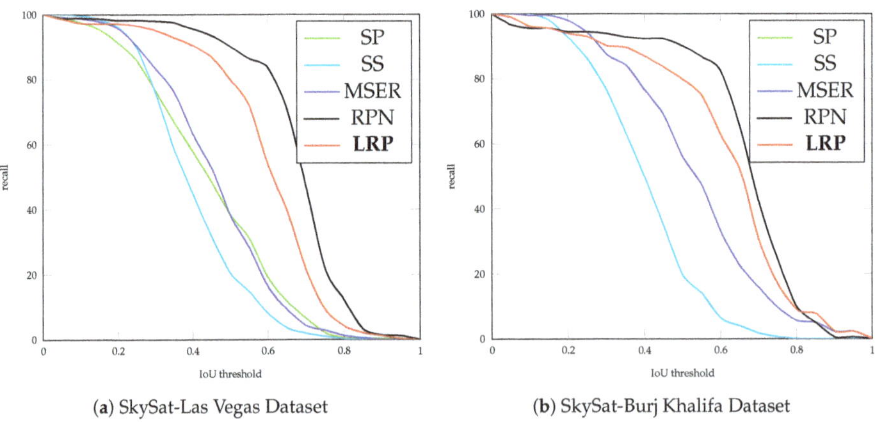

(a) SkySat-Las Vegas Dataset **(b)** SkySat-Burj Khalifa Dataset

Figure 9. Recall rates over different IoU thresholds.

In addition to aforementioned single-frame-based detection approach, we also compare our approach with three popular background subtraction-based approaches —Gaussian Mixture Model (GMM) [46], GMMv2 [14] and Visual Background Extractor (ViBe) [16] approaches (A post-processing is applied to all these background subtraction-based approaches for removing extremely small or large blobs.). Their performance are compared in terms of recall, precision and F_1 scores at IoU = 0.5. Compared with these background subtraction-based approaches, Fast-RCNN-LRP that uses our region proposals generates better F_1 scores, and the background subtraction-based approaches suffer from poor precision, as shown in Table 5.

Table 4. Detection performance evaluation.

Method	SkySat-Las Vegas				SkySat-Burj Khalifa			
	Rcll	Prcn	F_1	AP	Rcll	Prcn	F_1	AP
Fast R-CNN-SP	34.32%	35.53%	34.91%	29.20%	46.41%	31.82%	37.75%	35.30%
Fast R-CNN-SS	14.32%	19.57%	16.54%	7.43%	16.02%	12.78%	14.22%	5.90%
Fast R-CNN-MSER	30.45%	31.16%	30.80%	20.21%	41.44%	47.17%	44.12%	33.96%
Fast R-CNN-Top-hat-Ostu	1.82%	8.08%	2.97%	1.15%	26.52%	26.23%	26.37%	13.37%
Fast R-CNN-LRP	58.18%	43.91%	50.05%	**49.48**%	64.09%	42.49%	51.10%	**50.57**%
Faster R-CNN (Finetuned from Fast R-CNN-LRP)	59.32%	55.53%	56.31%	46.46%	62.43%	46.12%	53.05%	45.15%

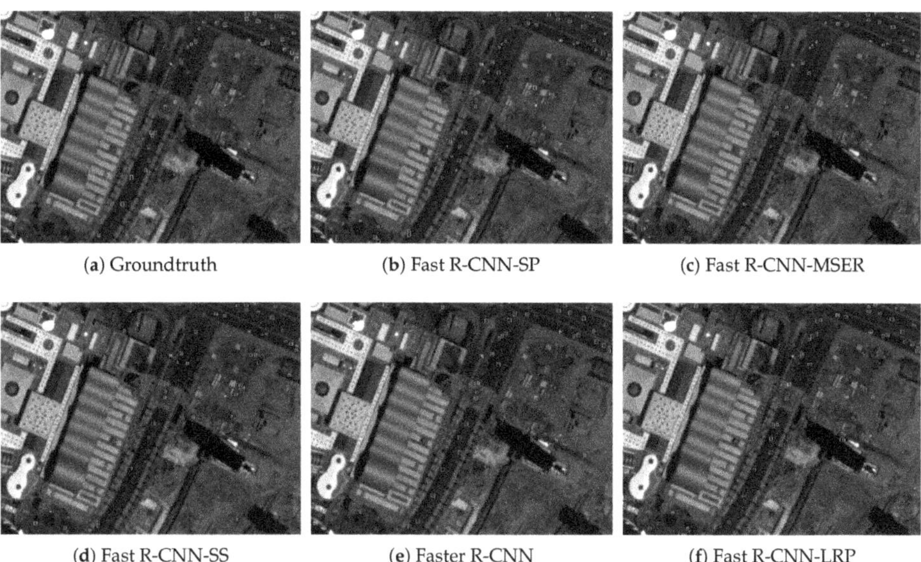

(a) Groundtruth (b) Fast R-CNN-SP (c) Fast R-CNN-MSER

(d) Fast R-CNN-SS (e) Faster R-CNN (f) Fast R-CNN-LRP

Figure 10. Visualization on detection results by selected approaches on SkySat-Burj Khalifsa dataset.

Table 5. Detection results comparisons.

Dataset	Method	Rcll	Prcn	F_1
SkySat-Las Vegas	GMM	45.8%	49.6%	47.6%
	GMMv2	64.7%	26.7%	37.8%
	ViBe	58.0%	16.7%	25.9%
	Fast-RCNN-LRP	58.18%	43.91%	50.05%
SkySat-Burj Khalifa	GMM	33.5%	56.7%	42.1%
	GMMv2	70.1%	37.7%	49.0%
	ViBe	74.6%	22.0%	34.0%
	Fast-RCNN-LRP	64.09%	42.49%	51.10%

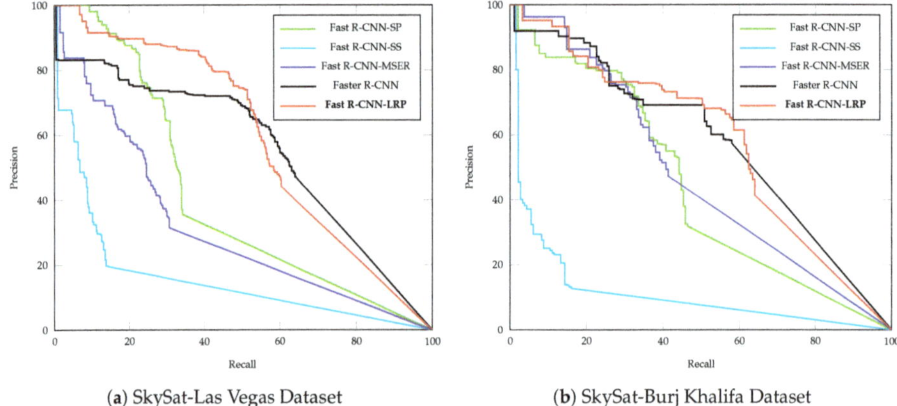

(**a**) SkySat-Las Vegas Dataset   (**b**) SkySat-Burj Khalifa Dataset

Figure 11. Precision-recall curve.

4. Discussion and Conclusions

Region proposal extraction is a valuable step to make target detection efficient. However, it is challenging to generate a small number of region proposals without missing any targets. This is more difficult when the targets are small and dim, such as those presented in satellite videos, due to their limited spatial resolution.

To address the degraded performance of current region proposal extraction methods for satellite videos, we proposed a novel region proposal approach (LRP), in which possible locations of targets are searched in semantic regions by coarse-scale segmentation and a Histogram Mixture Model (HistMM) is proposed to select region proposals with high likelihood from them.

The proposed LRP achieves improved recall rates of the targets with an acceptable increase in time cost, when compared with saliency object-based region proposal approaches, such as Superpixels (SP), Selective Search (SS), Maximally Stable Extremal Regions (MSER) and Top-hat-Otsu. Although the Region Proposal Network (RPN) recalls more targets with less time cost, it requires sufficient training samples or finetuning from a pre-trained model, such as the one obtained from LRP. Another advantage of the proposed LRP is that its training procedure only relies on positive training samples, even when a limited number of training samples is available.

With the improved recall rates by LRP, the detection performance by it with a slim Fast R-CNN is also superior to other saliency object-based region proposal approaches. The detection results are comparable with those by a finetuned Faster R-CNN model from our Fast R-CNN model. Compared with those background subtraction techniques, the proposal LRP approach outperforms them in term of precision, as fewer false alarms are generated.

As more satellite video data are available, more extensive testing can be conducted in the future study. In addition, the approach proposed in this manuscript is developed and tested on a panchromatic video data without color information. It may be extended to multi-channel data in the future research and improved detection performance can be expected.

Author Contributions: Conceptualization, J.Z.; methodology, J.Z. and X.J.; software, J.Z.; validation, J.Z.; writing–original draft preparation, J.Z. and X.J.; visualization, X.J.; supervision, X.J. and J.H.; project administration, X.J. and J.H.; funding acquisition, X.J.

Funding: This research received no external funding.

Acknowledgments: This work is partially supported by China Scholarship Council. The authors would like to thank Planet Team for providing the data in this research [47].

Conflicts of Interest: The authors declare no conflict of interest.

Remote Sens. **2019**, *11*, 2372

References

1. Luo, Y.; Zhou, L.; Wang, S.; Wang, Z. Video Satellite Imagery Super Resolution via Convolutional Neural Networks. *IEEE Geosci. Remote Sens. Lett.* **2017**, *14*, 2398–2402. [CrossRef]
2. Xiao, A.; Wang, Z.; Wang, L.; Ren, Y. Super-Resolution for "Jilin-1" Satellite Video Imagery via a Convolutional Network. *Sensors* **2018**, *18*, 1194. [CrossRef] [PubMed]
3. Wang, X.; Hu, R.; Wang, Z.; Xiao, J. Virtual Background Reference Frame Based Satellite Video Coding. *IEEE Signal Process. Lett.* **2018**, *25*, 1445–1449. [CrossRef]
4. Xiao, J.; Zhu, R.; Hu, R.; Wang, M.; Zhu, Y.; Chen, D.; Li, D. Towards Real-Time Service from Remote Sensing: Compression of Earth Observatory Video Data via Long-Term Background Referencing. *Remote Sens.* **2018**, *10*, 876. [CrossRef]
5. Du, B.; Sun, Y.; Cai, S.; Wu, C.; Du, Q. Object Tracking in Satellite Videos by Fusing the Kernel Correlation Filter and the Three-Frame-Difference Algorithm. *IEEE Geosci. Remote Sens. Lett.* **2018**, *15*, 168–172. [CrossRef]
6. Zhang, J.; Jia, X.; Hu, J.; Tan, K. Satellite Multi-Vehicle Tracking under Inconsistent Detection Conditions by Bilevel K-Shortest Paths Optimization. In Proceedings of the 2018 International Conference on Digital Image Computing: Techniques and Applications (DICTA), Canberra, Australia, 10–13 December 2018; pp. 1–8.
7. Yang, T.; Wang, X.; Yao, B.; Li, J.; Zhang, Y.; He, Z.; Duan, W. Small moving vehicle detection in a satellite video of an urban area. *Sensors* **2016**, *16*, 1528. [CrossRef] [PubMed]
8. Mou, L.; Zhu, X.X. Spatiotemporal scene interpretation of space videos via deep neural network and tracklet analysis. In Proceedings of the 2016 IEEE International Geoscience and Remote Sensing Symposium (IGARSS), Beijing, China, 10–15 July 2016; pp. 1823–1826.
9. Cristani, M.; Farenzena, M.; Bloisi, D.; Murino, V. Background subtraction for automated multisensor surveillance: A comprehensive review. *EURASIP J. Adv. Signal Process.* **2010**, *2010*, 343057. [CrossRef]
10. Piccardi, M. Background subtraction techniques: A review. In Proceedings of the 2004 IEEE International Conference on Systems, Man and Cybernetics, The Hague, Netherlands, 10–13 October 2004; Volume 4, pp. 3099–3104.
11. Reilly, V.; Idrees, H.; Shah, M. Detection and tracking of large number of targets in wide area surveillance. In Proceedings of the European Conference on Computer Vision, lHeraklion, Crete, Greece, 5–11 September 2010; Springer: Berlin, Germany, 2010; pp. 186–199.
12. Xiao, J.; Cheng, H.; Sawhney, H.; Han, F. Vehicle detection and tracking in wide field-of-view aerial video. In Proceedings of the 2010 IEEE Computer Society Conference on Computer Vision and Pattern Recognition, San Francisco, CA, USA, 13–18 June 2010; pp. 679–684.
13. Sommer, L.W.; Teutsch, M.; Schuchert, T.; Beyerer, J. A survey on moving object detection for wide area motion imagery. In Proceedings of the 2016 IEEE Winter Conference on Applications of Computer Vision (WACV), Lake Placid, NY, USA, 7–10 March 2016; pp. 1–9.
14. Zivkovic, Z. Improved adaptive Gaussian mixture model for background subtraction. In Proceedings of the Proceedings of the 17th International Conference on Pattern Recognition (ICPR 2004), Cambridge, UK, 26 August 2004; Volume 2, pp. 28–31.
15. Pollard, T.; Antone, M. Detecting and tracking all moving objects in wide-area aerial video. In Proceedings of the 2012 IEEE Computer Society Conference on Computer Vision and Pattern Recognition Workshops (CVPRW), Providence, RI, USA, 16–21 June 2012; pp. 15–22.
16. Barnich, O.; Van Droogenbroeck, M. ViBe: A universal background subtraction algorithm for video sequences. *IEEE Trans. Image Process.* **2011**, *20*, 1709–1724. [CrossRef]
17. Xiang, X.; Zhai, M.; Lv, N.; El Saddik, A. Vehicle counting based on vehicle detection and tracking from aerial videos. *Sensors* **2018**, *18*, 2560. [CrossRef]
18. Kang, K.; Ouyang, W.; Li, H.; Wang, X. Object detection from video tubelets with convolutional neural networks. In Proceedings of the IEEE Conference on Computer Vision and Pattern Recognition, Las Vegas, NV, USA, 26 June–1 July 2016; pp. 817–825.
19. Cheng, G.; Han, J. A survey on object detection in optical remote sensing images. *ISPRS J. Photogramm. Remote Sens.* **2016**, *117*, 11–28. [CrossRef]
20. Zhang, W.; Sun, X.; Fu, K.; Wang, C.; Wang, H. Object detection in high-resolution remote sensing images using rotation invariant parts based model. *IEEE Geosci. Remote Sens. Lett.* **2014**, *11*, 74–78. [CrossRef]

21. Cheng, G.; Han, J.; Guo, L.; Liu, Z.; Bu, S.; Ren, J. Effective and efficient midlevel visual elements-oriented land-use classification using VHR remote sensing images. *IEEE Trans. Geosci. Remote Sens.* **2015**, *53*, 4238–4249. [CrossRef]

22. Alexe, B.; Deselaers, T.; Ferrari, V. Measuring the objectness of image windows. *IEEE Trans. Pattern Anal. Mach. Intell.* **2012**, *34*, 2189–2202. [CrossRef] [PubMed]

23. Cheng, M.M.; Zhang, Z.; Lin, W.Y.; Torr, P. BING: Binarized normed gradients for objectness estimation at 300fps. In Proceedings of the IEEE Conference on Computer Vision and Pattern Recognition, Columbus, OH, USA, 23–28 June 2014; pp. 3286–3293.

24. Zitnick, C.L.; Dollár, P. Edge boxes: Locating object proposals from edges. In Proceedings of the European Conference on Computer Vision, Zurich, Switzerland, 6–12 September 2014; Springer: Berlin, Germany, 2014; pp. 391–405.

25. Gokberk Cinbis, R.; Verbeek, J.; Schmid, C. Segmentation driven object detection with fisher vectors. In Proceedings of the IEEE International Conference on Computer Vision, Sydney, NSW, Australia, 1–8 December 2013; pp. 2968–2975.

26. Uijlings, J.R.; Van De Sande, K.E.; Gevers, T.; Smeulders, A.W. Selective search for object recognition. *Int. J. Comput. Vis.* **2013**, *104*, 154–171. [CrossRef]

27. Manen, S.; Guillaumin, M.; Van Gool, L. Prime object proposals with randomized prim's algorithm. In Proceedings of the IEEE International Conference on Computer Vision, Sydney, NSW, Australia, 1–8 December 2013; pp. 2536–2543.

28. Rantalankila, P.; Kannala, J.; Rahtu, E. Generating object segmentation proposals using global and local search. In Proceedings of the IEEE Conference on Computer Vision and Pattern Recognition, Columbus, OH, USA, 23–28 June 2014; pp. 2417–2424.

29. Endres, I.; Hoiem, D. Category-independent object proposals with diverse ranking. *IEEE Trans. Pattern Anal. Mach. Intell.* **2014**, *36*, 222–234. [CrossRef] [PubMed]

30. Carreira, J.; Sminchisescu, C. CPMC: Automatic object segmentation using constrained parametric min-cuts. *IEEE Trans. Pattern Anal. Mach. Intell.* **2011**, *34*, 1312–1328. [CrossRef] [PubMed]

31. Pont-Tuset, J.; Arbelaez, P.; Barron, J.T.; Marques, F.; Malik, J. Multiscale combinatorial grouping for image segmentation and object proposal generation. *IEEE Trans. Pattern Anal. Mach. Intell.* **2016**, *39*, 128–140. [CrossRef] [PubMed]

32. Matas, J.; Chum, O.; Urban, M.; Pajdla, T. Robust wide-baseline stereo from maximally stable extremal regions. *Image Vis. Comput.* **2004**, *22*, 761–767. [CrossRef]

33. Teutsch, M.; Krüger, W.; Beyerer, J. Evaluation of object segmentation to improve moving vehicle detection in aerial videos. In Proceedings of the 2014 11th IEEE International Conference on Advanced Video and Signal Based Surveillance (AVSS), Seoul, Korea, 26–29 August 2014; pp. 265–270.

34. Zheng, Z.; Zhou, G.; Wang, Y.; Liu, Y.; Li, X.; Wang, X.; Jiang, L. A novel vehicle detection method with high resolution highway aerial image. *IEEE J. Sel. Top. Appl. Earth Obs. Remote Sens.* **2013**, *6*, 2338–2343. [CrossRef]

35. Szegedy, C.; Reed, S.; Erhan, D.; Anguelov, D.; Ioffe, S. Scalable, high-quality object detection. *arXiv* **2014**, arXiv:1412.1441.

36. Ren, S.; He, K.; Girshick, R.; Sun, J. Faster r-cnn: Towards real-time object detection with region proposal networks. In Proceedings of the Advances in Neural Information Processing Systems, Montreal, QC, Canada, 7–12 December 2015; pp. 91–99.

37. Zhang, F.; Du, B.; Zhang, L.; Xu, M. Weakly supervised learning based on coupled convolutional neural networks for aircraft detection. *IEEE Trans. Geosci. Remote Sens.* **2016**, *54*, 5553–5563. [CrossRef]

38. Xia, G.S.; Bai, X.; Ding, J.; Zhu, Z.; Belongie, S.; Luo, J.; Datcu, M.; Pelillo, M.; Zhang, L. DOTA: A large-scale dataset for object detection in aerial images. In Proceedings of the IEEE Conference on Computer Vision and Pattern Recognition, Salt Lake City, UT, USA, 18–23 June 2018; pp. 3974–3983.

39. Felzenszwalb, P.F.; Huttenlocher, D.P. Efficient graph-based image segmentation. *Int. J. Comput. Vis.* **2004**, *59*, 167–181. [CrossRef]

40. Long, Y.; Gong, Y.; Xiao, Z.; Liu, Q. Accurate object localization in remote sensing images based on convolutional neural networks. *IEEE Trans. Geosci. Remote Sens.* **2017**, *55*, 2486–2498. [CrossRef]

41. Zivkovic, Z.; van der Heijden, F. Recursive unsupervised learning of finite mixture models. *IEEE Trans. Pattern Anal. Mach. Intell.* **2004**, *26*, 651–656. [CrossRef] [PubMed]

42. Hosang, J.; Benenson, R.; Dollár, P.; Schiele, B. What makes for effective detection proposals? *IEEE Trans. Pattern Anal. Mach. Intell.* **2016**, *38*, 814–830. [CrossRef]

43. Maas, A.L.; Hannun, A.Y.; Ng, A.Y. Rectifier nonlinearities improve neural network acoustic models. In Proceedings of the ICML Workshop on Deep Learning for Audio, Speech and Language Processing, Atlanta, GA, USA, 16 June 2013.

44. Krizhevsky, A.; Sutskever, I.; Hinton, G.E. Imagenet classification with deep convolutional neural networks. In Proceedings of the Advances in Neural Information Processing Systems, Lake Tahoe, NV, USA, 3–8 December 2012; pp. 1097–1105.

45. Ioffe, S.; Szegedy, C. Batch Normalization: Accelerating Deep Network Training by Reducing Internal Covariate Shift. In Proceedings of the International Conference on Machine Learning, Lille, France, 6–11 July 2015; pp. 448–456.

46. KaewTraKulPong, P.; Bowden, R. An improved adaptive background mixture model for real-time tracking with shadow detection. In *Video-Based Surveillance Systems*; Springer: Berlin, Germany, 2002; pp. 135–144.

47. Team, P. Application Program Interface: In Space for Life on Earth. San Francisco, CA. Available online: https://api.planet.com (accessed on 31 August 2019).

Article

Efficient Object Detection Framework and Hardware Architecture for Remote Sensing Images

Lin Li [1,2,*], Shengbing Zhang [1] and Juan Wu [3]

1 School of Computer Science and Engineering, Northwestern Polytechnical University, Xi'an 710072, China; zhangsb@nwpu.edu.cn
2 Fourth Design Department, Beijing Institute of Microelectronics Technology, Beijing 100076, China
3 School of Animation and Software, Xi'an Vocational and Technical College, Xi'an 710077, China; wujuan0226@outlook.com
* Correspondence: lilin0106@mail.nwpu.edu.cn; Tel.: +86-29-65685192

Received: 18 August 2019; Accepted: 11 October 2019; Published: 13 October 2019

Abstract: Object detection in remote sensing images on a satellite or aircraft has important economic and military significance and is full of challenges. This task requires not only accurate and efficient algorithms, but also high-performance and low power hardware architecture. However, existing deep learning based object detection algorithms require further optimization in small objects detection, reduced computational complexity and parameter size. Meanwhile, the general-purpose processor cannot achieve better power efficiency, and the previous design of deep learning processor has still potential for mining parallelism. To address these issues, we propose an efficient context-based feature fusion single shot multi-box detector (CBFF-SSD) framework, using lightweight MobileNet as the backbone network to reduce parameters and computational complexity, adding feature fusion units and detecting feature maps to enhance the recognition of small objects and improve detection accuracy. Based on the analysis and optimization of the calculation of each layer in the algorithm, we propose efficient hardware architecture of deep learning processor with multiple neural processing units (NPUs) composed of 2-D processing elements (PEs), which can simultaneously calculate multiple output feature maps. The parallel architecture, hierarchical on-chip storage organization, and the local register are used to achieve parallel processing, sharing and reuse of data, and make the calculation of processor more efficient. Extensive experiments and comprehensive evaluations on the public NWPU VHR-10 dataset and comparisons with some state-of-the-art approaches demonstrate the effectiveness and superiority of the proposed framework. Moreover, for evaluating the performance of proposed hardware architecture, we implement it on Xilinx XC7Z100 field programmable gate array (FPGA) and test on the proposed CBFF-SSD and VGG16 models. Experimental results show that our processor are more power efficient than general purpose central processing units (CPUs) and graphics processing units (GPUs), and have better performance density than other state-of-the-art FPGA-based designs.

Keywords: object detection; remote sensing image; deep learning; convolutional neural networks (CNNs); hardware architecture; processor

---

1. Introduction

Object detection in high resolution optical remote sensing images is to determine if a given aerial or satellite image contains one or more objects belonging to the class of user focused and locate the position of each predicted object in the image [1]. As an important research topic of remote sensing images analysis, object detection in remote sensing images is widely applied to military reconnaissance, intelligent transportation, urban planning, and other domains [2–5]. In recent years, with the development of optical remote sensing technology and space-borne intelligent information processing system, it is a trend to construct a system that combines remote sensing detection with information processing on satellite or aircraft. Efficient object detection in a remote sensing image and processing on a satellite or aircraft can not only reduce the amount of communication data, but also achieve efficient, flexible, and fast earth observation tasks. However, there are many challenges to detect the user-concerned objects quickly and accurately from the massive remote sensing data. Firstly, remote sensing images have ultra-high spatial resolution, which usually contains tens or hundreds of millions of pixels. Quickly and accurately detecting the user-focused objects from massive amounts of data is a challenging task. Secondly, objects in remote sensing images have multi-scale features. For example, objects such as a ground track field, bridge, etc. have hundreds of pixels, while small objects such as a vehicle, ship, etc. may only contain a few pixels. This feature makes accurate object detection in remote sensing images more difficult, especially for small objects. Thirdly, objects in a remote sensing image viewed from overhead have any orientation, while natural image sets are typically acquired horizontally. Therefore, models trained on natural image sets cannot be directly applied to remote sensing image object detection. In addition, the general-purpose processor that carries out the algorithm cannot meet the requirements of space-borne or airborne information processing with high performance and low energy consumption. Therefore, the design of efficient object detection algorithm framework and hardware architecture for remote sensing images has become an urgent problem to be solved for space-borne or airborne information processing.

There are many methods for object detection in remote sensing images after years of research and development. We summarize these methods into traditional object detection methods based on prior information and manual features and deep learning based object detection methods. The traditional object detection methods based on prior information and manual designed features regard object detection as a classification problem composed of feature extraction and object classification, which includes template matching-based object detection methods, knowledge-based object detection methods, object-based image analysis-based (OBIA-based) object detection methods, and machine learning object detection methods based on prior information and manual designed features [1]. The template matching-based object detection methods are divided into two steps. Firstly, the template is trained from existing data by hand-crafting or statistical methods, and then the similarity measurement is performed on the pre-processed input image to complete the detection. The template matching-based object detection approaches are usually divided into a rigid template and deformable template according to the template type selected by the user [1,6,7]. The knowledge-based object detection methods setup knowledge and rules based on geometric information and context information, and generate hypotheses on the input image and convert the object detection problem into a hypothesis testing problem [8–10]. The OBIA-based object detection methods accomplish the detection task by segmenting the input image and classifying the object, wherein the scale of the image segmentation directly affects the detection result [11–13]. The machine learning object detection methods based on prior information and manual designed features are typically divided into two stages: feature extraction stage and object classification and recognition stage. In the feature extraction stage, selective search is usually used to extract handcrafted features, such as scale-invariant feature transform (SIFT) [14], histogram of oriented gradients (HOG) [15], bag-of-words (BoW) feature [16], texture features [17], and so on. In the stage of object classification and recognition, classifiers often include: support vector machine (SVM) [18], AdaBoost [19], deformable parts model (DPM) [20], condition random field (CRF) [21], sparse coding-based classifier [22] and artificial neural network (ANN) [23], and so on. However most of these methods mentioned above rely on the prior information and manual designed features, and it is difficult to efficiently achieve object detection tasks under massive remote sensing data.

In recent years, deep learning technology has achieved great success in computer vision applications, and the deep learning based object detection methods have become the mainstream in the field of image recognition. The deep convolutional neural network AlexNet proposed by Krizhevsky, A. et al. not only won the 2012 ImageNet Large Scale Visual Recognition Challenge (ILSVR 2012), but also set off a wave of deep learning research [24]. Since then, many researchers have proposed a variety of excellent deep learning algorithms, and deep convolutional neural networks have become the best deep learning algorithm in the field of image recognition. Researchers have also begun to apply deep learning technology to object recognition in remote sensing images. For example, Cheng, G. et al. applied the rotation-invariant convolutional neural networks (RICNN) to object detection in very high resolution (VHR) optical remote sensing images [25]. Wang, G. et al. proposed the infrastructure target detection of remote sensing image based on residual networks [26]. Although these object detection methods based on convolutional neural networks perform well in remote sensing image object detection, they do not form a unified and efficient object framework. Currently, the deep learning object detection algorithm framework based on convolutional neural network has made great progress. These mainstream algorithm frameworks include region proposal-based two-stage object detection algorithm and regression-based one stage object detection algorithm [27]. The region proposal-based algorithm generates a series of region proposals according to selective search (SS), Bing or edge boxes methods firstly, and then extracts the features by the deep neural networks, and implements object classification and boundary regression based on these features. For example, Girshick, R. et al. proposed region-based convolutional neural networks (R-CNN) [28], which combines object candidate regions and deep learning for object detection. Then he proposed the more efficient fast R-CNN algorithm [29], which overcomes the shortcomings of R-CNN's redundant operation when extracting features. Subsequently, Ren, S. et al. proposed a faster R-CNN algorithm [30], which uses region proposal networks (RPN) to extract object candidate regions, and integrates the entire object candidate region extraction, feature extraction, object recognition, and detection into a deep neural network framework. In order to solve the multi-scale detection problem, Lin, T.Y. et al. introduced the feature pyramid network (FPN) to improve the recognition efficiency of small objects [31]. However, the object detection algorithm framework based on the region proposal is not very efficient because it takes more time to extract the candidate region. The regression-based object detection algorithm has no candidate region extraction step, which combines all recognition and detection steps in a deep neural network, and has high detection and recognition efficiency. For this type of algorithm framework, the you only look once (YOLO) algorithm framework proposed by Redmon, J. et al. requires only a single network to evaluate the entire image to obtain the target bounding box and category [32]. The single shot multi-box detector (SSD) algorithm framework proposed by Liu, W. et al. introduces an anchor mechanism based on YOLO, which detects the object on the different scales and improves accuracy without affecting process speed [33]. In general, the deep learning based object detection methods have made great progress in the accuracy and efficiency of object detection compared to the traditional methods. At present, these algorithms are widely used in remote sensing image object detection. However, these algorithms are not satisfactory for the detection of small objects in the images, and they need to be improved. Meanwhile, for the airborne or space-borne application environment, not only the object detection accuracy but also the computation complexity and model size need to be considered.

The rapid development of deep learning technology is inseparable from the support of high performance hardware computing systems. The performance of deep learning algorithms is not only related to its own structure, but also depends on the hardware architecture of computing system that carries out the algorithm. Currently, the training and inference of deep learning algorithms mainly depends on general-purpose processors, such as central processing units (CPUs) and graphics processing units (GPUs). Although the general-purpose CPUs have higher flexibility and better parallel computing power, the deep learning algorithm does not achieve better execution efficiency. GPUs are widely used in training and inference of deep learning algorithms because of its unique many-core architecture and superior parallel computing power, but it cannot obtain good performance-power ratio due to high power consumption. In some specific applications, not only high processing performance but also stricter power consumption requirements are required. For example, in embedded application scenarios or on satellite or aircraft information processing systems, general-purpose CPUs and GPUs will not be able to accommodate the application requirements in such situations. Therefore, application-oriented domain-specific architecture (DSA) is the solution to overcome such problems currently [34]. In recent years, many researchers have proposed different hardware architectures for their respective application scenarios. Farabet, C. et al. proposed CNP [35] with parallel vector computing architecture for low-power lightweight unmanned aircraft vehicles (UAVs) or robots, and a scalable dataflow 2-D grid hardware architecture Neuflow [36] optimized for the computation of general-purpose vision algorithms. Peemem, M. et al. proposed hierarchical memory-centric accelerator architecture to improve the performance of convolutional operations and reduce the overhead of memory access [37]. Alwani, M. et al. reduced the transfer of off-chip feature map data by modifying the order of input data and fusing multiple continuous convolutional layer processing [38]. Chen, T. et al. presented a high-throughput algorithm accelerator DianNao based on adder tree structure for large-scale convolutional neural network (CNN) and deep neural network (DNN) [39]. Du, Z. et al. proposed ShiDianNao based on 2-D mesh topology structure for image recognition applications near to sensors, and reduced memory usage through weight sharing [40]. Zhang, C. et al. designed a CNN accelerator based on the adder tree structure by quantitative analysis of memory bandwidth required for throughput [41]. Google has introduced a high-performance tensor processing unit (TPU) for data centers based on 2-D systolic array architecture [42]. Li, L. et al. designed a co-processor with 2-D mesh topology structure for image recognition by optimization calculation of algorithm [43]. Chang, J.W. et al. proposed a deconvolutional neural networks accelerator (DCNN) with 2-D mesh architecture for super-resolution images [44]. These hardware architectures are designed for a specific application scenario and are mainly used to accelerate the calculation of the deep learning algorithms. The processing elements (PEs) in these processors are typically organized in 1-D or 2-D topology structure. These processors only implement parallel computation of synapses and neurons, which compute each feature map one by one. However, the feature map in the image object detection is 3-D, but the current design do not consider the parallel calculation of feature map, so there is still potential for mining parallelism. Meanwhile, the storage organization and data reuse need to be considered in the architecture to adapt to large-scale algorithms and parameters, which is very important for computing and storage dual-intensive remote sensing images object detection applications.

In order to adapt to the characteristics of object detection in remote sensing images and tackle the problems of the algorithm framework and hardware architecture mentioned above, we propose an efficient context-based feature fusion SSD (CBFF-SSD) algorithm framework. Subsequently, we have designed hardware architecture of deep learning processor with multiple 2-D mesh architecture supporting feature maps parallel processing by analyzing and optimizing the calculation of each layer in the deep learning algorithm framework. Finally, the efficiency and performance of the algorithm framework and processor are evaluated by multiple experiments and indicators. The main contributions of this paper are summarized as follows:

1. We propose a context-based feature fusion SSD (CBFF-SSD) framework for object detection in remote sensing images. MobileNet is used as the backbone network in the algorithm framework to reduce the amount of calculation and parameters, which makes the algorithm more efficient. Two feature fusion units and seven feature maps are used in the algorithm to enhance the detection of multi-scale objects and small objects, and improve the detection accuracy.
2. We analyze and optimize the calculation of each layer in the algorithm framework, which makes it easy to implement in hardware, and lays a foundation for the design of time-division multiplexing processing unit in the subsequent hardware architecture of deep learning processor.
3. We propose efficient hardware architecture of deep learning processor with multiple 2-D mesh topology oriented to image object recognition, which can simultaneously calculate multiple output feature maps. Hierarchical on-chip storage organization makes the neurons and weights to be efficiently delivered to the neural processing units (NPUs). The parallel architecture, the hierarchical storage organization, and the register designed in the PE effectively realize the sharing and reuse of the calculation data.

4. We evaluate the performance and efficiency of the algorithm framework and hardware architecture based on several experiments and evaluation indicators. The performance of the proposed algorithm framework is compared with the several popular algorithms on the NWPU-VHR-10 dataset. We realize the proposed hardware architecture of the deep learning processor on the field programmable gate array (FPGA), and then evaluate the processing performance and compared with the CPU, GPU, and the current popular deep learning processors. The experimental results confirmed the effectiveness and superiority of the proposed algorithm framework and hardware architecture of deep learning processor.

The rest of this paper is organized as follows. Section 2 proposes the context-based feature fusion SSD (CBFF-SSD) framework. The calculation of each layer in deep learning algorithm framework and optimization are described in Section 3. Section 4 introduces the details of the hardware architecture of deep learning processor. Section 5 presents the experimental results and analysis. The experimental results are discussed in the Section 6. Finally, the conclusions are drawn in Section 7.

2. Context-Based Feature Fusion SSD Framework

2.1. Related Works

The deep learning algorithm based on the convolutional neural network model has achieved excellent results in image object detection applications, which not only improves the accuracy of recognition, but also improves the efficiency of recognition. In particular, the recent rapid development of the region proposal-based objects detection algorithm framework and the regression-based objects detection algorithm framework are particularly outstanding. Since remote sensing images have ultra-high resolution, diverse object sizes and directions, including small targets, and diverse shooting angles, it is full of challenges to quickly and accurately detect the user-focused object from massive remote sensing data. For the application of remote sensing image object detection, many researchers have conducted a lot of research based on the two popular algorithm frameworks.

For examples, Han, X. et al. proposed a highly efficient and robust integrated geospatial object detect framework based on the faster region-based convolutional neural network (Faster R-CNN), which realized the integrated procedure by sharing features between the region proposal generation stage and the object detection stage [45]. Zhu, M. et al. proposed an effective airplane detection method in remote sensing images based on Faster R-CNN and multiplayer feature fusion, which solved the problem of insufficient representation ability of weak and small objects and overlapping detection boxes in airplane object detection [3]. Etten, A.V. presented a rapid multi-scale object detection method based on YOLO and DarkNet for the detection of small objects in large satellite imagery [46]. In order to solve the detection of small and dense objects, Zhang, X. et al. proposed an effective region-based VHR remote sensing imagery object detection framework named double multi-scale feature pyramid network (DM-FPN) [47].

In summary, the above algorithms are improved for the small objects detection in remote sensing images, and have achieved good object detection results. However, efficient on-board remote sensing image object detection not only needs to consider the accuracy of detection, but also needs to think about the efficiency of calculations such as computational complexity and the number of parameters.

2.2. CBFF-SSD Framework

Based on the in-depth analysis of the characteristics and challenges of object detection in a remote sensing image on a satellite or aircraft, we proposed a context-based feature fusion SSD (CBFF-SSD) algorithm framework shown in Figure 1. The whole algorithm framework was designed based on the SSD framework [33]. This is because on the one hand, the regression-based object detection framework is considered to be more efficient in image object detection, and the other is because the algorithm framework is more suitable for multi-scale object detection. Different from the SSD framework, the backbone network uses the MobileNet [48] instead of VGGNet. The lightweight MobileNet uses depth-wise separable convolution to effectively reduce the amount of calculation and parameters of the algorithm, which is more conducive to efficient object detection in embedded applications, especially in a space-borne or airborne application environment. We compared the number of parameters and the amount of calculations of the proposed CBFF-SSD algorithm framework and the SSD algorithm in Table 1. It can be seen that the number of parameters of the proposed CBFF-SSD algorithm framework was 56.09% of the SSD algorithm, and the calculation amount was only 17.56% of the SSD algorithm. By reducing the size of the model parameters and computational complexity, the proposed algorithm framework was more effective in object detection.

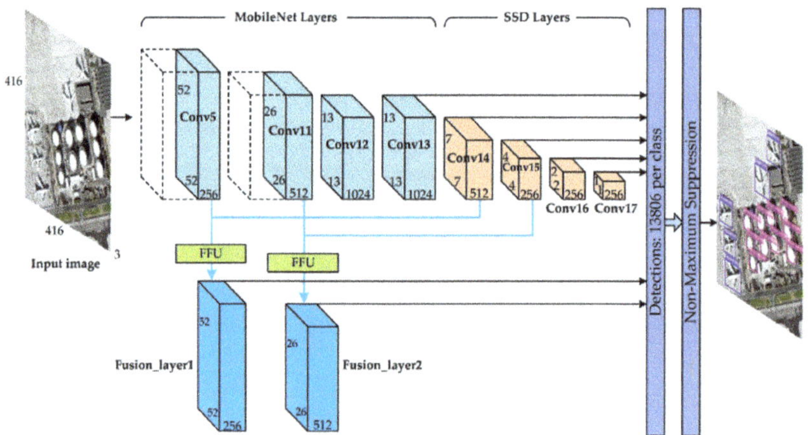

Figure 1. The overall structure of the proposed context-based feature fusion single shot multi-box detector (CBFF-SSD) algorithm framework.

Table 1. Comparison of the number of parameters and calculation amount of the SSD and the proposed CBFF-SSD framework.

Framework	Input Size	Parameters	Mult_Adds
SSD	$300 \times 300 \times 3$	26.28 Million	31.37 Billion
CBFF-SSD	$416 \times 416 \times 3$	14.74 Million	5.51 Billion

The high-level features of the convolutional network were rich in semantics and suitable for detecting large objects. After layer-by-layer down-sampling, the features lose too much detail information, and it is often important to small object detection. The feature rich in semantic information was mapped back to the lower layer features with larger resolution and richer detail information, and they were fused in an appropriate way to improve the effect of small object detection. Therefore, in the algorithm framework, the low-level feature map Conv5 ($52 \times 52 \times 256$) was added, and the Conv14 and Conv15 were respectively up-sampled and fused with Conv5 and Conv11 layers to improve the precision of small object detection.

Figure 1 shows the architecture details of the CBFF-SSD algorithm framework. The Fusion_layer1, Fusion_layer2, Conv13, Conv14, Conv15, Conv16, and Conv17 were used to predict both location and confidences. In the deep learning algorithm, the resolution of the input image and the number of detection boxes for each class of object affected the accuracy of the detection. Based on the SSD framework, we adjusted the input resolution from 300×300 to 416×416, and increased the number of detection boxes from 8732 to 13806 to ensure the recognition accuracy of the proposed algorithm framework.

2.3. Feature Fusion Unit (FFU)

There are two feature fusion units in the proposed algorithm framework as shown in Figure 1. The structure of the feature fusion unit is shown in Figure 2. The design of feature fusion unit is inspired by the design of the deconvolution module in the deconvolutional single shot detector (DSSD) [49]. The high-level feature maps are processed by deconvolution to the same size and channel as the lower-level feature maps. Then they are fused by element wise addition.

Taking the feature fusion layer 1 as an example, in order to fuse the feature maps of the Conv14 and Conv5, it is necessary to up-sample the resolution of the Conv14 layer by eight times. Specifically, for the Conv14 layer, we designed three deconvolution layers with stride 2 to achieve up-sampling. Since the feature maps in the feature fusion unit are computationally intensive, we also applied depth-wise separable convolutions here to reduce parameters and computational complexity. The deconvolution layer was followed by depth-wise separable convolution. The depth-wise separable convolution was composed of 3 × 3 depth-wise convolutional layer, batch normalization, rectified linear unit (ReLU) layer, 1 × 1 point-wise convolutional layer, batch normalization layer, and ReLU layer. The Conv5 layer underwent a depth-wise separable convolution module. After the normalization layer, we fused them by element-wise addition, and finally passed the ReLU to complete the fusion.

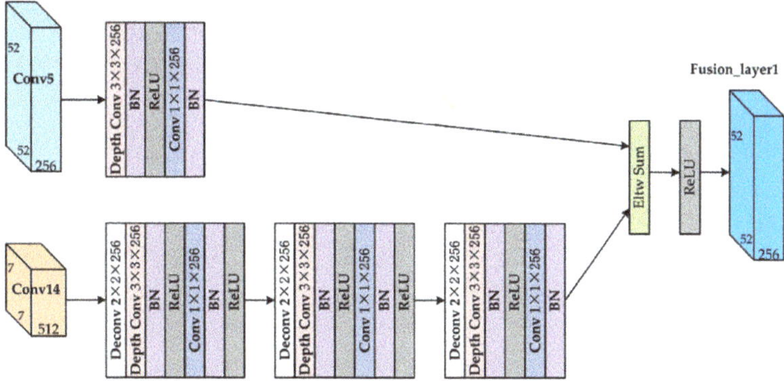

Figure 2. The structure of feature fusion unit (FFU).

Fusion layer 2 used the same calculation method, and only the channel was adjusted. Only minor modifications were required for models with different input resolutions.

2.4. Training

We used the same training strategy as SSD [33]. During training, a set of default boxes was matched to the ground truth boxes. For each ground truth box, we matched it to the default box with the Jaccard overlap higher than a threshold (e.g., 0.5). This was more conducive to predicting multiple bounding boxes with high confidence for overlapped objects. We selected the non-matched default boxes with top loss value as the negative samples so that the ratio of negative and positive was 3:1.

The training objective was for multiple object categories. We set x to be an indicator for matching the default box to the ground truth box of category p, which equaled to 1 or 0. The c, l, and g represent confidences, the predicted box, and ground truth box respectively. The overall objective loss function is a weighted sum of the localization loss L_{loc} and the confidence loss L_{conf}, which is defined as:

$$L(x, c, l, g) = \frac{1}{N}(L_{conf}(x, c) + \alpha L_{loc}(x, l, g)), \tag{1}$$

where N is the number of matched default boxes, and the weight term α is set to 1 by cross validation. The localization loss L_{loc} is a smooth L1 loss between the predicted box (l) and the ground truth box (*g*) parameters. The confidence loss L_{conf} is the Softmax loss over multiple classes' confidences (*c*). The confidence loss L_{conf} and the localization loss L_{loc} are defined the same as SSD [33].

Seven feature maps were used to predict both location and confidences in the proposed CBFF-SSD framework. Specifically, we adopted four default boxes at each feature map for Fusion_layer1, Conv16, and Conv17, and used six default boxes at each feature map location for all other layers. We made a minor change on the scale of default boxes, the lowest layer had a scale of 0.15 and the highest layer had a scale of 0.9. The aspect ratio setting of default boxes were selected as 1, 2, 3, 1/2, and 1/3.

We also used the data augmentation strategy that was consistent with SSD to make the framework more robust to various input object sizes and shapes. These methods included random cropping, flipping, photometric distortion, and random expansion augmentation trick, which are very helpful for detecting small objects.

3. Calculation of Each Layer in the Deep Learning Algorithm Framework and Optimization

Based on the analysis of many deep learning algorithms used in image object detection and the framework proposed in Section 2, it can be seen that although the structures of these algorithms were different, they were all designed based on the deep convolutional neural network. These object detection algorithms are composed of some basic calculation layers, including the convolutional layer, deconvolutional layer, pooling layer, nonlinear activation function layer, normalization layer, element-wise sum layer, full connection layer, and Softmax layer. The hardware architecture suitable for algorithm computing was abstracted based on the analysis and optimization of each layer in the following.

3.1. Convolutional Layer and Deconvolutional Layer

The convolutional layer is composed of several convolution kernels, which is used to extract various features from the input feature maps. When calculating, the size of input feature maps f_{in} is defined as $W \times H \times C$; the kernels are expressed as $K_x \times K_y \times C \times M$ (M is the number of kernels, which is also equal to the number of output feature maps).

The output neuron N at position (x, y) of output feature map f_{out} is computed with:

$$N_{x,y}^{f_{out}} = \sum_{f_{in}=0}^{C-1} \sum_{j=0}^{K_y-1} \sum_{i=0}^{K_x-1} W_{i,j}^{f_{in},f_{out}} * N_{x*S_x+i,y*S_y+j}^{f_{in},f_{out}} + Bias^{f_{in},f_{out}}, \tag{2}$$

where W and $Bias$ represent the kernels and bias parameters between input feature map f_{in} and output feature map f_{out} respectively, and S_x and S_y are the sliding steps when the image convoluted in the x-direction and y-direction.

The standard convolutional formula is shown as Equation (2), which can be transformed into depth-wise convolution and 1×1 point-wise convolution by modifying the kernels. The depth-wise convolution and point-wise convolution are the two key computation layers of depth-wise separable convolution, which is widely used in my proposed framework. When the kernel channels C are equal to 1, and the number of kernels M are equal to the number of channels of input feature maps, the kernels can be expressed as $K_x \times K_y \times M$, the standard convolution has been transformed into depth-wise convolution. When the kernel size K_x and K_y are both equal to 1, the standard convolution has been transformed into 1×1 point-wise convolution. It can be seen from the above Equation that the basic calculations of the convolutional layer are multiplication and addition.

Deconvolutional in image object detection typically refer to transposed convolution or dilated convolution, which is used to up-sample the result of the convolutional layer back to the resolution of the original image. The calculations of deconvolutional are similar to the convolutional layer. The output neuron can be calculated according to the Equation (2), and its basic calculation is also composed of multiplication and addition.

3.2. Pooling Layer

The pooling layer is also called the down-sampling layer, which reduces the amount of data onto feature maps by maximizing or averaging the neuron in each pooling window. The calculation of pooling not only retains the main features, simplifies the computational complexity of network, but also effectively controls the risk of over-fitting of deep neural networks. Maximum pooling and average pooling are two commonly used pooling methods. In the pooling calculation of two-dimensional image feature map, the pooling window is defined as $P_x \times P_y$, and the input feature map f_{in} and output feature f_{out} are one-to-one correspondence.

The maximum pooling formula for the output neuron N at position (x, y) of output feature f_{out} is:

$$N_{x,y}^{f_{out}} = \max_{0 \leq i \leq P_x-1, 0 \leq j \leq P_y-1} N_{x+i,y+j}^{f_{in}}, \tag{3}$$

where the maximum pooling is done by successively comparing the maximum values of neuron in the pooling windows, which can be achieved by a hardware comparator. The average pooling formula for the output neuron N at position (x, y) of output feature f_{out} is:

$$N_{x,y}^{f_{out}} = \frac{\sum_{i=0}^{P_x-1} \sum_{j=0}^{P_y-1} N_{x+i,y+j}^{f_{in}}}{P_x * P_y}. \tag{4}$$

The calculation of average pooling is realized by the accumulating the neurons in the pooling window and dividing by the size of the pooling window. Since the structure of the neural network is determined, the $1/(P_x \times P_y)$ in formula (4) can be regarded as a coefficient, that multiplied by each neuron in the pooling window and accumulated to realize the calculation of average pooling. This method saves hardware resources by eliminating divisions in average pooling calculations and converting them into multiplication and addition operations.

3.3. Nonlinear Activation Function Layer

The nonlinear activation functions are widely used in neural networks, which not only make the neural network have nonlinear learning and an expression ability by layered nonlinear mapping compound, but also enhance the ability of the network to represent the high-level semantics of data. Commonly used activation functions include Sigmoid, Tanh, ReLU, and so on. The Sigmoid and Tanh functions are usually approximated by piecewise linear interpolation method, which is composed of multiplication and addition.

Recently, the ReLU has been widely used in many current deep learning algorithms and neural networks because it can effectively alleviate over-fitting and is less prone to gradient loss. Its calculation formula is:

$$ReLU(N) = \begin{cases} N, N > 0 \\ 0, N \leq 0 \end{cases}, \tag{5}$$

where N is the neuron. Since the signed fixed-point number is adopted in the hardware architecture design of deep learning processor, the calculation for ReLU can be completed directly by judging the sign bit of neuron.

In this paper, we drew on an efficient hardware pipeline architecture proposed by Li, L. et al. to realize the calculation of the activation function, which is another result of our work [50].

3.4. Normalization Layer

The normalization operation in the neural network is to solve the problem that the distribution of the data in the middle layer changes during the training process to prevent the gradient from disappearing or exploding and speed up the training. Krizhevsky, A. et al. used a local response normalization (LRN) operation in AlexNet to reduce the error rates of top-1 and top-5 by 1.4% and

1.2% [24]. Du, Z. et al. added local response normalization (LRN) and local contrast normalization (LCN) in the design of ShiDianNao, which improved the recognition accuracy, but increased the computational and hardware complexity [40]. The batch normalization proposed by Ioffe, S. et al. in 2015 is widely used in the deep neural network, which effectively accelerates the speed of training and convergence [51]. The batch normalization is calculated as a separate layer in the neural network forward inference process. The batch normalization formula for the neuron N located at the (x, y) position of output feature map f_{out} is as follows:

$$N_{x,y}^{f_{out}} = (N_{x,y}^{f_{in}} - \frac{mean_{x,y}^{f_{in}}}{scale factor}) / \sqrt{\frac{Variance_{x,y}^{f_{in}}}{scale factor} + \varepsilon},\tag{6}$$

where *mean* and *Variance* are the mean and variance of the input feature map fin respectively, and *scalefactor* is the scaling factor. These three parameters are learned by network training. The ε is a small constant, usually taken as 0.00001. It can be intuitively seen that the batch normalization calculation includes complex operations such as division and square root, and we will optimize the calculation using the parameter preprocessing strategy. Therefore, the Equation (6) can be converted into the Equation (7) shown below,

$$\begin{cases} N_{x,y}^{f_{out}} = BN\_a_{x,y}^{f_{in}} * N_{x,y}^{f_{in}} + BN\_b_{x,y}^{f_{in}} \\ BN\_a_{x,y}^{f_{in}} = 1 / \sqrt{\frac{Variance_{x,y}^{f_{in}}}{scale factor} + \varepsilon} \\ BN\_b_{x,y}^{f_{in}} = -\frac{mean_{x,y}^{f_{in}}}{scale factor} * BN\_a_{x,y}^{f_{in}} \end{cases},\tag{7}$$

where *BN_a* and *BN_b* are new parameters obtained by parameter preprocessing. Thus, complex batch normalization calculations are converted into multiplication and addition operations, thereby simplifying the complexity of the hardware structure.

3.5. Element-Wise Sum Layer

The calculation of element-wise sum is introduced when performing feature fusion, and this layer fuses feature maps of the same size produced on different paths. The main calculation of this layer is the addition of neurons at the corresponding positions of input feature maps, which can be implemented by hardware adder.

3.6. Full Connection Layer

The full connected layer is used to combine the features extracted from previous layers of the network, which is usually at the top of the neural network and used as a classifier. Although the full connect layer is no longer used in some current object detection algorithms, in order not to lose generality, the calculations of this layer are listed here. The input feature maps f_{in} is a vector of $1 \times 1 \times C$, and the output feature maps f_{out} is a vector of $1 \times 1 \times M$. The output neuron N at position $(1, 1)$ of output feature map f_{out} is computed with:

$$N_{1,1}^{f_{out}} = \sum_{f_{in}=0}^{C-1} W_{1,1}^{f_{in},f_{out}} * N_{1,1}^{f_{in}} + Bias^{f_{in},f_{out}},\tag{8}$$

where W and *Bias* represent the kernels and bias parameters between input feature map f_{in} and output feature map f_{out} respectively. The calculation of the full connect layer is similar to that of the convolutional layer, which is composed of multiplication and addition.

3.7. Softmax Layer

The Softmax layer is used for the output of the multi-classification neural network, which maps the M-dimensional vector V into an M-dimensional vector S with a range between (0,1) and a cumulative sum equal to 1. The calculation of the Softmax layer is:

$$S_i = \frac{e^{V_i}}{\sum_{j=1}^{M} e^{V_j}} (i = 1, 2, \ldots, M), \tag{9}$$

where V and S are both M-dimensional vectors. The Softmax layer also includes exponential calculation. Consequently, we drew on an efficient hardware pipeline architecture based on piecewise linear interpolation proposed by Li, L. et al. [50], which shares the same hardware as the calculation of the activation function.

Based on the characteristics of neural network layer-by-layer calculation, this section analyzed and optimized the calculation of each layer, and summarized the basic calculation of deep learning processor. The algorithm also includes some calculations that are difficult to implement in hardware. We would implement them in software, such as the calculation of the default boxes, non-maximum suppression (NMS), and so on. By considering the complexity of the hardware implementation, the same or a similar calculation were used at each layer as much as possible, laying the foundation for the design of the time-division multiplex hardware architecture in the next section.

4. Hardware Architecture of Deep Learning Processor

In order to adapt to the high performance and low power consumption environment on the satellite or aircraft, we used the dedicated deep learning processor to replace the traditional CPUs or GPUs for object detection algorithm processing. Due to various considerations, the deep learning hardware architecture we designed only supports the calculation of the algorithm inference stage, and the training and the acquisition of parameters are realized by the offline mode.

Based on the characteristics of algorithmic hierarchical computing and the analysis and optimization of the calculation of each layer in deep learning algorithm in Section 3, we designed the hardware architecture of deep learning processor shown in Figure 3. Our deep learning processor consists of the following main components: memory controller interface, feature maps buffer, parameters buffer, instruction memory, decoder, buffer controller and processing data setup module, neural processing engine, and activation pipeline module.

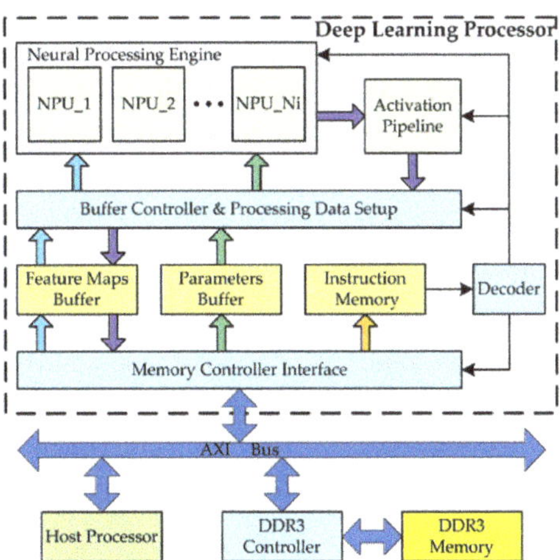

Figure 3. The overall diagram of the designed hardware architecture of the deep learning processor.

The memory controller interface interacts with the outside system through the advanced extensible interface (AXI) bus. It stores the received image data, parameters, and instructions into the feature maps buffer, the parameters buffer, and instruction memory respectively, and returns the calculation result of the deep learning processor to the system. The feature maps buffer is used to store input and output feature maps. The parameters buffer is used to store the weight and bias parameters obtained by offline training. The instructions of deep learning processor are stored in the instruction memory. After the instruction is decoded, the control signals are respectively transmitted to the respective functional modules. The buffer controller and processing data setup module readout feature maps data and parameters, and then sends them to the neural processing engine after organization, and stores the calculated output feature maps data of the activation pipeline module into the feature maps buffer. The neural processing engine performs basic neuron operations such as multiplication, addition and comparison. The activation pipeline module is used to implement an approximate calculation of the nonlinear activation function.

4.1. Parallel Computing Architecture

In the image object detection application, the input and output feature maps are both three-dimensional. According to the characteristics of the feature map, we naturally thought of parallel computing for three different dimensions in calculation. However, in the current research, most designs were performed in parallel for a two-dimensional neuron of an output feature map until this one was finished and next one began [35–44]. Therefore, we designed multiple neural processing units (NPUs) in the neural processing engine to realize parallel computing of multiple output feature maps. The hardware architecture of neural processing engine we designed is shown in Figure 4.

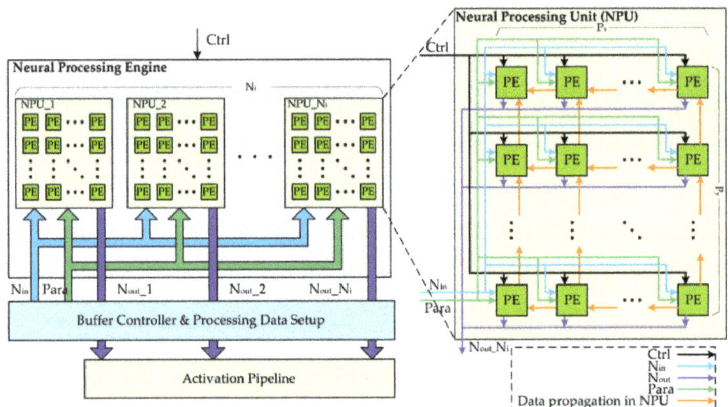

Figure 4. The hardware architecture of the neural processing engine. The left part is the neural processing engine consisting of multiple neural processing units (NPUs) and its data flow. The right part is the structure and data flow of a neural processing unit (NPU) composed of multiple processing elements (PEs).

We designed N_i neural processing units (NPUs) in the neural processing engine, which can simultaneously calculate the N_i output feature maps. The buffer controller and processing data setup module provides neurons and synapses for each neural processing unit for calculation. It reads data from the feature maps buffer and parameters buffer and reorganizes it for distribution to the neural processing units. After the calculation is completed, the neural processing engine outputs to the activation pipeline module for processing, and then writes the results to the feature maps buffer-by-buffer controller. The neural processing unit (NPU) consists of $P_x \times P_y$ processing elements (PEs) arranged in 2-D format. As can be seen from the right part of Figure 4, the neuron data calculated by the PEs could come not only from the buffer controller but also from right or bottom PEs. The data propagation between the PEs achieved data reuse, and reduced the bandwidth of reading the feature maps buffer.

The data propagation path and the computation path were designed in the PE structure shown in Figure 5 to implement data reuse and neuron calculation functions respectively. The data propagation path was used for local transfer of neuron data between PEs, thereby realizing data reuse. Inspired by the calculation process of the sliding window during image convolutional operations, the input of this path included three ways of reading from the buffer controller and reading from the right or bottom PE. We set two sets of shifter registers in this path, which were row shift register (Row_Shifter) and column shift register (Col_Shifter). These registers were used to implement temporary storage and propagation of the reused neuron data. Based on the analysis of many current mainstream deep learning algorithms and the object detection algorithm proposed in Section 2, the stride of convolutional operations generally did not exceed 2. Therefore, in each set of shift registers we set up two serially connected 16-bit registers to form a queue. The specific use of these two registers was detailed in the Section 4.3 data sharing and reuse. The hardware structure of the computation path included a multiplexer, multiplier, adder, comparator, and register. The computation path mainly completed the calculation of the PE, which supported multiplication, addition and comparison operations, and implemented all the calculation of the algorithm layers except for activation function and Softmax. When the computation path worked, the neuron data came from the PE_Reg, and the parameters were selected according to the calculation of each layer. For example, weight and bias were used for the calculation of convolutional layer, and two pre-processed parameters BN_a and BN_b for the computation of the normalization layer.

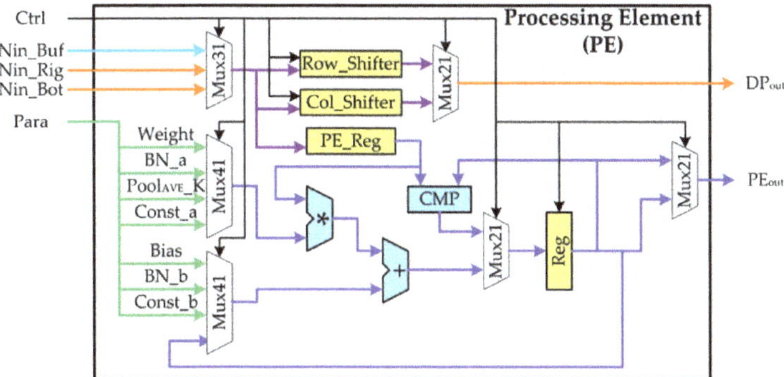

Figure 5. The structure of the processing element.

4.2. Hierarchical Storage Organization

The application of object detection in a remote sensing image is not only computation intensive, but also storage-intensive. Therefore, the efficient calculation of the neural processing engine is inseparable from the efficient organization and timely delivery of neuron data. The hierarchical storage organization structure as show in Figure 6 was adopted to store the feature maps in our design.

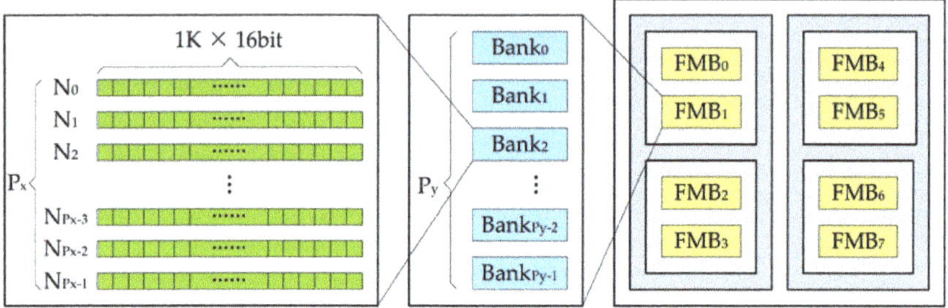

Figure 6. The hierarchical storage organization of the feature maps buffer.

We can see from the Figure 6, the structure of the feature map buffer was related to the number of PEs in the NPU. We designed P_x independent 16-bit width memory in one bank, where P_x is the number of PEs in a row of the NPU. Designing with independent memory made it easy for us to address a single neuron. In the upward level of storage, we designed P_y banks, where P_y is the number of PEs in a column of the NPU. This design allowed us to not only address a single neuron, but also read $P_x \times P_y$ neurons in one clock period. In order to achieve greater memory bandwidth, we could also design the separate memory similar to the multi-port format of the register file. The feature map buffer was divided into eight blocks, each of which contained P_y banks. Each of the two memory blocks constituted a set of Ping-Pong memories for alternately storing input or output feature maps. When a set of memory (e.g., FMB_0 and FMB_1) is the input feature map buffer, another set (e.g., FMB_2 and FMB_3) is used to store the output feature map. The input and output characteristics of the buffer were alternately changed, and the current output feature map buffer was the input feature map buffer calculated by the next layer of the algorithm. The other half of the buffer was used as a temporary buffer to handle data interactions of large algorithm when the feature maps size was larger than the on-chip buffer capacity. At this time, the two halves of the buffer acted as Ping-Pong memory to achieve alternate storage of the feature maps.

In order to effectively read out neuron data and reorganize it, the buffer controller reads the feature map buffer in four ways:

- Read $P_x \times P_y$ neurons from P_y banks.
- Read P_x neurons from 1 bank.
- Read P_y neurons from P_y banks with given stride.
- Read a single neuron.

These four reading modes can complete the reading of the neuron data in each layer of the deep learning algorithm.

The parameters buffer mainly stores the weights and bias data of the convolutional layer and the preprocessing parameters of the normalization layer. Due to the parameter sharing feature in the convolutional network, the structure of the parameters buffer is slightly different from the feature map buffer. The parameters buffer contains two levels storage structure. We designed N_i independent memories in its first level storage for a bank, where N_i is the number of NPUs. The second level storage contains four independent banks, which form two sets of Ping-Pong memory. In order to adapt to the large-scale model, the two sets of buffer were alternately used for the storage of current parameters and the interaction of subsequent parameters. The parameters are sequentially stored in the first-level independent memory according to the dimensions of the output feature map. The instruction buffer was designed as a FIFO (first input first output) with a width of 128 bits and a depth of 2 k. The very long instruction word (VLIW) instructions are sequentially stored in it. Since the structure of the instruction buffer is simple, it will not be described more here.

4.3. Data Sharing and Reuse

In order to reduce the bandwidth of the read buffer and perform calculations efficiently, and to make full use of the weight sharing of the convolutional neural network, we took a lot of work in the design of the deep learning processor. Taking the convolutional operation with the largest amount of calculation in the object detection algorithm as an example, since the N_i NPUs share the input neurons, only the $P_x \times P_y$ neurons data are read out and broadcast to the N_i NPUs for calculation in the first cycle. At the same time, the weights are shared within the NPU, and only the N_i weights are read from the parameters buffer and sent to the corresponding NPU for calculation. In the subsequent cycles of convolutional operation, not only data sharing but also data reuse based on inter-PE neuron propagation was adopted. The specific process of data reuse based on inter-PE neuron propagation in the convolutional operation is show in Figures 7 and 8.

We can see from Figure 7 that when stride was equal to 1, neuron reuse used only one row register and one column register for 3×3 convolutional operations. The row and column shift registers in the PE could be configured by the control signal to write neuron data directly to the output stage registers while masking the first stage registers. When Stride was equal to 1, the data could be reused for all cycles of the convolutional operation except for the first cycle. Specifically, $(K_x - 1) \times K_y$ row propagations were performed, and $(K_y - 1)$ column propagations were performed, where K_x and K_y was the kernel size.

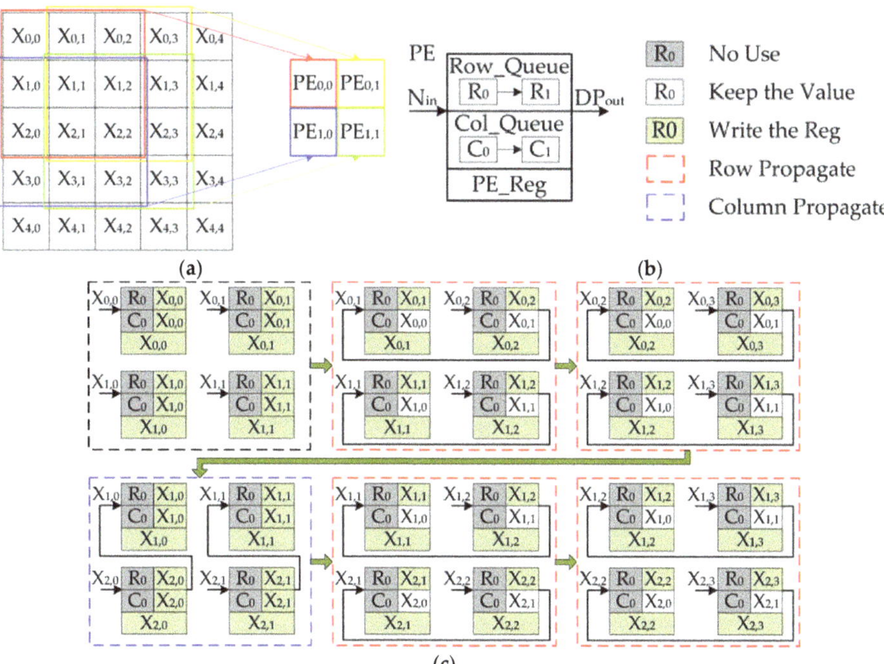

Figure 7. The neuron propagation between PEs and data reuse during the convolutional operation when stride = 1. (**a**) The mapping between neurons and PEs. (**b**) PE model, register operations, and neuron propagation between PEs. (**c**) The neuron propagation and register operation of the neuron between PEs in first six cycles during the convolutional operation.

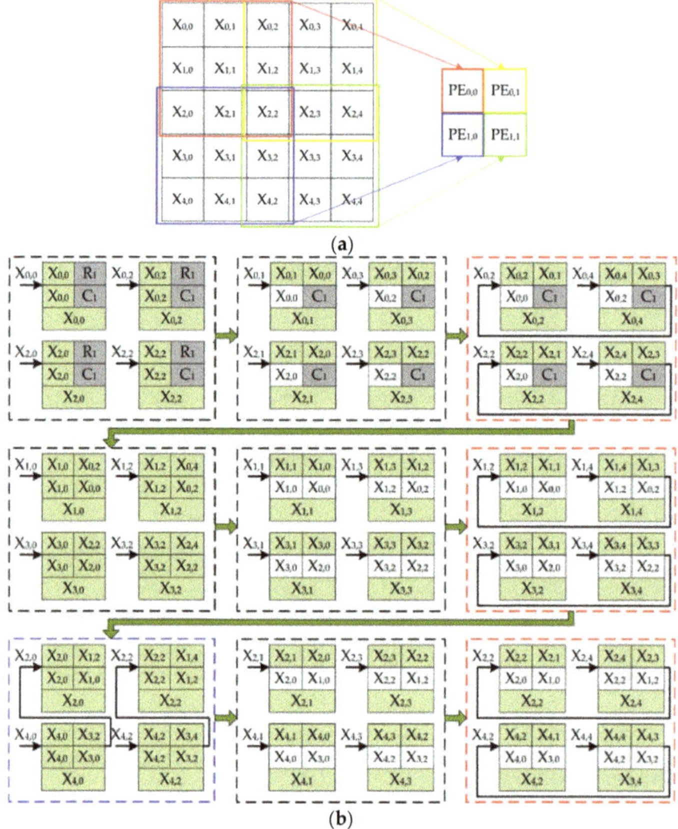

Figure 8. The neuron propagation between PEs and data reuse during the convolutional operation when stride = 2. (**a**) The mapping between neurons and PEs. (**b**) The specific operation of the PE registers and neuron propagation between PEs.

Based on the analysis of many current mainstream deep learning algorithms and the object detection algorithm proposed in Section 2, the stride of convolutional operations generally did not exceed 2. Therefore, in each set of shift registers in PE we set up to two serially connected 16-bit registers to form a queue. We can see from Figure 8, for a convolutional operation with a stride equal to 2, only four 16-bit registers were needed to achieve data reuse, which saved hardware resources compared to the design of FIFO for six 16-bit storage cells in [40]. When stride was equal to 2, taking 3×3 convolution as an example, data reuse could be performed in only four cycles. Fortunately, this situation was a small percentage of the calculation in the algorithm.

Data sharing and reuse were mainly applied to the convolutional layer, the normalized layer, and the fully connected layer. The remaining layers required a larger buffer read bandwidth because there was no overlapping of data.

5. Experiments and Results

In order to evaluate the efficiency and accuracy of our proposed algorithm framework and performance of the deep learning processor hardware architecture, we tested them separately by several experiments and evaluation indicators.

5.1. Experimental Settings

The evaluation experiment was divided into two parts. First, we used the publicly available remote sensing image dataset to evaluate the performance of the proposed algorithm framework in an open source deep learning framework. Then, we would implemented the hardware architecture of the deep learning processor on the FPGA and test its processing performance. The dataset, experiment environment, and test procedure used in the experiments are detailed below.

5.1.1. Dataset Description

We evaluated the performance of the proposed algorithm framework on the Northwestern Polytechnical University very high resolution remote sensing image dataset with 10 classes objects (NWPU VHR-10), which was constructed by Cheng, G. [1,25]. The resolution of these remote sensing images is between 0.5–2 m, and the average size is about 600×800 [25,45]. This dataset contained in total 800 VHR remote sensing images, which was divided into two parts, one was the positive image set and the other was the negative image set. The positive image set contained 650 remote sensing images, which were manually annotated with 757 airplanes, 302 ships, 655 storage tanks, 390 baseball diamonds, 524 tennis courts, 150 basketball courts, 163 ground track fields, 224 harbors, 124 bridges, and 477 vehicles. The negative image set was not used in this paper.

In our work, in order to obtain better detection results, we took the data augmentation on the dataset by cropping, flipping, rotating, scaling, and chromatic spatial transformation. During training, we randomly selected 20% as training set, 20% as validation set, and 60% as the test set.

5.1.2. Experiment Environment

To evaluate the performance of the proposed algorithm framework and the hardware architecture of deep learning processor, we leveraged the popular open source Caffe framework [52] and FPGA. For the evaluation of algorithm performance, we implemented the proposed algorithm framework on the Caffe, which executed on a 64-bit Ubuntu 16.04 PC with Intel i7-7700 CPU, 16GB memory and NVIDIA GeForce GTX1070Ti GPU. For the evaluation of the deep learning processor performance, we described the proposed hardware architecture using Verilog HDL and implemented it on Xilinx Zynq-XC7Z7100 FPGA.

5.1.3. Test Procedure

In the performance evaluation of the algorithm, we implemented the proposed CBFF-SSD algorithm framework on the Caffe, and then trained it with GPU. We used the pre-trained MobileNet-SSD model on VOC0712 [53] dataset for CBFF-SSD training, and then fine-tuned our model on NWPU VHR-10 dataset. We set batch size to 8 for 416×416 input during training, and set the learning rate at 10^{-3} for the first 60 k iterations, then decreased it to 10^{-4} for the next 40 k iterations, and 10^{-5} for the last 20 k iterations. The momentum and weight decay were set to 0.9 and 0.0005 respectively by using stochastic gradient descent (SGD). The performance of the proposed algorithm framework was compared with the newly trained CNN [25], rotation-invariant CNN (RICNN) [25], R-P-Faster R-CNN [45], and SSD [33]. For a fair and accuracy comparison, the detection accuracy, computational time, and the precision-recall curves (PRCs) were taken as the evaluation indexes.

For the evaluation of the performance of the hardware architecture of the deep learning processor, the proposed hardware architecture was implemented on the Xilinx XC7Z100 FPGA. In the specific implementation of the deep learning processor, we designed 16 NPUs to form the neural processing engine, and designed 8×8 PEs in each NPU. The feature map buffer contained eight buffer blocks, each contained eight banks, each bank with a capacity of 8×16 bit. The multiplier and adder in the processor were implemented by embedded digital signal processing slice (DSPs) in the FPGA. The feature maps buffer, parameter maps buffer, instruction buffer, and the look up table (LUT) in the activation function pipeline were all implemented by block random access memory (BRAM). The resource utilization of the deep learning processor is shown in Table 2.

Table 2. Resource utilization of the deep learning processor.

Resource	BRAM (18 k)	DSP	FF	LUT
Available	1510	2020	554,800	277,400
Utilization	912	1152	156,238	136,892
Utilization (%)	60	57	28	49

The performance of the proposed hardware architecture of deep learning processor was compared with CPU and GPU platform, and the other state-of-the-art processor based on FPGA implementations. The technology, frequency, power, performance, power efficiency, and performance density were taken as the evaluation indexes.

5.2. Evalutaion Indicators

In order to quantitatively evaluate the performance of the proposed algorithm framework and the deep learning processor, the widely utilized evaluation indicators of average precision (AP), mean average precision (mAP), average running time per image, and precision–recall curves (PRCs) were adopted for the object detection algorithm framework, and the Giga operations per second (GOP/s), power consumption, power efficiency, and performance density were adopted for the deep learning processor.

5.2.1. Average Precision

The average precision (AP) is the average of the precision over the interval from recall = 0 and recall =1, which is also equal to the area under the precision-recall curve. The higher the AP value, the better the performance of the algorithm. The mean average precision (mAP) is another indicator, which reflects the average of the average precision (AP) of all categories in the dataset. The AP and mAP reflect the performance of the algorithm from the detection accuracy. The average running time per image reflects the execution efficiency of the algorithm, but it depends on the hardware that carries the algorithm.

5.2.2. Precision–Recall Curve

The precision and recall are two important indicators that evaluate the performance of the object detection algorithm from other perspectives. They are formulated as follows:

$$Precision = \frac{TP}{TP + FP}. \tag{10}$$

$$Recall = \frac{TP}{TP + FN}. \tag{11}$$

The precision refers to the proportion of true positive (TP) in all data that are predicted to be positive. The recall reflects to the proportion of data predicted to be true positive (TP) to all positive data. When the area overlap between the predicted box and the ground-truth box exceeds a threshold (e.g., 0.5), the detection map is considered to be a TP. Otherwise, the detection map is considered to be a false positive (FP). In addition, the detection is considered to be a false negative (FN) when the predicted boxes that overlap with ground-truth box but does not have the maximum overlap value. The PRCs describes the relationship between the precision and recall, the larger the area under the curve, the better the performance of the detector.

5.2.3. Processor Performance

The performance evaluation indicators of the deep learning processor mainly include the Giga operations per second (GOP/s), power consumption, power efficiency, and performance density. It reflects the processing performance of the deep learning processor. The power consumption is another indicator of the processor, which is particularly sensitive to embedded systems, especially for the on-board information processing system in my work. The power efficiency is the ratio of processing performance to power consumption, which reflects the processing performance of the same energy consumption. The performance density is defined as the number of arithmetic operations performed by one DSP slice in one cycle, which can better reflect the computing performance of deep learning processors with different hardware architectures based on FPGA implementation [56].

5.3. Test Results of Object Detection Algorithm Framework

Detection examples of proposed CBFF-SSD algorithm framework on the NWPU VHR-10 dataset are shown in Figure 9. The qualitative detection results of proposed CBFF-SSD algorithm framework for the ten categories of aircraft, ship, storage tank, baseball diamond, tennis court, basketball court, ground track field, harbor, bridge, and vehicle are shown in Figure 9, respectively. It can be seen from Figure 9 that the proposed CBFF-SSD algorithm framework shows good detection performance both for large objects such as storage tank, baseball diamond, tennis court, basketball court, ground track field, harbor, and bridge, as well as small objects such as airplane, ship, and vehicle.

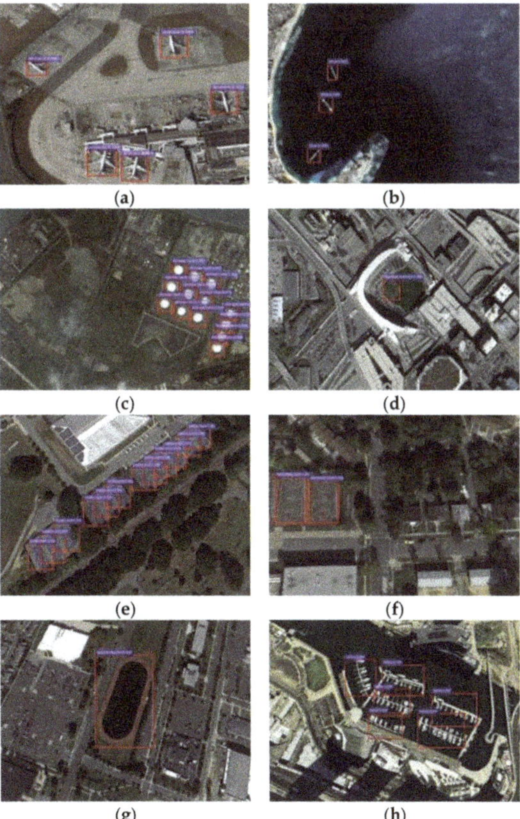

Figure 9. *Cont.*

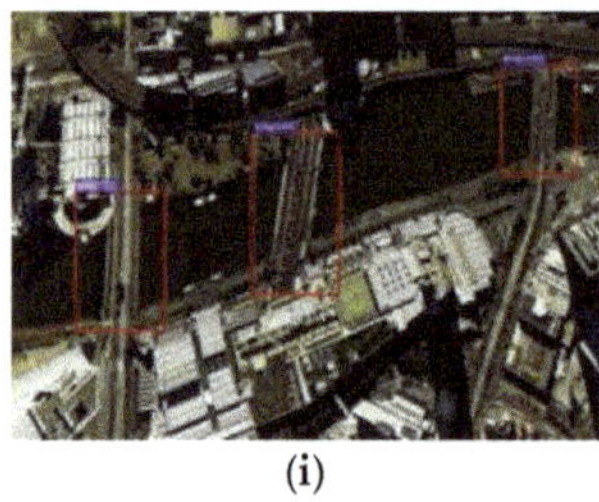 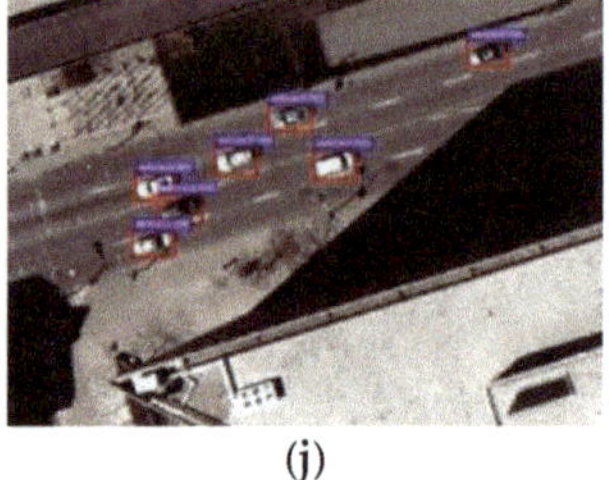

(i) (j)

Figure 9. Object detection examples on NWPU VHR-10 dataset with the proposed CBFF-SSD algorithm framework. (**a**) Airplane; (**b**) ship; (**c**) storage tank; (**d**) baseball diamond; (**e**) tennis court; (**f**) basketball court; (**g**) ground track field; (**h**) harbor; (**i**) bridge and (**j**) vehicle.

In Table 3 and Figure 10, we quantitatively compared the performance of five different object detection algorithms by the AP value, mAP value, average running time per image, and PRCs, respectively. In each row of Table 3, the bold number indicates the best test result. It can be seen from the comparison of the test data of the five algorithms in Table 3 that the proposed algorithm had an advantage in the average accuracy of the detection of six classes of objects such as airplane, ship, tennis court, basketball court, ground track field, and vehicle. The detection accuracy of storage tank was lower than the newly trained CNN and RICNN algorithms. The proposed algorithm was slightly inferior to the SSD algorithm in the detection of tennis courts, harbor, and bridge. The average precision of the test data also verified that the proposed CBFF-SSD algorithm framework was more effective for detecting small objects such as an airplane, ship, and vehicle. In Table 3, it can be seen that the proposed CBFF-SSD algorithm obtained the best mean AP value of 0.9142 among all the object detection algorithms. It can also be seen from Table 3 that the proposed CBFF-SSD algorithm framework detected an image with an average running time of 0.0133s, which was faster than other algorithms. This is due to the efficient of regression-based SSD object detection framework and the reduced computational complexity of lightweight MobileNet backbone network. In addition, it was also inseparable from the support of high-performance GPU evaluation platform.

Table 3. Performance comparison of different algorithms.

	Newly Trained CNN [25]	RICNN [25]	R-P-Faster R-CNN [45]	SSD [33]	CBFF-SSD
Airplane	0.7014	0.8835	0.9060	0.9565	**0.9693**
Ship	0.6370	0.7734	0.7620	0.9356	**0.9426**
Storage tank	0.8433	**0.8527**	0.4030	0.6087	0.8095
Baseball diamond	0.8361	0.8812	0.9080	**0.9939**	0.9909
Tennis court	0.3546	0.4083	0.7970	0.8765	**0.9150**
Basketball court	0.4680	0.5845	0.7740	0.9200	**0.9264**
Ground track field	0.8120	0.8673	0.8800	0.9864	**0.9882**
Harbor	0.6228	0.6860	0.7620	**0.9460**	0.9159
Bridge	0.4538	0.6151	0.5750	**0.9704**	0.8968
Vehicle	0.4480	0.7110	0.6660	0.7447	**0.7878**
Mean AP	0.6177	0.7263	0.7430	0.8939	**0.9142**
Average running time per image (second)	8.770	8.770	0.150	0.0217	**0.0133**

As can be seen form Figure 10, the proposed CBFF-SSD algorithm framework shows better detection performance in most classes, especially in the classes of airplane, ship, tennis court, basketball court, ground track field, and vehicle. However, it was slightly inferior to the other algorithms in the classes of storage tank, tennis courts, harbor, and bridge.

By jointly analyzing the AP values, mAP value, average running time per image, the recall rate, and the PRCs, it can be seen that the proposed CBFF-SSD algorithm was superior to other algorithms in most classes of detection accuracy and detection efficiency.

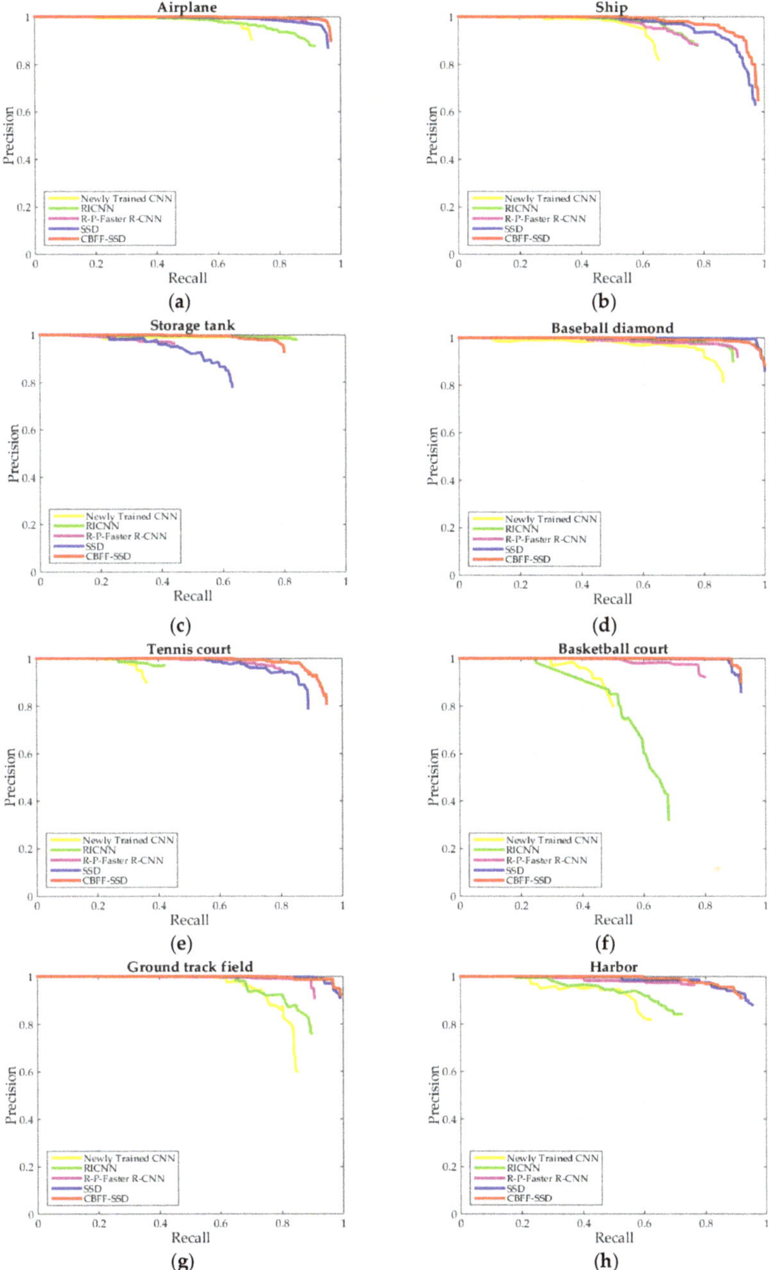

Figure 10. *Cont.*

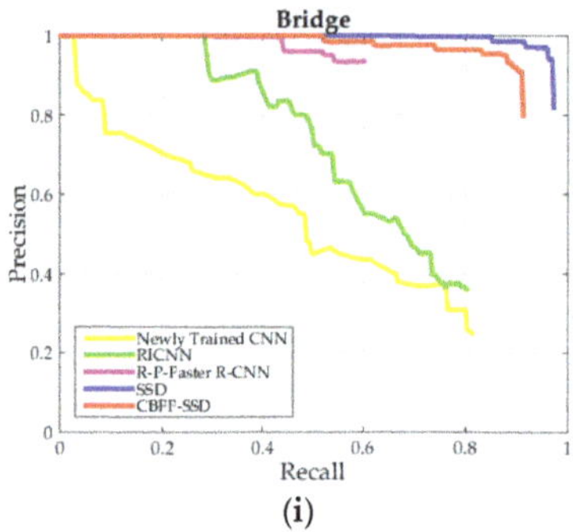
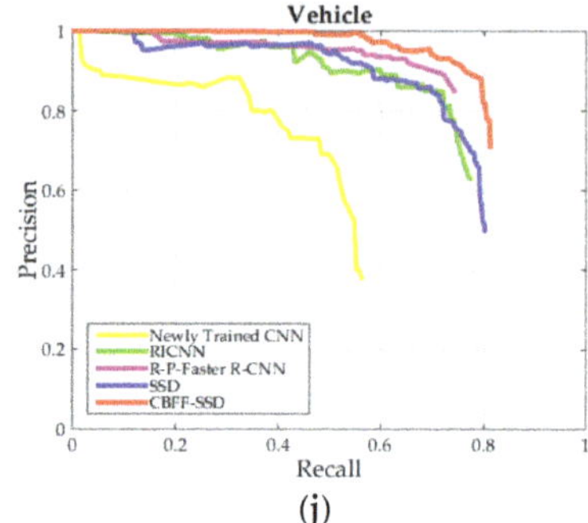

Figure 10. The precision–recall curves (PRCs) of the proposed method and other state-of-the-art methods for 10 classes object in NWPU VHR-10 dataset. (**a**) Airplane; (**b**) ship; (**c**) storage tank; (**d**) baseball diamond; (**e**) tennis court; (**f**) basketball court; (**g**) ground track field; (**h**) harbor; (**i**) bridge; and (**j**) vehicle.

5.4. Performance Evaluation of the Deep Learning Processor

In order to evaluate the performance of the processor fairly, we chose the proposed CBFF-SSD remote sensing object detection algorithm as a benchmark to compare the performance of the CPU, GPU and our deep learning processor when executing the algorithm. The evaluation results of the proposed processor, CPU, and GPU are shown in Table 3. The performance evaluation of CPU and GPU were based on the open source Caffe framework [52], which could choose the platform (CPU or GPU) to run. The CPU platform was Intel i7-7700 CPU @ 3.60 GHz with 16 GB DRAM. The GPU platform was NVIDIA GTX1070Ti GPU with 8 GB memory. The thermal design power (TDP) values of the CPU and GPU were 65 W and 180 W, separately. The total power of our deep learning processor was 19.52 W according to the power report by the Vivado design suite. We can see from Table 4 that the NVIDIA GTX1070Ti GPU had a great leading advantage in terms of computing performance, and its throughput was 1452 GOP/s. Our deep learning processor achieved the best power efficiency among all the platforms, reaching 23.20 GOP/s/W. The power efficient of our deep learning processor was 29.74 times that of CPU and 2.87 times that of GPU.

Table 4. Evaluation results on CPU, GPU, and our processor.

Platform	CPU	GPU	FPGA
Vendor	Intel i7-7700	NVIDIA GTX1070Ti	Xilinx XC7Z100
Technology (nm)	14	16	28
Frequency (MHz)	3600	1607	200
Power (W)	65	180	19.52
Benchmarks	CBFF-SSD	CBFF-SSD	CBFF-SSD
Latency (ms)	382.15	13.27	42.59
Performance (GOP/s)	51	1452	452.8
Power Efficiency (GOP/s/W)	0.78	8.07	23.20

In order to further evaluate the performance of the proposed deep learning processor, we compared it with the state-of-the-art FPGA-based deep learning processor or algorithm accelerator. In Table 5, we compared the performance of deep learning processors implemented on different FPGA platforms. For the sake of fairness, VGG16 was also selected as a benchmark for evaluating. The performance of the deep learning processor based on FPGA was closely related to the on-chip resources used. Therefore, it was inaccurate to simply evaluate the processor performance by GOP/s. We introduced the performance density indicator, which is related to the GOP/s, the on-chip DSPs resources used, and the operating frequency [56]. Performance density can better reflect the computing performance of deep learning processors with different hardware architectures based on FPGA implementation. Note that a MAC (multiply and accumulate) was counted as two operations. As can be seen from Table 5, our processor operated at 200 MHz, which was higher than other implements. Our processor used 1152 on-chip DSPs in the implementation, achieving 452.8 GOP/s processing performance, and its performance density was 1.97 OP/DSP/cycle. Although our processor did not achieve the highest value of performance term, its performance density was the best among all the processors compared [54–56].

Table 5. Comparisons with previous implementations.

	Ref. [54]	Ref. [55]	Ref. [56]	Ours
FPGA Chip	Xilinx XC7Z045	Arria10 GX1150	Xilinx XC7VX690T	Xilinx XC7Z100
Frequency (MHz)	150	150	120	200
Precision	16-bit fixed	8–16 fixed	fixed	16-bit fixed
CNN Model	VGG16	VGG16	VGG16	VGG16
DSPs	780	3036	3595	1152
Performance (GOP/s)	187.80	645.25	691.6	452.8
Performance Density (OP/DSP/cycle)	1.61	1.42	1.60	1.97

6. Discussion

We adopted the NWPU VHR-10 dataset to train, validate, and test the proposed CBFF-SSD algorithm framework, which achieved considerable results in the object detection of very high resolution optical remote sensing images. In order to fully verify the effectiveness of the algorithm proposed in this paper and the deep learning hardware architecture, we carried out experiments on small object detection, large-scale remote sensing image object detection, and deep learning processor performance comparison, and analyzed the experimental results.

6.1. Small Objects Detection Result and Analysis

The algorithm proposed in this paper was optimized on the SSD [33] algorithm. The superiority in computational efficiency was verified by the average running time of each image in Table 3. In order to show the effect of the proposed CBFF-SSD algorithm on small object detection, we compared the detection of small objects of different classes by the two algorithm and the results are shown in Figure 11.

By comparing the results of the two algorithms in Figure 11 on the detection of small objects such as the storage tank, airplane, and ship, it can be seen that both SSD algorithm and CBFF-SSD algorithm had achieved good results for the detection of storage tank. For small and dense objects detection such as airplane and ships, the CBFF-SSD algorithm was superior to the SSD algorithm. The experimental results verified the effectiveness of the feature fusion unit that adds context feature fusion to the SSD algorithm structure.

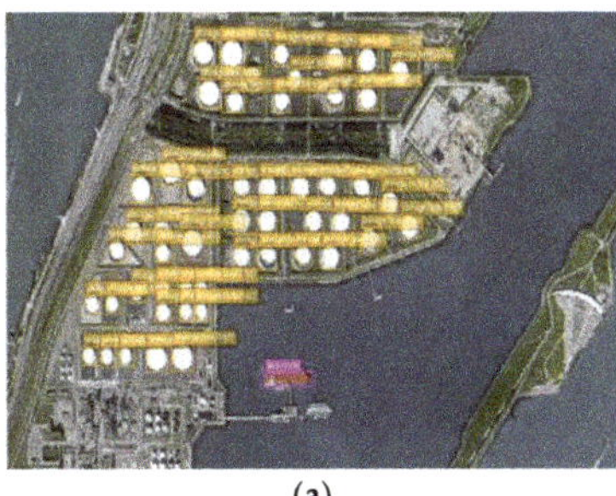

(a)

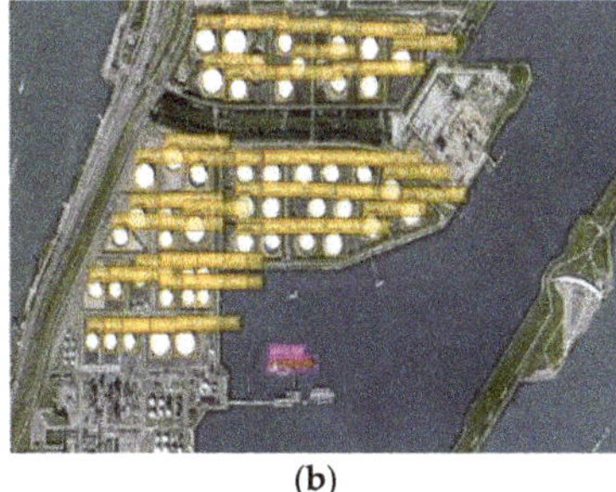

(b)

Figure 11. *Cont.*

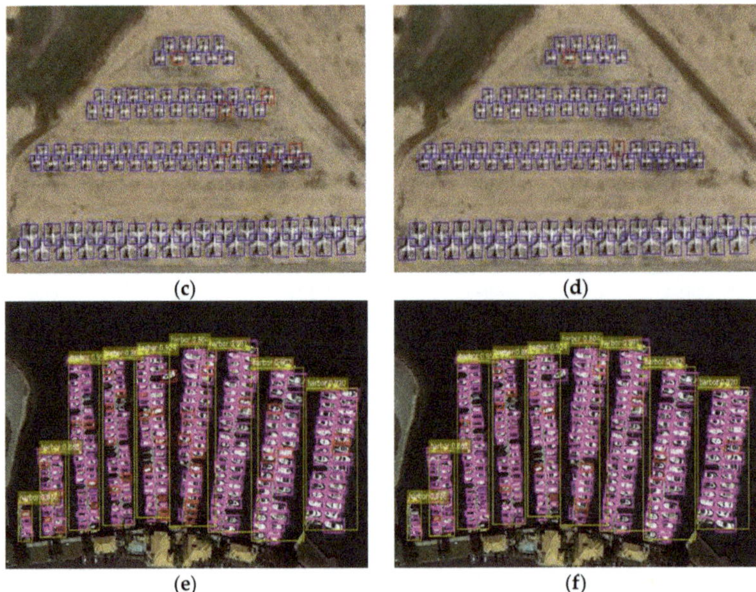

Figure 11. Small object detection results comparison. (a,c,e) are the detection results of the SSD algorithm; (b,d,f) are the detection results of our proposed CBFF-SSD algorithm. The red rectangle indicates incorrect detection.

6.2. Large-Scale Remote Sensing Image Object Detection Results and Analysis

In order to show the overall detection effect, we performed inferences on large-scale remote sensing images and the results are shown in Figure 12.

We tested the proposed algorithm and deep learning processor through large-scale remote sensing images. After the remote sensing image was calculated by the processor, the coordinates, categories and confidence were marked by the software. It should be noted that for the testing of large-scale remote sensing images, we could not directly scale them as input images. We processed the original image by dividing and setting the overlap rate of 10%, and adopted non-maximum suppression (NMS) to process the detection results of the overlap part.

It can be seen from the experimental results that the proposed algorithm and deep learning processor had a good object detection effect for large-scale remote sensing images. However, it can be seen from the experimental results that illumination and shadow had a certain influence on the object detection. Meanwhile, the detection results of objects that did not appear in the training set were unsatisfactory (for example, helicopters without wings). These problems could be improved by training a large number of samples.

6.3. Discussion on Processor Implementation and Model Compression Methods

In the current research, there were many implementations of deep learning processors, such as the general-purpose processor, FPGAs, application specific integrated circuits (ASICs), etc. We compared the different implementations in Table 6.

By comparison we could see that the ASIC-implemented deep learning processors had higher operating frequency, better processing performance, and lower power consumption. This provided us with ideas and directions for the next step in improving the performance of our proposed deep learning processor.

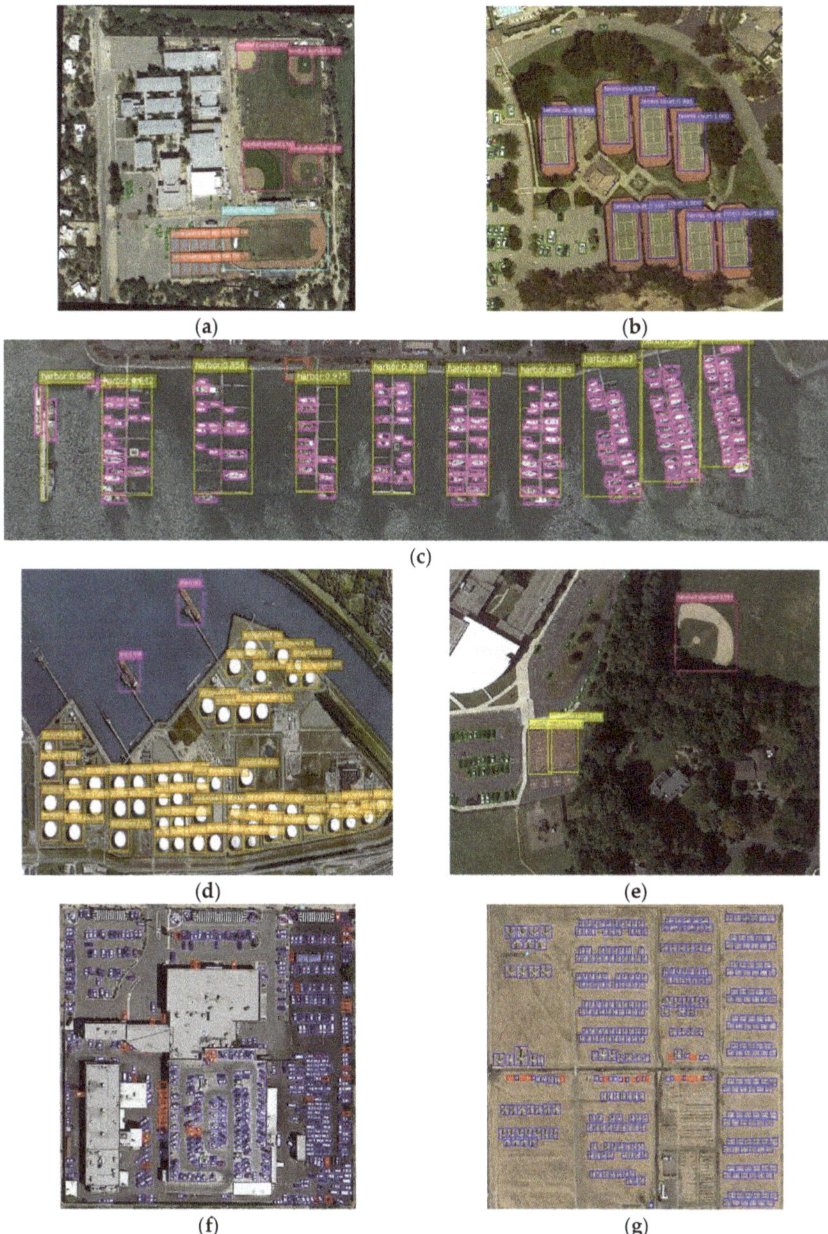

Figure 12. Detection results on large-scale remote sensing images. (**a**) Baseball diamond, ground track filed, tennis court, and vehicle; (**b**) vehicle, and tennis court; (**c**) harbor, and ship; (**d**) storage tank, ship; (**e**) baseball diamond, basketball court, and vehicle; (**f**) vehicle; and (**g**) airplane.

We were currently using a refined model way to compress model parameters and reduce the amount of computation. It can be seen from Table 1 that the number of parameters of the proposed CBFF-SSD algorithm framework was 56.09% of the SSD algorithm, and the calculation amount was only 17.56% of the SSD algorithm. There are also model compression methods such as weight pruning, weight sparse, processing data quantification, and so on. In particular, the method of processing data quantization can achieve better model compression without changing the model structure. This method includes binary neural networks [57], quantized neural networks (n-bits) [58], integer CNNs [59], and so on. Google's TPU [42] uses 8-bit integer data for superior processing performance and negligible error relative to floating-point production in the engineering applications. Our current model compression strategy could be combined with strategies for processing data quantification to achieve better model compression. This will be the direction of our next step in optimizing the algorithm model.

Table 6. Comparison of deep learning processors with different implementations.

	Jetson AGX Xavier [60]	TPU [42]	ShiDianNao [40]	Ours
Implementations	ASIC (ARM+GPU)	ASIC	ASIC	FPGA
Frequency (MHz)	1370	700	1000	200
Precision	8-bit integer	8-bit integer	16-bit fixed	16-bit fixed
Performance (GOP/s)	22,000	92,000	194	452.8
Power (W)	10/15/30	75	0.320	19.52

7. Conclusions

In this paper, an efficient context-based feature fusion SSD (CBFF-SSD) framework and hardware architecture of deep learning processor with multi-processor clusters were proposed to object detection in remote sensing images on space-borne or airborne. The design of the CBFF-SSD framework fully considers small object detection, detection accuracy, and efficiency. The deep learning processor uses multiple neural processing units (NPUs) composed of 2-D processing elements (PEs) to simultaneously calculate multiple output feature maps. The parallel architecture, hierarchical on-chip storage organization, and the register designed in the PE make the calculation of the processor more efficient.

A comparison test with five algorithms on the NWPU VHR-10 dataset shows that our algorithm framework had an advantage in the average accuracy of the detection of six classes of objects, and it was superior to other algorithms in terms of the mean AP value and average running timer per image. The effectiveness of the proposed CBFF-SSD algorithm was verified by small object detection experiments and large-scale remote sensing image object detection experiments. The proposed processor implemented in FPGA was compared with CPU, GPU, and other FPGA-based deep learning processor. In the comparative test of general-purpose processors, our deep learning processor achieved the best power efficiency, which was 29.74 times that of CPU and 2.87 times that of GPU. In comparison with the state-of-the-art FPGA-based deep learning processors, although our processor did not achieve the highest value of performance term, its performance density was the best among all the processors compared.

In the future, we would improve our algorithm frameworks in terms of identifying the classes of the object and accuracy to realize more effective network for remote sensing images object detection. We would use a strategy for processing data quantification to optimize the proposed algorithm framework. Furthermore, we would implement the proposed hardware architecture of deep learning processor in application specific integrated circuit (ASIC) way to achieve higher computing performance and lower power consumption.

Author Contributions: Conceptualization, L.L. and S.Z.; methodology, L.L.; software, L.L. and J.W.; validation, L.L. and J.W.; formal analysis, L.L.; investigation, L.L. and J.W.; resources, S.Z.; data curation, L.L.; writing—original draft preparation, L.L.; writing—review and editing, S.Z. and J.W.; supervision, S.Z.

Funding: This research received no external funding.

Acknowledgments: This research is supported by Northwestern Polytechnical University and Beijing Institute of Microelectronics Technology. The authors would like to thank all the teachers and colleagues who provided inspirations and helpful suggestions.

Conflicts of Interest: The authors declare no conflict of interest.

References

1. Cheng, G.; Han, J. A survey on object detection in optical remote sensing images. *ISPRS J. Photogramm. Remote Sens.* **2016**, *117*, 11–28. [CrossRef]
2. Xu, Y.; Zhu, M.; Li, S. End-to-end airport detection in remote sensing images combining cascade region proposal networks and multi-threshold detection networks. *Remote Sens.* **2018**, *10*, 1156. [CrossRef]
3. Zhu, M.; Xu, Y.; Ma, S.; Li, S.; Ma, H.; Han, Y. Effective airplane detection in remote sensing images based on multilayer feature fusion and improved nonmaximal suppression algorithm. *Remote Sens.* **2019**, *11*, 1062. [CrossRef]
4. Leitloff, J.; Hinz, S.; Stilla, U. Vehicle detection in very high resolution satellite images of city areas. *IEEE Trans. Geosci. Remote Sens.* **2010**, *48*, 2795–2806. [CrossRef]
5. He, H.; Yang, D.; Wang, S.C.; Wang, S.Y.; Li, Y. Road extraction by using atrous spatial pyramid pooling integrated encoder-decoder network and structural similarity loss. *Remote Sens.* **2019**, *11*, 1015. [CrossRef]
6. Zhang, J.; Lin, X.; Liu, Z.; Shen, J. Semi-automated road tracking by template matching and distance transformation in urban areas. *Int. J. Remote Sens.* **2011**, *32*, 8331–8347. [CrossRef]
7. Liu, G.; Sun, X.; Fu, K.; Wang, H. Interactive geospatial object extraction in high resolution remote sensing images using shape-based global minimization active contour model. *Pattern Recog. Lett.* **2013**, *34*, 1186–1195. [CrossRef]
8. Ok, A.O.; Senaras, C.; Yuksel, B. Automated detection of arbitrarily shaped buildings in complex environments from monocular VHR optical satellite imagery. *IEEE Trans. Geosci. Remote Sens.* **2013**, *51*, 1701–1717. [CrossRef]
9. Leninisha, S.; Vani, K. Water flow based geometric active deformable model for road network. *ISPRS J. Photogramm. Remote Sens.* **2015**, *102*, 140–147. [CrossRef]
10. Peng, J.; Liu, Y. Model and context-driven building extraction in dense urban aerial images. *Int. J. Remote Sens.* **2005**, *26*, 1289–1307. [CrossRef]
11. Hussain, M.; Chen, D.; Cheng, A.; Wei, H.; Stanley, D. Change detection from remotely sensed images: From pixel-based to object-based approaches. *ISPRS J. Photogramm. Remote Sens.* **2013**, *80*, 91–106. [CrossRef]
12. Mishra, N.B.; Crews, K.A. Mapping vegetation morphology types in a dry savanna ecosystem: Integrating hierarchical object-based image analysis with Random Forest. *Int. J. Remote Sens.* **2014**, *35*, 1175–1198. [CrossRef]
13. Feizizadeh, B.; Tiede, D.; Rezaei Moghaddam, M.H.; Blaschke, T. Systematic evaluation of fuzzy operators for object-based landslide mapping. *South East. Eur. J. Earth Obs. Geomat.* **2014**, *3*, 219–222.
14. Lowe, D.G. Object recognition from local scale-invariant features. In Proceedings of the 7th IEEE International Conference on Computer Vision, Kerkyra, Greece, 20–27 September 1999; pp. 1150–1157.
15. Dalal, N.; Triggs, B. Histograms of oriented gradients for human detection. In Proceedings of the IEEE Conference on Computer Vision and Pattern Recognition, San Diego, CA, USA, 21–23 September 2005; pp. 886–893.
16. Sun, H.; Sun, X.; Wang, H.; Li, Y.; Li, X. Automatic target detection in high-resolution remote sensing images using spatial sparse coding bag-of-words model. *IEEE Geosci. Remote Sens. Lett.* **2012**, *9*, 109–113. [CrossRef]
17. Zhu, C.; Zhou, H.; Wang, R.; Guo, J. A novel hierarchical method of ship detection from spaceborne optical image based on shape and texture features. *IEEE Trans. Geosci. Remote Sens.* **2010**, *48*, 3446–3456. [CrossRef]
18. Mountrakis, G.; Im, J.; Ogole, C. Support vector machines in remote sensing: A review. *ISPRS J. Photogramm. Remote Sens.* **2011**, *66*, 247–259. [CrossRef]
19. Collins, M.; Schapire, R.E.; Singer, Y. Logistic regression, adaboost and bregman distances. *Mach. Learn.* **2002**, *48*, 253–285. [CrossRef]
20. Ali, A.; Olaleye, O.G.; Bayoumi, M. Fast region-based DPM object detection for autonomous vehicles. In Proceedings of the 2016 IEEE 59th International Midwest Symposium on Circuits and Systems, Abu Dhabi, United Arab Emirates, 16–19 October 2016; pp. 1–4.
21. Wegner, J.D.; Haensch, R.; Thiele, A.; Soergel, U. Building detection from one orthophoto and high-resolution InSAR data using conditional random fields. *IEEE J. Sel. Top. Appl. Earth Observ. Remote Sens.* **2011**, *4*, 83–91. [CrossRef]

22. Cheng, G.; Han, J.; Zhou, P.; Yao, X.; Zhang, D.; Guo, L. Sparse coding based airport detection from medium resolution Landsat-7 satellite remote sensing images. In Proceedings of the 2014 3rd International Workshop on Earth Observation and Remote Sensing Applications, Changsha, China, 11–14 June 2014; pp. 226–230.

23. Mokhtarzade, M.; Zoej, M.V. Road detection from high-resolution satellite images using artificial neural networks. *Int. J. Appl. Earth Observ. Geoinform.* **2007**, *9*, 32–40. [CrossRef]

24. Krizhevsky, A.; Sutskever, I.; Hinton, G.E. ImageNet classification with deep convolutional neural networks. In Proceedings of the 25th International Conference on Neural Information Processing Systems, Lake Tahoe Nevada, NV, USA, 3–6 December 2012; pp. 1097–1105.

25. Cheng, G.; Zhou, P.; Han, J. Learning rotation-invariant convolutional neural networks for object detection in VHR optical remote sensing images. *IEEE Trans. Geosci. Remote Sens.* **2016**, *54*, 7405–7415. [CrossRef]

26. Wang, G.; Chen, J.; Gao, F.; Wu, J. Research on the infrastructure target detection of remote sensing image based on deep learning. *Radio Eng.* **2018**, *48*, 219–224.

27. Jiao, L.; Zhao, J.; Yang, S.; Liu, F. *Deep Learning, Optimization and Recognition*, 1st ed.; Tsinghua University Press: Beijing, China, 2017; pp. 341–367.

28. Girshick, R.; Donahue, J.; Darrelland, T.; Malik, J. Rich feature hierarchies for object detection and semantic segmentation. In Proceedings of the 2014 IEEE Conference on Computer Vision and Pattern Recognition, Los Alamitos, CA, USA, 23–28 June 2014; pp. 580–587.

29. Girshick, R. Fast R-CNN. In Proceedings of the 2015 IEEE International Conference on Computer Vision, Santiago, Chile, 7–13 December 2015; pp. 1440–1448.

30. Ren, S.; He, K.; Girshick, R.; Sun, J. Faster R-CNN: Towards real-time object detection with region proposal networks. *IEEE Trans. Pattern Anal. Mach. Intell.* **2015**, *39*, 1137–1149. [CrossRef] [PubMed]

31. Lin, T.Y.; Dollar, P.; Girshick, R.; He, K.; Hariharan, B.; Belongie, S. Feature pyramid networks for object detection. In Proceedings of the 30th IEEE Conference on Computer Vision and Pattern Recognition, Honolulu, HI, USA, 21–26 July 2016; pp. 936–944.

32. Redmon, J.; Divvala, S.; Girshick, R.; Farhadi, A. You Only Look Once: Unified, Real-Time Object Detetction. *arXiv* **2015**, arXiv:1506.02640.

33. Liu, W.; Anguelov, D.; Erhan, D.; Szegedy, C.; Reed, S.; Fu, C.Y.; Berg, A.C. SSD: Single Shot MultiBox Detector. In Proceedings of the 14th European Conference on Computer Vision, Amsterdam, The Netherlands, 8–16 October 2016; pp. 21–37.

34. Hennessy, J.L.; Patterson, D.A. *Computer Architecture: A Quantitative Approach*, 6th ed.; Morgan Kaufman: Cambridge, MA, USA, 2019; pp. 539–617.

35. Farabet, C.; Poulet, C.; Han, J.Y.; Lecun, Y. CNP: An FPGA based processor for convolutional networks. In Proceedings of the 2009 International Conference on Field Programmable Logic and Applications, Prague, Czech Republic, 31 August–2 September 2009; pp. 32–37.

36. Farabet, C.; Martini, B.; Corda, B.; Akselrod, P.; Culurciello, E.; Lecun, Y. NeuFlow: A runtime reconfigurable dataflow processor for vision. In Proceedings of the 2011 IEEE Computer Society Conference on Computer Vision and Pattern Recognition Workshops, Colorado Springs, CO, USA, 20–25 June 2011; pp. 109–116.

37. Peemen, M.; Setio, A.A.A.; Mesman, B.; Corporaal, H. Memory-centric accelerator design for convolutional neural networks. In Proceedings of the 2013 IEEE 31st International Conference on Computer Design, Asheville, NC, USA, 6–9 October 2013; pp. 13–19.

38. Alwani, M.; Chen, H.; Ferdman, M.; Milder, P. Fused-layer CNN accelerators. In Proceedings of the 2016 49th Annual IEEE/ACM International Symposium on Microarchitecture, Taipei, Taiwan, 15–19 October 2016; pp. 1–12.

39. Chen, T.; Du, Z.; Sun, N.; Wang, J.; Wu, C.; Chen, Y.; Temam, O. DianNao: A small-footprint high-throughput accelerator for ubiquitous machine-learning. *ACM Sigplan Not.* **2014**, *49*, 269–284.

40. Du, Z.; Fasthuber, R.; Chen, T.; Ienne, P.; Li, L.; Luo, T.; Feng, X.B.; Chen, Y.J.; Temam, O. ShiDianNao: Shifting vision processing closer to the sensor. *SIGARCH Comput. Archit. News* **2015**, *43*, 92–104. [CrossRef]

41. Zhang, C.; Li, P.; Sun, G.; Guan, Y.; Xiao, B.; Cong, J. Optimizing FPGA-based accelerator design for deep convolutional neural networks. In Proceedings of the 2015 ACM/SIGDA International Symposium on Field-Programmable Gate Arrays, Monterey, CA, USA, 22–24 February 2015; pp. 161–170.

42. Jouppi, N.P.; Young, C.; Patil, N.; Patterson, D.; Agrawal, G.; Bajwa, R.; Bates, S.; Bhatia, S.; Boden, N.; Borchers, A.; et al. In-datacenter performance analysis of a tensor processing unit. *SIGARCH Comput. Archit. News* **2017**, *45*, 1–12. [CrossRef]

Remote Sens. **2019**, *11*, 2376

43. Li, L.; Zhang, S.B.; Wu, J. Design and realization of deep learning coprocessor oriented to image recognition. In Proceedings of the 2017 17th IEEE International Conference on Communication Technology, Chengdu, China, 27–30 October 2017; pp. 1553–1559.

44. Chang, J.W.; Kang, K.W.; Kang, S.J. An energy-efficient FPGA-based deconvolutional neural networks accelerator for single image super-resolution. *IEEE Trans. Circuits Sys. Video Tech.* **2018**. [CrossRef]

45. Han, X.; Zhong, Y.; Zhang, L. An efficient and robust integrated geospatial object detection framework for high spatial resolution remote sensing imagery. *Remote Sens.* **2017**, *9*, 666. [CrossRef]

46. Etten, A.V. You Only Look Twice: Rapid Multi-Scale Object Detection in Satellite Imagery. *arXiv* **2018**, arXiv:1805.09512.

47. Zhang, X.; Zhu, K.; Chen, G.; Tan, X.; Zhang, L.; Dai, F.; Liao, P.; Gong, Y. Geospatial object detection on high resolution remote sensing imagery based on double multi-scale feature pyramid network. *Remote Sens.* **2019**, *11*, 755. [CrossRef]

48. Howard, A.G.; Zhu, M.; Chen, B.; Kalenichenko, D.; Wang, W.; Weyand, T.; Andreetto, M.; Adam, H. MobileNets: Efficient Convolutional Neural Networks for Mobile Vision Applications. *arXiv* **2017**, arXiv:1704.04861.

49. Fu, C.Y.; Liu, W.; Ranga, A.; Tyagi, A.; Berg, A.C. DSSD: Deconvolutional Single Shot Detector. *arXiv* **2017**, arXiv:1701.06659.

50. Li, L.; Zhang, S.B.; Wu, J. An efficient hardware architecture for activation function in deep learning processor. In Proceedings of the 2018 3rd IEEE International Conference on Image, Vision and Computing, Chongqing, China, 27–29 June 2018; pp. 911–918.

51. Ioffe, S.; Szegedy, C. Batch normalization: Accelerating deep network training by reducing internal covariate shift. In Proceedings of the 32nd International Conference on Machine Learning, Lille, France, 6–11 July 2015; pp. 448–546.

52. Jia, Y.; Shelhamer, E.; Donahue, J.; Karayev, S.; Long, J.; Girshick, R.; Guadarrama, S.; Darrell, T. Caffe: Convolutional architecture for fast feature embedding. In Proceedings of the 22nd ACM International Conference on Multimedia, Orlando, FL, USA, 3–7 November 2014; pp. 675–678.

53. Everingham, M.; Gool, L.; Williams, L.K.; Winn, J.; Zisserman, A. The pascal visual object classes (VOC) challenge. *IJCV* **2010**, *88*, 303–338. [CrossRef]

54. Qiu, J.; Wang, J.; Yao, S.; Guo, K.; Li, B.; Zhou, E.; Yu, J.; Tang, T.; Xu, N.; Song, S. Going deeper with embedded FPGA platform for convolutional neural network. In Proceedings of the 2016 ACM/SIGDA International Symposium on Field-Programmable Gate Arrays, Monterey, CA, USA, 21–23 February 2016; pp. 26–35.

55. Ma, Y.; Cao, Y.; Vrudhula, S.; Seo, J.S. Optimizing loop operation and dataflow in FPGA acceleration of deep convolutional neural networks. In Proceedings of the 2016 ACM/SIGDA International Symposium on Field-Programmable Gate Arrays, Monterey, CA, USA, 22–24 February 2017; pp. 45–54.

56. Liu, Z.; Chow, P.; Xu, J.; Jiang, J.; Dou, Y.; Zhou, J. A uniform architecture design for accelerating 2D and 3D CNNs on FPGAs. *Electronics* **2019**, *8*, 65. [CrossRef]

57. Courbariaux, M.; Bengio, Y.; David, J.P. Binaryconnect: Training deep neural networks with binary weights during propagations. In Proceedings of the 28th International Conference on Neural Information Processing Systems, Montreal, Canada, 7–12 December 2015; pp. 3123–3131.

58. Hubara, I.; Courbariaux, M.; Soudry, D.; El-Yaniv, R.; Bengio, Y. Quantized neural networks: Training neural networks with low precision weights and activations. *J. Mach. Learn. Res.* **2018**, *18*, 1–30.

59. Jacob, B.; Kligys, S.; Chen, B.; Zhu, M.; Tang, M.; Howard, A.; Adam, H.; Kalenichenko, D. Quantization and training of neural networks for efficient integer-arithmetic-only inference. In Proceedings of the IEEE Conference on Computer Vision and Pattern Recognition, Salt Lake City, UT, USA, 18–23 June 2018; pp. 2704–2713.

60. Jetson AGX Xavier. Available online: https://developer.nvidia.com/embedded/jetson-agx-xavier (accessed on 12 December 2018).

 *remote sensing*

Article

Unsupervised Saliency Model with Color Markov Chain for Oil Tank Detection

Ziming Liu [1,2,3], Danpei Zhao [1,2,3,*], Zhenwei Shi [1,2,3] and Zhiguo Jiang [1,2,3]

[1]   Image Processing Center, School of Astronautics, Beihang University, Beijing 100191, China;
       zm_liu@buaa.edu.cn (Z.L.); shizhenwei@buaa.edu.cn (Z.S.); jiangzg@buaa.edu.cn (Z.J.)
[2]   Beijing Key Laboratory of Digital Media, Beihang University, Beijing 100191, China
[3]   Key Laboratory of Spacecraft Design Optimization and Dynamic Simulation Technologies,
       Ministry of Education, Beijing 100191, China
*   Correspondence: zhaodanpei@buaa.edu.cn; Tel.: +86-134-2626-3900

Received: 4 April 2019; Accepted: 3 May 2019; Published: 7 May 2019

Abstract: Traditional oil tank detection methods often use geometric shape information. However, it is difficult to guarantee accurate detection under a variety of disturbance factors, especially various colors, scale differences, and the shadows caused by view angle and illumination. Therefore, we propose an unsupervised saliency model with Color Markov Chain (US-CMC) to deal with oil tank detection. To avoid the influence of shadows, we make use of the CIE Lab space to construct a Color Markov Chain and generate a bottom-up latent saliency map. Moreover, we build a circular feature map based on a radial symmetric circle, which makes true targets to be strengthened for a subjective detection task. Besides, we combine the latent saliency map with the circular feature map, which can effectively suppress other salient regions except for oil tanks. Extensive experimental results demonstrate that it outperforms 15 saliency models for remote sensing images (RSIs). Compared with conventional oil tank detection methods, US-CMC has achieved better results and is also more robust for view angle, shadow, and shape similarity problems.

Keywords: oil tank detection; unsupervised saliency model; Color Markov Chain; bottom-up and top-down

1. Introduction

With the rapid development of remote sensing applications, the detection of valuable remote sensing targets has become a hot issue in the field of remote sensing images (RSIs) and computer vision. As an important energy storage device, oil tanks have become one of the key targets for remote sensing reconnaissance or exploration systems.

In RSIs, an oil tank is usually round in shape and painted in white or other light colors, the arrangement rule of which is very random. Due to many factors such as illumination, position, viewing angle, and imaging quality, the edges of oil tanks become fuzzy, and their colors are not uniform. They may also have a certain degree of geometric deformation. Moreover, the background of large area RSIs becomes complicated and oil tank targets are relatively small. These complex situations have brought great difficulties to the detection and identification of oil tanks. Therefore, how to accurately and completely detect an oil tank is the most important question. Based on the above, in this paper, we mainly resolve how to accurately and completely detect a tank target under the interference of complex ground objects.

In recent years, many oil tank detection methods have been proposed, which include the template matching method [1,2], geometric shape method [3–7], saliency detection method [8,9], and machine learning method [10–12]. The template matching method requires lots of calculations. Furthermore, the template selection is susceptible to many factors such as scale and rotation. Ref [2] combines an improved Hough transform algorithm with Canny and a fast template matching algorithm. Through template matching to locate oil tanks, the detection rate is often low. Many conventional methods for oil tank detection are based on geometric shape, such as the standard circular Hough transform proposed by Duda [13]. Ref [3] employs Hough circle detection method with scale invariance to improve efficiency of detection. Ref [4] proposes an improved fuzzy Hough transform, which avoids the occurrence of peak diffusion and false peaks, thus improving the detection results. Han [5] uses a depth-first map search strategy, grouping the detected circles according to the special distribution of the oil tank and then eliminating false alarms. Ref [6] applies semantic analysis to retrieve the oil tank area and combines this with the Hough transform to detect oil tanks in the optical satellite images. However, this method is only used for specific images including big targets and is not universal. In case of unsupervised detection, the Hough transform relies on the color gradient of the image and clear object boundaries. When the background is very complicated, the detection result is not often satisfied. In addition to this, the methods above pay more attention to the bottom characteristics of the oil tank and almost ignore the influence of the background, resulting in a high probability of misses and false detections. In terms of shape information, a new method of detecting oil tanks has appeared in recent years. Ok [7] proposes a detection method based on circular radial symmetry by calculating the boundary gradient direction and the center of the circle to obtain a very good detection result. But oil tanks cannot be detected completely when oil tanks are small or have shadows, three-dimensional structures, and low contrast. In the synthetic aperture radar(SAR) images, [1] detects the oil tanks by using a template to combine circular shadows and high-brightness parts of the light. Ref [14] proposes a method based on multidimensional feature vector clustering to search for oil tank targets in SAR images. With regard to supervised methods, [10–12] use convolutional neural networks(CNNs) to extract the depth feature of the oil tank in the network and then classified the final results. However, CNN requires a large number of samples for training. To save training time and solve the problems such as shadow, shape interference, and certain angle due to the latitude of the earth, we consider using human visual perception to provide complete and accurate results for oil tank detection, so we choose saliency methods.

In recent years, saliency has become a very popular area due to its ability to highlight salient areas of the image faster and better, just to provide the interest candidate areas for object detection. Traditional bottom-up saliency models, such as [15–20], utilize bottom feature mining. Most of these models take advantage of the color contrast differences in the images, and the resulting resolution is often low and only applies to natural images, not RSIs. Ref [21] obtains the saliency result by calculating the reconstruction errors of sparse and dense graphs, then uses K-means clustering and an object-biased Gaussian model to optimize the result. Ref [22–25] apply boundary connectivity to obtain background information and use it to find out if super-pixels are connected to the background, then use saliency optimization to get better results. The saliency methods above rely too much on the color information of the boundary. When the target exists at the boundary, they often cause false detection. As for top-down saliency models, in [26,27] for example, they add a bootstrap learning algorithm and hierarchical cellular automata to detect the saliency of the image. With the popularity of deep convolutional neural networks, many saliency detection methods have begun to use CNNs for feature extraction. Some people also think that CNNs can provide a lot of help for saliency detection. Ref [28] employs VGG-net to extract advanced features and combines high-level features with low-level features for detection. By using two deep neural networks, [29] proposes a saliency detection model that combines local estimation and global search. However, the saliency methods above only consider natural images in terms of space and color. Although saliency models have been applied to the field of RSIs, there is still a lack of reasonable use of saliency in oil tank detection, which often leads to a large number of false and misdetections. Ref [8] makes use of a saliency model and Hough transform, combined with support vector machine(SVM), for oil tank detection. But this method only works for larger oil tanks, and its detection rate is relatively low. Ref [9] used a saliency region detection method with an Otsu threshold to detect oil tanks; however, the lack of shape guidance led to the omission of oil tanks that are not salient in the RSI.

Though the methods above use saliency to detect oil tanks, they have missed some of the oil tanks and only have good results under certain conditions. At present, oil tank detection still has many problems, such as when oil tanks present three-dimensional structures, similar shape interference, and shadow interference, and it is impossible to accurately and effectively detect the oil tank area. To solve above problems, we propose an unsupervised saliency model with Color Markov Chain (US-CMC). US-CMC not only utilizes bottom-up low-level color features that can highlight the oil tank areas and eliminate the interference of shadows, but also introduces top-down characteristics of the shape to the saliency model, which can effectively eliminate surrounding similar colors or view angle interference. By fusing bottom-up and top-down feature cues, US-CMC has better performance and robustness than other oil tank detection methods.

The main contributions of our approach are summarized in three aspects below:

1. Aiming at the difficult problems of oil tank detection, we propose an unsupervised saliency model with Color Markov Chain (US-CMC). US-CMC makes use of the CIE Lab space to construct a Color Markov Chain for effectively reducing the influence of the shadow. Moreover, the constraint matrix is constructed to suppress the interference of non-oil-tank circle areas. By using the linear interpolation process, the oil tank targets with variable view angles can also be detected completely.
2. Different from the previous methods using circular features, US-CMC transforms circular radial symmetry features into a circular gradient map and then generates a series of confidence values for the circular region.
3. We employ an unsupervised framework, which can avoid the extra time cost in large sample training. Furthermore, US-CMC can restrain other salient regions apart from oil tanks by combining bottom-up latent saliency maps with top-down circular feature maps. Consequently, our model can not only quickly locate the oil tank targets, but can also maintain the detection accuracy.

2. Proposed Method

2.1. Bottom-Up Latent Saliency Map Based on Color Markov Chain

As shown in Figure 1, the US-CMC model is mainly divided into the following steps. Firstly, we use the SLIC [30] method to segment the image into super-pixels, and then we use a Color Markov Chain to obtain a coarse saliency region. Then we calculate the color and position matrix of super-pixel blocks and use them to obtain the saliency image. Secondly, we use the radial symmetry method to get information about the circular shapes in the image. Thirdly, The final grayscale image is obtained by Bayesian integration. Finally, the binary detection result is obtained by GrabCut and post-processing. Next, we will introduce each step in subsections.

Because oil tanks are distributed targets, just like the absorption nodes in the Absorbing Markov Chain, each oil tank can be regarded as an absorbed node, and the super-pixels along the image side are regarded as the start nodes to absorb the entire super-pixels. Thus, we use Absorbing Markov Chain to help us with oil tank detection. a Markov chain is a relatively common and familiar random process. A Markov chain containing an absorption state is called an Absorbing Markov Chain. Given a series of data $S = \{s_1, s_2, \ldots, s_l\}$, the process starts from one of these states and moves continuously from one state to another. If the chain is currently in state s_i, the probability of moving to state s_j is called the transition probability, represented by P_{ij}. Therefore, the Markov chain can be determined by the transfer matrix P. For any Absorbing Markov Chain with k absorption states and m transient states, the canonical form of the above transition matrix P can be obtained as follows:

$$P = \begin{bmatrix} Q & R \\ 0 & I \end{bmatrix},$$
(1)

where Q is the probability transfer matrix of the transient state, the element in R means the transition probability between the transition and the absorption state, and I is the identity matrix of $k \times k$. We can get another matrix based on the matrix Q:

$$N = (I - Q)^{-1}. \tag{2}$$

The element N_{ij} in N gives the expected number of times that the process changes to the transient state s_j if it starts at the transient state s_i. The final absorption probability matrix is as follows:

$$B = NR. \tag{3}$$

In the past, the saliency model based on an Absorbing Markov Chain used the absorption time to determine the saliency value. What we propose is a Color Markov Chain model that judges the saliency value based on the contrast of the image in CIE Lab color. The Color Markov Chain is a random walk model that is used to detect saliency regions in the image. Due to the segment representation by the super-pixel method, we can identify the image as $G(V, E)$. The vertex V is represented by a super-pixel, and E is a set of undirected edges containing the connectivity between two super-pixels. The edge connecting the two nodes i and j is denoted as e_{ij}, and w_{ij} represents the weight of the edge e_{ij} based on the similarity between the features defined on the nodes i and j. We use the CIE Lab color space to define the color characteristics of each super-pixel node because the CIE Lab color model describes how the color is displayed based on a human's perception of color. The weight relationship expression between adjacent nodes i and j is:

$$w_{ij} = exp(-\frac{\|c_i - c_j\|^2}{2\sigma^2}). \tag{4}$$

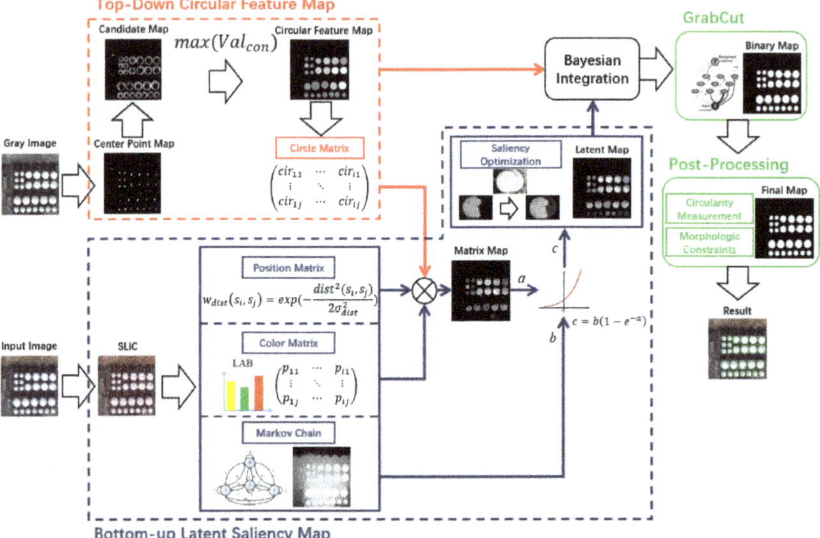

Figure 1. Flowchart of the unsupervised saliency model with Color Markov Chain (US-CMC) method. First, a bottom-up latent saliency model combined with a circle matrix was applied to get the latent map. At the same time, the radial symmetry method was used to get the circular feature map then combine both maps by using Bayesian integration. Then, after One Cut for GrabCut and post-processing, the final result was obtained.

We take all three channels of the CIE Lab color in the super-pixel block as the feature quantity of the two nodes i and j, which are represented by c_i and c_j, and $\|c_i - c_j\|^2 = (c_{i_L} - c_{j_L})^2 + (c_{i_a} - c_{j_a})^2 + (c_{i_b} - c_{j_b})^2$, where c_{i_L}, c_{i_a}, and c_{i_b} are the values of three channels in c_i. c_{j_L}, c_{j_a}, and c_{j_b} are the values of three channels in c_j. The super-pixel result is shown in Figure 2b. σ is a constant used to control the weight; generally we set σ^2 to 0.05. Then we get the affinity matrix $\mathbf{W} = [w_{ij}]_{mm}$ of the undirected graph $G(V, E)$. Then we use the edge of the selected image as the absorption node to start the absorb process. At the same time, another affinity matrix $\mathbf{T} = [t_{ij}]_{mk}$ between the transient node i and the absorption node j is established, where t_{ij} represents the correlation between nodes i and j in the image. $t_{ij} = exp(-\frac{\|c_i - c_j\|^2}{2\sigma^2})$. The σ is the same as in (4). We find the final correlation matrix $D = \{d_{11}, d_{22}, \ldots, d_{mm}\}$, where $d_{ii} = \sum_{j=1}^{m} w_{ij} + \sum_{j=1}^{k} t_{ij}$. Then make $Q = D^{-1}\mathbf{W}$, $R = D^{-1}\mathbf{T}$. The final absorption probability matrix B is obtained according to the above formula. Then define the saliency of each node as the dissimilarity with the image boundary. For the transient state s_i in the Color Markov Chain, the probability b_{ij} absorbed by the absorption state s_j actually represents the relationship between them. For each node i in the image, we sort the absorption probability values b_{ij} of all boundary nodes $j(j \in \{1, 2, \ldots, k\})$ in the image boundary in descending order:

$$b_{i1} \geq b_{i2} \geq \cdots \geq b_{ik}. \tag{5}$$

We take the first r column of b_{ij}, r is the number of columns in b_i (where $\frac{1}{2}k \leq r \leq k$), which is $bg(i) = \sum_{j=1}^{r} b_{ij}$ to represent the similarity between node i and the boundary node. And we present $fg(i) = \sum_{j=r}^{k} b_{ij}$ as the dissimilarity between node i and boundary nodes. So we get the final saliency value for each node:

$$Markov(i) = e^{(1 - bg(i))} \cdot fg(i). \tag{6}$$

In this way, we get a coarse saliency map based on the Color Markov Chain, which is the foreground probability distribution map. The results are shown in Figure 2d. Although it is not very accurate, it suppresses most of the background from the map. It also removes some of the shadow interference. Next we will use the color, position, and shape information of the map to construct a function to extract the features and then will highlight the salient areas we are really interested in.

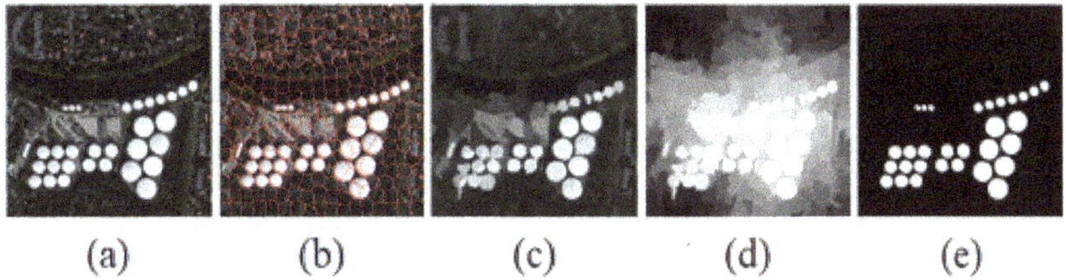

(a)   (b)   (c)   (d)   (e)

Figure 2. An example of the process of a Color Markov Chain. (**a**) is the input image, (**b**) is the super-pixel segmentation, (**c**) is the result after averaging, (**d**) is the result of the Color Markov Chain, and (**e**) is the ground truth.

When the image is divided by super-pixel, the image is labeled as $S = \{s_1, s_2, \ldots, s_n\}$, where n represents the number of super-pixels. In the previous process of the Color Markov Chain, we have formed an undirected graph $G(V, E)$, so that we can calculate the Euclidean distance of each super-pixel in the CIE Lab color space as the weight value of the edges. We define it as $Lab_{ij}(s_i, s_j)$. Then, the Euclidean distance $dist(s_i, s_j)$ between the center points of the super-pixel blocks s_i and s_j is obtained, and the weight formula of the Euclidean distance is proposed:

$$w_{dist}(s_i, s_j) = exp(-\frac{dist^2(s_i, s_j)}{2\sigma_{dist}^2}). \tag{7}$$

We set σ_{dist} to 0.25 according to the convention. Finally, we can get the contrast determination formula between two super-pixel blocks:

$$Ctr(s_i, s_j) = Lab_{ij}(s_i, s_j)w_{dist}(s_i, s_j)d_{cir}(s_i, s_j),\tag{8}$$

where $d_{cir}(s_i, s_j)$ is the circle matrix, which is the Euclidean distance of the average gray value in a super-pixel block between the super-pixels s_i and s_j in the circular feature map. The Formula (8) is also the basic formula for contrast optimization of the Color Markov Chain. In this paper, considering the characteristics of remote sensing object detection and the shape characteristics of the oil tank, we should also combine shape features for processing when we optimize the contrast of the Color Markov Chain. For shape features, we extract the circular features by calculating the radial symmetry of the gradients in the image. The method of radial symmetry will be explained in the next section. We integrate the result of the contrast determination formula into the Color Markov Chain as follows:

$$Sal(s_i, s_j) = d_{markov}(s_i, s_j) \cdot (1 - exp(-Ctr(s_i, s_j))),\tag{9}$$

where $d_{markov}(s_i, s_j)$ is the Euclidean distance of saliency values between each super-pixel block in $Markov(s_i)$. Then we sum $Sal(s_i, s_j)$ to get the saliency value of each super-pixel block itself:

$$FinalSal(s) = \sum_{i=1}^{n} Sal(s, s_i).\tag{10}$$

The relevant results are shown in Figure 3c. The calculation of the saliency map in the above formula simply uses the method of multiplication and weighted summation, combining various low-level features and clues, and then optimizes the results of the Color Markov Chain. However, this kind of optimization is not enough, obviously. Through experiments, we find that some of the background areas are still not removed. Furthermore, we will optimize the results of the saliency map.

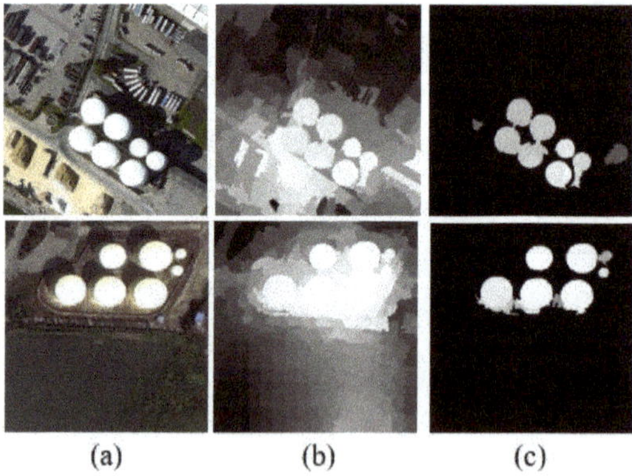

(a)   (b)   (c)

Figure 3. (**a**) is the input image, (**b**) is the result of Color Markov Chain, and (**c**) is the result after contrast constraint.

2.2. Saliency Map Optimization and Background Suppression

Since the background of the original input image is very complicated, there is a significant disadvantage in some saliency maps where the background of the saliency map is not sufficiently suppressed. To solve this problem, we propose the following two methods. First, we group the nodes of the input image into K clusters by a K-means clustering algorithm in the CIE Lab color space, and then the saliency of each node can be corrected by simple interpolation of nodes in the same cluster. The saliency optimization of the node s_i can be implemented by the following formula:

$$S_f(s_i) = \alpha \cdot FinalSal(s_i) + (1 - \alpha)\frac{\sum_{j=1}^{m} D_{ij} \cdot FinalSal(s_j)}{\sum_{j=1}^{m} D_{ij}}, \tag{11}$$

where m is the number of nodes in the cluster. $D_{ij} = e^{-\frac{\|c_i - c_j\|^2}{2\sigma_I}}$, and σ_I is the sum of the variances in each feature dimension of the CIE Lab space. The term on the left side of Equation (11) represents the saliency of the initial optimization of the node s_i, while the $FinalSal(s_i)$ and $FinalSal(s_j)$ on the right side of the formula are the original saliency results of the nodes s_i and s_j, respectively. The above parameter α is set to 0.5 according to experience.

Although the above interpolation method effectively highlights the foreground of some saliency maps, some parts are still not well suppressed. To further solve this problem, we introduce a simple piecewise function to remove the part that is not saliency, or large, error regions. The function is defined as:

$$f(s_i) = \begin{cases} s_i, & s_i > \theta \\ \frac{s_i^3}{\theta^2}, & s_i \leq \theta \end{cases}, \tag{12}$$

where θ is the threshold that defines the saliency range, set to 0.6 according to experience, and s_i is the super-pixel. This way we get the final image of the latent map, as shown in Figure 4c.

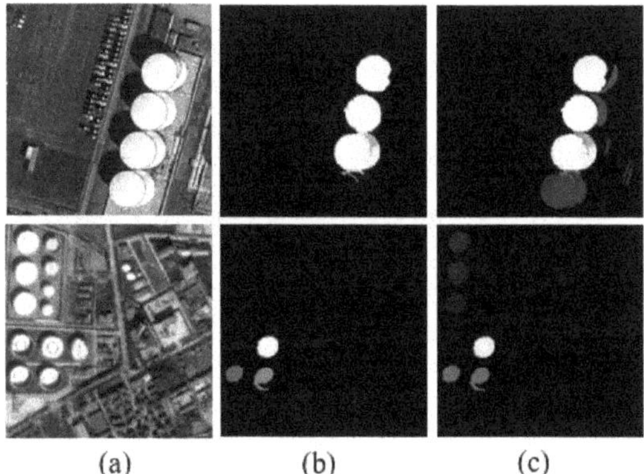

(a)   (b)   (c)

Figure 4. (**a**) is the input image, (**b**) is the result after the above saliency model, and (**c**) is the latent map.

2.3. Top-Down Circular Feature Map Based on Circular Confidence Value

Oil tanks are often constructed of metal and are cylindrical in shape. In RSIs, the shape of the oil tank often appears round. Therefore, the result of target detection can be obtained by detecting circle areas in the image. However, the satellite may be at a certain angle to the surface of the earth. Especially in some areas with high latitudes, the final result of detection is probably not a regular circle. Therefore, the traditional Hough transform circle detection method is not able to detect most of the oil tanks that appear and even has a large number of misdetection problems. Aiming at this problem, we propose a circle detection method based on improved fast radial symmetric transformation, which can be applied to the above situation and has strong robustness. Though target detection only based on shape is not enough, it gives us a new way to think about the problem. The process is as follows:

Given a radius interval R, let the radius of oil tank $r' \in R$. Then, after obtaining the gradient of the image using Sobel transform, we can get the magnitude, which is called the "impact", caused by a series of radii r and also calculate the direction of the gradient to determine whether the radius "fits" the gradient to form a circle. In theory, if the pixel is located just at the boundary of the circle with radius r', then at $r' \in R$, the effect of these pixels will be enriched at the center of the circle so that the approximate center position can be obtained. We can get the confidence value Val_{con} of each center point by adding all the magnitudes together. Then we find the point with $max(Val_{con})$ and use this point as a center to form a circle with r' as its radius. We set the confidence value of the center as the gray value of the entire circle after being normalized. Then we can obtain the average distance $d_{cir}(s_i, s_j)$ between the super-pixels s_i and s_j in the circular feature map. Figure 5 shows the circular feature map generation process.

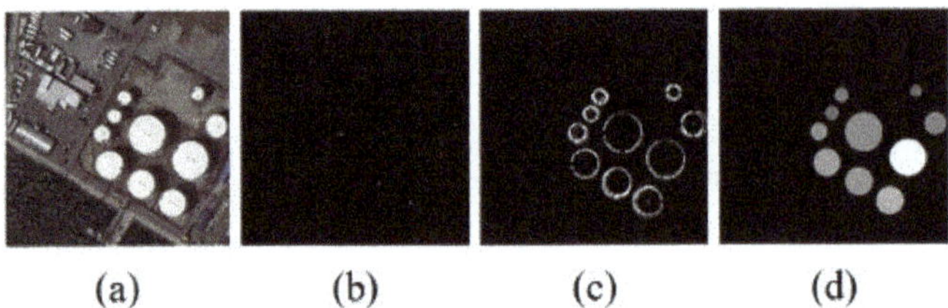

(a)   (b)   (c)   (d)

Figure 5. (**a**) is the input image, (**b**) is the center point map, (**c**) is the candidate map, and (**d**) is the final circular feature map.

2.4. Fusion of Color Saliency Map and Circular Feature Map

After extracting the shape and color features of the target, we need to integrate these two different modules to achieve a better result. Therefore, we introduce a Bayesian integration function to fuse the results of the above two modules.

We set the latent map to S_1 and the result of the circular feature map to S_2. We first define one of the pictures $S_i(i = \{1,2\})$ as the foreground map and the other picture $S_j(j \neq i, j = \{1,2\})$ as the background map. We calculate the possibilities of both so that we can integrate more information from different saliency maps. First, we use the average gray value of the image as the threshold and then we divide the graph S_i by it. The images are divided into $Fore_i$ and $Back_i$, respectively, where $Fore_i$ is the foreground area and $Back_i$ is the background area. In each region, we calculate the likelihood by comparing the foreground and background regions of S_j and S_i at pixel z. The formula is as follows:

$$p(S_j(z) \mid Fore_i) = \frac{N_{b_{Fore_i(S_j(z))}}}{N_{Fore_i}}, \tag{13}$$

$$p(S_j(z) \mid Back_i) = \frac{N_{b_{Back_i(S_j(z))}}}{N_{Back_i}}, \tag{14}$$

where N_{Fore_i} is the number of nonzero pixels of the foreground region in the image i, and $N_{b_{Fore_i(S_j(z))}}$ represents the number of pixels whose saliency regions fall into the foreground bin $b_{Fore_i(S_j(z))}$, which contains $S_j(z)$. Similarly, N_{Back_i} represents the number of pixels of the background area i. The formula for calculating the posterior probability using S_i is:

$$p(Fore_i \mid S_j(z)) = \frac{S_i(z)p(S_j(z) \mid Fore_i)}{S_i(z)p(S_j(z) \mid Fore_i) + (1 - S_i(z))p(S_j(z) \mid Back_i)}. \tag{15}$$

We can also use the formula in (16) to get another posterior probability using S_j.

$$p(Fore_j \mid S_i(z)) = \frac{S_j(z)p(S_i(z) \mid Fore_j)}{S_j(z)p(S_i(z) \mid Fore_j) + (1 - S_j(z))p(S_i(z) \mid Back_j)}. \tag{16}$$

We then use these two posterior probabilities to calculate the final saliency map, which is as follows:

$$S_{map}(S_1(z), S_2(z)) = p(Fore_1 \mid S_2(z)) + p(Fore_2 \mid S_1(z)). \tag{17}$$

We should notice that Bayesian integration enforces the two graphs as prior and they cooperate with each other in an efficient manner; then the final result uniformly highlights the salient objects. So we add the two images together so that we can highlight the foreground of the two images at the same time. But the final result we need to get is a binary map. In the current situation, it is difficult to find a suitable threshold to convert the final saliency map into a binary map. The idea of setting a fixed threshold is too simple and easy to get false detections. So we introduce the One Cut for GrabCut [31] method to implement the binarization process. It is an improved version of GraphCut that sets an energy minimization function to calculate the threshold in each image, while also considering the completeness of the segmentation results. Minimized energy can be obtained by the following formula:

$$E_{sal}(S) = \sum_{p \in \Omega} m(S_{map}) - S_{map}(z) \cdot bin_z + |\delta S| - Penalty, \tag{18}$$

where S_{map} is the saliency map we get, and $S_{map}(z)$ refers to the saliency value of pixel z. bin_z indicates whether this pixel belongs to the foreground, 1 means the foreground, and 0 means the background. $Penalty = \|\theta^s - \theta^{\bar{s}}\|$ is the background and foreground overlap penalty, θ^s and $\theta^{\bar{s}}$ are histograms inside object S and background $\bar{S}$, respectively, and $|\delta S|$ is a smooth term. With One Cut, we can get the binarized form of each map.

2.5. Post-Processing

The post-processing part was introduced because we have obtained the binary map results for the oil tank area, but the result we need to get is a more accurate one. Because Grabcut brings problems such as some small noises, we extract the area information and shape information from all of the noises separately and compare it with the area of the oil tank, then propose a method to remove the noise. We first remove the regions with less than 40 pixels and then solve the circularity for all the remaining connected domain. The circularity formula is as follows:

$$\tau = \frac{4\pi \cdot Area(i)}{PM^2(i)}, \tag{19}$$

where $Area(i)$ is the area of the connected domain i, and $PM(i)$ is the perimeter of it. Then we will get the circularity τ and can remove all the parts that do not conform to the circle, and the remaining regions we see as the final oil tank area, so we get the final result.

3. Experimental Results And Discussion

We created a dataset for detection that is comprised of a total of 240 multi-resolution images from Google Earth. The resolution of the test images is 400×400. All test images contain at least two or more oil tanks for testing. The full dataset has a total of more than 2200 oil tanks of various colors and luminances. The scales of the oil tanks are from 7 m to 40 m. In addition, we manually mark all of the ground truth to ensure that the ground truth fits perfectly with the oil tank area. At the same time, in order to make the detection task more difficult, we also added nearly 35 test images with similar color and shape interferences for detection. Here is our experiment:

3.1. Parameter Selection

Before comparing the saliency models and oil tank detection methods, we need to use experiments to verify the parameters. For all of the parameters such as σ^2, σ_{dist}, α, and θ in our text, we supplanted the parameter analysis through a P–R curve to verify the performance of each parameter on our results.

As shown in Figure 6, the change of four parameters does not have a significant effect on the P–R curve. Therefore, these four parameter selections are appropriate.

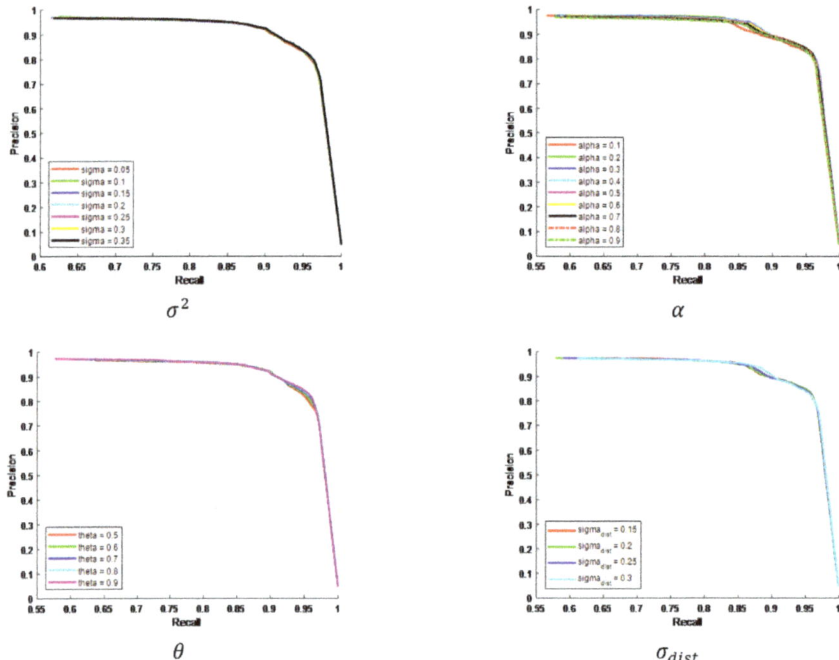

Figure 6. The P–R curve result of four parameters.

3.2. Comparison with Saliency Models

In order to highlight the superiority of our method, we first compare the US-CMC model with 15 currently advanced saliency models, namely wCtr [22], SF [20], GS [24], MR [23], BL [26], DSR [21], FT [17], GC [32], HS [33], RCRR [34], AMC [35], MST [25], LPS [36], SCA [27], Itti [15]. We will use three indicators, mean absolute error(MAE), precision–recall curve(P–R Curve), and F-measure to evaluate the advantages and disadvantages of each model. The mean absolute error formula and the F-measure formula are as follows:

$$MAE(X,h) = \frac{1}{W \times H} \sum_{x=1}^{W} \sum_{y=1}^{H} |S(x,y) - GT(x,y)|,$$

(20)

$$F_{measure} = \frac{(1 + \beta^2) \times Precision \times Recall}{\beta^2 Precision + Recall},$$

(21)

When the test results are evaluated using the F-Measure in this paper, the value of β^2 is set to 0.3. Since the P–R curve needs to ensure that the final image is a grayscale image, we compare the S_{map} results with other saliency models.

As shown in Table 1, Figures 7 and 8, our model is the best of all the saliency detection models and far better than all of the others. This is mainly due to the following reasons: Firstly, previous saliency models are generally designed to solve the problem of natural images. For RSIs, due to their complex background information, saliency models can only judge whether the region is salient by relying on the color information and the position of the parts of interest in the image. Although the color of the oil tank is mostly light, it is still easy to misjudge just according to color information because other light-colored natural objects or buildings are also regarded as oil tanks. At the same time, it is difficult to accurately identify the oil tank because of the similarity in color and texture between oil tank and background. Since it combines a bottom-up low-level feature with a top-down circular feature map, the US-CMC model can be more accurate in identifying the oil tank target. Therefore, our model has better detection performance for oil tanks than the traditional saliency models.

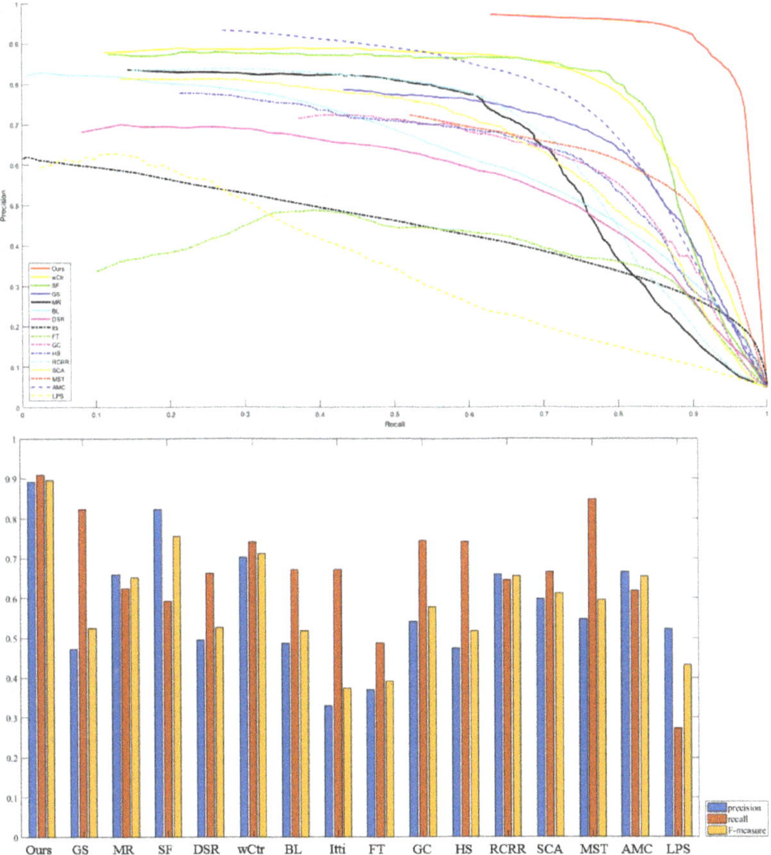

Figure 7. The result of the P–R curve and F-measure in the experiment.

Table 1. The MAE result.

Ours	wCtr [22]	SF [20]	GS [24]	MR [23]
0.009276	0.043808	0.028008	0.116892	0.093963
BL [26]	**DSR** [21]	**FT** [17]	**GC** [32]	**HS** [33]
0.166116	0.082601	0.093677	0.106212	0.189343
RCRR [34]	**AMC** [35]	**MST** [25]	**LPS** [36]	**SCA** [27]
0.089126	0.082153	0.078306	0.066183	0.109985

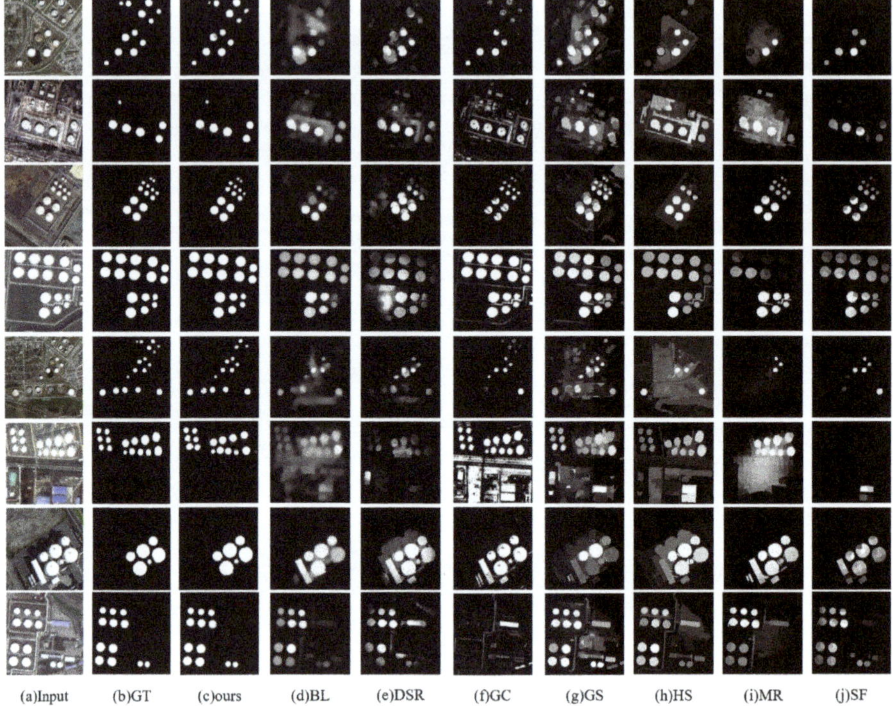

(a)Input   (b)GT   (c)ours   (d)BL   (e)DSR   (f)GC   (g)GS   (h)HS   (i)MR   (j)SF

Figure 8. Comparison of results with other saliency models.

3.3. Comparison with Oil Tank Detection Methods

After comparing it with the saliency detection models, in order to verify the effectiveness of our proposed method, we also need to compare it with the current advanced oil tank detection methods. We compare two different methods of oil tank detection. The first [10] is a supervised algorithm combined with CNN feature extractor and SVM classifier, and the other [7] is an oil tank detection method based entirely on geographic shape information. We randomly select 190 of the 240 images to train the CNN network, and then use the remaining images to compare the effects of the three oil tank detection methods. To evaluate the test results, we define the result of true positives when the Intersection over Union(IoU) value is greater than 0.7 and define false positives as the result of IoU values less than 0.3. This will prevent some unsuccessfully detected oil tanks from being labeled with both false and missed inspections. The test results are shown in Table 2.

In Table 2, we can see that our method is the best in terms of both the precision rate and the recall rate compared with the other state-of-art detection methods. Although Zhang's method combines feature extraction and supervised learning with CNN, it requires a large dataset as training set, which undoubtedly consumes human and material resources for marking ground truth and lacks efficiency. In the face of a small training dataset, there is no superiority of the supervised algorithm. The excellent performance of our method shows that even without effective learning, there are still very good performances. As shown in Figure 9c, Ok's method relies only on the circle shape information obtained from the gradient of the image. It has its own limitations. When the oil tank has a circular shadow, or other non-oil tank areas have a circular gradient, it is easy to lead to false detection. At the same time, many external factors such as illumination will change the gradient of the oil tank contour, and because of its lack of guidance and assistance from other features, they always misjudge the actual size of the oil tank, resulting in false detection results.

(a)Input   (b)ours   (c)OK   (d)MR   (e)SF   (f)GC   (g)HS

Figure 9. Comparison of final test results. Green is the positive detection area, blue is the false detection area, and red is the missed detection area.

Table 2. IoU test results for three oil tank detection methods.

Methods	Precision	Recall	F-Measure
Ours	**0.928**	**0.946**	**0.932**
Ok's method [7]	0.923	0.909	0.920
Zhang's method [10]	0.898	0.710	0.846

Compared with the radial symmetry detection method separately, we find that our method performs better than the radial oil tank detection method proposed by Ok in the following three cases. Figures 10–12 show that our method can overcome these influences well when the oil tank is at a certain angle, or when there are certain non-tank circular zone disturbances and shadow problems caused by the sunlight.

Firstly, our method is based on a US-CMC model. In the process of the Color Markov Chain, due to the color difference with the main part of the oil tank and the similarity with the background, shadows are more easily absorbed. Therefore, the saliency value of a shadow is generally low, while the saliency value of an oil tank's body is relatively high. With further color constraints, the shadow has been largely eliminated. So this method is robust to the presence of shadows. At the same time, when the oil tank is at a certain angle to the satellite, according to Figure 10, we can see that the body part of the oil tank can be detected well after color interpolation. Therefore, even if the oil tank has a certain angle due to the latitude of the Earth, the oil tank can still be detected completely, while using only one single feature such as the circle is unachievable. Finally, when similar shapes occur, as shown in the second column of Figure 11, there is a roundabout in the input image, which is identified as the oil tank by Ok's method. However, our method determines this area to be background by comparing with the surrounding area. The input image in the fourth column has an incomplete ring next to the oil tank, which is identified as a true target by Ok's method. It is obvious that there are great limitations in oil tank detection by relying only on shape detection. Our method can solve such limitations and improve the accuracy of the whole process of detection.

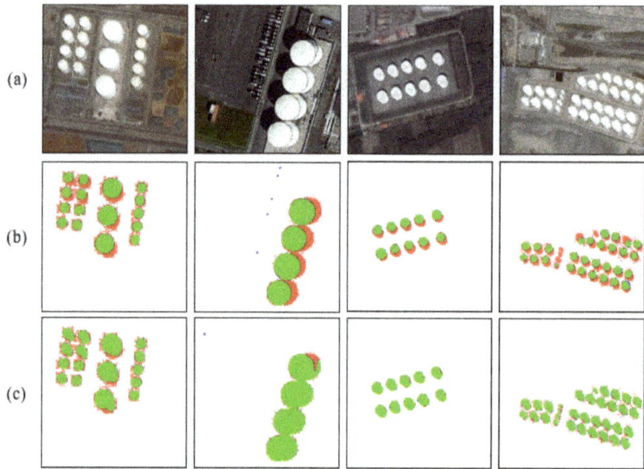

Figure 10. Comparison of two detection methods when the oil tank is at a certain angle. (**a**) is the input image, (**b**) is the method from Ok, (**c**) is our method.

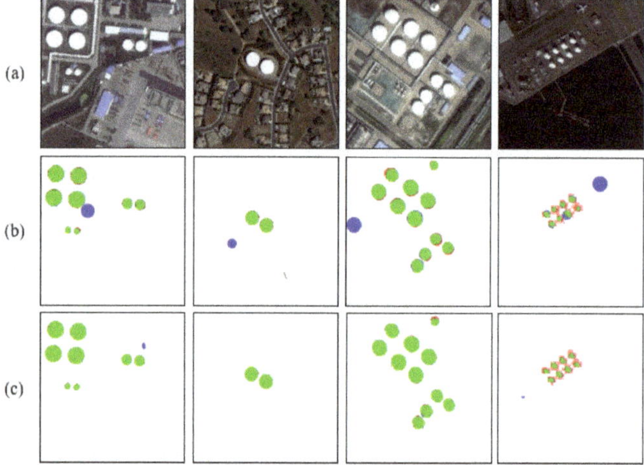

Figure 11. Comparison of two detection methods when similar shapes occur. (**a**) is the input image, (**b**) is the method from Ok, (**c**) is our method.

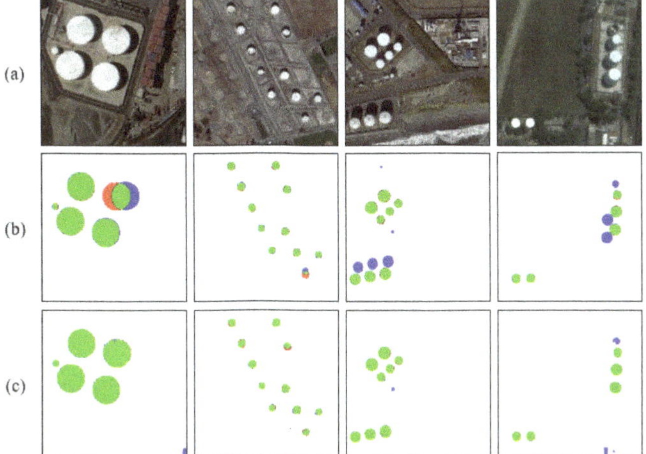

Figure 12. Comparison of two detection methods when shadow interference occurs. (**a**) is the input image, (**b**) is the method from Ok, (**c**) is our method.

3.4. Robustness for Our Methods

In this section, we will show that our method is good not only when the oil tank is at a certain angle, or has certain non-tank circular zone disturbances and shadow problems caused by the sunlight, but also has excellent results when oil tanks are in misty and dusty condition, as well as in higher resolution visible light remote sensing images.

The principle of our algorithm is based on color, gradient, and shape information so that the target existing in the image can be detected. In addition to against disturbances caused by shadows and sensor inclination angle, our method also works well in misty and dusty environments. From the result shown in Figure 13, it can be seen that our method is also excellent. In terms of high resolution, our method can also achieve better results and is robust in high-resolution remote sensing images. The results can be seen in Figure 14. The input images in Figure 14 are all four times bigger than the images in the test dataset.

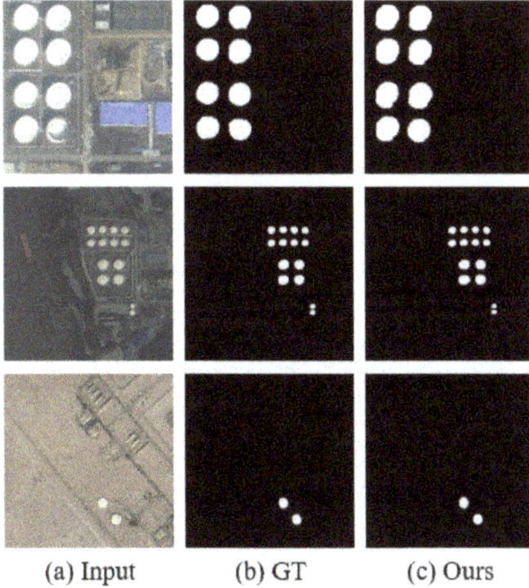

Figure 13. Our method works in misty and dusty condition. (**a**) is the input image, (**b**) is the ground truth, (**c**) is our method.

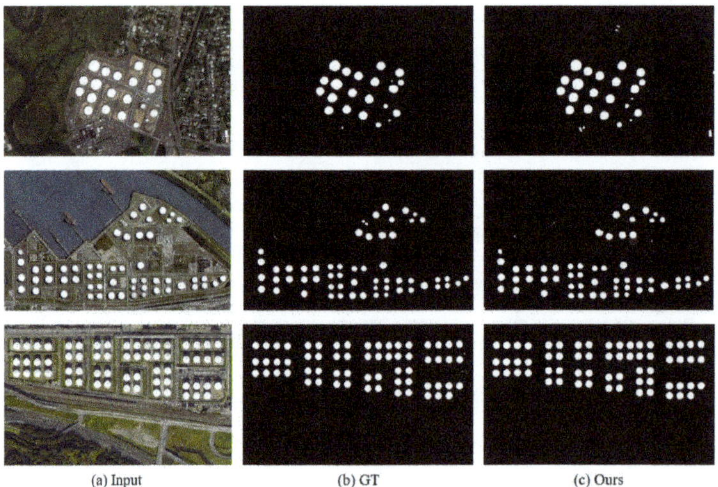

(a) Input   (b) GT   (c) Ours

Figure 14. Our methods works in high resolution images. (**a**) is the input image, (**b**) is the ground truth, (**c**) is our method.

4. Conclusions

According to the characteristics of oil tanks in RSIs, in this paper, we propose an unsupervised oil tank detection method that takes advantage of low-level saliency to highlight latent target areas and introduce a circular feature map to the saliency model to suppress the background. Compared with other saliency models, our model is designed for oil tanks, and can better eliminate the interference of color and texture in similar areas. Our method is also simpler and faster than the learning-based detection method because it does not need sample collection or a training process. Compared with the geometry-based detection method, we incorporate shape information into the saliency model, both using color features to extract the potential target regions and shape information to eliminate the interference of similar regions. Consequently, the US-CMC has outstanding performance in terms of precision rate and recall rate under the conditions with shadows, view angle, and shape interference. Next, we plan to apply the Markov chain to higher-resolution remote sensing images and try to use shape information to absorb and reconstruct the super-pixel blocks in order to obtain better results.

Author Contributions: Z.L. designed and implemented the whole detection model and drafted the manuscript. D.Z. proposed the direction of research and designed the experiment. Z.S. and Z.J. reviewed and edited the manuscript.

Funding: This research was supported by the National Key R&D Program of China under Grant 2017YFC 1405600, the National Natural Science Foundation of China under Grant 61671037, and the Aeronautical Science Foundation of China under Grant 2017ZC51046.

Acknowledgments: The authors would like to thank the editors and anonymous reviewers for their valuable comments, which are very helpful in revising this paper.

Conflicts of Interest: The authors declare no conflict of interest.

References

1. Xu, H.; Chen, W.; Sun, B.; Chen, Y.; Li, C. Oil tank detection in synthetic aperture radar images based on quasi-circular shadow and highlighting arcs. *J. Appl. Remote Sens.* **2014**, *8*, 397–398. [CrossRef]
2. Zhang, W.; Hong, Z.; Chao, W.; Tao, W. Automatic oil tank detection algorithm based on remote sensing image fusion. In Proceedings of the IEEE International Geoscience and Remote Sensing Symposium (IGARSS), Seoul, Korea, 25–29 July 2005; pp. 3956–3958.
3. Atherton, T.J.; Kerbyson, D.J. Size invariant circle detection. *Image Vis. Comput.* **1999**, *17*, 795–803. [CrossRef]

4. Li, B.; Yin, D.; Yuan, X.; Li, G. Oilcan recognition method based on improved Hough transform. *Opto-Electron. Eng.* **2008**, *35*, 30–44.

5. Han, X.; Fu, Y.; Li, G. Oil depots recognition based on improved Hough transform and graph search. *J. Electron. Inf. Technol.* **2011**, *33*, 66–72. [CrossRef]

6. Zhu, C.; Liu, B.; Zhou, Y.; Yu, Q.; Liu, X.; Yu, W. Framework design and implementation for oil tank detection in optical satellite imagery. In Proceedings of the IEEE International Geoscience and Remote Sensing Symposium (IGARSS), Munich, Germany, 22–27 July 2012; pp. 6016–6019.

7. Ok, A.O.; Baseski, E. Circular oil tank detection from panchromatic satellite images: A new automated approach. *IEEE Geosci. Remote Sens. Lett.* **2015**, *12*, 1347–1351. [CrossRef]

8. Cai, X.; Sui, H.; Lv, R.; Song, Z. Automatic circular oil tank detection in high-resolution optical image based on visual saliency and hough transform. In Proceedings of the 2014 IEEE Workshop on Electronics, Computer and Applications(IWECA), Ottawa, ON, Canada, 8–9 May 2014; pp. 408–411.

9. Yao, Y.; Jiang, Z.; Zhang, H. Oil tank detection based on salient region and geometric features. In Proceedings of the SPIE/COS Photonics Asia, Beijing, China, 9 October 2014.

10. Zhang, L.; Shi, Z.; Wu, J. A hierarchical oil tank detector with deep surrounding features for high-resolution optical satellite imagery. *IEEE J. Sel. Top. Appl. Earth Obs. Remote Sens.* **2015**, *8*, 4895–4909. [CrossRef]

11. Wang, Q.; Zhang, J.; Hu, X.; Wang, Y. Automatic Detection and Classification of Oil Tanks in Optical Satellite Images Based on Convolutional Neural Network. In Proceedings of the International Conference on Image and Signal Processing(ICISP), Trois-Rivières, QC, Canada, 30 May–1 June 2016; pp. 304–313.

12. Cheng, G.; Zhou, P.; Han, J. Learning rotation-invariant convolutional neural networks for object detection in vhr optical remote sensing images. *IEEE Trans. Geosci. Remote Sens.* **2016**, *54*, 7405–7415. [CrossRef]

13. Duda, R.O.; Hart, P.E. Use of the Hough transformation to detect lines and curves in pictures. *Commun. ACM* **1972**, *15*, 11–15. [CrossRef]

14. Zhang, L.; Wang, S.; Liu, C.; Wang, Y. Saliency-Driven Oil Tank Detection Based on Multidimensional Feature Vector Clustering for SAR Images. *IEEE Geosci. Remote Sens. Lett.* **2014**, 1–5. [CrossRef]

15. Itti, L.; Koch, C.; Niebur, E. A model of saliency-based visual attention for rapid scene analysis. *IEEE Trans. Pattern Anal. Mach. Intell.* **1998**, *20*, 1254–1259. [CrossRef]

16. Hou, X. Zhang, L. Saliency detection: A spectral residual approach. In Proceedings of the IEEE Conference on Computer Vision and Pattern Recognition (CVPR), Minneapolis, MN, USA, 18–23 June 2007.

17. Achanta, R.; Hemami, S.; Estrada, F.; Susstrunk, S. Frequency-tuned salient region detection. In Proceedings of the IEEE Conference on Computer Vision and Pattern Recognition (CVPR), Miami, FL, USA, 20–25 June 2009; pp. 1597–1604.

18. Goferman, S.; Zelnik-Manor, L.; Tal, A. Context-aware saliency detection. In Proceedings of the IEEE Conference on Computer Vision and Pattern Recognition (CVPR), San Francisco, CA, USA, 13–18 June 2010; pp. 2376–2383.

19. Cheng, M.; Zhang, G.; Mitra, N.J.; Huang, X.; Hu, S. Global contrast based salient region detection. In Proceedings of the IEEE Conference on Computer Vision and Pattern Recognition (CVPR), Colorado Springs, CO, USA, 20–25 June 2011; pp. 409–416.

20. Perazzi, F.; Krahenbuhl, P.; Pritch, Y.; Hornung, A. Saliency filters: Contrast based filtering for salient region detection. In Proceedings of the IEEE Conference on Computer Vision and Pattern Recognition (CVPR), Providence, RI, USA, 16–21 June 2012; pp. 733–740.

21. Lu, H.; Li, X.; Zhang, L.; Ruan, X.; Yang, M.H. Dense and sparse reconstruction error based saliency descriptor. *IEEE Trans. Image Process.* **2016**, *25*, 1592–1603. [CrossRef] [PubMed]

22. Zhu, W.; Liang, S.; Wei, Y.; Sun, J. Saliency optimization from robust background detection. In Proceedings of the IEEE Conference on Computer Vision and Pattern Recognition (CVPR), Columbus, OH, USA, 23–28 June 2014; pp. 2814–2821.

23. Yang, C.; Zhang, L.; Lu, H.; Ruan, X.; Yang, M.H. Saliency Detection via Graph-Based Manifold Ranking. In Proceedings of the IEEE International Conference on Computer Vision and Pattern Recognition (CVPR), Portland, OR, USA, 23–28 June 2013; pp. 3166–3173.

24. Wei, Y.; Wen, F.; Zhu, W.; Sun, J. Geodesic Saliency Using Background Priors. In Proceedings of the European Conference on Computer Vision (ECCV), Florence, Italy, 7–13 October 2012; pp. 29–42.

25. Tu, W.; He, S.; Yang, Q.; Chien, S. Real-Time Salient Object Detection with a Minimum Spanning Tree. In Proceedings of the IEEE Conference on Computer Vision and Pattern Recognition (CVPR), Las Vegas, NV, USA, 27–30 June 2016; pp. 2334–2342.

26. Tong, N.; Lu, H.; Ruan, X.; Yang, M. Salient Object Detection via Bootstrap Learning. In Proceedings of the IEEE Conference on Computer Vision and Pattern Recognition (CVPR), Boston, MA, USA, 7–12 June 2015; pp. 1884–1892.

27. Qin, Y.; Lu, H.; Xu, Y.; Wang, H. Saliency Detection via Cellular Automata. In Proceedings of the IEEE Conference on Computer Vision and Pattern Recognition (CVPR), Boston, MA, USA, 7–12 June 2015; pp. 110–119.

28. Lee, G.; Tai, Y.; Kim, J. Deep Saliency with Encoded Low Level Distance Map and High Level Features. In Proceedings of the IEEE Conference on Computer Vision and Pattern Recognition (CVPR), Las Vegas, NV, USA, 27–30 June 2016; pp. 660–668.

29. Wang, T.; Zhang, L.; Wang, S.; Lu, H.; Yang, G.; Ruan, X.; Borji, A. Detect Globally, Refine Locally: A Novel Approach to Saliency Detection. In Proceedings of the IEEE Conference on Computer Vision and Pattern Recognition (CVPR), Salt Lake City, UT, USA, 18–22 June 2018; pp. 4321–4329.

30. Achanta, R.; Shaji, A.; Smith, K.; Lucchi, A.; Fua, P.; Süsstrunk, S. SLIC superpixels compared to state-of-the-art superpixel methods. *IEEE Trans. Pattern Anal. Mach. Intell.* **2012**, *34*, 2274–2282. [CrossRef] [PubMed]

31. Tang, M.; Gorelick, L.; Veksler, O.; Boykov, Y. GrabCut in One Cut. In Proceedings of the IEEE International Conference on Computer Vision (ICCV), Sydney, Australia, 1–8 December 2013; 1769–1776.

32. Cheng, M.; Warrell, J.; Lin, W.; Zheng, S.; Vineet, V.; Crook, N. Efficient Salient Region Detection with Soft Image Abstraction. In Proceedings of the International Conference on Computer Vision (ICCV), Sydney, Australia, 1–8 December 2013; pp. 1529–1536.

33. Yan, Q.; Xu, L.; Shi, J.; Jia, J. Hierarchical Saliency Detection. In Proceedings of the IEEE Conference on Computer Vision and Pattern Recognition (CVPR), Portland, OR, USA, 23–28 June 2013; pp. 1155–1162.

34. Yuan, Y.; Li, C.; Kim, J.; Cai, W.; Feng, D.D. Reversion Correction and Regularized Random Walk Ranking for Saliency Detection. *IEEE Trans. Image Process.* **2018**, *27*, 1311–1322. [CrossRef] [PubMed]

35. Sun, J.; Lu, H.; Liu, X. Saliency Region Detection Based on Markov Absorption Probabilities. *IEEE Trans. Image Process.* **2015**, *24*, 1639–1649. [CrossRef] [PubMed]

36. Li, H.; Lu, H.; Lin, Z.; Shen, X.; Price, B. Inner and Inter Label Propagation: Salient Object Detection in the Wild. *IEEE Trans. Image Process.* **2015**, *24*, 3176–3186. [CrossRef] [PubMed]

Article

Affine-Function Transformation-Based Object Matching for Vehicle Detection from Unmanned Aerial Vehicle Imagery

Shuang Cao [1,†], Yongtao Yu [2,†], Haiyan Guan [1,*], Daifeng Peng [1] and Wanqian Yan [3]

[1] School of Remote Sensing & Geomatics Engineering, Nanjing University of Information Science & Technology, Nanjing 210044, China

[2] Faculty of Computer and Software Engineering, Huaiyin Institute of Technology, Huaian 223003, China

[3] School of Geographical Science, Nanjing University of Information Science & Technology, Nanjing 210044, China

[*] Correspondence: guanhy.nj@nuist.edu.cn; Tel.: +86-25-5873-1152

[†] S.C. and Y.Y. contributed equally to this paper.

Received: 10 June 2019; Accepted: 17 July 2019; Published: 19 July 2019

Abstract: Vehicle detection from remote sensing images plays a significant role in transportation related applications. However, the scale variations, orientation variations, illumination variations, and partial occlusions of vehicles, as well as the image qualities, bring great challenges for accurate vehicle detection. In this paper, we present an affine-function transformation-based object matching framework for vehicle detection from unmanned aerial vehicle (UAV) images. First, meaningful and non-redundant patches are generated through a superpixel segmentation strategy. Then, the affine-function transformation-based object matching framework is applied to a vehicle template and each of the patches for vehicle existence estimation. Finally, vehicles are detected and located after matching cost thresholding, vehicle location estimation, and multiple response elimination. Quantitative evaluations on two UAV image datasets show that the proposed method achieves an average completeness, correctness, quality, and F1-measure of 0.909, 0.969, 0.883, and 0.938, respectively. Comparative studies also demonstrate that the proposed method achieves compatible performance with the Faster R-CNN and outperforms the other eight existing methods in accurately detecting vehicles of various conditions.

Keywords: vehicle detection; object matching; superpixel segmentation; unmanned aerial vehicle; remote sensing imagery

---

1. Introduction

Periodically and effectively monitoring traffic conditions is greatly important for transportation management department to conduct traffic controls and make road plans. Accurate traffic monitoring can help to avoid potential traffic disasters and alleviate traffic congestions. Traditionally, traffic monitoring is basically performed through on-site surveillances of traffic police or using traffic cameras installed along roads. To monitor the traffic condition over a large area, the monitoring data from different observation sites should be collected manually or digitally and further merged to carry out post analysis. Therefore, such means are labor-intensive and inefficient to some extent. With the advent and rapid advance of remote sensing techniques, the acquisition of high-resolution and rich-detail remote sensing images can be easily and quickly accomplished using satellite sensors and unmanned aerial vehicles (UAV). Satellite images have a large perspective and can cover an extensive area of interest, as well as collecting a series of data over a long period of time [1]. Comparatively, benefiting from high portability, low-cost platform, and flying flexibility, UAV systems can quickly reach the surveillance area and capture images with different levels of details [2]. Thus, due to the advantages of high convenience, low cost, and abundant information, remote sensing sensors and their resultant images have been applied to various traffic-related applications. Consequently, extensive studies have also been conducted for information extraction and interpretation from remote sensing images, such as road segmentation [3,4], road feature extraction [5,6], vehicle detection [7,8], and traffic monitoring [9,10].

Among the wide range of traffic-related applications, vehicle detection plays a significant role in intelligent transportation and has attracted increasing attention in recent years. The vehicle detection results can be used for controlling traffic flows, planning road networks, estimating parking situations, tracking specific targets, and analyzing economic levels of cities and living standards of citizens. Consequently, a great effort has been paid for vehicle detection using remote sensing images and a great number of achievements have been made in the literature. The existing approaches for vehicle detection from remote sensing images can be simply categorized into implicit model-based methods [11,12] and explicit model-based methods [13,14]. Implicit model-based methods typically characterize intensity or texture features in the vicinity of individual pixels or pixel clusters. The detection of vehicles is performed by evaluating the features surrounding the target region. In contrast, explicit model-based methods usually depict a vehicle using a box, a wireframe representation, or a morphological model. The detection of vehicles is performed by a top-down matching scheme or a classification-oriented strategy. However, automated and accurate detection and localization of vehicles from remote sensing images is still facing great challenges because of orientation variations, within-class dissimilarities and between-class similarities in texture and geometry, partial occlusions caused by trees and buildings, and illumination condition variations.

To explore distinct feature representations of vehicles or its local parts towards vehicle detection, a great number of strategies have been proposed in the literature. Niu [15] developed a semi-automatic framework to detect vehicles based on a geometric deformable model. By minimizing the objective function that connects the optimization problem with the propagation of regular curves, the geometric deformable model obtained a promising vehicle detection rate. Kembhavi et al. [11] combined the histograms of oriented gradients (HOG) features, color probability maps, and pairs of pixels to capture the statistical and structural features of vehicles and their surroundings. Vehicle detection was performed through partial least squares regression. To achieve invariant feature characterization, Bag-of-Words model was explored and used by Zhou et al. [8] to detect vehicles. In this method, local steering kernel descriptor and orientation aware scanning were introduced to localize vehicle positions in the image. Similarly, orientation aware vehicle detection was also designed by Zhou et al. [16]. Wan et al. [17] presented a cascaded vehicle detection framework consisting of affine invariant interest point detection, bag-of-words feature encoding, and large-margin dimensionality reduction. Xu et al. [18] proposed to detect vehicles using a hybrid scheme integrating the Viola-Jones and linear support vector machines (SVM) with HOG features. Later on, to solve the sensitivity of Viola-Jones to in-plane rotations of objects, an enhanced version of Viola-Jones through road orientation adjustment was presented by Xu et al. [19] for vehicle detection. A segment-before-detect pipeline was suggested by Audebert et al. [20] to detect vehicles through semantic segmentation of images. In this method, a semantic map was constructed to segment vehicle instances by extracting connected components. By using integral channel features in a soft-cascade structure, Liu and Mattyus [21] designed a fast binary detector to conduct vehicle detection. The output of the binary detector was further fed into a multiclass classifier for orientation and type analysis. Recently, disparity maps [22], hard example mining [23], catalog-based approach [24], and expert features [25] have also been studied for vehicle detection from remote sensing images.

To tackle occlusions and complicated scenarios towards accurate vehicle detection, machine learning based methods and classification-based methods have been intensively exploited in recent years. Generally, such methods use extracted features to train different classifiers, which convert the vehicle detection task into a binary classification problem. Cao et al. [26] proposed to detect vehicles using exemplar-SVMs classifiers with a hard negative example selection scheme. The features used for training the classifiers were extracted through a deep convolutional neural network. To handle the difficulty of labelling sufficient training instances, weakly supervised, multi-instance discriminative learning and transfer learning were also explored by Cao et al. [7,27]. In their implementations, weakly labelled instances and across domain samples were selected for SVM classifiers training. Similarly, SVM classifier trained with deep features was also adopted by Ammour et al. [28] to detect vehicles. Sparse representation was introduced to assist high-performance classifier construction towards vehicle detection [13,29]. The feature encoded dictionaries created through sparse representation were applied to distinct training sample selection. Considering both local and global structures of vehicles, Zhang et al. [30] trained a part detector and a root detector using front windshield samples and entire vehicle samples, respectively. The root detector localized a potential vehicle candidate, while the part detector scanned within the candidate to remove false alarms. To well handle illumination, rotation, and scale variations, Bazi and Melgani [31] designed a convolutional SVM network. The convolutional SVM network was constructed based on a set of alternating convolutional and reduction layers that were terminated by a linear SVM classification layer. Elmikaty and Stathaki [32] proposed a combination of two subsystems, namely window-evaluation and window-classification systems, to achieve robust detection of vehicles. The window-evaluation subsystem used a Gaussian-mixture-model classifier to extract regions of interest, whereas the window-classification subsystem adopted an SVM classifier to distinguish descriptors related to vehicles. In addition, multi-source data fusion strategies have also been explored and applied to vehicle detection recently [33,34].

Deep learning techniques [35–37] have shown their superior advantages in mining hierarchical, high-level, distinctive feature representations. They have been widely used in a variety of applications, such as image segmentation [38,39], object detection [40,41], classification [42,43], image registration [44], etc. Consequently, vehicle detection by using deep learning techniques has also been intensively studied [45]. Mou and Zhu [46] proposed a semantic boundary-aware multitask learning network to detect and segment vehicle instances. In this method, through residual learning, a fully convolutional network was constructed to encode multilevel contextual features. To effectively generate and select representative training samples, Wu et al. [47] presented a superpixel segmentation and convolutional neural network (CNN) iteration strategy. Patches were generated based on the centers of segmented superpixels. The CNN used as a feature extractor was iteratively refined through a training sample iterative selection strategy. Tang et al. [48] combined region convolutional neural networks (R-CNNs) and hard negative example mining to improve vehicle detection performance. To accurately extract vehicle-like targets, a hyper region proposal network was constructed with a combination of hierarchical feature maps. Similarly, Deng et al. [49] adopted coupled R-CNNs to detect vehicles. Schilling et al. [50] designed a multi-branch CNN model containing two CNN branches, respectively, for vehicle detection and segmentation purposes. Zhong et al. [51] constructed a cascaded CNN model consisting of two independent CNNs. The first CNN was applied to generate vehicle-like regions from multi-feature maps, whereas the second CNN functioned to extract features and make decisions. To solve the problem of vehicle scale variations and the production limitation of training samples, Yang et al. [52] suggested using a multi-perspective CNN that was trained with different initial receptive fields. Utilizing a regression-based CNN model, Tang et al. [53] proposed an oriented single shot multi-box detector aiming at detecting vehicles with arbitrary orientations. On the whole, deep learning techniques have achieved plentiful breakthroughs on vehicle detection tasks. However, the performance of the deep learning-based methods suffered greatly from the sufficient number of labelled training samples and the rational selection of representative training samples.

In this paper, we propose an affine-function transformation-based object matching framework for vehicle detection from UAV images. The proposed method can effectively deal with vehicles with varying conditions: such as scale variations, orientation variations, shadows, and partial occlusions. The contributions of this paper include: (1) an affine-function transformation-based object matching framework is designed for vehicle detection; (2) a successive convexification scheme is proposed to obtain tight transformation parameters. For a test image, first, superpixel segmentation strategy is adopted to generate meaningful and non-redundant patches. Then, object matching is carried out between a vehicle template and each of the patches. Finally, after matching cost thresholding, vehicle location estimation, and multiple detection results elimination, vehicles are detected and located in the image.

The remainder of this paper is organized as follows. Section 2 details the affine-function transformation-based object matching framework and the methodology for vehicle detection. Section 3 reports and discusses the experimental results. Finally, Section 4 gives the concluding remarks.

2. Methodology

A detailed vehicle detection workflow is illustrated in Figure 1. As shown in Figure 1, for a test image, we first over-segment it into a group of superpixels using the simple linear iterative clustering (SLIC) superpixel segmentation method [54]. Then, centered at each superpixel, a patch is generated with a size of $n_p \times n_p$ pixels. Then, to estimate the existence of vehicles from these patches, we proposed an affine-function transformation-based object matching method, in which both the template and each of the patches, a collection of scale-invariant feature transform (SIFT) feature points are generated and characterized with SIFT feature vectors, and then a vehicle template is selected for conducting matching between the template and each of the generated patches. Compared to traditional methods that usually adopt a sliding window strategy to generate a group of candidate regions for individual vehicle detection [8], we, in this paper, the SLIC superpixel segmentation method to generate meaningful and non-redundant patches as operating units for individual vehicle detection. The SLIC superpixel segmentation method is detailed in the literature [55]. In the following subsections, we focus on the description of the affine-function transformation-based object matching framework, followed by an optimal matching processing by using a successive convexification scheme in Section 2.2.

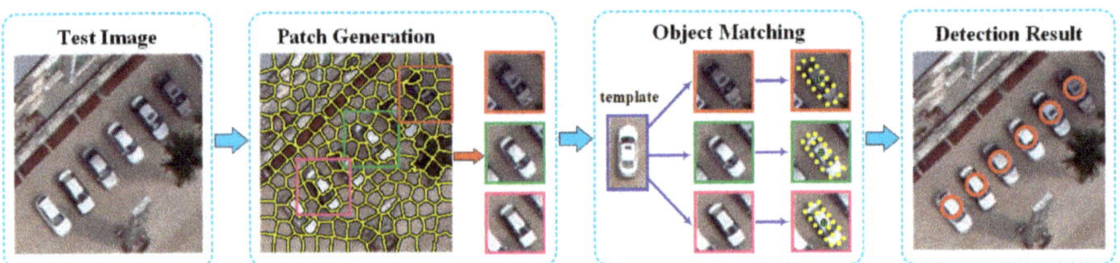

Figure 1. Illustration of the vehicle detection workflow using the proposed affine-function transformation-based object matching framework.

2.1. Affine-Function Transformation Based Object Matching

The problem of object matching can be defined as matching a group of template feature points, representing a specific object of interest, to another group of scene feature points, representing a scene containing an instance of the object of interest (See Figure 2). Each feature point has a unique location and is depicted with a feature vector that characterizes the local appearance around that location. The matched scene feature points should preserve similar local features and relative spatial relationships of the template feature points. Most of existing object matching techniques dedicate to seek for point-to-point matching results, which might show low performance when dealing with occlusions. In contrast, we propose an affine-function transformation-based object matching framework, whose objective is to determine each template feature point's optimal transformation parameters (not point-to-point matching) so that the matching location (which may not be a specific scene feature point) of each template feature point is close to a scene feature point with similar local appearance and geometric structure.

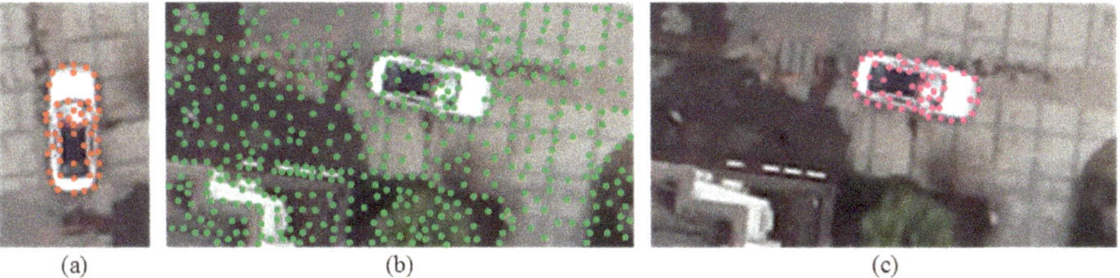

(a)   (b)   (c)

Figure 2. Illustration of object matching. (**a**) A group of template feature points representing a vehicle, (**b**) a group of scene feature points representing a scene containing a vehicle instance, and (**c**) the matched scene feature points.

(1) Affine-function transformation

Denote n_t and n_s as the numbers of template feature points and scene feature points, respectively. Let $\mathbf{P} = \{p_i = [x_{p_i}, y_{p_i}]^{\mathrm{T}} | i = 1, 2, \ldots, n_t\}$ and $\mathbf{Q} = \{q_j = [x_{q_j}, y_{q_j}]^{\mathrm{T}} | j = 1, 2, \ldots, n_s\}$ be the sets of template feature points and scene feature points, respectively. Then, our object matching objective is to optimize the transformation parameters of each template feature point in $\mathbf{P}$ based on the scene feature points in $\mathbf{Q}$. Define $T_i(\mathbf{\Phi}_i) : \mathbf{R}^n \to \mathbf{R}^2, i = 1, 2, \ldots, n_t$ as an affine transformation function that transforms the ith template feature point p_i into a location in the scene with transformation parameters $\mathbf{\Phi}_i \in \mathbf{R}^n$. The result of $T_i(\mathbf{\Phi}_i)$ is the corresponding matching location of template feature point p_i in the scene. In this paper, we define the affine transformation function as follows:

$$T_i(\mathbf{\Phi}_i) = \begin{bmatrix} \alpha & \beta & \phi \\ \gamma & \delta & \varphi \end{bmatrix} \begin{bmatrix} x_{p_i} \\ y_{p_i} \\ 1 \end{bmatrix} + \begin{bmatrix} \xi_i \\ \vartheta_i \end{bmatrix} \tag{1}$$

where $T_i(\mathbf{\Phi}_i) : \mathbf{R}^8 \to \mathbf{R}^2$ computes the matching location of template feature point p_i under an affine transformation with parameters $\mathbf{\Phi}_i = [\alpha, \beta, \gamma, \delta, \phi, \varphi, \xi_i, \vartheta_i]^{\mathrm{T}} \in \mathbf{R}^8$. We define a separate affine transformation function for each template feature point. The matching location of a template feature point p_i is computed by its corresponding function $T_i(\mathbf{\Phi}_i)$. In Equation (1), $[\alpha, \beta, \gamma, \delta, \phi, \varphi]^{\mathrm{T}}$ are the global affine transformation parameters that are shared by all template feature points, whereas $[\xi_i, \vartheta_i]^{\mathrm{T}}$ are the local translation parameters for only template feature point p_i. Therefore, different template feature points might have different versions of $[\xi_i, \vartheta_i]^{\mathrm{T}}$. As illustrated in Figure 3, the local translation parameters allow small local deformations between the template feature points and their matched locations.

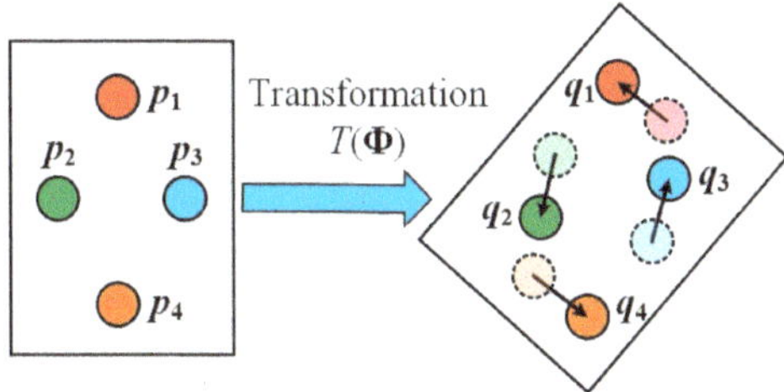

Figure 3. Illustration of the affine transformation model. *q1* to *q4* are, respectively, the corresponding matched locations of *p1* to *p4* after applying the affine transformation function.

(2) Dissimilarity measure

According to the object matching principles, one objective is to match each template feature point p_i to the corresponding location $T_i(\mathbf{\Phi}_i)$ in the scene with the constraint that the local appearances of p_i and $T_i(\mathbf{\Phi}_i)$ should be similar. Therefore, we define a dissimilarity measure function $diss_i(q) : \mathbf{R}^2 \to \mathbf{R}^1, i = 1, 2, \ldots, n_t$, respectively, for each template feature point to measure the local appearance dissimilarities between template feature point p_i and its corresponding matched location q in the scene. Generally, two feature points having similar local appearances will result in a low dissimilarity measure value.

To solve the object matching problem, our overall objective is to determine the optimal transformation parameters $\mathbf{\Phi}_1, \mathbf{\Phi}_2, \ldots, \mathbf{\Phi}_{n_t}$ for template feature points $p_1, p_2, \ldots, p_{n_t}$ to minimize the following objective function:

$$
\begin{aligned}
&\underset{\mathbf{\Phi}_1, \mathbf{\Phi}_2, \ldots, \mathbf{\Phi}_{n_t}}{\text{minimize}} \sum_{i=1}^{n_t} diss_i(T_i(\mathbf{\Phi}_i)) + R(\mathbf{\Phi}_1, \mathbf{\Phi}_2, \ldots, \mathbf{\Phi}_{n_t}), \\
&\text{subject to } C_j(\mathbf{\Phi}_1, \mathbf{\Phi}_2, \ldots, \mathbf{\Phi}_{n_t}) \leq 0, j = 1, 2, \ldots, n_c
\end{aligned}
\tag{2}
$$

where $diss_i(T_i(\mathbf{\Phi}_i))$ computes the local appearance dissimilarity between template feature point p_i and its corresponding matching location $T_i(\mathbf{\Phi}_i)$ in the scene. $R(\mathbf{\Phi}_1, \mathbf{\Phi}_2, \ldots, \mathbf{\Phi}_{n_t})$ denotes a convex relaxation term regularizing the transformation parameters. $C_j(\mathbf{\Phi}_1, \mathbf{\Phi}_2, \ldots, \mathbf{\Phi}_{n_t}) \leq 0, j = 1, 2, \ldots, n_c$ defines a series of convex constraints. Here, n_c is the number of convex constraints. By such a definition, the overall objective function in Equation (2) can be effectively solved through convex optimization techniques. Next, we focus on the design of the dissimilarity measure function.

Recall that each feature point is associated with a location, as well as a feature vector characterizing the local appearance around that location. In this paper, each feature point is described using a scale-invariant feature transform (SIFT) vector [56]. Let $Cost_{i,j}, i = 1, 2, \ldots, n_t, j = 1, 2, \ldots, n_s$ denote the feature dissimilarity between a template feature point p_i and a scene feature point q_j. Then, we define $Cost_{i,j}$ as the square root of the χ^2 distance [57] between the SIFT feature vectors of p_i and q_j as follows:

$$
Cost_{i,j} = \sqrt{\sum_k \frac{\left(F_{p_i}^k - F_{q_j}^k\right)^2}{F_{p_i}^k + F_{q_j}^k}}
\tag{3}
$$

where $F_{p_i}^k$ and $F_{q_j}^k$ are the kth channels of the SIFT feature vectors of feature points p_i and q_j, respectively. Then, for each template feature point $p_i, i = 1, 2, \ldots, n_t$, we define a discrete version of the dissimilarity measure function $Diss_i(q_j) : \mathbf{Q} \to \mathbf{R}^1, i = 1, 2, \ldots, n_t$ as follows:

$$
Diss_i(q_j) = Cost_{i,j}, j = 1, 2, \ldots, n_s, q_j \in \mathbf{Q}
\tag{4}
$$

The domain of this function indicates that a template feature point p_i can be only matched to a certain scene feature point q_j with the feature dissimilarity measure determined by function $Diss_i(q_j)$. Minimizing $Diss_i(q_j)$ still results in a point-to-point matching pattern, which violates our objective to optimize the affine transformation parameters to compute the matching locations. Moreover, the discrete function $Diss_i(q_j)$ is non-convex. Therefore, adopting $Diss_i(q_j)$ as the dissimilarity measure in Equation (2) to minimize the overall objective function is difficult and cannot effectively obtain optimal solutions.

(3) Convex dissimilarity measure

To solve the aforementioned problem, we relax each discrete function $Diss_i(q_j)$ and construct a continuous and convex dissimilarity measure function $diss_i(q)$, which can be effectively optimized through convex optimization techniques. To this end, for each template feature point $p_i, i = 1, 2, \ldots, n_t$, we organize all the scene feature points together with their feature dissimilarities $Cost_{i,j}, j = 1, 2, \ldots, n_s$ as a set of three-dimensional (3D) points $\{[x_{q_j}, y_{q_j}, Cost_{i,j}]^T | j = 1, 2, \ldots, n_s\}$, whose first two dimensions are the location of a scene feature point and the third dimension is the corresponding feature dissimilarity. As illustrated in Figure 4, we give an example of the feature dissimilarities, viewed as a 3D point set, of the scene feature points associated with a template feature point. Obviously, this is actually the discrete version of the dissimilarity measure function $Diss_i(q_j)$.

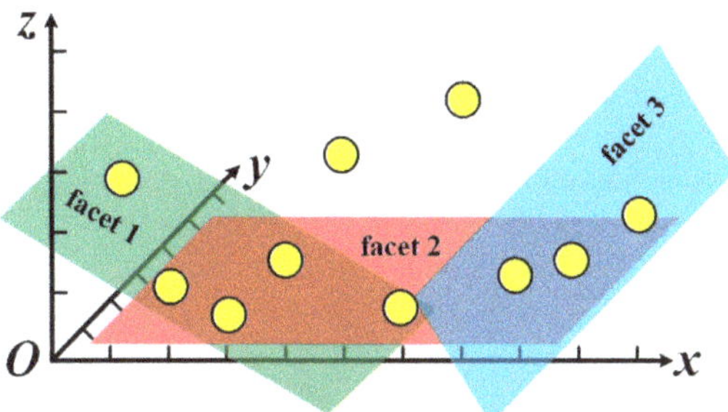

Figure 4. Illustration of the feature dissimilarity measures viewed as a 3D point set and the constructed convex dissimilarity measure function (facets).

We construct the convex dissimilarity measure function $diss_i(q)$ based on the lower convex hull of the 3D point set associated with template feature point p_i with respect to the feature dissimilarity dimension. As shown in Figure 3, the facets are the lower convex hull of the 3D point set. Denote $\{z = a_k x + b_k y + c_k | k = 1, 2, \ldots, n_f\}$ as the plane functions defining the n_f facets on the lower convex hull. $[a_k, b_k, c_k]^T$ are the plane parameters of the kth plane. Then, we define the continuous convex dissimilarity measure function as follows:

$$diss_i\left([x, y]^T\right) = \max_k (a_k x + b_k y + c_k), k = 1, 2, \ldots, n_f \tag{5}$$

where $[x, y]^T$ can be any location in the scene domain. In other words, by such a relaxation, template feature point p_i can be matched to any location $[x, y]^T$ in the scene, not necessarily being a specific scene feature point. To effectively minimize Equation (5), we convert it into an equivalent linear programming problem:

$$\underset{x,y}{\text{minimize}}\, diss_i\left([x, y]^T\right) \Leftrightarrow \begin{array}{l} \underset{x,y,u_i}{\text{minimize}}\, u_i \\ \text{subject to } a_k x + b_k y + c_k \leq u_i, k = 1, 2, \ldots, n_f \end{array} \tag{6}$$

where u_i is an auxiliary variable representing the upper bound of $diss_i\left([x, y]^T\right)$. Equation (6) can be efficiently optimized using convex optimization techniques.

In order to use Equation (6) to minimize $diss_i(T_i(\Phi_i))$ in the overall objective function in Equation (2), we rewrite the affine transformation function $T_i(\Phi_i)$ into $T_i(\Phi_i) = [f_i(\Phi_i), g_i(\Phi_i)]^T$, where $f_i(\Phi_i) = \alpha x_{p_i} + \beta y_{p_i} + \phi + \xi_i$ and $g_i(\Phi_i) = \gamma x_{p_i} + \delta y_{p_i} + \varphi + \vartheta_i$ are affine functions that computes the x and y components of the matching location of template feature point p_i. By substituting x and y in Equation (6) with $f_i(\Phi_i)$ and $g_i(\Phi_i)$, we obtain the following convex optimization model which is equivalent to minimizing $diss_i(T_i(\Phi_i))$ with respect to transformation parameters Φ_i:

$$\underset{\Phi_i}{\text{minimize}}\, diss_i(T_i(\Phi_i)) \Leftrightarrow \begin{array}{l} \underset{\Phi_i,u_i}{\text{minimize}}\, u_i \\ \text{subject to } a_k f_i(\Phi_i) + b_k g_i(\Phi_i) + c_k \leq u_i, k = 1, 2, \ldots, n_f \end{array} \tag{7}$$

Then, summing up all the minimization terms $diss_i(T_i(\mathbf{\Phi}_i)), i = 1, 2, \ldots, n_t$ results in our overall objective function with respect to optimizing the affine transformation parameters $\mathbf{\Phi}_1, \mathbf{\Phi}_2, \ldots, \mathbf{\Phi}_{n_t}$ with convex constraints defined in Equation (7):

$$\underset{\mathbf{\Phi}_1,\mathbf{\Phi}_2,\ldots,\mathbf{\Phi}_{n_t}}{\text{minimize}} \sum_{i=1}^{n_t} diss_i(T_i(\mathbf{\Phi}_i)) + \lambda \sum_{i=1}^{n_t} \left\| \begin{bmatrix} \xi_i \\ \vartheta_i \end{bmatrix} \right\|_2^2 \tag{8}$$

where the regularization term functions to penalize local deformations of the matching locations in the scene. It indicates that the local deformations of the matching locations should not be too large. λ is a parameter that weights the dissimilarity measure term and the regularization term.

When partial occlusions of an object of interest exist in the scene, directly optimizing Equation (8) may degrade the performance of the proposed affine-function transformation-based object matching framework. To solve this problem, we assign a weight factor $w_i, i = 1, 2, \ldots, n_t$ for each template feature point $\boldsymbol{p}_i$ to describe its distinctiveness and contribution to the matching. Then, we obtain the final overall objective function with convex constraints defined in Equation (7) as follows:

$$\underset{\mathbf{\Phi}_1,\mathbf{\Phi}_2,\ldots,\mathbf{\Phi}_{n_t}}{\text{minimize}} \sum_{i=1}^{n_t} w_i \cdot diss_i(T_i(\mathbf{\Phi}_i)) + \lambda \sum_{i=1}^{n_t} \left\| \begin{bmatrix} \xi_i \\ \vartheta_i \end{bmatrix} \right\|_2^2 \tag{9}$$

2.2. Successive Convexification Scheme for Solving the Objective Function

Recall that the continuous convex dissimilarity measure function $diss_i([x, y]^{\mathrm{T}}) : \mathbf{R}^2 \to \mathbf{R}^1$ is constructed by relaxing the discrete dissimilarity measure function $Diss_i(q) : \mathbf{Q} \to \mathbf{R}^1$ based on the lower convex hull. If the feature descriptions of feature points are distinctive, the dissimilarity measures, computed using $Diss_i(q)$, between a template feature point and all the scene feature points differ significantly. Therefore, the lower convex hull relaxation provides a satisfactory lower bound to the discrete measure function $Diss_i(q)$. However, when features are not distinctive, the lower convex hull might not generate a very tight lower bound to $Diss_i(q)$. To solve this problem, we propose a successive convexification scheme, similar to that adopted by Jiang et al. [58], to iteratively optimize the overall objective function to obtain a tighter solution.

Initially, we assign an identical weight factor $w_i = 1, i = 1, 2, \ldots, n_t$ to all template feature points. In each iteration of the convexification, a trust region is defined for each template feature point. Only the scene feature points within the trust region can be used to construct the convex dissimilarity measure functions. In the first iteration, we fix the weight factors $w_i = 1, i = 1, 2, \ldots, n_t$ and define the entire scene as the trust region for each template feature point $\boldsymbol{p}_i$, as illustrated by $D_i^{(1)}$ in Figure 5a. That is, initially, all scene feature points are used to construct the convex dissimilarity measure functions. Then, these convex dissimilarity measure functions are applied to the overall objective function in Equation (9) to optimize the affine transformation parameters $\mathbf{\Phi}_1, \mathbf{\Phi}_2, \ldots, \mathbf{\Phi}_{n_t}$. The corresponding matching locations of template feature points are computed by $T_1(\mathbf{\Phi}_1), T_2(\mathbf{\Phi}_2), \ldots, T_{n_t}(\mathbf{\Phi}_{n_t})$. Afterwards, we adjust the weight factors $w_i, i = 1, 2, \ldots, n_t$ to deal with partial occlusions. If the dissimilarity measure value $diss_i(T_i(\mathbf{\Phi}_i))$ between template feature point $\boldsymbol{p}_i$ and its matching location $T_i(\mathbf{\Phi}_i)$ is high, w_i is decreased by Δw (i.e., $w_i = w_i - \Delta w$) to degrade the contribution of $\boldsymbol{p}_i$. Otherwise, if the dissimilarity measure value is low, w_i is increased by Δw (i.e., $w_i = w_i + \Delta w$) to upgrade the contribution of $\boldsymbol{p}_i$. In this way, the actual matching locations $T_i(\mathbf{\Phi}_i)$ occluded by other objects in the scene will be considered less to optimize the overall objective function.

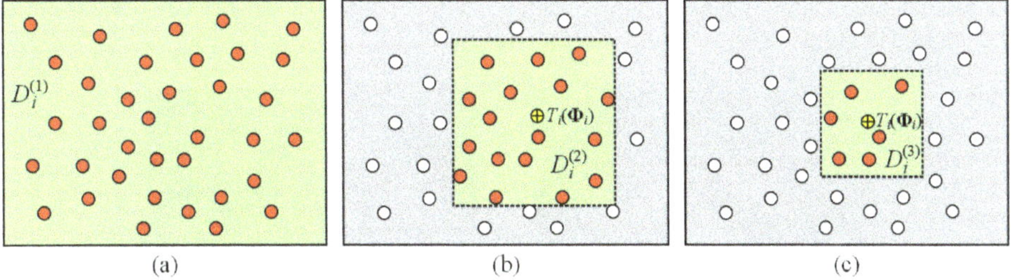

Figure 5. Illustration of the successive convexification scheme. (**a**) In the first iteration, all the scene feature points (red dots) are used to construct the convex dissimilarity measure function, (**b**) in the second iteration, only the scene feature points in the trust region (red dots) are used to construct the convex dissimilarity measure function, and (**c**) similar operations as those in the second iteration are performed in the latter iterations.

In the second iteration, we fix the weight factors $w_i, i = 1, 2, \ldots, n_t$ and define a shrunken trust region centered at the matching location $T_i(\Phi_i) = [f_i(\Phi_i), g_i(\Phi_i)]^T$ for each template feature point p_i. Only the scene feature points located within the trust region are used to construct the convex dissimilarity measure functions (See Figure 5b). Mathematically, the trust region of p_i in the second iteration is defined as follows:

$$D_i^{(2)} = \left\{ [x, y]^T \in \mathbf{R}^2 | f_i(\Phi_i) - \frac{L^{(2)}}{2} \le x \le f_i(\Phi_i) + \frac{L^{(2)}}{2}, g_i(\Phi_i) - \frac{L^{(2)}}{2} \le y \le g_i(\Phi_i) + \frac{L^{(2)}}{2} \right\} \qquad (10)$$

where $L^{(2)} = \min(H_s, W_s)/2$ is the side length of the trust region in the second iteration. Here, H_s and W_s are the height and width of the scene, respectively. Then, we apply the convex dissimilarity measure functions constructed using the scene feature points in trust regions $D_i^{(2)}, i = 1, 2, \ldots, n_t$ to optimize the overall objective function to obtain a set of tighter affine transformation parameters $\Phi_1, \Phi_2, \ldots, \Phi_{n_t}$. The tighter matching locations are represented by $T_1(\Phi_1), T_2(\Phi_2), \ldots, T_{n_t}(\Phi_{n_t})$. Afterwards, we adjust the weight factors $w_i, i = 1, 2, \ldots, n_t$ using the same principle as described in the first iteration.

The same optimization operations are performed in the subsequent iterations with smaller and smaller trust regions that consider fewer and fewer scene feature points (See Figure 5c). Specifically, in the kth iteration, the trust region is defined as follows:

$$D_i^{(k)} = \left\{ [x, y]^T \in \mathbf{R}^2 | f_i(\Phi_i) - \frac{L^{(k)}}{2} \le x \le f_i(\Phi_i) + \frac{L^{(k)}}{2}, g_i(\Phi_i) - \frac{L^{(k)}}{2} \le y \le g_i(\Phi_i) + \frac{L^{(k)}}{2} \right\} \qquad (11)$$

where $L^{(k)} = L^{(k-1)}/2$ is the side length of the trust region in the kth iteration. Generally, four iterations are enough. Through the proposed successive convexification scheme, we can obtain a tighter matching result with satisfactory consideration of handling partial occlusions.

After optimizing the overall objective function in Equation (9) through the proposed successive convexification scheme, we obtain two results: a set of affine transformation parameters $\Phi_1, \Phi_2, \ldots, \Phi_{n_t}$ and a matching cost (i.e., the value of the overall objective function). The corresponding matching locations in the patch can be computed by $T_1(\Phi_1), T_2(\Phi_2), \ldots, T_{n_t}(\Phi_{n_t})$, and the matching cost is used to estimate the existence of a vehicle in the patch. If the matching cost lies below a predefined threshold, we confirm that there is a vehicle instance in the patch. Then, as illustrated in Figure 6a, the location of the vehicle is estimated as the geometric centroid of the matching locations $T_1(\Phi_1), T_2(\Phi_2), \ldots, T_{n_t}(\Phi_{n_t})$. However, as shown in Figure 6b, a vehicle instance might exist in multiple patches by using the superpixel segmentation-based patch generation strategy. Consequently, multiple locations are estimated for a single vehicle instance. In fact, these locations associated with a vehicle instance exhibit a cluster form and are extremely close to each other. Thus, we further adopt a non-maximum suppression process [55] to eliminate the repetitive detection results. The final vehicle detection result is illustrated in Figure 5.

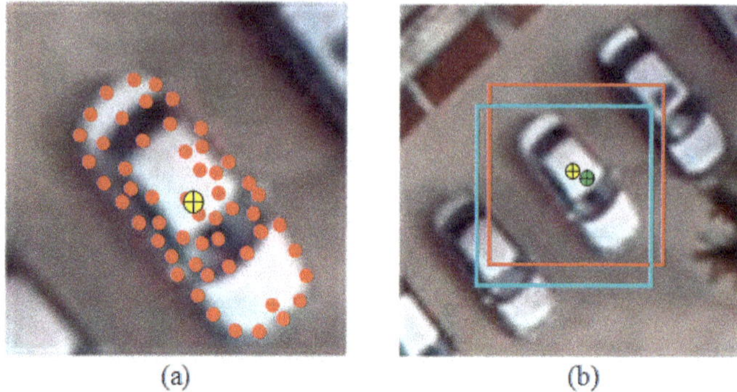

(a)   (b)

Figure 6. Illustrations of (**a**) the vehicle location is estimated as the centroid of the matching locations (yellow dot), and (**b**) a vehicle existing in two patches generates two locations.

3. Results and Discussions

3.1. Study Areas and Datasets

In this paper, we tested our proposed vehicle detection method on the UAV images. The UAV images used in this study were captured using the DJI Phantom 4 Pro UAV system (See Figure 7a). This is a quadrotor aircraft mounted with a one-inch high-resolution Exmor R CMOS image sensor. The maximum measuring frequency is 20 Hz. The detailed specification of the DJI Phantom 4 Pro UAV system is listed in Table 1. While surveying, we set the image capture mode to be "BURST mode" (a continuous shooting mode) with an image capture interval of 2 seconds. Therefore, a total number of 30 images with a size of 5472 × 3648 pixels were captured every minute.

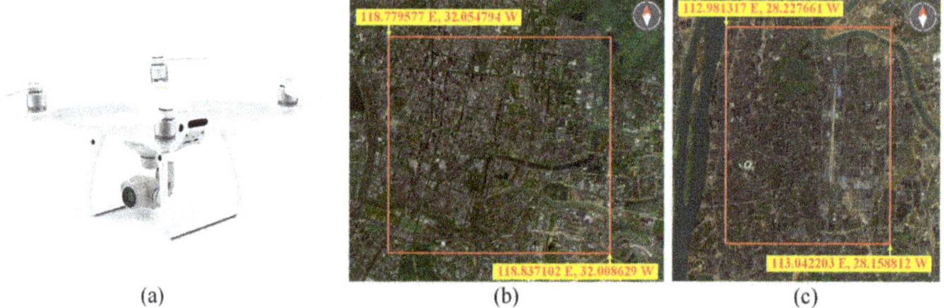

(a)   (b)   (c)

Figure 7. Illustrations of (**a**) the DJI Phantom 4 Pro Unmanned Aerial Vehicle (UAV) system, (**b**) the Nanjing study area, and (**c**) the Changsha study area.

In this study, two UAV image datasets were collected in two different urban areas for evaluating the proposed vehicle detection method. As shown in Figure 7b, the first dataset was collected in the urban area of City Nanjing, Jiangsu Province, China in 2016. The surveying area had a size of about 5.0×5.5 km^2. While surveying, the UAV system was flying at a height of about 150 m with a horizontal flight speed of about 15 m/s. A total number of 30,728 UAV images were collected to form the first dataset. As shown in Figure 7c, the second dataset was collected in the urban area of City Changsha, Hunan Province, China, in 2017. The surveying area had a size of about 7.5×6.0 km^2. While surveying, the UAV system was flying at a height of about 150 m with a horizontal flight speed of about 16 m/s. A total number of 44,180 UAV images were captured to form the second dataset.

Table 1. Specification of the DJI Phantom 4 Pro Unmanned Aerial Vehicle (UAV) system.

Parameter	Value
Manufacturer	DJI
Weight	1388 g
Maximum horizontal flight speed	72 km/h
Maximum take-off altitude	6000 m
Maximum flight height	500 m
Maximum flight time	30 min
Maximum wind speed tolerable	10 m/s
Field of view	front-back: 70o, left-right: 50o
Measuring frequency	20 Hz
Image sensor	1 inch Exmor R CMOS sensor, 20M pixels
Image size	5472 × 3648 pixels
Maximum control distance	7000 m

3.2. Robustness Evaluation

In the UAV images, different vehicles exhibit different sizes and orientations. Some are even partially occluded by other high-rise objects (e.g., trees and buildings). The illumination conditions and the vibrations of the surveying platform also affect the quality of the captured images. Thus, the proposed vehicle detection method should have the capability to effectively deal with the aforementioned circumstances. In this section, we evaluated the robustness of the proposed affine-function transformation-based object matching framework to scale variations, orientation variations, partial occlusions, and noise contaminations. To this end, we manually created a scene test dataset containing 500 scenes, each of which contains a vehicle instance. A subset of the test scene dataset is shown in Figure 8b. The same vehicle template shown in Figure 8a was used to conduct matching in all the experiments. For each matching experiment between the template and a scene, a group of feature points were extracted and described using the SIFT features. Then, we applied the affine-function transformation-based object matching framework to a pair of template and scene to conduct matching. After matching, if the matching cost was below a predefined threshold, the scene was regarded as correctly matched. Otherwise, it was regarded as a bad matching.

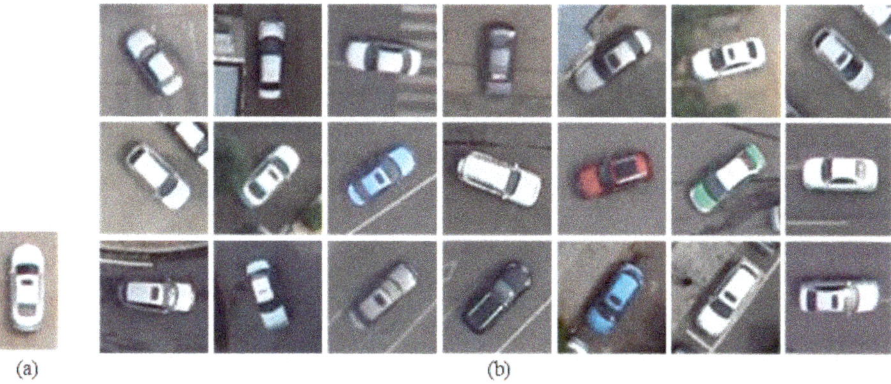

(a) (b)

Figure 8. Illustrations of (**a**) the vehicle template, and (**b**) a subset of the scene dataset used for robustness evaluation.

To examine the properties of the proposed object matching framework to scale variations, we transformed each scene with the following scale factors: 0.4, 0.5, 0.6, 0.8, 1.0, 1.2, 1.4, 1.5, 1.6, 1.8, and 2.0. Therefore, a scale variation test dataset containing 5500 scenes was created for robustness evaluation. Figure 9a shows an example of a scene transformed with different scales. Then, we applied the affine-function transformation-based object matching framework to the template shown in Figure 8a and each scene in the scale variation test dataset to conduct matching. The matching results are detailed in Table 2. We used the matching rate, which was defined as the proportion of correctly matched scenes, to analyze the matching performance in different scales. As shown in Table 2, when the scenes were enlarged by scale factors ranging from 1.2 to 2.0, as well as shrunk by scale factors 0.6 and 0.8, the matching performance was unaffected. This is because when a scene is enlarged or shrunk slightly, the features of the feature points and their relative position relationships are almost unchanged. Thus, the proposed matching framework obtained stable performance. However, when the scenes were shrunk by scale factors 0.4 and 0.5, some of the scenes were not correctly matched, resulting in a decrease of the matching rate. In fact, when a scene is shrunk greatly, the local descriptions of the feature points will change greatly. In addition, the distinctiveness between the feature points will diminish. Moreover, some adjacent feature points might be merged into one feature point. Thus, the matching performance was degraded. Fortunately, in actual UAV images captured with a bird view, the scale variations among vehicles are not very big, since the sizes of vehicles do not change dramatically. Therefore, the matching performance is hardly affected.

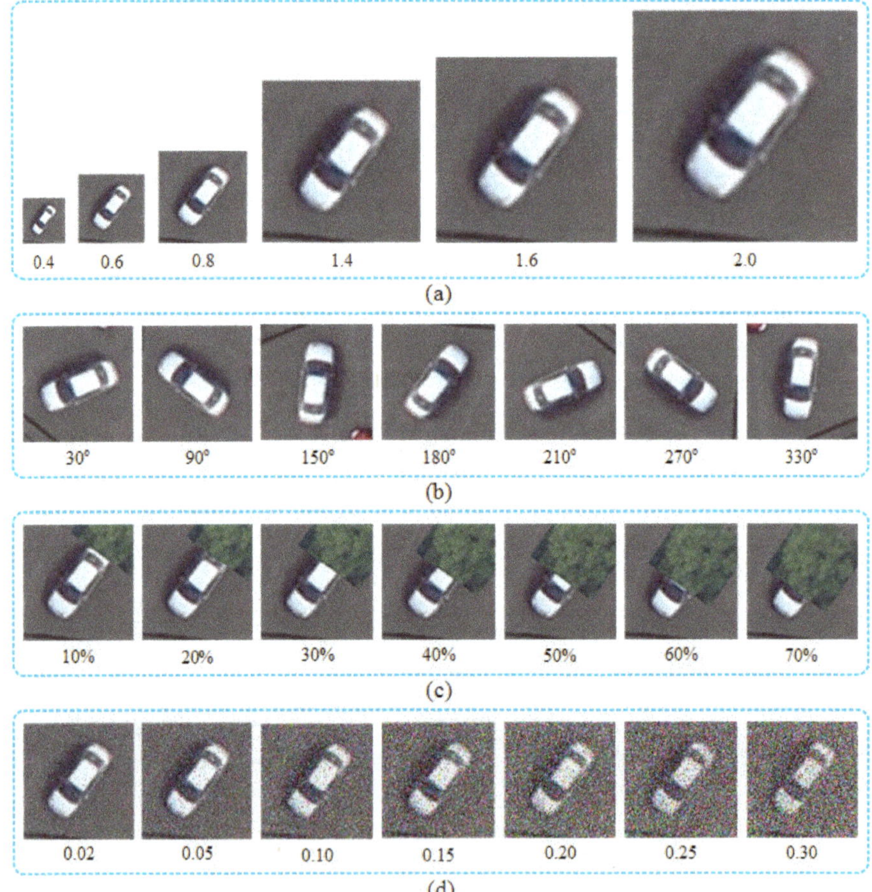

Figure 9. Illustrations of (**a**) a scene sample transformed with different scales, (**b**) a scene sample rotated with different angles, (**c**) a scene sample occluded with different proportions, and (**d**) a scene contaminated with different levels of salt and pepper noises.

Table 2. Vehicle matching results on the scale variation test dataset.

Scale	0.4	0.5	0.6	0.8	1.0	1.2	1.4	1.5	1.6	1.8	2.0
Matching rate	0.94	0.98	1.0	1.0	1.0	1.0	1.0	1.0	1.0	1.0	1.0

To test the robustness of the proposed object matching framework to rotation variations, we successively rotated each scene clockwise with an angle interval of 30 degrees. Therefore, a rotation variation test dataset including 6000 scenes was created for robustness evaluation. Figure 9b presents an example of a scene rotated with different angles. Then, object matching was carried out between the template and each of the rotated scenes. The vehicle matching results in different rotations are listed in Table 3. As reflected in Table 3, the matching rates were the same and unaffected by the rotations of scenes. This is because, when a scene is rotated, the features of the feature points, as well as their relative position relationships, still maintain without any modifications. Thus, the proposed matching framework performed equally under different rotation variations. This property is very useful for handling real world scenes, since vehicles always exhibit with different orientations.

Table 3. Vehicle matching results on the rotation variation test dataset.

Rotation	0°	30°	60°	90°	120°	150°	180°	210°	240°	270°	300°	330°
Matching rate	1.0	1.0	1.0	1.0	1.0	1.0	1.0	1.0	1.0	1.0	1.0	1.0

To assess the performance of the proposed object matching framework to occlusion variations, we manually masked the vehicle instance in each scene with the following proportions of occlusions: 0%, 10%, 20%, 30%, 40%, 50%, 60%, 70%, 80%, and 90%. Therefore, an occlusion variation test dataset containing 5000 scenes was created for robustness evaluation. Figure 9c shows an example of a vehicle instance occluded with different proportions. Then, object matching was performed on each pair of the template and an occluded scene. Table 4 details the vehicle matching results. As shown in Table 4, when the vehicles were partially occluded slightly (occlusion proportions ranging from 10% to 40%), the proposed matching framework performed effectively. All the scenes were correctly matched. This is benefited from the introduction of the weight factors in the overall objective function for evaluating the contributions of different feature points to a matching. When half part of a vehicle instance was occluded, the matching performance was slightly affected but still satisfactory. However, when the vehicle instances were occluded significantly, the matching performance dramatically decreased. This is because, when a vehicle instance is occluded significantly, the number of feature points contributing to the matching become very less. The majority of the matching locations computed through the affine transformation parameters are not correct, resulting in high dissimilarity measure values.

Table 4. Vehicle matching results on the occlusion variation test dataset.

Occlusion Proportion	0%	10%	20%	30%	40%	50%	60%	70%	80%	90%
Matching rate	1.0	1.0	1.0	1.0	1.0	0.98	0.82	0.51	0.32	0.17

To evaluate the robustness of the proposed object matching framework to noises, we superimposed each scene with different levels of "salt and pepper" noises. We tested the following noise densities in our experiments: 0.00, 0.02, 0.05, 0.08, 0.10, 0.12, 0.15, 0.18, 0.20, 0.22, 0.25, 0.28, and 0.30. Therefore, a noise contamination dataset including 6500 scenes was constructed for robustness evaluation. Then, we applied the proposed object matching framework to the template and each of the noise-contaminated scenes to conduct matching. Table 5 presents the vehicle matching results analyzed using the matching rate. As reflected in Table 5, the proposed object matching framework showed superior performance when the scenes were contaminated with low densities of noises. This is benefitted from the robustness of the SIFT feature descriptor, which has excellent properties of noise resistance. However, when the scenes were contaminated with high levels of noises, the feature descriptions of the feature points would be influenced, resulting in feature difference between a template feature point and its matching location. Thus, the dissimilarity measures between the template feature points and their matching locations became higher.

Table 5. Vehicle matching results on the noise contamination dataset.

Noise Density	0.00	0.02	0.05	0.08	0.10	0.12	0.15	0.18	0.20	0.22	0.25	0.28	0.30
Matching rate	1.0	1.0	1.0	1.0	1.0	1.0	1.0	1.0	0.97	0.93	0.88	0.81	0.78

3.3. Vehicle Detection

To evaluate the performance of our proposed vehicle detection method, we applied it to the Nanjing and Changsha UAV image datasets aforementioned in Section 3.1. For a test image, first, we over-segmented it into a series of superpixels using the SLIC superpixel segmentation method. Then, meaningful and non-redundant patches were generated centered at each superpixel. In the UAV images, the length of a vehicle is approximately 54 pixels. Thus, in order to enclose an entire vehicle instance in a patch with some relaxations, we set the patch size to be 70×70 pixels. Next, object matching between the vehicle template shown in Figure 8a and each of the generated patches was carried out using the proposed affine-function transformation-based object matching framework. To successively optimize the overall objective function to obtain a group of tight transformation parameters, we performed four iterations of the successive convexification process and configured the weight adjustment factor as $\Delta w = 0.2$. After matching cost thresholding, vehicle location estimation, and multiple detection results elimination, we obtained the final vehicle detection result.

Table 6 lists the vehicle detection results, as well as the ground truths, on the two UAV image datasets. As reflected in Table 6, for each of the datasets, the majority of vehicles were correctly detected and only a small number of false alarms were generated. However, the number of false alarms took a very small proportion and was acceptable. For visual inspections, Figures 10 and 11 illustrate a subset of the vehicle detection results on the two UAV image datasets. As shown in these figures, the vehicles exhibiting with different colors, different sizes, different orientations, different illumination conditions, different densities, and different levels of occlusions were effectively detected by the proposed vehicle detection method. Specifically, as shown in Figure 11a, for a scene with very high density of vehicles, the proposed method still obtained promising vehicle detection results. Figure 11b shows a scene containing vehicles covered with large areas of shadows. These shadows might affect the appearance and the saliency of the vehicles. Fortunately, benefiting from the use of the SIFT features, which has a strong property of invariance to illumination variations, these vehicles were correctly detected by using the proposed method. However, as shown by the vehicle marked by a yellow box labeled with #1 in Figure 10, it was covered with a severe shadow. The vehicle was almost hidden in the background. Thus, our proposed method failed to detect it because of extremely low distinctiveness of feature points. As shown in Figure 11c,d, some vehicles were partially occluded by high-rise buildings and overhead trees. Since in our proposed affine-function transformation-based object matching framework, occlusion is considered and handled by assigning each template feature point with a weight factor, which is successively adjusted to degrade the contributions of occluded matching positions in the successive convexification process. Therefore, our proposed method still achieved promising performance on such occluded vehicles. However, as shown by the vehicles marked by yellow boxes labeled with #2, #3, and #4 in Figure 10 and the vehicles marked by yellow boxes in Figure 11c,d, these vehicles were occluded severely, resulting in very high matching costs. Therefore, they were failed to be detected. In addition, as shown by the vehicle marked by a yellow box labeled with #5, it was entirely covered with a cloth. Its appearance feature being a vehicle almost disappeared. Thus, it was also undetected. Moreover, due to the high similarities of some real-world objects (e.g., air conditioner external units) to the vehicles, they were falsely detected as vehicles caused by low matching costs.

Table 6. Vehicle detection results and quantitative evaluations on the Nanjing and Changsha Unmanned Aerial Vehicle (UAV) image datasets.

Dataset	Ground Truth	Detection Results		Quantitative Evaluations			
		Vehicles	False Alarms	Completeness	Correctness	Quality	F1-Measure
Nanjing	672,184	613,032	17,659	0.912	0.972	0.889	0.941
Changsha	896,722	813,327	28,626	0.907	0.966	0.879	0.936
Average	1568,906	1426,359	46,285	0.909	0.969	0.883	0.938

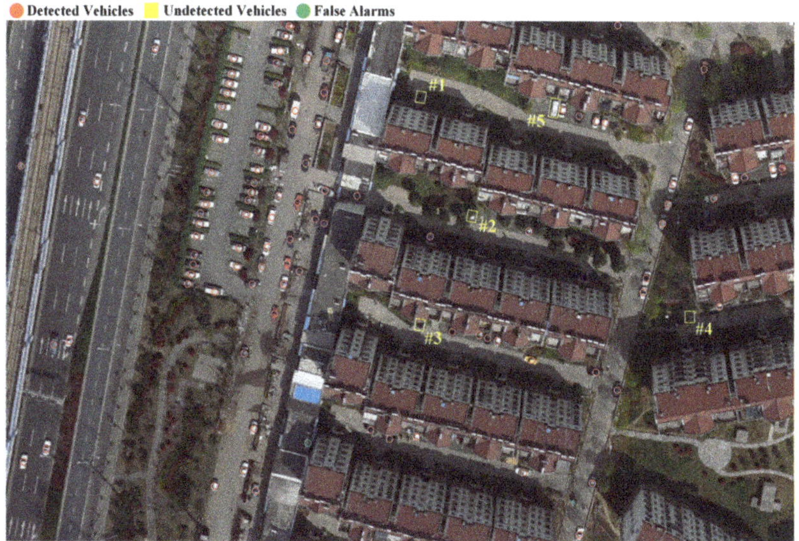

Figure 10. Illustration of a subset of vehicle detection results on a Unmanned Aerial Vehicle (UAV) image.

To quantitatively evaluate the accuracy and correctness of the vehicle detection results on the two UAV image datasets, we adopted the following four quantitative measures: completeness, correctness, quality, and F1-measure [41]. Completeness assesses the proportion of correctly detected vehicles with respect to the ground truth. Correctness evaluates the proportion of correctly detected vehicles with respect to all the detected instances. Quality and F1-measure reflect the overall performance. They are defined as follows:

$$completeness = \frac{TP}{TP + FN} \tag{12}$$

$$correctness = \frac{TP}{TP + FP} \tag{13}$$

$$quality = \frac{TP}{TP + FN + FP} \tag{14}$$

$$F_1 - measure = \frac{2 \cdot completeness \cdot correctness}{completeness + correctness} \tag{15}$$

where TP, FN, and FP are the numbers of correctly detected vehicles, undetected vehicles, and falsely detected non-vehicle objects, respectively. The quantitative evaluation results using these four measures are listed in Table 6. The proposed vehicle detection method achieved a completeness, correctness, quality, and F1-measure of 0.912, 0.972, 0.889, and 0.941, respectively, on the Nanjing UAV image dataset. For the Changsha UAV image dataset, a completeness, correctness, quality, and F1-measure of 0.907, 0.966, 0.879, and 0.936, respectively, were obtained. On the whole, through visual inspections and quantitative evaluations, we confirmed that the proposed vehicle detection method performed effectively and was feasible for vehicle detection from UAV images.

The proposed vehicle detection method was tested on a cloud computing platform with eight 16-GB GPUs, one 16-core CPU, and a memory size of 64 GB. In practice, for a test image, after patch generation, the generated patches were distributed to the eight GPUs for parallel processing. The processing time of the proposed method was also recorded to analyze its computational performance. On average, the proposed method achieved a processing speed of 31 patches per second on a GPU. Thus, by adopting the parallel processing strategy, 248 patches were under processing every second.

Figure 11. Illustrations of vehicle detection results on Unmanned Aerial Vehicle (UAV) images under challenging scenarios. (**a**) high density of vehicles, (**b**) vehicles covered with shadows, (**c**) vehicles occluded by high-rise buildings, and (**d**) vehicles occluded by overhead trees.

3.4. Comparative Studies

To further compare the performance of the proposed method in this paper and other existing vehicle detection methods, a set of comparative experiments were conducted with the following nine existing vehicle detection methods: coupled region-based convolutional neural networks (CR-CNN) [49], hard example mining based convolutional neural networks (HEM-CNN) [23], affine invariant description and large-margin dimensionality reduction based method (AID-LDR) [17], bag-of-words and orientation aware scanning based method (BoW-OAS) [8], Viola-Jones based method (VJ) [18], enhanced Viola-Jones based method (EVJ) [19], fast binary detector based method (FBD) [21], YOLOv3 [59], and Faster R-CNN [36]. In the CR-CNN method, first, vehicle candidate regions are extracted based on a vehicle proposal network; then, a coupled region-based CNN is performed on the candidate regions to detect vehicles. For the HEM-CNN method, to train an effective CNN model, hard example mining is applied to the stochastic gradient descent to select informative training samples; then, the CNN model is used for vehicle detection based on a sliding window strategy. Both of the AID-LDR and BoW-OAS methods adopt the bag-of-words model to represent the statistical feature of a vehicle. The detection of vehicles is achieved through a sliding window-based classification process. For the VJ and EVJ methods, Viola–Jones object detection scheme is proposed to detect vehicles. To well handle vehicles of varying orientations, a road orientation adjustment method is adopted to make sure that roads and on-road vehicles are aligned with the horizontal direction. In the FBD method, a fast binary detector using integral channel features is designed to detect vehicles. YOLOv3 is a one-stage object detection network which accomplishes feature extraction and object prediction in a single network. In contrast, Faster R-CNN is a two-stage object detection framework composed of a region proposal network and an object detection network. The region proposal network generates a group of object proposals, which are further identified by the object detection network to verify the objects of interest.

We applied these nine methods to the Nanjing and Changsha UAV image datasets to evaluate their performances on vehicle detection. Quantitative evaluations using completeness, correctness, quality, and F1-measure were also carried out on the detection results. The detailed detection results and quantitative evaluations of different methods are listed in Table 7. As reflected by the overall evaluations of quality and F1-measure, the HEM-CNN and AID-LDR methods obtained relatively lower performances on the two datasets; whereas the YOLOv3 and Faster R-CNN methods obtained the best performance. In addition, the BoW-OAS and EVJ methods obtained similar performances. By analyzing the number of correctly detected vehicles with respect to the ground truth and the number of correctly detected vehicles with respect to the detected objects, the YOLOv3 and Faster R-CNN methods outperformed the other seven methods with higher completeness and correctness values. This is because, these two methods adopt deep learning techniques to exploit high-level features of vehicles. Thus, they showed superior performance than the other methods. Comparatively, by using region proposal mechanism, Faster R-CNN performed a little better than YOLOv3. However, the AID-LDR and BoW-OAS methods generated more false alarms, thereby resulting in relatively lower correctness values than the other methods. This is because the AID-LDR and BoW-OAS methods adopt mid-level statistical features of vehicles represented using the bag-of-words model. According to the statistical property, the bag-of-words representation can only characterize the existence of some features; however, the relative relationships of these features cannot be reflected. As shown in Figure 12, Figure 12a is a patch containing a normal vehicle; Figure 12b and c are generated by cutting Figure 12a into four parts and making some transformations and combinations on these parts. Apparently, Figure 12b,c cannot be considered as a vehicle. However, the bag-of-words representations of these three patches are almost similar. They are equally detected as normal vehicles. Therefore, more false alarms were detected by the AID-LDR and BoW-OAS methods. Compared with these nine methods, our proposed method obtained compatible performance with the Faster R-CNN and outperformed the other eight methods. Through comparative studies, our proposed method can effectively tackle various scene conditions and obtained advantageous performance in accurately detecting vehicles from UAV images.

Table 7. Vehicle detection results and quantitative evaluations of different methods.

Method	Dataset	Detection Results		Quantitative Evaluations			
		Vehicles	False Alarms	Completeness	Correctness	Quality	F1-Measure
CR-CNN	Nanjing	564,635	32,231	0.840	0.946	0.802	0.890
	Changsha	752,350	54,027	0.839	0.933	0.791	0.884
HEM-CNN	Nanjing	545,142	32,951	0.811	0.943	0.773	0.872
	Changsha	719,172	49,996	0.802	0.935	0.760	0.863
AID-LDR	Nanjing	576,062	48,734	0.857	0.922	0.799	0.888
	Changsha	739,796	71,383	0.825	0.912	0.764	0.866
BoW-OAS	Nanjing	602,277	48,833	0.896	0.925	0.835	0.910
	Changsha	782,839	70,856	0.873	0.917	0.809	0.894
VJ	Nanjing	576,734	40,094	0.858	0.935	0.810	0.895
	Changsha	745,176	57,815	0.831	0.928	0.781	0.877
EVJ	Nanjing	590,850	35,050	0.879	0.944	0.835	0.910
	Changsha	772,078	56,332	0.861	0.932	0.810	0.895
FBD	Nanjing	580,767	41,706	0.864	0.933	0.814	0.897
	Changsha	759,524	66,944	0.847	0.919	0.788	0.882
YOLOv3	Nanjing	605,638	20,021	0.901	0.968	0.875	0.933
	Changsha	806,153	36,222	0.899	0.957	0.864	0.927
Faster R-CNN	Nanjing	612,360	16,993	0.911	0.973	0.889	0.941
	Changsha	811,533	27,695	0.905	0.967	0.878	0.935

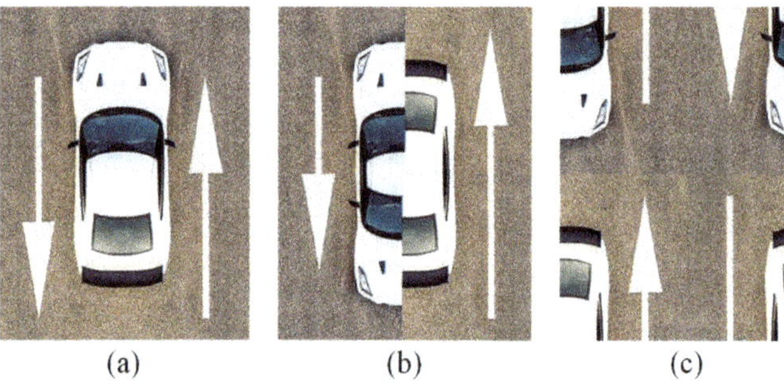

(a)   (b)   (c)

Figure 12. Illustration of three patches having almost the similar bag-of-words representations. (**a**) a complete vehicle, (**b**) and (**c**) transformed vehicles on (**a**).

4. Conclusions

In this paper, we have proposed an affine-function transformation-based object matching framework for detecting vehicles from UAV images. The proposed method has advantageous properties to tackle scale variations, orientation variations, illumination variations, and partial occlusions of vehicles. For a test image, to generate meaningful and non-redundant patches, an SLIC-based superpixel segmentation strategy is adopted for patch generation. Then, the affine-function transformation-based object matching framework is applied to a vehicle template and each of the generated patches for vehicle existence estimation. Finally, after matching cost thresholding, vehicle location estimation, and multiple response elimination, vehicles are accurately detected and located in the image. The proposed method has been tested on two UAV image datasets for performance evaluation on vehicle detection. Quantitative evaluations confirmed that an average completeness, correctness, quality, and F1-measure of 0.909, 0.969, 0.883, and 0.938, respectively, were achieved towards vehicle detection. Visual inspections also showed the robustness of the proposed method in handling various vehicle conditions. In addition, comparative studies with nine existing vehicle

Remote Sens. **2019**, *11*, 1708

detection methods demonstrated that the proposed method obtained compatible performance with the Faster R-CNN and outperformed the other eight methods in detecting vehicles from UAV images.

Author Contributions: C.S. and Y.Y. conceived of and designed the experiments and performed the experiments; H.G. and Y.Y. analyzed the data; H.G., D.P., and W.Y. revised the paper. H.G. and Y.Y. did the field works and preprocessed the data.

Funding: This work was supported in part by the National Natural Science Foundation of China under Grants 61603146, 41671454, and 41801386 in part by the Natural Science Foundation of Jiangsu Province under Grant BK20160427 and BK20180797, and in part by the Natural Science Research in Colleges and Universities of Jiangsu Province under Grant 16KJB520006.

Acknowledgments: The authors would like to acknowledge the anonymous reviewers for their valuable comments.

Conflicts of Interest: The authors declare no conflict of interest.

References

1. Liu, W.; Yamazaki, F.; Vu, T.T. Automated vehicle extraction and speed determination from QuickBird satellite images. *IEEE J. Sel. Top. Appl. Earth Obs. Remote Sens.* **2011**, *4*, 75–82. [CrossRef]
2. Zhou, H.; Kong, H.; Wei, L.; Creighton, D.; Nahavandi, S. On detecting road regions in a single UAV image. *IEEE Trans. Intell. Transp. Syst.* **2017**, *18*, 1713–1722. [CrossRef]
3. Li, M.; Stein, A.; Bijker, W.; Zhan, Q. Region-based urban road extraction from VHR satellite images using binary partition tree. *Int. J. Appl. Earth Obs. Geoinf.* **2016**, *44*, 217–225. [CrossRef]
4. Mokhtarzade, M.; Zoej, V.M.J. Road detection from high-resolution satellite images using artificial neural networks. *Int. J. Appl. Earth Obs. Geoinf.* **2007**, *9*, 32–40. [CrossRef]
5. Jin, H.; Feng, Y. Automated road pavement marking detection from high resolution aerial images based on multi-resolution image analysis and anisotropic Gaussian filtering. In Proceedings of the 2010 2nd International Conference on Signal Processing Systems, Dalian, China, 5–7 July 2010; pp. 337–341.
6. Pan, Y.; Zhang, X.; Cervone, G.; Yang, L. Detection of asphalt pavement potholes and cracks based on the unmanned aerial vehicle multispectral imagery. *IEEE J. Sel. Top. Appl. Earth Obs. Remote Sens.* **2018**, *11*, 3701–3712. [CrossRef]
7. Cao, L.; Luo, F.; Chen, L.; Sheng, Y.; Wang, H.; Wang, C.; Ji, R. Weakly supervised vehicle detection in satellite images via multi-instance discriminative learning. *Pattern Recognit.* **2017**, *64*, 417–424. [CrossRef]
8. Zhou, H.; Wei, L.; Lim, C.P.; Creighton, D.; Nahavandi, S. Robust vehicle detection in aerial images using bag-of-words and orientation aware scanning. *IEEE Trans. Geosci. Remote Sens.* **2018**, *56*, 7074–7085. [CrossRef]
9. Eslami, M.; Faez, K. Automatic traffic monitoring using satellite images. In Proceedings of the 2010 2nd International Conference on Computer Engineering and Technology, Chengdu, China, 16–18 April 2010; pp. 130–135.
10. Khalil, M.; Li, J.; Sharif, A.; Khan, J. Traffic congestion detection by use of satellites view. In Proceedings of the 2017 14th International Computer Conference on Wavelet Active Media Technology and Information Processing (ICCWAMTIP), Chengdu, China, 15–17 December 2017; pp. 278–280.
11. Kembhavi, A.; Harwood, D.; Davis, L.S. Vehicle detection using partial least squares. *IEEE Trans. Pattern Anal. Mach. Intell.* **2011**, *33*, 1250–1265. [CrossRef]
12. Zheng, Z.; Zhou, G.; Wang, Y.; Liu, Y.; Li, X.; Wang, X.; Jiang, L. A novel vehicle detection method with high resolution highway aerial image. *IEEE J. Sel. Top. Appl. Earth Obs. Remote Sens.* **2013**, *6*, 2338–2343. [CrossRef]
13. Chen, Z.; Wang, C.; Luo, H.; Wang, H.; Chen, Y.; Wen, C.; Yu, Y.; Cao, L.; Li, J. Vehicle detection in high-resolution aerial images based on fast sparse representation classification and multiorder feature. *IEEE Trans. Intell. Transp. Syst.* **2016**, *17*, 2296–2309. [CrossRef]
14. Holt, A.C.; Seto, E.Y.; Rivard, T.; Gong, P. Object-based detection and classification of vehicles from high-resolution aerial photography. *Photogramm. Eng. Remote Sens.* **2009**, *75*, 871–880. [CrossRef]
15. Niu, X. A semi-automatic framework for highway extraction and vehicle detection based on a geometric deformable model. *ISPRS J. Photogramm. Remote Sens.* **2006**, *61*, 170–186. [CrossRef]
16. Zhou, H.; Wei, L.; Creighton, D.; Nahavandi, S. Orientation aware vehicle detection in aerial images. *Electron. Lett.* **2017**, *53*, 1406–1408. [CrossRef]

17. Wan, L.; Zheng, L.; Huo, H.; Fang, T. Affine invariant description and large-margin dimensionality reduction for target detection in optical remote sensing images. *IEEE Geosci. Remote Sens. Lett.* **2017**, *14*, 1116–1120. [CrossRef]

18. Xu, Y.; Yu, G.; Wang, Y.; Wu, X.; Ma, Y. A hybrid vehicle detection method based on Viola-Jones and HOG+SVM from UAV images. *Sensors* **2016**, *16*, 1325. [CrossRef] [PubMed]

19. Xu, Y.; Yu, G.; Wu, X.; Wang, Y.; Ma, Y. An enhanced Viola-Jones vehicle detection method from unmanned aerial vehicles imagery. *IEEE Trans. Intell. Transp. Syst.* **2017**, *18*, 1845–1856. [CrossRef]

20. Audebert, N.; Saux, B.L.; Lefèvre, S. Segment-before-detect: Vehicle detection and classification through semantic segmentation of aerial images. *Remote Sens.* **2017**, *9*, 368. [CrossRef]

21. Liu, K.; Mattyus, G. Fast multiclass vehicle detection on aerial images. *IEEE Geosci. Remote Sens. Lett.* **2015**, *12*, 1938–1942.

22. Tuermer, S.; Kurz, F.; Reinartz, P.; Stilla, U. Airborne vehicle detection in dense urban areas using HOG features and disparity maps. *IEEE J. Sel. Top. Appl. Earth Obs. Remote Sens.* **2013**, *6*, 2327–2337. [CrossRef]

23. Koga, Y.; Miyazaki, H.; Shibasaki, R. A CNN-based method of vehicle detection from aerial images using hard example mining. *Remote Sens.* **2018**, *10*, 124. [CrossRef]

24. Moranduzzo, T.; Melgani, F. Detecting cars in UAV images with a catalog-based approach. *IEEE Trans. Geosci. Remote Sens.* **2013**, *52*, 6356–6367. [CrossRef]

25. Razakarivony, S.; Jurie, F. Vehicle detection in aerial imagery: A small target detection benchmark. *J. Vis. Commun. Image Represent.* **2016**, *34*, 187–203. [CrossRef]

26. Cao, L.; Jiang, Q.; Cheng, M.; Wang, C. Robust vehicle detection by combining deep features with exemplar classification. *Neurocomputing* **2016**, *215*, 225–231. [CrossRef]

27. Cao, L.; Wang, C.; Li, J. Vehicle detection from highway satellite images via transfer learning. *Inf. Sci.* **2016**, *366*, 177–187. [CrossRef]

28. Ammour, N.; Alhichri, H.; Bazi, Y.; Benjdira, B.; Alajlan, N.; Zuair, M. Deep learning approach for car detection in UAV imagery. *Remote Sens.* **2017**, *9*, 312. [CrossRef]

29. Chen, Z.; Wang, C.; Wen, C.; Teng, X.; Chen, Y.; Guan, H.; Luo, H.; Cao, L.; Li, J. Vehicle detection in high-resolution aerial images via sparse representation and superpixels. *IEEE Trans. Geosci. Remote Sens.* **2016**, *54*, 103–116. [CrossRef]

30. Zhang, J.; Tao, C.; Zou, Z. An on-road vehicle detection method for high-resolution aerial images based on local and global structure learning. *IEEE Geosci. Remote Sens. Lett.* **2017**, *14*, 1198–1202. [CrossRef]

31. Bazi, Y.; Melgani, F. Convolutional SVM networks for object detection in UAV imagery. *IEEE Trans. Geosci. Remote Sens.* **2018**, *56*, 3107–3118. [CrossRef]

32. Elmikaty, M.; Stathaki, T. Car detection in aerial images of dense urban areas. *IEEE Trans. Aerosp. Electron. Syst.* **2018**, *54*, 51–63. [CrossRef]

33. Liu, Y.; Monteiro, S.T.; Saber, E. Vehicle detection from aerial color imagery and airborne LiDAR data. In Proceedings of the 2016 IEEE International Geoscience and Remote Sensing Symposium (IGARSS), Beijing, China, 10–15 July 2016; pp. 1384–1387.

34. Schilling, H.; Bulatov, D.; Middelmann, W. Object-based detection of vehicles using combined optical and elevation data. *ISPRS J. Photogramm. Remote Sens.* **2018**, *136*, 85–105. [CrossRef]

35. Krizhevsky, A.; Sutskever, I.; Hinton, G.E. ImageNet classification with deep convolutional neural networks. In Proceedings of the Advances in Neural Information Processing Systems, Lake Tahoe, NV, USA, 3–6 December 2012; pp. 1097–1105.

36. Ren, S.; He, K.; Girshick, R.; Sun, J. Faster R-CNN: Towards real-time object detection with region proposal networks. *IEEE Trans. Pattern Anal. Mach. Intell.* **2017**, *39*, 1137–1149. [CrossRef] [PubMed]

37. Salakhutdinov, R.; Tenenbaum, J.B.; Torralba, A. Learning with hierarchical-deep models. *IEEE Trans. Pattern Anal. Mach. Intell.* **2013**, *35*, 1958–1971. [CrossRef] [PubMed]

38. Chen, G.; Zhang, X.; Wang, Q.; Dai, F.; Gong, Y.; Zhu, K. Symmetrical dense-shortcut deep fully convolutional networks for semantic segmentation of very-high-resolution remote sensing images. *IEEE J. Sel. Top. Appl. Earth Obs. Remote Sens.* **2018**, *11*, 1633–1644. [CrossRef]

39. Kemker, R.; Salvaggio, C.; Kanan, C. Algorithms for semantic segmentation of multispectral remote sensing imagery using deep learning. *ISPRS J. Photogramm. Remote Sens.* **2018**, *145*, 60–77. [CrossRef]

40. Ding, P.; Zhang, Y.; Deng, W.J.; Jia, P.; Kuijper, A. A light and faster regional convolutional neural network for object detection in optical remote sensing images. *ISPRS J. Photogramm. Remote Sens.* **2018**, *141*, 208–218. [CrossRef]

41. Yu, Y.; Guan, H.; Zai, D.; Ji, Z. Rotation-and-scale-invariant airplane detection in high-resolution satellite images based on deep-Hough-forests. *ISPRS J. Photogramm. Remote Sens.* **2016**, *112*, 50–64. [CrossRef]

42. Han, W.; Feng, R.; Wang, L.; Cheng, Y. A semi-supervised generative framework with deep learning features for high-resolution remote sensing image scene classification. *ISPRS J. Photogramm. Remote Sens.* **2018**, *145*, 23–43. [CrossRef]

43. Zhang, C.; Pan, X.; Li, H.; Gardiner, A.; Sargent, I.; Hare, J.; Atkinson, P.M. A hybrid MLP-CNN classifier for very fine resolution remotely sensed image classification. *ISPRS J. Photogramm. Remote Sens.* **2018**, *140*, 133–144. [CrossRef]

44. Wang, S.; Quan, D.; Liang, X.; Ning, M.; Guo, Y.; Jiao, L. A deep learning framework for remote sensing image registration. *ISPRS J. Photogramm. Remote Sens.* **2018**, *145*, 148–164. [CrossRef]

45. Sommer, L.; Schuchert, T.; Beyerer, J. Comprehensive analysis of deep learning based vehicle detection in aerial images. *IEEE Trans. Circuits Syst. Video Tech.* **2018**, in press. [CrossRef]

46. Mou, L.; Zhu, X.X. Vehicle instance segmentation from aerial image and video using a multitask learning residual fully convolutional network. *IEEE Trans. Geosci. Remote Sens.* **2018**, *56*, 6699–6711. [CrossRef]

47. Wu, D.; Zhang, Y.; Chen, Y.; Zhong, S. Vehicle detection in high-resolution images using superpixel segmentation and CNN iteration strategy. *IEEE Geosci. Remote Sens. Lett.* **2019**, *16*, 105–109. [CrossRef]

48. Tang, T.; Zhou, S.; Deng, Z.; Zou, H.; Lei, L. Vehicle detection in aerial images based on region convolutional neural networks and hard negative example mining. *Sensors* **2017**, *17*, 336. [CrossRef]

49. Deng, Z.; Sun, H.; Zhou, S.; Zhao, J.; Zou, H. Toward fast and accurate vehicle detection in aerial images using coupled region-based convolutional neural networks. *IEEE J. Sel. Top. Appl. Earth Obs. Remote Sens.* **2017**, *10*, 3652–3664. [CrossRef]

50. Schilling, H.; Bulatov, D.; Niessner, R.; Middelmann, W.; Soergel, U. Detection of vehicles in multisensor data via multibranch convolutional neural networks. *IEEE J. Sel. Top. Appl. Earth Obs. Remote Sens.* **2018**, *11*, 4299–4316. [CrossRef]

51. Zhong, J.; Lei, T.; Yao, G. Robust vehicle detection in aerial images based on cascaded convolutional neural networks. *Sensors* **2017**, *17*, 2720. [CrossRef]

52. Yang, C.; Li, W.; Lin, Z. Vehicle object detection in remote sensing imagery based on multi-perspective convolutional neural network. *ISPRS Int. J. Geo-Inf.* **2018**, *7*, 249. [CrossRef]

53. Tang, T.; Zhou, S.; Deng, Z.; Lei, L.; Zou, H. Arbitrary-oriented vehicle detection in aerial imagery with single convolutional neural networks. *Remote Sens.* **2017**, *9*, 1170. [CrossRef]

54. Achanta, R.; Shaji, A.; Smith, K.; Lucchi, A.; Fua, P.; Süsstrunk, S. SLIC superpixels compared to state-of-the-art superpixel methods. *IEEE Trans. Pattern Anal. Mach. Intell.* **2012**, *34*, 2274–2282. [CrossRef]

55. Neubeck, A.; Van Gool, L. Efficient non-maximum suppression. In Proceedings of the 18th International Conference on Pattern Recognition (ICPR'06), Hong Kong, China, 20–24 August 2006; pp. 850–855.

56. Lowe, D.G. Distinctive image features from scale-invariant keypoints. *Int. J. Comput. Vis.* **2004**, *60*, 91–110. [CrossRef]

57. Wahl, E.; Hillenbrand, U.; Hirzinger, G. Surflet-pair-relation histograms: A statistical 3D-shape representation for rapid classification. In Proceedings of the Fourth International Conference on 3-D Digital Imaging and Modeling, 2003. 3DIM 2003, Banff, UK, 6–10 October 2003; pp. 474–481.

58. Jiang, H.; Drew, M.S.; Li, Z. Matching by linear programming and successive convexification. *IEEE Trans. Pattern Anal. Mach. Intell.* **2007**, *29*, 959–975. [CrossRef]

59. Redmon, J.; Farhadi, A. YOLOv3: An incremental improvement. *arXiv* **2018**, arXiv:1804.02767v1.

 *remote sensing*

Article

Hyperspectral Anomaly Detection via Dictionary Construction-Based Low-Rank Representation and Adaptive Weighting

Yixin Yang, Jianqi Zhang *, Shangzhen Song and Delian Liu *

School of Physics and Optoelectronic Engineering, Xidian University, Xi'an 710071, China;
yxyang618@163.com (Y.Y.); szsong118@163.com (S.S.)
* Correspondence: jqzhang@mail.xidian.edu.cn (J.Z.); dlliu@xidian.edu.cn (D.L.);
  Tel.: +86-135-7259-1055 (J.Z.); +86-133-8921-2012 (D.L.)

Received: 13 December 2018; Accepted: 17 January 2019; Published: 19 January 2019

Abstract: Anomaly detection (AD), which aims to distinguish targets with significant spectral differences from the background, has become an important topic in hyperspectral imagery (HSI) processing. In this paper, a novel anomaly detection algorithm via dictionary construction-based low-rank representation (LRR) and adaptive weighting is proposed. This algorithm has three main advantages. First, based on the consistency with AD problem, the LRR is employed to mine the lowest-rank representation of hyperspectral data by imposing a low-rank constraint on the representation coefficients. Sparse component contains most of the anomaly information and can be used for anomaly detection. Second, to better separate the sparse anomalies from the background component, a background dictionary construction strategy based on the usage frequency of the dictionary atoms for HSI reconstruction is proposed. The constructed dictionary excludes possible anomalies and contains all background categories, thus spanning a more reasonable background space. Finally, to further enhance the response difference between the background pixels and anomalies, the response output obtained by LRR is multiplied by an adaptive weighting matrix. Therefore, the anomaly pixels are more easily distinguished from the background. Experiments on synthetic and real-world hyperspectral datasets demonstrate the superiority of our proposed method over other AD detectors.

Keywords: anomaly detection; hyperspectral imagery; low-rank representation; dictionary construction; HSI reconstruction; sparse coding; adaptive weighting

---

1. Introduction

In contrast to color and multispectral imagery, hundreds of narrow and contiguous spectral bands covering a wide range of wavelengths contained in hyperspectral imagery provide abundant spatial and spectral information about Earth observations [1,2]. Since each material has unique electromagnetic reflection characteristics at different wavelengths, their spectral information can be used for target detection [3]. According to the availability of prior knowledge about the target signatures, target detection can be divided into two categories: supervised and unsupervised [4]. Unsupervised target detection, known as anomaly detection (AD), has attracted a lot of attention over the last 20 years because it does not require any prior information about the spectral characteristics of targets that are usually difficult to obtain [5]. Moreover, it does not need radiation calibration and atmospheric absorption compensation [6].

Anomalies refer to the small objects with low probability of occurrence and whose spectra are significantly different from the main background. AD can be regarded as a binary classification problem designed to separate the background class and the anomaly class automatically [7]. In recent years, many AD methods have been proposed, and among them, the Reed-Xiaoli (RX) detector is the most well-known method based on statistical modeling [8]. It uses the probability density functions of the multivariate normal distribution to measure the probability of the detected pixel to be background, and its solution is the Mahalanobis distance between the spectrum of the detected pixel and the background. It has two versions: global RX (GRX) and local RX (LRX). Specifically, GRX estimates background statistics from the full image scene, whereas the background in LRX is estimated from the local neighborhood of the detected pixel using a dual-window strategy [9]. However, the background composition of HSI is usually complicated and nonhomogeneous in practical, so a single multivariate normal distribution is generally unsuitable for describing the background [10]. Moreover, the anomaly contamination in background statistics (background mean and covariance matrix) is another potential problem with RX. Based on these two shortcomings, several improved RX-based AD methods have been proposed. For example, the Gaussian mixture model-based detector [11] uses a mixture of multivariate Gaussian distributions to model the multimode background to capture the complexity of the background. The cluster-based anomaly detector (CBAD) [12] applies a clustering technique to divide the dataset into some homogeneous clusters and then implements RX on each cluster. The subspace RX (SSRX) [13] performs RX on a finite number of principal components obtained by principal component analysis (PCA), thereby reducing computational cost and improving the separability of background and anomalies. Due to the rich nonlinear information among the inter-bands of HSI, kernel-RX (KRX) [14] and support vector data description (SVDD) [15] are applied to project the original data into an infinite high-dimensional space through a kernel function. Cluster KRX (CKRX) [16], as an improved version of KRX, groups background pixels into clusters and then applies a fast eigendecomposition algorithm to generate anomaly indexes. It significantly reduces computation time by replacing each pixel with its cluster center. There are some AD methods trying to mitigate anomaly contamination for a pure estimation of the background. For example, the random-selection-based anomaly detector (RSAD) [17] applies a selection procedure several times to choose some representative background pixels. The blocked adaptive computationally efficient outlier nominator (BACON) detector [18] uses the subsets of the entire HSI to iteratively update a stable and robust background to suppress anomaly contamination in the background estimation.

With the development of representation theory in recent years, some representation-based methods have been successfully applied to AD. They sidestep the difficulty of modeling the complicated distribution of background in statistics-based methods. The sparse representation-based detector (SRD) [19] assumes that the spectrum of a pixel can be sparsely represented by a linear combination of a few sparse coefficients with respect to a background dictionary, and the reconstruction error is used to measure the anomaly response. The collaborative representation-based detector (CRD) [20] is based on the fact that background pixels can be well approximated by their spatial neighborhoods, whereas anomalies cannot. In addition to CRD, there are some other methods to incorporate spatial or feature information into detection and classification. In [21], during the recovery of sparse vector in sparse representation, two different approaches are proposed to incorporate the contextual information of HSI to improve the classification performance. In [22], the joint sparsity model is extended to a feature space induced by a nonlinear kernel function for improving the discrimination between background and targets. In this case, the spectral, spatial, and feature information are jointly used.

Recently, low-rank-based methods have drawn much attention and been applied to AD. It exploits the intrinsic low-rank property of background and the sparse property of anomalies [23]. It also does not require modeling the distribution of complex background. For instance, robust principal component analysis (RPCA) [24] performs detection by decomposing HSI data into a low-rank background matrix and a sparse anomaly matrix. However, the sparse matrix obtained is always contaminated by isolated noise, thus causing some false alarm points [25]. As an improvement, low-rank and sparse matrix decomposition (LRaSMD) [26] extracts noise from the valuable signals, and then further separates the low-rank background and sparse anomalies. The anomaly detector in [27] first extracts some source components by using the unmixing operation, and then identifies the components that are sparse and have the largest accumulated distance from other components. The optimization problem is converted to a low-rank matrix decomposition problem and can be solved. Low-rank representation (LRR) [28] assumes that the HSI data lie in multiple subspaces and requires a dictionary to span the data space to separate the background and anomalies. Due to the mixed property of real-world datasets, the pixels of an HSI are usually drawn from multiple subspaces. Therefore, compared with RPCA and LRaSMD, LRR is theoretically more suitable for real HSI datasets by imposing l_{21} constraint on the sparse component [25]. In addition, the l_{21} constraint makes the background component unaffected by the column-wise sparse anomalies [29]. Some advanced LRR-based AD methods have been proposed in recent years and they improve the detection performance of LRR from different aspects. For example, the anomaly detector based on low-rank and learned dictionary (LRALD) in [23] constructs a dictionary from the whole image with a random selection process and then performs LRR. The abundance- and dictionary-based low-rank decomposition (ADLR) in [25] applies spectral unmixing to obtain some abundance maps that contains more distinctive features, and then constructs a dictionary based on the mean shift clustering, and finally performs LRR. The low-rank and sparse representation-based detector (LRASR) in [28] improves LRR through a sparsity-inducing regularization term and a cluster-based dictionary construction strategy. It can be found that all these methods build a reasonable dictionary and try to make the anomalies easier to be recognized. Dictionary construction is an important process in many HSI problems and there are many ways to implement it. [30] proposes an AD method based on sparse presentation through constructing multiple dictionaries to learn discriminative features. In each category, the representative spectra that can significantly enhance the difference between background and anomalies are selected.

In the original model of LRR, the entire input dataset is used as the dictionary to span the data space. However, due to the anomaly contamination in this dictionary, sparse anomalies cannot be effectively separated from the background component [23]. In addition, the heavy computational burden caused by large data size is also an important issue. In the LRR model based on randomly selected dictionary, there is no guarantee that the selected dictionary atoms contain all background categories. In this paper, taking into account the above issues, a novel AD algorithm via dictionary construction-based LRR and adaptive weighting is proposed. To better represent the background subspace and separate the anomaly component from the background, a background dictionary construction strategy based on the usage frequency of each dictionary atom for HSI reconstruction is adopted in LRR. To cover all background classes in the dictionary, the *K*-means clustering is first executed to divide the data into several clusters. Then, we estimate the background pixels in each cluster. It is based on the observation that if an atom has a high usage frequency for HSI reconstruction, it is more likely to be a background pixel [31]. Therefore, from the perspective of the usage frequency of the dictionary atoms used for HSI reconstruction in each cluster, we can obtain a reasonable estimation of the background dictionary, which can exclude anomaly contamination and contain all background categories. Furthermore, for further enhancing the response difference between the anomaly pixels and the background pixels, an adaptive weighting method based on the reconstruction residual of the entire data with respect to the background dictionary constructed above is proposed. The final anomaly response of each pixel is calculated by multiplying the value obtained through LRR by the weight. Compared with the existing LRR-based detectors, our proposed algorithm avoids the randomness brought by the random selection process (compared with LRALD), does not damage the physical structure of HSI (compared with ADLR), and needs less computation time than LRASR, which adds a sparsity-inducing regularization term to LRR. In addition, the distinction between background and anomalies can be significantly improved by our adaptive weighting method, which has not been used in other LRR-based algorithms. The main contributions of our proposed algorithm for AD can summarized as follows:

(1) Use of the LRR model. First, the LRR model is highly consistent with the hyperspectral AD problem and is therefore used in this paper. Second, the real-world HSIs are usually lying in multiple subspaces due to the presence of mixed pixels caused by insufficient sensor resolution [25]. The LRR model assumes that the data are in multiple subspaces by imposing l_{21} constraint on the sparse component, so it is suitable for real data. Third, the l_{21} constraint also makes the background unaffected by the column-wise sparse anomalies [29].

(2) Background dictionary construction strategy. To better separate the sparse anomalies from the background component and reduce the computational burden, a novel background dictionary is constructed by analyzing the usage frequency of the dictionary atoms for HSI reconstruction in each cluster. The dictionary is an excellent representation of the background subspace since it excludes anomaly contamination and covers all background categories. Therefore, the sparse component containing most of the anomaly information is extracted accurately.

(3) Adaptive weighting method. To further enhance the diversity between the background pixels and the anomaly pixels, an adaptive weighting method is introduced in our proposed algorithm by reusing the constructed background dictionary. By multiplying the results of LRR by the weights, the background and anomalies are more easily distinguished in the final detection map.

The rest of this paper is organized as follows. In Section 2, we briefly review the LRR model and its solution. In Section 3, the background dictionary construction strategy and adaptive weighting method in our proposed algorithm are described in detail. In Section 4, experimental results and analysis based on synthetic and real-world HSI datasets are provided. Finally, Section 5 concludes this paper.

2. Low-Rank Representation and Its Solution

In this section, we briefly introduce the consistency of the LRR model and the hyperspectral AD theory. Then the solution of LRR is provided. It plays a significant role in our proposed algorithm.

2.1. LRR Model for AD

There are several typical characteristics in HSIs. (1) Unlike anomaly pixels, there are strong correlations among the background pixels, i.e., the spectrum of a background pixel can be represented as a linear combination of some other background pixels [32]. (2) Anomalies occupy only a few pixels with a low probability of occurrence, that is, they are sparse spatially [33]. (3) Due to the limitation of the resolution of hyperspectral sensors, there are many mixed pixels in the real-world HSIs. Since the spectrum of each mixed pixel can be represented as a mixture of some pure materials (endmembers) and each endmember can be described in a subspace, all pixels in the HSI can be drawn from multiple subspaces [34]. The LRR model takes into account the above characteristics of HSI and is therefore very suitable for AD. The model of LRR is as follows:

$$\min_{\mathbf{S},\mathbf{E}} \|\mathbf{S}\|_* + \lambda \|\mathbf{E}\|_{2,1} \quad s.t. \ \mathbf{X} = \mathbf{DS} + \mathbf{E}, \tag{1}$$

where the HSI matrix $\mathbf{X}$ is decomposed into a background component $\mathbf{DS}$ and an anomaly component $\mathbf{E}$. $\mathbf{D}$ is the dictionary spanning the data space, and $\mathbf{S}$ is called the low-rank representation of $\mathbf{X}$ with respect to $\mathbf{D}$. $\|\cdot\|_*$ is the nuclear norm, which is a good alternative to the rank function because of the convex optimization problem it causes. It attempts to find the lowest-rank representation of all data jointly by imposing low-rank constraint on the representation coefficient matrix $\mathbf{S}$, instead of the background itself. $\|\cdot\|_{2,1}$ is the l_{21}-norm used to encourage the sparse nature of $\mathbf{E}$, indicating that the anomalies are column-wise sparse, i.e., sample-specific. $\mathbf{E}$ is obtained by the residual of the data and the recovered background component. It contains most of the anomaly information and can therefore be used for AD. $\lambda > 0$ is the tradeoff parameter used to balance these two parts. LRR assumes that the data are drawn from multiple subspaces corrupted by anomalies and tries to find the lowest-rank representation of all data jointly to recover the underlying multiple subspaces. In the original LRR model, the entire input matrix $\mathbf{X}$ is used as the dictionary $\mathbf{D}$ to span the data space.

The difference between the PCA model and the LRR model is illustrated in Figure 1. As we can see, with the l_{21} constraint, LRR assumes that the data lie in multiple subspaces, while the data in PCA are drawn from a single subspace because of the l_1 constraint. Due to the presence of mixed pixels, multiple subspaces can better describe the real HSI data. In addition, the l_{21} constraint on the sparse component in LRR indicates that the anomalies are column-wise sparse, i.e., sample-specific. It means that most of the data vectors are clean and a few of them are corrupted, which ensures that the background spectra are not affected by the anomalies. On the contrary, in RPCA and LRaSMD [35,36], the anomalies are entry-wise sparse, and all the spectra of background component can be affected by the nonzero anomalies due to the l_1 constraint on the anomaly component. Moreover, LRR can also exclude the noise that is normally randomly distributed in each band from the anomaly component. After comparison, we find that LRR can better separate a sparse component as pure as possible from the background. The models and characteristics of RPCA, LRaSMD and LRR are summarized in Table 1. The advantages of LRaSMD over RPCA is that it considers the additive noise in the dataset and thus avoids the isolated noise being detected as anomalies [37]. In Section 4.2, we will experimentally demonstrate that the l_{21} constraint is superior to the l_1 constraint for the LRR model.

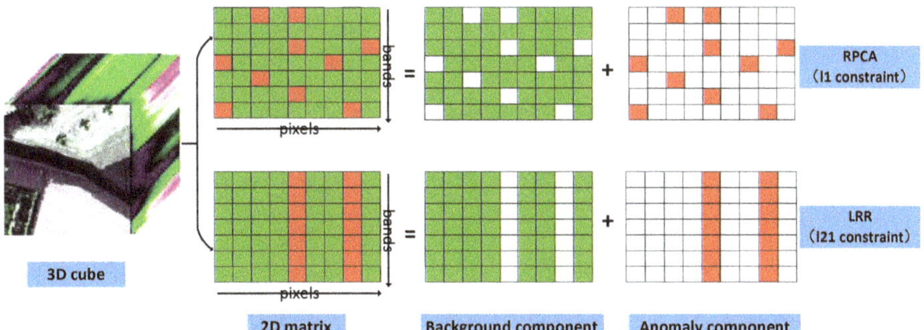

Figure 1. Difference between the l_1 constraint in RPCA and the l_{21} constraint in LRR. Each square represents the digital number of a pixel in a band. The greens correspond to the backgrounds and the reds correspond to the anomalies.

Table 1. Comparison of models and characteristics of RPCA, LRaSMD and LRR.

Methods	Models	Theories	Characteristics	
RPCA	$\mathbf{X} = \mathbf{L} + \mathbf{E}$	Abstract the low-rank component as $\mathbf{L}$ and the sparse component as $\mathbf{E}$		
LRaSMD	$\mathbf{X} = \mathbf{L} + \mathbf{E} + \mathbf{N}$	Consider the additional noise; Abstract the low-rank component as $\mathbf{L}$ and the sparse component as $\mathbf{E}$ with predefined $\mathrm{rank}(\mathbf{L})$ and $\mathrm{card}(\mathbf{E})$	Single subspace assumption	l_1 constraint on $\mathbf{E}$
LRR	$\mathbf{X} = \mathbf{DS} + \mathbf{E}$	Recover the background component $\mathbf{DS}$ using the lowest representation of all data jointly	Multiple subspaces assumption	$l_{2,1}$ constraint on $\mathbf{E}$

2.2. Solution of LRR

To solve the problem in Equation (1), we introduce an auxiliary variable $\mathbf{J}$ to make the objective function separable [38]. The optimization problem is converted to:

$$\min_{\mathbf{S},\mathbf{E},\mathbf{J}} \|\mathbf{J}\|_* + \lambda \|\mathbf{E}\|_{2,1} \quad s.t. \quad \mathbf{X} = \mathbf{DS} + \mathbf{E}, \quad \mathbf{S} = \mathbf{J}, \tag{2}$$

Then, the following Lagrange function can be obtained:

$$\min_{\mathbf{S},\mathbf{E},\mathbf{J},\mathbf{Y}_1,\mathbf{Y}_2} \|\mathbf{J}\|_* + \lambda \|\mathbf{E}\|_{2,1} + \mathrm{tr}\left[\mathbf{Y}_1^T(\mathbf{X} - \mathbf{DS} - \mathbf{E})\right] + \mathrm{tr}\left[\mathbf{Y}_2^T(\mathbf{S} - \mathbf{J}))\right]$$
$$+ \frac{\mu}{2}(\|\mathbf{X} - \mathbf{DS} - \mathbf{E}\|_F^2 + \|\mathbf{S} - \mathbf{J}\|_F^2), \tag{3}$$

where $\mathbf{Y}_1$ and $\mathbf{Y}_2$ are Lagrange multipliers, $\mu > 0$ is the penalty parameter. The equation can be solved by inexact Augmented Lagrange Multiplier (ALM) via alternatively updating one variable when the others are fixed [39]. The solution of LRR is outlined in Algorithm 1.

Algorithm 1. Solving LRR by Inexact ALM for AD

Input: dataset matrix: $\mathbf{X}$; dictionary matrix: $\mathbf{D}$; tradeoff parameter: $\lambda > 0$
Initialize: $\mathbf{S} = \mathbf{J} = \mathbf{E} = \mathbf{Y}_1 = \mathbf{Y}_2 = \mathbf{0}$, $\mu = 10^{-6}$, $\mu_{\max} = 10^{10}$, $\rho = 1.1$, $\varepsilon = 10^{-8}$
While not converged do
1. Update $\mathbf{J}$ and fix the others: $\mathbf{J} = \arg \min \frac{1}{\mu}\|\mathbf{J}\|_* + \frac{1}{2}\|\mathbf{J} - (\mathbf{S} + \mathbf{Y}_2/\mu)\|_F^2$
2. Update $\mathbf{S}$ and fix the others: $\mathbf{Z} = (\mathbf{D}^T\mathbf{D} + \mathbf{I})^{-1}\left[\mathbf{D}^T\mathbf{X} - \mathbf{D}^T\mathbf{E} + \mathbf{J} + (\mathbf{D}^T\mathbf{Y}_1 - \mathbf{Y}_2)/\mu\right]$
3. Update $\mathbf{E}$ and fix the others: $\mathbf{E} = \arg \min \frac{\lambda}{\mu}\|\mathbf{E}\|_{2,1} + \frac{1}{2}\|\mathbf{E} - (\mathbf{X} - \mathbf{DS} - \mathbf{Y}_1/\mu)\|_F^2$
4. Update the Lagrange multipliers: $\mathbf{Y}_1 = \mathbf{Y}_1 + \mu(\mathbf{X} - \mathbf{DS} - \mathbf{E}))$, $\mathbf{Y}_2 = \mathbf{Y}_2 + \mu(\mathbf{S} - \mathbf{J}))$
5. Update the tradeoff parameter μ: $\mu = \min(\rho\mu, \mu_{\max})$
6. Check the convergence conditions: $\|\mathbf{X} - \mathbf{DS} - \mathbf{E}\|_\infty < \varepsilon$ and $\|\mathbf{S} - \mathbf{J}\|_\infty < \varepsilon$, where $\|\cdot\|_\infty$ is the infinite norm.
end while
Output: the optimal solution of $\mathbf{S}$ and $\mathbf{E}$

In Algorithm 1, the sub-problems in step 1 and step 3 are respectively solved by the singular value thresholding operation [40] and the l_{21} minimization operation [38].

Finally, the anomaly response of pixel $\mathbf{x}$ is calculated by the l_2-norm of the corresponding column of $\mathbf{E}$, i.e.,

$$v(\mathbf{x}) = \|\mathbf{E}_i(:)\|_2 \quad (1 \leq i \leq N), \tag{4}$$

where $\mathbf{E}_i(:)$ is the corresponding column of pixel $\mathbf{x}$ in $\mathbf{E}$, and N is the number of pixels in $\mathbf{X}$.

3. Proposed Method

The LRR model has high consistency with the hyperspectral AD problem because it can effectively capture the low-rank representation of all data jointly and mine the sparse component contained in the dataset for AD [38]. However, in LRR, the entire input dataset or randomly selected data are usually used as the dictionary, where the former will bring a large computational burden and an unsatisfactory separation of sparse anomalies from the background component, while the latter cannot ensure that all background material categories are covered in the dictionary [28]. In this case, to achieve a better separation performance between the background component and the anomaly component with a low computational complexity, a background dictionary that excludes anomaly contamination and contains all background categories is required. In Section 3.1, we propose a novel background dictionary construction strategy based on the usage frequency of the dictionary atoms for HSI reconstruction in each cluster. In addition, for further enhancing the response difference between the background pixels and the anomaly pixels, an adaptive weighting method based on the reconstruction residual of the entire data with respect to the constructed dictionary is introduced in Section 3.2.

3.1. Background Dictionary Construction Strategy

To contain all background categories in the dictionary, the K-means clustering algorithm is first used to divide the data into K clusters, where the value of K can be estimated a priori by the HySime algorithm [41]. A complex background consisting of many types of background materials should have a larger K, and the value of K we choose should be larger than the true number of background categories in the scene to cover all background materials. After performing K-means clustering on dataset $\mathbf{X}$, we obtain K clusters $\{\mathbf{X}_1, \mathbf{X}_2, ..., \mathbf{X}_K\}$. For each cluster $\mathbf{X}_i (1 < i < K)$, we randomly select M percent of the pixels to form the dictionary $\mathbf{B}$ to sparsely reconstruct each sample in $\mathbf{X}_i$, and then the sparse reconstruction coefficients are obtained by using the sparse coding method [31]. Specifically, the spectrum of pixel $\mathbf{x}$ is assumed to be approximately represented as a linear combination of only a few atoms in $\mathbf{B}$, i.e.,

$$\mathbf{x} = \mathbf{B}\boldsymbol{\alpha} + \mathbf{r}, \tag{5}$$

where $\mathbf{x}$ is a sample in $\mathbf{X}_i$, $\boldsymbol{\alpha}$ is the reconstruction coefficient vector where most of the entries are zero, and $\mathbf{r}$ is the residual vector. Given a fixed dictionary $\mathbf{B}$, $\boldsymbol{\alpha}$ can be obtained by solving the following optimization problem:

$$\min \|\mathbf{x} - \mathbf{B}\boldsymbol{\alpha}\|_2 \quad s.t. \quad \|\boldsymbol{\alpha}\|_0 < K_0, \tag{6}$$

where $\|\cdot\|_0$ denotes the l_0-norm and K_0 is the upper bound of the sparsity level. The sparse coding method provides the optimal solution of $\boldsymbol{\alpha}$ using greedy pursuit algorithms, such as matching pursuit (MP) [42] and orthogonal matching pursuit (OMP) [43], where OMP is superior to MP due to its fewer iterations and better convergence. For cluster $\mathbf{X}_i$, the sparse coefficient vector $\boldsymbol{\alpha}$ for each sample is obtained, constituting the sparse coefficient matrix $\mathbf{A}_i$.

We focus on $\mathbf{A}_i$ and then count the usage frequency of each atom in $\mathbf{B}$ for reconstructing $\mathbf{X}_i$. For a pixel in $\mathbf{X}_i$, some dictionary atoms in $\mathbf{B}$ participate in its reconstruction while the others do not. As mentioned above, background dominates the scene while the anomalies occupy only a few pixels with a low probability of occurrence. From this point of view, we can conclude that if a dictionary atom is used frequently for reconstruction, it contains more background information and is more likely to be a background pixel [31]. In contrast, the rarely used atoms are anomaly pixels with high probability. In this case, in cluster $\mathbf{X}_i$, assuming $\mathbf{b}_j$ is the jth atom of $\mathbf{B}$, its usage frequency f_j for reconstructing $\mathbf{X}_i$ is defined as:

$$f_j = \frac{\sum_{k=1}^{N_i} |\alpha_{j,k}|}{\|\mathbf{A}_i\|_1}, \tag{7}$$

where $\|\cdot\|_1$ denotes the l_1-norm, which is the sum of the absolute values of all elements in a matrix. N_i is the number of pixels in $\mathbf{X}_i$. The numerator in Equation (7) is the sum of the reconstruction coefficients of atom $\mathbf{b}_j$ used to reconstruct all pixels in $\mathbf{X}_i$, and the denominator is the sum of all entries in $\mathbf{A}_i$. Then, we choose P atoms corresponding to the first P largest usage frequency to constitute the background pixels we estimate in $\mathbf{X}_i$.

The above procedure is repeated in each cluster with the same M and P. The estimated background pixels in all clusters are summarized, constituting the estimated background pixels in the whole image. Figure 2 shows an illustration of the background dictionary construction strategy. The constructed background dictionary, which effectively excludes possible anomalies and contains all background categories in the scene, is finally used for LRR. It is worth nothing that since sparse coding requires an over-complete dictionary, in each cluster, the number of atoms randomly selected for HSI reconstruction should be larger than the dimension H of the dataset. If the total number of pixels in a cluster is less than H, then this cluster should be ignored and skipped because it may belong to the anomalies due to its small size and we have set K larger than the true number of background material categories.

In Section 4.3, we will compare the dictionary we construct with two other commonly used dictionaries, including the dictionary using the entire input data and the dictionary with randomly selected atoms, to demonstrate the advantages of our proposed dictionary construction strategy in terms of detection performance and computation time.

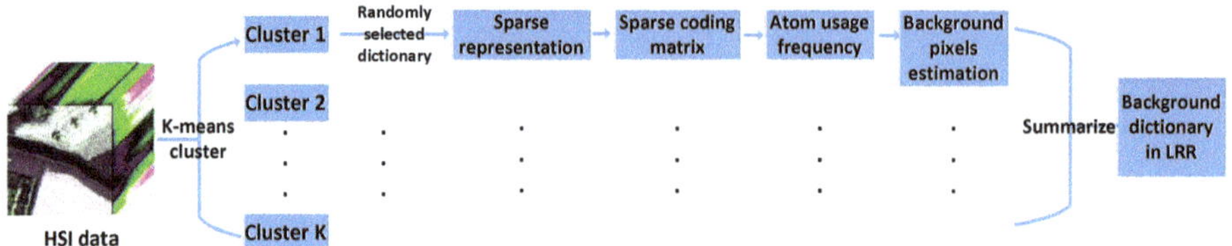

Figure 2. Illustration of the background dictionary construction strategy.

3.2. Adaptive Weighting Method

After performing LRR on an HSI based on our constructed background dictionary, the anomaly response of each pixel is calculated using the sparse component obtained. However, the response difference between anomaly pixels and background pixels can be further enhanced to improve the discrimination degree between them. Fortunately, through implementing sparse reconstruction on the entire dataset based on the background dictionary constructed in Section 3.1, the resulting reconstruction residuals provide an effective way to assign adaptive weight values to different pixels according to their likelihood of being background pixels or anomalies. It is well known that the background in HSI is highly correlated and the spectrum of a background pixel can be represented by a linear combination of some other background pixels, while the anomalies cannot. That is to say, compared with anomaly pixels, the background pixels can be better sparsely reconstructed by the background dictionary $\mathbf{D}$ [44]. Similarly, the sparse coefficient vector can be solved by the OMP algorithm [43]. Therefore, the following reconstruction residual can be used to assign an adaptive weight to each pixel:

$$\xi(\mathbf{x}) = \|\mathbf{x} - \mathbf{D}\boldsymbol{\beta}\|_2, \tag{8}$$

where $\mathbf{x}$ is an arbitrary test pixel in $\mathbf{X}$, $\mathbf{D}$ is the background dictionary constructed in Section 3.1, and $\boldsymbol{\beta}$ is the sparse coefficient vector of $\mathbf{x}$ with respect to $\mathbf{D}$. Obviously, an anomaly pixel will obtain a larger residual while the residual for a background pixel will be small. In this case, the response difference between the background pixels and anomalies is enhanced, which will further improve the AD performance. The final anomaly response of each pixel is calculated by multiplying the weight defined in Equation (8) by the anomaly value obtained through LRR, i.e.,

$$v'(\mathbf{x}) = \xi(\mathbf{x}) \cdot v(\mathbf{x}), \tag{9}$$

3.3. Overview of the Proposed Algorithm

According to the consistency of the LRR model and the AD theory, the detection algorithm proposed in this paper is based on the LRR model, which can effectively mine the hidden lowest-rank structure in the data and extract the sparse component for AD [38]. A background dictionary construction strategy is applied to better depart the sparse anomalies from the background component. An adaptive weighting method is introduced for further enhancing the response difference between the background pixels and the anomaly pixels. Our proposed method is called the hyperspectral AD algorithm via dictionary construction-based LRR and adaptive weighting (DCLaAW). The main steps of DCLaAW are summarized as Algorithm 2, and the corresponding schematic flowchart is given in Figure 3.

Algorithm 2. Hyperspectral AD via the proposed DCLaAW

Input: HSI data: **X**; parameters: $K, M, P, \lambda > 0$
1. Divide **X** into K clusters using K-means clustering.
2. for $i = 1 : K$
   (1) Randomly select M percent of the pixels in this cluster as the dictionary atoms for HSI reconstruction.
   if $L < H$ (L is the number of pixels in this cluster, and H is the number of bands of **X**)
      ignore and skip this cluster.
   end
   (2) Perform sparse coding to obtain the sparse coefficient matrix **A**.
   (3) Count the usage frequency f of each atom in the dictionary based on **A**.
   (4) Choose P pixels corresponding to the first P largest f as the background pixels we estimate.
   end
3. Summarize the estimated background pixels in all clusters to constitute the background dictionary **D** for LRR.
4. Perform LRR using Algorithm 1 to obtain the anomaly component **E**, and then calculate the response value v of each pixel.
5. Create the weight matrix based on the reconstruction residuals of **X** with respect to **D**.
6. Multiply v by the weight to obtain the final anomaly response value of each pixel.
Output: Anomaly response values of **X**

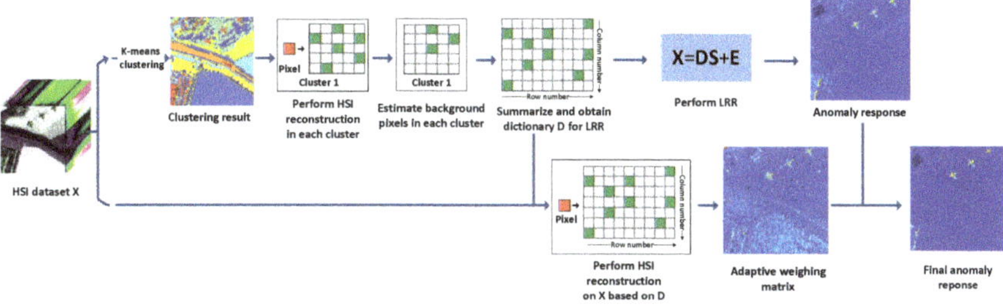

Figure 3. Schematic flowchart of the proposed DCLaAW algorithm for hyperspectral anomaly detection.

4. Experiments and Analysis

In this section, the effectiveness and superiority of our proposed DCLaAW are evaluated on both synthetic and real-world datasets. The AD performance is assessed by four commonly used indexes, including color detection map, ROC (receiver operating characteristic) curve [45], AUC (area under curve) value [46], and background-anomaly separation map. The superiority of the l_{21} constraint in LRR, the effectiveness of both the dictionary construction strategy and the adaptive weighting method are illustrated in Section 4.2, Section 4.3 and Section 4.4, respectively. In Section 4.5, we compare the detection performance of DCLaAW with that of eight existing state-of-the-art anomaly detectors in detail. Then, the sensitivity of the detection performance of DCLaAW to the relevant parameters is analyzed in Section 4.6. In Section 4.7, we provide a comparison between the LRR, the sparsity formulation, and the L2 formulation to further demonstrate the superiority of our proposed algorithm. All the experiments are implemented on a personal computer with an Intel Core i3 3.70-GHz central processing unit, 8GB memory, and 64-bit Windows 7. MATLAB 2016a provides the simulation and computing platform.

4.1. Dataset Description

The synthetic dataset is generated based on a real-world dataset collected by the HyMap airborne hyperspectral imaging sensor from a small town of Cook City, MT, USA [47]. It has an area of 280×800 pixels and 126 spectral bands with wavelengths ranging from 450 to 2500 nm. After removing the bands corresponding to water absorption regions and low signal-to-noise ratio, 120 bands are retained. A sub-region with a size of 230×240 pixels on the right side of the scene is chosen to form the simulated image, where the background types mainly conclude trees, grasses, and rocks. Based on the linear mixing model (LMM), a synthetic subpixel anomaly with spectrum $\mathbf{x}$ and a specified abundance fraction α is generated by fractionally implanting a desired target with spectrum $\mathbf{t}$ in a given background pixel with spectrum $\mathbf{b}$ [48], as follows:

$$\mathbf{x} = \alpha \cdot \mathbf{t} + (1 - \alpha) \cdot \mathbf{b}, \tag{10}$$

The implanted target corresponds to a vehicle with distinctive spectral characteristics outside the scene. In this experiment, 30 anomalies are synthesized and distributed in 5 rows and 6 columns. In each row, the abundance fraction α remains unchanged and the sizes of anomalies are $1 \times 1, 1 \times 1$, $3 \times 3, 3 \times 3, 5 \times 5$, and 5×5 from left to right. In each column, the abundance fraction α are 0.1, 0.3, 0.5, 0.8, and 1.0 from top to bottom. The pseudo-color image, the ground-truth map, and the spectral curves of the implanted target and main backgrounds are shown in Figure 4a–c, respectively.

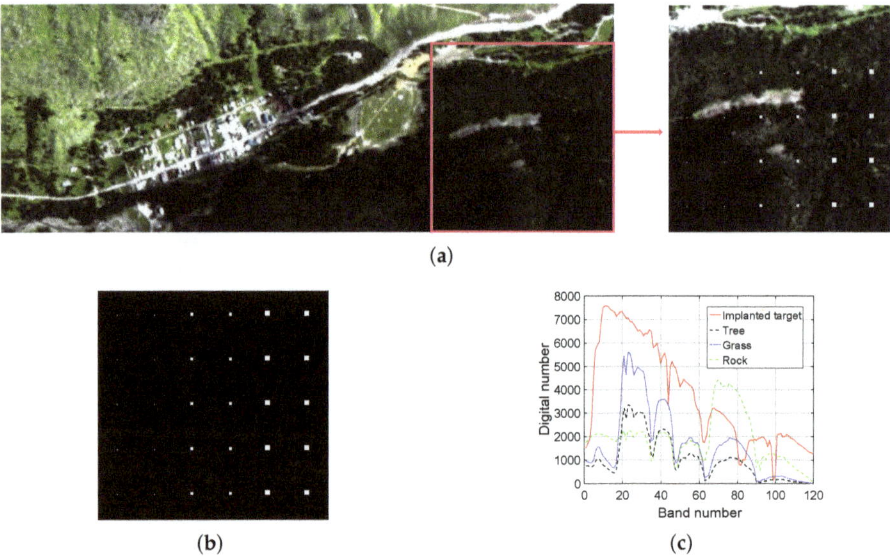

Figure 4. Synthetic dataset. (**a**) Pseudo-color image of the scene; (**b**) Ground-truth map; (**c**) Spectral curves of implanted target and main backgrounds.

The first real-world dataset was collected by the Airborne Visible/Infrared Imaging Spectrometer (AVIRIS) sensor from the San Diego airport area, San Diego, CA, USA [49]. It has a spatial resolution of approximately 3.5m and 224 spectral bands spanning a wavelength range of 0.37 to 2.51 um. After removing the bands corresponding to water absorption regions and low signal-to-noise ratio, 189 bands are retained. A sub-region with a size of 100×100 pixels is chosen for this experiment, where the background types mainly include parking apron, road, roofs, and shadow. Three aircraft, occupying 58 pixels in the image, are considered as anomalies in this experiment. The pseudo-color image, the ground-truth map, and the spectral curves of mean anomalies and main backgrounds are shown in Figure 5a–c, respectively.

The second real-world hyperspectral dataset was collected by the Hyperspectral Digital Imagery Collection Experiment (HYDICE) remote sensor. It covers a suburban residential area with 10 nm spectral resolution and 210 spectral bands ranging from 0.4 to 2.5 um [50]. After removing the bands corresponding to water absorption regions and low signal-to-noise ratio, 160 bands are retained. A sub-region with a size of 80×100 pixels is chosen for this experiment, where the background types mainly include parking lot, water, soil and two roads. Some synthetic vehicles, containing 21 pixels, are the anomalies in this experiment. The pseudo-color image, the ground-truth map, and the spectral curves of mean anomalies and main backgrounds are shown in Figure 6a–c, respectively.

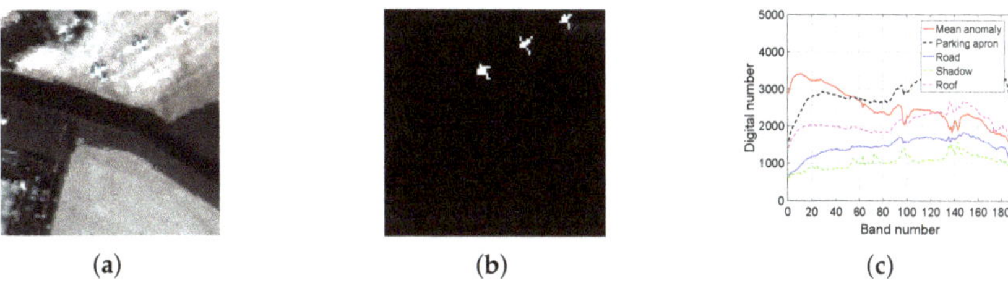

Figure 5. San Diego dataset. (**a**) Pseudo-color image; (**b**) Ground-truth map; (**c**) Spectral curves of mean anomalies and main backgrounds.

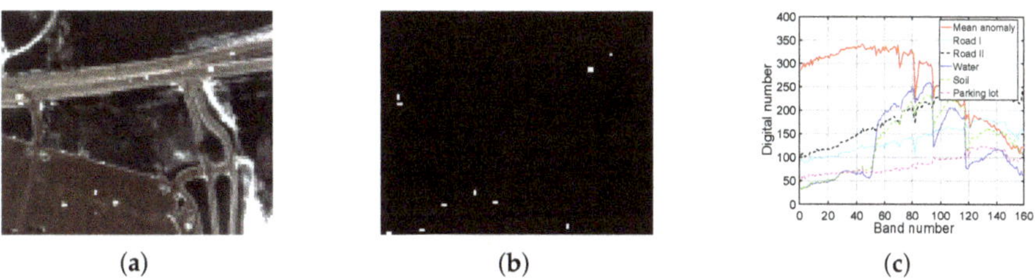

Figure 6. Urban dataset. (**a**) Pseudo-color image; (**b**) Ground-truth map; (**c**) Spectral curves of mean anomalies and main backgrounds.

4.2. Superiority of the l_{21} Constraint for LRR

As described in Section 2.1, for the sparse component in LRR, the l_{21} constraint is theoretically more suitable to discriminate the background and anomalies than the l_1 constraint. In this section, to experimentally demonstrate the superiority of the l_{21} constraint, the detection performance of LRR under l_{21} constraint is compared with that under l_1 constraint. To compare only the effects of different constraints in the performance of LRR, the optimal background dictionary is adopted while the adaptive weighting is not implemented. Here we present the experimental results for the San Diego dataset, and the other two datasets can get the similar conclusions. The detection maps obtained by LRR with different constraints are shown in Figure 7, and the corresponding AUC values and calculation times (in seconds) are listed in Table 2. The ROC curve plots the relationship between the false alarm rate (FAR) and the detection rate (DR), where the FAR is generally measured by a base 10 logarithmic scale to better illustrate the details. The closer the ROC curve is to the upper left corner of the coordinate plane, the better the performance of the corresponding detector. The AUC value represents the whole area under the ROC curve, so a larger AUC value usually means a better detection performance. For each constraint, the sensitivity of the obtained AUC value to the number of dictionary atoms is shown in Figure 8.

As shown in Figure 7, the detection map obtained by the l_1 constraint has significantly more false alarm points than the l_{21} constraint. This is mainly because the l_1 constraint finds the entry-wise sparse points, which are usually sparse in a certain band, not in all bands. This results in the background pixels that are sparse in only a band being extracted into the sparse component, and further leads to serious false alarms in the detection result. From Table 2, we see that the l_{21} constraint achieves a slightly larger AUC value, consistent with the observation in the detection maps. In addition, the l_{21} constraint requires less computation time than the l_1 constraint and is therefore more practical. After several experiments, we find that the l_1 constraint requires 240 iterations in one experiment, while the l_{21}

constraint requires only 152 iterations. Figure 8 shows that the LRR with l_{21} constraint is more robust to the number of dictionary atoms. Therefore, after comprehensive consideration, we believe that the l_{21} constraint is superior to the l_1 constraint both theoretically and experimentally.

(a)

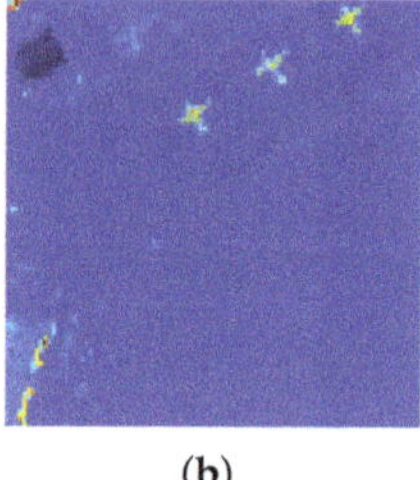

(b)

Figure 7. Color detection maps obtained by LRR with different constraints for the San Diego dataset. (a) l_1 constraint; (b) l_{21} constraint.

Table 2. Performance comparison of different constraints for the San Diego dataset.

Constraint	l_1 Constraint	l_{21} Constraint
AUC value	0.9936	0.9949
Computation time (s)	72.923	49.260

Figure 8. AUC values achieved by LRR under different numbers of dictionary atoms for the San Diego dataset. (a) l_1 constraint; (b) l_{21} constraint.

4.3. Effectiveness of the Background Dictionary Construction Strategy

In this section, our proposed background dictionary construction strategy is compared with two other commonly used LRR dictionaries, including the dictionary using the entire input data and the dictionary with randomly selected atoms, to demonstrate the superiority of our dictionary in terms of detection performance and computation time. In the original LRR, the entire input matrix is used as the dictionary to span the data space. In the randomly selected dictionary-based LRR, atoms in the dictionary are randomly selected from the entire dataset. In this comparison, for the sake of fairness, the number of randomly selected atoms is set equal to the number of atoms in DCLaAW. To make objective comparisons only for different dictionaries, we do not implement weighting operation when performing DCLaAW in this section. When the original LRR is executed on a large data, an error occurs due to "out of memory". Therefore, in this part, a sub-region taken from the upper right corner of the San Diego image is used as the toy dataset to perform the experiment. The ground-truth map, and the color detection maps achieved by these three different algorithms are shown in Figure 9 for intuitive comparisons. The ROC curves of each algorithm and their corresponding AUC values are plotted in Figure 10 for quantitative comparisons. In addition, the computation times of each algorithm are listed in Table 3 for a practical comparison.

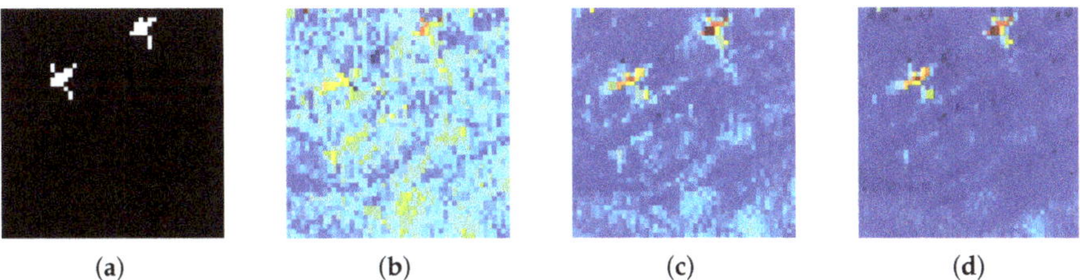

(**a**) (**b**) (**c**) (**d**)

Figure 9. Color detection maps obtained by LRR using different dictionaries for the toy dataset. (**a**) Ground-truth map; (**b**) Original LRR; (**c**) LRR using randomly selected dictionary; (**d**) LRR using our dictionary.

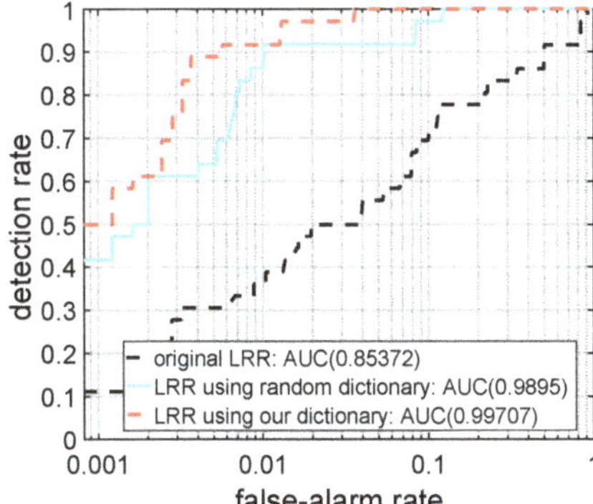

Figure 10. ROC curves and AUC values achieved by LRR using different dictionaries for the toy dataset.

Table 3. Computation times of LRR using different dictionaries for the toy dataset.

Time(s)	Original LRR	LRR Using Random Dictionary	LRR Using Our Dictionary
Toy Dataset	1340.505	9.471	11.057

As shown in Figure 9, the LRR algorithm using our dictionary achieves the best detection map in terms of background suppression and anomaly highlighting. For the original LRR, since the whole dataset, as the dictionary for LRR, cannot separate the background component and the anomaly component very well, it is difficult to identify the anomalous aircraft in the detection map. For LRR based on randomly selected dictionary, random selection cannot avoid anomalies being selected as dictionary atoms, and it is difficult to ensure that each background category is covered. Therefore, the background component extracted by it cannot adequately describe the real background. Our proposed background dictionary construction strategy can guarantee the exclusion of anomaly contamination and the inclusion of all background categories in the background dictionary to a considerable extent, thus providing the best detection map. From Figure 10, we can see that the ROC curve obtained by the LRR algorithm using our dictionary is basically always above that obtained by the LRR using the other two dictionaries. Consistently, the AUC value achieved by LRR using our dictionary is the largest. In addition, Table 3 shows that the time taken to execute the original LRR is long, so it is impractical to use it to process the real-world HSI datasets. Although the computational cost of the LRR using our dictionary is slightly larger than that of the LRR using a random dictionary, it is within an acceptable range.

4.4. Effectiveness of the Adaptive Weighting

After performing LRR based on the background dictionary we construct, the adaptive weighting method described in Section 3.2 is implemented to further increase the diversity between the background pixels and the anomaly pixels. The weighting effect can be clearly reflected by the detection map and the background-anomaly separation map. Here, to demonstrate the effectiveness of our proposed adaptive weighting method, the detection result obtained by DCLaAW with adaptive weighting is compared with that obtained by DCLaAW without adaptive weighting. The detection maps and normalized background-anomaly separation maps for the three datasets are shown in Figures 11 and 12, respectively. The background-anomaly separation map is a graph used to evaluate the separation performance of background pixels and anomaly pixels. It normalizes the detection result to 0-1 and uses a green box and a red box to represent the compactness and tendency of the distribution of backgrounds and anomalies, respectively. The central mark of each box is the median, the bottom and top edges refer to the lower quartile and the upper quartile, and the whisker are the extreme values within 1.5 times the interquartile range from the end of the box. Therefore, a larger gap between two boxes means a better separation between background and anomalies.

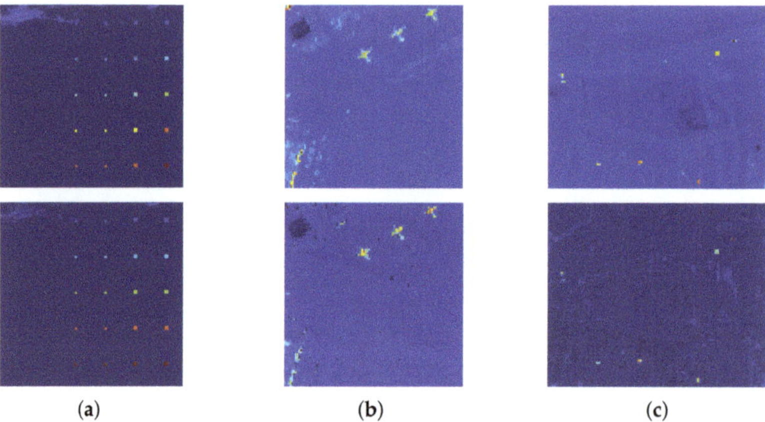

$$\text{(a)} \qquad\qquad \text{(b)} \qquad\qquad \text{(c)}$$

Figure 11. Effect of adaptive weighting on the detection map obtained by DCLaAW for each dataset. For each dataset, the top is the DCLaAW without adaptive weighting and the bottom is the DCLaAW with adaptive weighting. (**a**) Synthetic dataset; (**b**) San Diego dataset; (**c**) Urban dataset.

From Figure 11, we can see that for the San Diego and Urban datasets, the response brightness of the background pixels through weighting is significantly lower than that without weighting. The anomalous are also brightened noticeably. For the Synthetic dataset, after weighting, the background materials in the upper left corner are suppressed and the response outputs of the anomalies in the third to fifth rows are greatly improved. This effect can be clearly observed through the background-anomaly separation map shown in Figure 12, where the gap between the background box and the anomaly box becomes larger after weighting, meaning an easier identification of anomalous objects from the background.

4.5. Detection Performance

Eight state-of-the-art anomaly detectors are used as the benchmarks to evaluate the detection performance of our proposed DCLaAW, including GRX [8], LRX [9], KRX [14], CKRX [16], SSRX [13], CRD [20], LRaSMD [36], and LRASR [28]. All compared detectors are implemented with their optimal parameters.

For the synthetic dataset, the color detection maps of all compared algorithms are shown in Figure 13 for an intuitive comparison. As shown, GRX obtains the worst detection map, where almost no anomalies can be successfully detected. LRX performs well for anomalies with an abundance fraction greater than 0.5 because of its advantages in dealing with local uniform background. For KRX, all the anomalies except for those in the first row are highlighted satisfactorily, but it is obvious that the background materials corresponding to the grasses and rocks in the scene are not well suppressed. For CKRX, all the anomalies are effectively highlighted, but the background materials, especially rocks and grasses, have undesirably high response values. For SSRX, anomalies with a large abundance fraction are well identified, but there are still some background materials with slightly high response. Both CRD and LRaSMD achieve a satisfactory extrusion for almost all anomalies, regardless of their sizes. However, they perform poorly for anomalies with an abundance fraction of 0.1 and there is some noise pollution scattered throughout the detection map of CRD. Compared with LRASR, our proposed DCLaAW achieves a better performance in anomaly highlighting and background suppression due to its more reasonable background dictionary construction strategy and adaptive weighing. All anomalies can be detected by DCLaAW, regardless of their sizes and abundance fractions. In general, besides DCLaAW, the detection map of LRaSMD is relatively good.

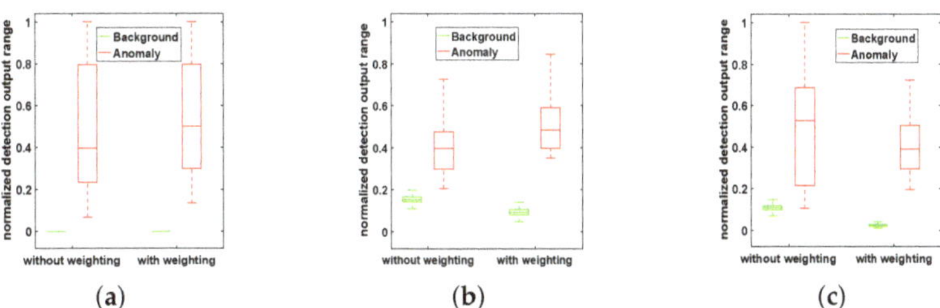

Figure 12. Effect of adaptive weighting on the background-anomaly separation map for each dataset. (**a**) Synthetic dataset; (**b**) San Diego dataset; (**c**) Urban dataset.

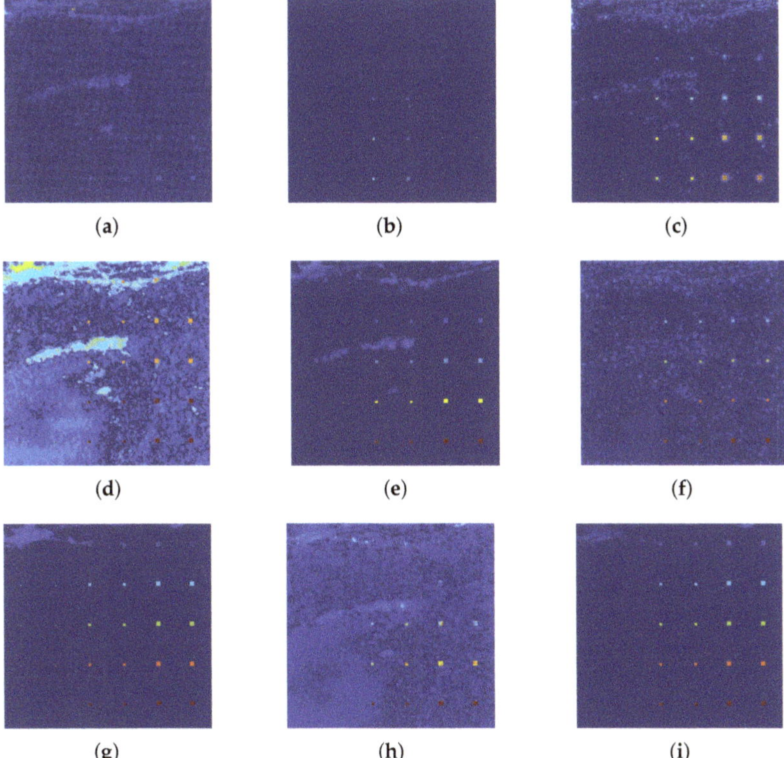

Figure 13. Color detection maps of all compared algorithms for the Synthetic dataset. (**a**) RX; (**b**) LRX; (**c**) KRX; (**d**) CKRX; (**e**) SSRX; (**f**) CRD; (**g**) LRaSMD; (**h**) LRASR; (**i**) DCLaAW.

Figure 14 provides the quantitative comparisons of these detectors for the synthetic dataset through the ROC curves and normalized background-anomaly separation maps. As shown in Figure 14a, our proposed DCLaAW obtains the best ROC curve with a DR close to 1 for all FARs. The ROC curves of LRaSMD and SSRX are slightly worse than that of DCLaAW, but still better than that of the other 6 detectors. The ROC curve of LRX approximates a straight line. Figure 14b shows the normalized background-anomaly separation maps of each detector. As shown, DCLaAW achieves the largest gap between the background box and the anomaly box with no overlap. In addition to DCLaAW, CKRX and LRaSMD can also satisfactorily separate anomalies from the background.

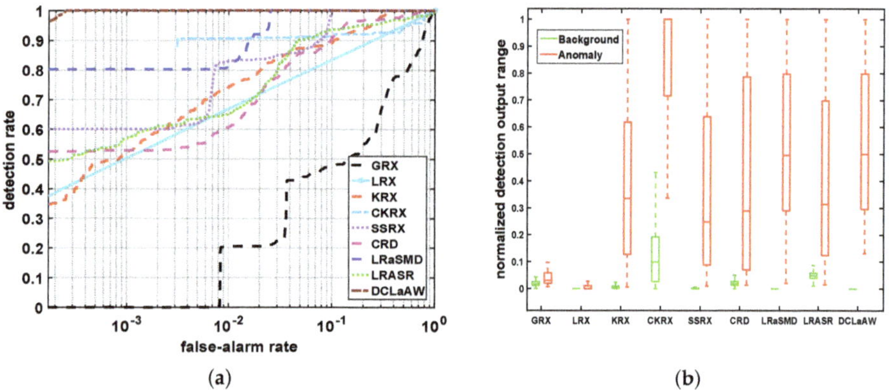

(a)   (b)

Figure 14. Quantitative comparisons of all compared algorithms for the synthetic dataset. (**a**) ROC curves; (**b**) Background-anomaly separation maps.

For the real-world San Diego dataset, the color detection maps of all compared algorithms are shown in Figure 15. We can see that neither GRX nor LRX can identify any anomalous aircraft from the background, thus providing the worst detection maps among all detectors. KRX achieves the most outstanding anomaly extrusion in all detectors, but there are some serious false alarms in the lower left and upper left corners. The anomaly extrusion of CKRX is satisfactory, but the background in the lower left corner needs to be further suppressed. For SSRX, when eliminating redundant background interference, some useful anomaly information is also removed by PCA, resulting in weak brightness of anomaly pixels in the detection map of SSRX, as shown in Figure 15e. The centers of the anomalous aircraft are well extruded by CRD, but the edges are ignored. LRaSMD achieves a very satisfactory background suppression for most of the background areas, but it is obvious that there are some high background responses in the lower left corner of the scene. For our proposed DCLaAW, all three aircraft are extracted from the background with very high brightness, and the background interference is well suppressed, demonstrating its superiority over LARSR which has relatively weak brightness in the anomaly pixels. Figure 16 presents the ROC curves and background-anomaly separation maps of these detectors. As shown in Figure 16a, DCLaAW obtains a DR greater than 0.3 when the FAR is approximately 0, and its DR is about 0.95 when the FAR is 0.007. Therefore, our proposed DCLaAW achieves the best detection performance among all detectors. The ROC curves of GRX and LRX are the worst, consistent with the conclusions of the above detection maps. Figure 16b illustrates that both LRX and LRaSMD successfully suppress the background to a very low and narrow range of brightness, but the anomalies in LRX are not well highlighted. The separation of CKRX is quite good, but the brightness of the background is too high. Although our proposed DCLaAW is not optimal for background suppression, it can obtain the maximum distance between the background box and the anomaly box, thus achieving the best background-anomaly separation performance.

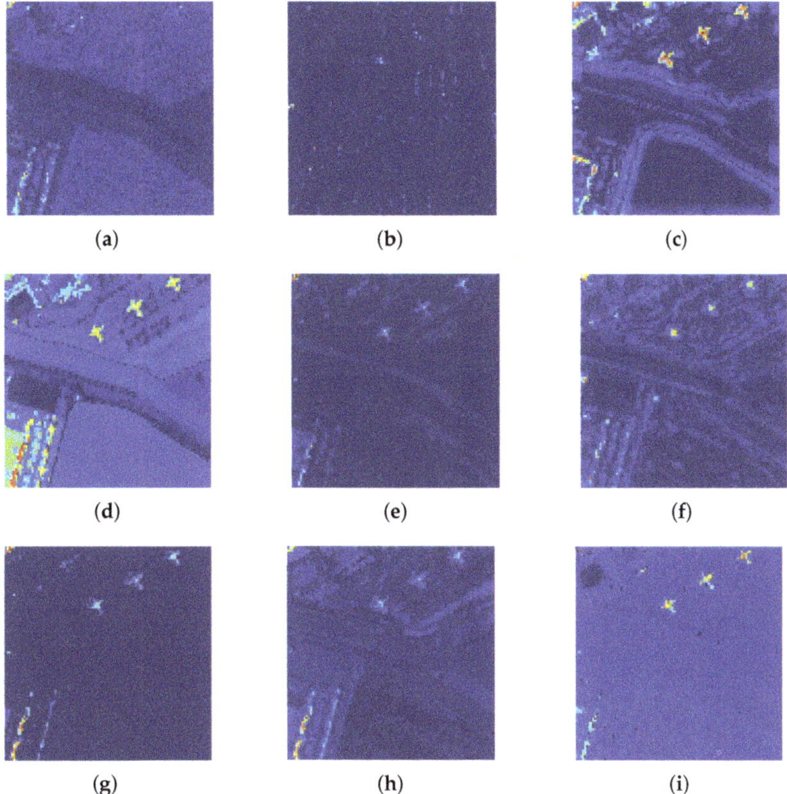

Figure 15. Color detection maps of all compared algorithms for the San Diego dataset. (**a**) RX; (**b**) LRX; (**c**) KRX; (**d**) CKRX; (**e**) SSRX; (**f**) CRD; (**g**) LRaSMD; (**h**) LRASR; (**i**) DCLaAW.

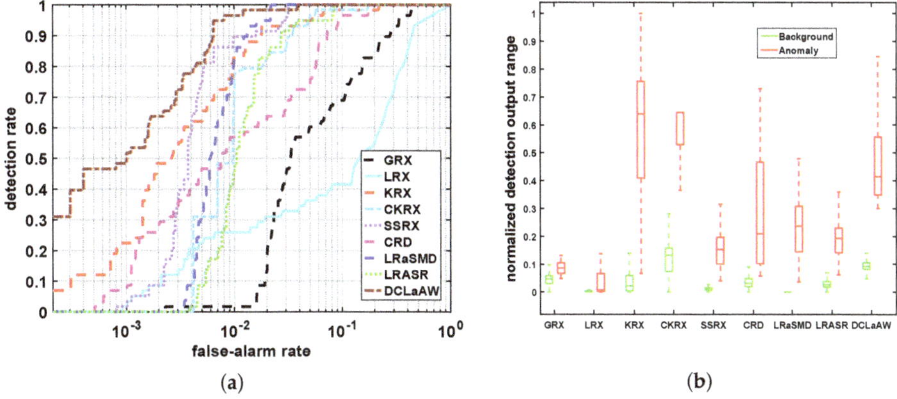

Figure 16. Quantitative comparisons of all compared algorithms for the San Diego dataset. (**a**) ROC curves; (**b**) Background-anomaly separation maps.

For the real-world Urban dataset, the detection maps are shown in Figure 17. As we can see, compared with GRX, LRX effectively eliminates some false alarm points in the scene. However, some anomalies are also suppressed undesirably by LRX. For KRX and CKRX, there are some background areas with high brightness, especially in the lower right corner of CKRX. For SSRX, almost all anomalies can be found, and its background suppression is much better than KRX and CKRX. For LRaSMD, the anomalies are well highlighted and most of the background areas in the scene are suppressed to a very low brightness. However, due to the presence of some background objects with sparse property, the detection map of LRaSMD may also contain some bright background responses, as shown in Figure 17g. Our proposed DCLaAW achieves an excellent anomaly extrusion from the background with almost no false alarms, and all background pixels are suppressed to a small interval. Figure 18 gives quantitative comparisons of these detectors by ROC curves and background-anomaly separation maps. It can be observed from Figure 18a that our DCLaAW obtains a DR greater than 0.6 when the false alarm is 0, and its FAR is the smallest compared with others when the DR reaches 1. The ROC curves of KRX and CKRX are the worst as they are close to the lower right corner of the coordinate plane. From Figure 18b, we can see that both LRX and LRaSMD achieve the best background suppression because their background boxes are very narrow, and their background values are close to 0. For DCLaAW, the gap between background and anomalies is the largest, meaning the best background-anomaly separation performance among all detectors.

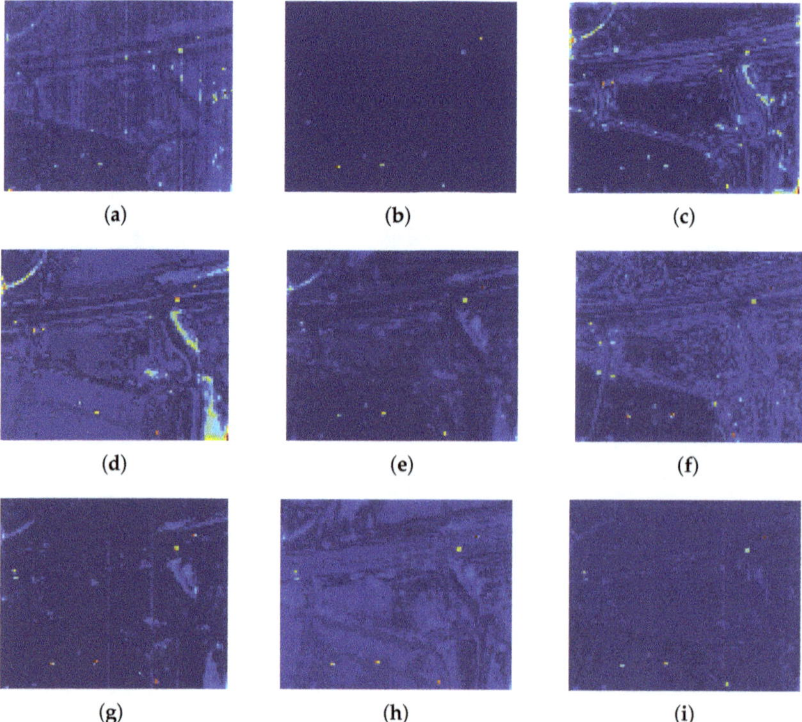

Figure 17. Color detection maps of all compared algorithms for the Urban dataset. (a) RX; (b) LRX; (c) KRX; (d) CKRX; (e) SSRX; (f) CRD; (g) LRaSMD; (h) LRASR; (i) DCLaAW.

In addition, the AUC values of all compared algorithms for each dataset are listed in Figure 19. It can be seen that DCLaAW obtains the largest AUC value for all three datasets, proving its advantages in AD. For the Urban dataset, all these detectors achieve an AUC value larger than 0.9, which mainly because this dataset has high anomaly fractions, relatively uniform background and weak anomaly contamination caused by small anomaly size.

Overall, our proposed DCLaAW generally performs best on both synthetic and real-world hyperspectral datasets. Compared with these compared algorithms, the main reasons for the superior performance of DCLaAW can be summarized as follows: (1) it requires no assumptions on the distribution of the background, which is the main limitation of the conventional probability distribution-based RX methods. (2) anomaly contamination in LRX and CRD is a major factor affecting their performances, which can lead to some false alarms and the missed detection of real anomalies. (3) for LRaSMD, because of the decomposition error, the sparse property of some background objects and the large upper bound of sparsity level, some background information is usually included in the extracted sparse component, which may result in the presence of some false alarms. (4) LRASR and DCLaAW, as improved versions of LRR, both construct a reliable background dictionary that can remove anomalies and contain all background categories. However, our proposed weighting strategy further enhances the response difference between the background pixels and the anomaly pixels, thus providing a better AD performance.

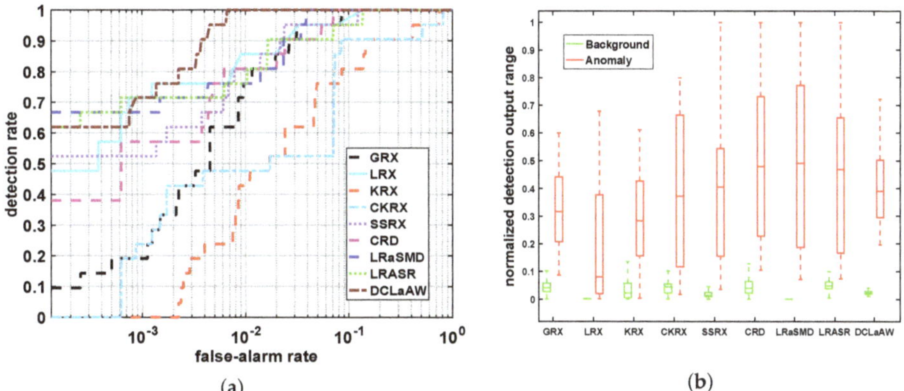

Figure 18. Quantitative comparisons of all compared algorithms for the Urban dataset. (**a**) ROC curves; (**b**) Background-anomaly separation maps.

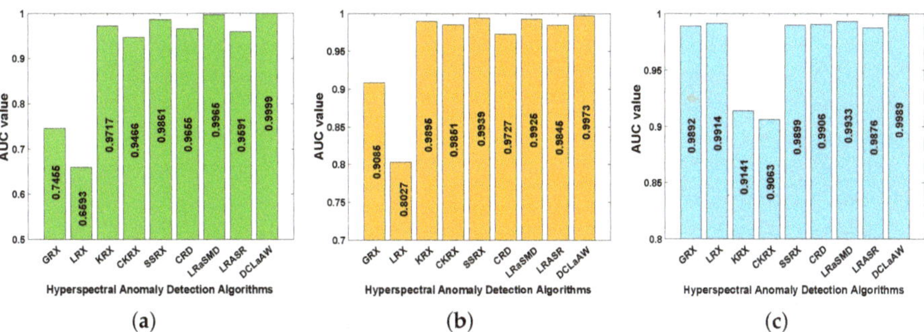

Figure 19. AUC values of all compared algorithms for each dataset. (**a**) Synthetic dataset; (**b**) San Diego dataset; (**c**) Urban dataset.

Furthermore, the computational costs of all these algorithms for each dataset are listed in Table 4 for a practical comparison. The computational cost of each algorithm refers to its runtime on our designated platform, and the number is in seconds. For the three datasets, although the detection performance of CKRX is slightly worse than KRX, its computation time is significantly less. The LRR-based algorithms, such as LRASR and DCLaAW, require more time to perform the detection operation than other algorithms. Due to the use of sparsity-inducing regularization term in LRASR, the computational cost of LRASR is slightly larger than that of our proposed DCLaAW. It is worth nothing that although our dictionary construction strategy greatly reduces the computation time of the original LRR algorithm, the main calculation of DCLaAW is still spent on the solution of LRR. Specifically, for the synthetic dataset, the San Diego dataset, and the Urban dataset, LRR accounts for 88.12%, 90.41% and 90.38% of the computational cost of DCLaAW, respectively.

Table 4. Computational costs of all compared detectors for each dataset.

Times (s)	GRX	LRX	KRX	CKRX	SSRX	CRD	LRaSMD	LRASR	DCLaAW
Synthetic Dataset	0.698	87.726	21.043	11.823	0.464	32.134	58.079	520.356	466.527
San Diego Dataset	0.157	48.108	10.218	1.946	0.161	9.953	16.885	62.877	60.015
Urban Dataset	0.143	20.930	2.561	1.677	0.155	2.552	10.919	58.111	55.787

4.6. Parameter Analysis

There are some important parameters in our proposed DCLaAW that may influence the detection performance, mainly including: (1) in the K-means clustering step: K is the number of clusters. (2) in the background dictionary construction step: M is the percentage of atoms selected for HSI reconstruction in each cluster; P is the number of pixels selected as the estimated background pixels in each cluster. (3) in the LRR step: λ is the tradeoff parameter. When we analyze the specified parameters, the other parameters are set to be optimal.

Firstly, we investigate the sensitivity of the detection performance of DCLaAW to K and M with the other parameters fixed. The AUC values are calculated when jointly taking K and M into consideration. Without loss of generality, K is set as {1, 2, 4, 6, 8, 10, 12, 14, 16, 20} and M is set as 10–100% with an interval of 10%. For each dataset, the AUC values obtained with different combinations of K and M are exhibited in Figure 20. It should be noted that since the sparse coding in each cluster requires an over-complete dictionary, we ignore and skip the clusters where the number of pixels is less than the number of dimensions of the dataset. As shown in Figure 20, it is clear that the AUC surfaces for the three datasets are similar, where the DCLaAW algorithm is more sensitive to the transformation of K than that of M. The detection performance of DCLaAW with small K is poor, mainly because the value of K is too small to enable the K-means clustering algorithm to segment the HSI dataset into a sufficient number of clusters. In this case, the constructed background dictionary for LRR cannot contain enough background categories and therefore cannot span the entire data space. The AUC value is relatively low when both K and M are very small. When K is in the range of 8–20 and M is in 30–100%, the AUC values are stable and satisfactory for all three datasets, demonstrating the robustness of DCLaAW to parameter K and M. For simplicity, in our experiments, we choose $K = 12$ and $M = 50\%$ for all the three datasets. It is worth noting that $K = 12$ is also slightly larger than the number of categories estimated by HySime and is therefore a reasonable choice.

Then, we investigate the influence of P on the detection performance of DCLaAW for each dataset. Since the value of K can significantly affect the variation of detection performance with P, here we jointly analyze K and P. The value of K is set as {1, 4, 8, 12, 16, 20} and P is in the range of 10–100 with an interval of 10. Since the background dictionary we use for HSI reconstruction in the weighting operation needs to be over-complete, the product of K and P should be larger than the dimension of the dataset. Therefore, when the product of K and P is lower than the dimension, we do not execute the adaptive weighting operation. It is worth mentioning that since we have made the dictionary for sparse coding in each cluster over-complete, we can ensure that the selection of P atoms in each cluster is sufficient, even if P takes the maximum value of 100. Figure 21a–c illustrate the change of AUC values with P under different K for each dataset.

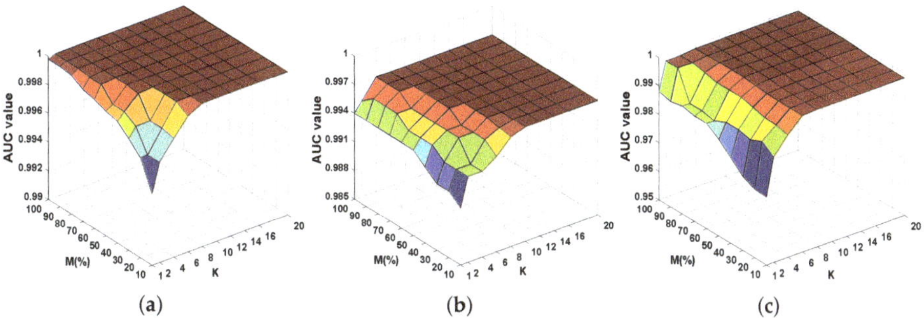

Figure 20. AUC illustration of DCLaAW with different combinations of K and M for each dataset. (**a**) Synthetic dataset; (**b**) San Diego dataset; (**c**) Urban dataset.

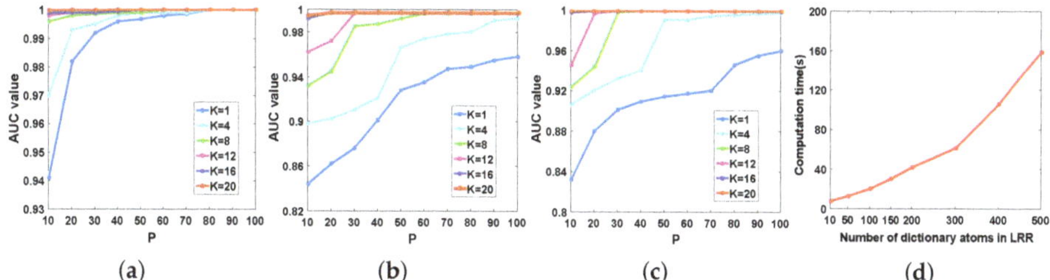

Figure 21. AUC values achieved by DCLaAW with different K and P for each dataset. (**a**) Synthetic dataset; (**b**) San Diego dataset; (**c**) Urban dataset; (**d**) Relationship between calculation time and the number of dictionary atoms used for LRR.

As shown in Figure 21a–c, we can see that for the three datasets, the changes of AUC exhibit similar characteristics. Specifically, on the one hand, an increased K means a better clustering result and a more comprehensive background dictionary, thus resulting in a more satisfactory detection performance. On the other hand, as P increases, more background dictionary atoms for LRR make the background space to be more adequately spanned and thus further lead to a larger AUC value. However, as P further increases, the AUC value will not increase anymore because the background space has been fully described. Here, we choose several representative K-curves to illustrate the details. For $K = 1$, the detection performance is very poor because the weighting method is not executed in this case and such a small K makes the background dictionary unable to contain enough background categories. For $K = 4$, there is a turning point where the AUC value increases rapidly. Prior to this point, the weighting strategy is not implemented. At this point, the weighting strategy optimizes the detection results. For $K = 20$, the weighting strategy is executed under all P values, so the detection performance is satisfactory. It is worth nothing that although a larger P and K can result in a larger AUC value, it also brings a greater computational cost. Though experiments, we plot the change of the calculation time of LRR with the number of atoms in the dictionary, as shown in Figure 21d, where the x-axis is the number of atoms in the dictionary for LRR and the y-axis is the calculation time. Therefore, the values of K and P should be chosen to be moderate after jointly considering the detection performance and time cost. For example, $K = 12$ and $P = 30$ is a good choice for all the three datasets.

Finally, the sensitivity of DCLaAW to the tradeoff parameter λ is analyzed. λ is chosen from $\{0.001, 0.005, 0.01, 0.02, 0.05, 0.1, 0.2, 0.3, 0.4, 0.5\}$, and the ROC curve is used as the evaluation measure. From the results shown in Figure 22, we can see that the variation trend of ROC curves with λ for the synthetic dataset is significantly different from that for the two real-world datasets. Specifically, for the synthetic dataset, as λ increases, the ROC curve initially becomes better and then reaches the best when λ is 0.02, and finally deteriorates as λ further increases. In the detection maps, λ larger than 0.2 will result in the appearance of false alarm points corresponding to the rocks and grasses in the scene. Differently, for the two real-world datasets, the ROC curves exhibit similar trends and are not sensitive to λ. To observe the details, we plot the AUC values as a function of λ, as shown in Figure 23. It reveals that for the two real-world datasets, all λ in the range of {0.001, 0.5} can achieve an AUC value larger than 0.994, demonstrating the robustness of DCLaAW to λ. For the synthetic dataset, when λ is less than 0.1, we can achieve an AUC value larger than 0.98. In our experiments, we choose $\lambda = 0.02$, $\lambda = 0.02$ and $\lambda = 0.4$ for the three datasets, respectively.

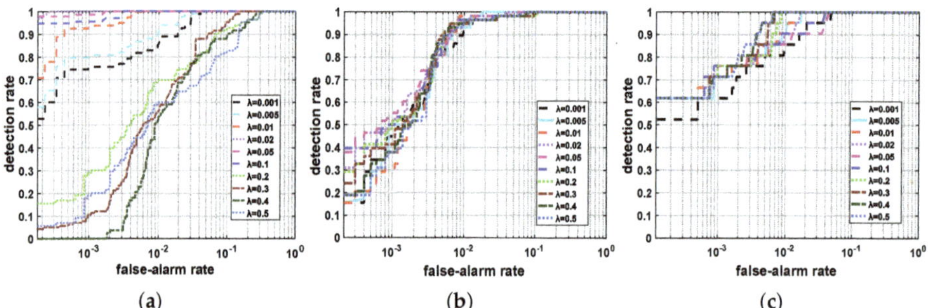

Figure 22. ROC curves achieved by DCLaAW with different λ for each dataset. (**a**) Synthetic dataset; (**b**) San Diego dataset; (**c**) Urban dataset.

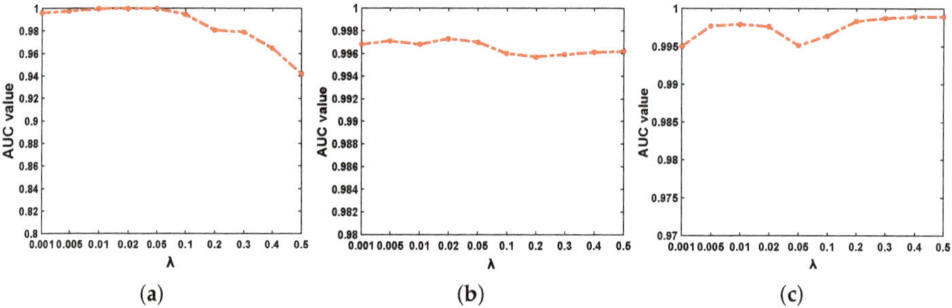

Figure 23. AUC values achieved by DCLaAW with different λ for each dataset. (**a**) Synthetic dataset; (**b**) San Diego dataset; (**c**) Urban dataset.

4.7. Comparison between Sparsity and l_2 Formulation

In our proposed algorithm, based on the constructed background dictionary, the LRR is used to separate the sparse anomaly component from the background for AD. As described in Section 3.2, since the background pixels can be reconstructed sparsely by the background dictionary very well, while the anomalies cannot, the reconstruction errors of the sparsity formulation can be used to assign anomaly responses to pixels. l_2 formulation, as a more commonly used approach, can theoretically also achieve AD based on the constructed dictionary. That is to say, the LRR, the sparsity formulation, and the l_2 formulation perform AD from different aspects, and they adopt different models. In this section, the AD performances of these three approaches are compared through experiments to demonstrate the superiority of our proposed algorithm.

The models of LRR and sparsity formulation are Equation (1) and Equation (6), respectively, both of which are used in our algorithm. l_2 formulation, whose regression is called ridge regression, is usually used to prevent data overfitting. In fact, with the l_2 formulation, the entries of the coefficient vector are close to 0, but not equal to 0, which is the main difference between it and the sparsity formulation. With the background dictionary $\mathbf{D}$, the l_2 formulation is as follows:

$$\min \|\mathbf{x} - \mathbf{D}\boldsymbol{\theta}\|_2^2 + \delta \|\boldsymbol{\theta}\|_2^2 , \tag{11}$$

where δ is the Lagrange multiplier. It can be found that for the l_2 formulation, each pixel is reconstructed by all atoms in the background dictionary. Differently, for the sparsity formulation, each pixel is sparsely reconstructed by a few atoms in the dictionary. The above optimization problem can be solved by making the derivative zero, and an analytical expression can be obtained. Without adaptive weighting, the optimal detection maps obtained by these three approaches are shown in Figure 24, and the corresponding AUC values are listed in Table 5. In addition, for each approach, the relationship between the calculation time (in seconds) and the number of dictionary atoms is shown in Figure 25 for practical comparisons. Here we only show the experimental results for the San Diego dataset, and the other two datasets can get the similar conclusions.

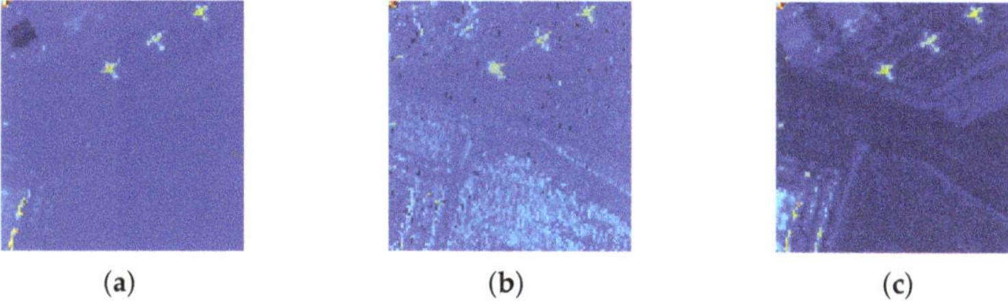

(a)   (b)   (c)

Figure 24. Detection maps obtained by different approaches for the San Diego dataset. (**a**) LRR with l_{21} constraint; (**b**) Sparsity formulation; (**c**) l_2 formulation.

Table 5. AUC values obtained by different approaches for the San Diego dataset.

Approach	LRR with l_{21} Constraint	Sparsity Formulation	l_2 Formulation
AUC value	0.9949	0.9922	0.9937

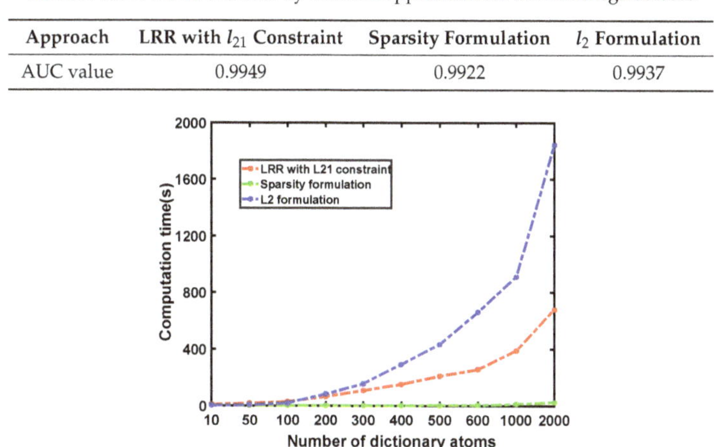

Figure 25. Relationship between computation time and the number of dictionary atoms for each approach for the San Diego dataset.

As shown in Figure 24, LRR achieves a uniform and satisfactory suppression for almost all background materials, thus obtaining the largest AUC value, as listed in Table 5. For the sparse formulation, the overall background brightness is too high and needs to be further suppressed. The l_2 formulation achieves the best suppression for most background areas, but the response values of the background objects in the upper left and lower left corners are quite high. Table 5 shows that the detection performance of LRR is the best, followed by the l_2 formulation, while the sparse formulation has the worst performance. However, in our algorithm, if we use the l_2 formulation instead of the sparsity formulation to adaptively weight, the final AUC value obtained by the LRR weighted by l_2 formulation is 0.9952, while the final AUC value obtained by the LRR weighted by sparsity formulation is 0.9973. The reason may be that for the l_2 formulation, the high background responses in the upper left and lower left corners make the FAR of the final detection result serious. For the sparsity formulation, although the overall background suppression in the weight map is not satisfactory, it is uniform. As a result, the final detection performance can be effectively improved. As can be seen from Figure 25, for the l_2 formulation, the AUC value increases rapidly as the number of dictionary atoms increases. Therefore, when the number of dictionary atoms is large, it is impractical to process HSI datasets using the l_2 formulation. In fact, the LRR with l_1 constraint has the longest computation time compared to these three approaches. After jointly considering the final detection performance and the calculation time, we use the sparsity formulation to weight the detection result of the LRR with l_{21} constraint, while the l_2 formulation is not adopted.

5. Conclusions

In this paper, a novel hyperspectral AD algorithm via DCLaAW is proposed. Based on the consistency of the LRR model and the hyperspectral AD problem, the LRR is used to mine the lowest-rank representation of all data jointly and extract the sparse component for AD. Considering the shortcomings of the conventional dictionaries for LRR and the fact that the background atoms participate more frequently in HSI reconstruction, a background dictionary construction strategy based on the usage frequency of the dictionary atoms for HSI reconstruction in each cluster is proposed. Such a background dictionary guarantees the exclusion of anomaly pixels and the inclusion of all background categories to a considerable extent, thus achieving a satisfactory separation between the anomaly component and the background component. In addition, to further enhance the response difference between the background pixels and the anomaly pixels, an adaptive weighting method based on the reconstruction error of the entire data with respect to the constructed background dictionary is proposed. The final anomaly value of each pixel is calculated by multiplying the weight value by the response value obtained through LRR.

Experiments on both synthetic and real-world datasets demonstrate the superiority of our proposed anomaly detection algorithm over the other eight state-of-the-art AD detectors. Moreover, the effectiveness of the dictionary construction strategy and the adaptive weighting method is proven by experiments. Finally, the influences of relevant parameters on the detection performance of our algorithm are analyzed in detail. Although our algorithm can greatly alleviate the computational burden of the original LRR, its calculation time is still larger than some other anomaly detectors. Therefore, computational complexity is the focus of future research.

Author Contributions: Conceptualization, Y.Y. and J.Z.; methodology, Y.Y. and D.L.; software, Y.Y.; data curation, Y.Y. and S.S; Investigation, Y.Y., J.Z., and S.S.; supervision, J.Z.; writing–original draft preparation, Y.Y.; writing–review and editing, Y.Y. and D.L.

Funding: This research was funded in part by the National Natural Science Foundation of China under grant number 61774120, in part by the Fundamental Research Funds for the Central Universities under grant number JBX170507, and in part by the 111 Project under grant number B17035.

Conflicts of Interest: The authors declare no conflict of interest.

References

1. Plaza, A.; Benediktsson, J.A.; Boardman, J.W.; Brazile, J.; Bruzzone, L.; Camps-Valls, G.; Chanussot, J.; Fauvel, M.; Gamba, P.; Gualtieri, A.; et al. Recent advances in techniques for hyperspectral image processing. *Remote Sens. Environ.* **2009**, *113*, S110–S122. [CrossRef]

2. Manolakis, D.; Shaw, G. Detection algorithms for hyperspectral imaging applications. *IEEE Signal Process. Mag.* **2002**, *19*, 29–43. [CrossRef]

3. Matteoli, S.; Diani, M.; Corsini, G. A total overview of anomaly detection in hyperspectral images. *IEEE Aerosp. Electron. Syst. Mag.* **2010**, *25*, 5–28. [CrossRef]

4. Bioucas-Dias, J.M.; Plaza, A.; Camps-Valls, G.; Scheunders, P.; Nasrabadi, N.; Chanussot, J. Hyperspectral remote sensing data analysis and future challenges. *IEEE Geosci. Remote Sens. Mag.* **2013**, *1*, 6–36. [CrossRef]

5. Li, F.; Zhang, X.; Zhang, L.; Jiang, D.; Zhang, Y. Exploiting Structured Sparsity for Hyperspectral Anomaly Detection. *IEEE Trans. Geosci. Remote Sens.* **2018**, *56*, 4050–4064. [CrossRef]

6. Nasrabadi, N.M. Hyperspectral target detection: An overview of current and future challenges. *IEEE Signal Process. Mag.* **2014**, *31*, 34–44. [CrossRef]

7. Shaw, G.; Manolakis, D. Signal processing for hyperspectral image exploitation. *IEEE Signal Process. Mag.* **2002**, *19*, 12–16. [CrossRef]

8. Reed, I.S.; Yu, X. Adaptive multiple-band cfar detection of an optical pattern with unknown spectral distribution. *IEEE Trans. Acoust. Speech Signal Process.* **1990**, *38*, 1760–1770. [CrossRef]

9. Borghys, D.; Kåsen, I.; Achard, V.; Perneel, C. Comparative evaluation of hyperspectral anomaly detectors in different types of background. In Proceedings of the Algorithms and Technologies for Multispectral, Hyperspectral, and Ultraspectral Imagery XVIII, Baltimore, MD, USA, 24 May 2012; International Society for Optics and Photonics: Bellingham, WA, USA, 2012.

10. Du, B.; Zhang, L. A discriminative metric learning based anomaly detection method. *IEEE Trans. Geosci. Remote Sens.* **2014**, *52*, 6844–6857.

11. Veracini, T.; Matteoli, S.; Diani, M.; Corsini, G. Fully unsupervised learning of gaussian mixtures for anomaly detection in hyperspectral imagery. In Proceedings of the 2009 Ninth International Conference on Intelligent Systems Design and Applications, Pisa, Italy, 30 November–2 December 2009; pp. 596–601.

12. Carlotto, M.J. A cluster-based approach for detecting man-made objects and changes in imagery. *IEEE Trans. Geosci. Remote Sens.* **2005**, *43*, 374–387. [CrossRef]

13. Schaum, A. Joint subspace detection of hyperspectral targets. In Proceedings of the 2014 IEEE Aerospace Conference, Big Sky, MT, USA, 6–13 March 2004.

14. Kwon, H.; Nasrabadi, N.M. Kernel rx-algorithm: A nonlinear anomaly detector for hyperspectral imagery. *IEEE Trans. Geosci. Remote Sens.* **2005**, *43*, 388–397. [CrossRef]

15. Banerjee, A.; Burlina, P.; Diehl, C. A support vector method for anomaly detection in hyperspectral imagery. *IEEE Trans. Geosci. Remote Sens.* **2006**, *44*, 2282–2291. [CrossRef]

16. Zhou, J.; Kwan, C.; Ayhan, B.; Eismann, M.T. A Novel Cluster Kernel RX Algorithm for Anomaly and Change Detection Using Hyperspectral Images. *IEEE Trans. Geosci. Remote Sens.* **2016**, *54*, 6497–6504. [CrossRef]

17. Du, B.; Zhang, L. Random-selection-based anomaly detector for hyperspectral imagery. *IEEE Trans. Geosci. Remote Sens.* **2011**, *49*, 1578–1589. [CrossRef]

18. Billor, N.; Hadi, A.S.; Velleman, P.F. Bacon: Blocked adaptive computationally efficient outlier nominators. *Comput. Stat. Data Anal.* **2000**, *34*, 279–298. [CrossRef]

19. Li, J.; Zhang, H.; Zhang, L.; Ma, L. Hyperspectral Anomaly Detection by the Use of Background Joint Sparse Representation. *IEEE J. Sel. Top. Appl. Earth Obs. Remote Sens.* **2015**, *8*, 2523–2533. [CrossRef]

20. Li, W.; Du, Q. Collaborative Representation for Hyperspectral Anomaly Detection. *IEEE Trans. Geosci. Remote Sens.* **2014**, *53*, 1463–1474. [CrossRef]

21. Chen, Y.; Nasrabadi, N.M.; Tran, T.D. Hyperspectral Image Classification Using Dictionary-Based Sparse Representation. *IEEE Trans. Geosci. Remote Sens.* **2011**, *49*, 3973–3985. [CrossRef]

22. Dao, M.; Kwan, C.; Koperski, K.; Marchisio, G. A Joint Sparsity Approach to Tunnel Activity Monitoring Using High Resolution Satellite Images. In Proceedings of the IEEE 8th Annua Ubiquitous Computing, Electronics & Mobile Communication Conference, New York, NY, USA, 19–21 October 2017; pp. 322–328.

23. Niu, Y.; Wang, B. Hyperspectral Anomaly Detection Based on Low-Rank Representation and Learned Dictionary. *Remote Sens.* **2016**, *8*, 289. [CrossRef]

24. Candes, E.J.; Li, X.; Ma, Y.; Wright, J. Robust principal component analysis? *J. ACM* **2011**, *58*, 11. [CrossRef]

25. Qu, Y.; Wang, W.; Guo, R.; Ayhan, B.; Kwan, C.; Vance, S.D.; Qi, H. Hyperspectral Anomaly Detection Through Spectral Unmixing and Dictionary-Based Low-Rank Decomposition. *IEEE Trans. Geosci. Remote Sens.* **2018**, *56*, 4391–4405. [CrossRef]

26. Zhu, L.; Wen, G. Low-Rank and Sparse Matrix Decomposition with Cluster Weighting for Hyperspectral Anomaly Detection. *Remote Sens.* **2018**, *10*, 707. [CrossRef]

27. Wang, W.; Li, S.; Ayhan, B.; Kwan, C. Identify Anomaly Component by Sparsity and Low Rank. In Proceedings of the 7th Workshop on Hyperspectral Image and Signal Processing: Evolution in Remote Sensing (WHISPERS), Tokyo, Japan, 2–5 June 2015.

28. Xu, Y.; Wu, Z.; Li, J.; Plaza, A.; Wei, Z. Anomaly detection in hyperspectral images based on low-rank and sparse representation. *IEEE Trans. Geosci. Remote Sens.* **2016**, *54*, 1990–2000. [CrossRef]

29. Sun, W.; Tian, L.; Xu, Y. A Randomized Subspace Learning Based Anomaly Detector for Hyperspectral Imagery. *Remote Sens.* **2018**, *10*, 417. [CrossRef]

30. Ma, D.; Yuan, Y.; Wang, Q. Hyperspectral Anomaly Detection via Discriminative Feature Learning with Multiple-Dictionary Sparse Representation. *Remote Sens.* **2018**, *10*, 745. [CrossRef]

31. Zhao, R.; Du, B.; Zhang, L. Hyperspectral anomaly detection via a sparsity score estimation framework. *IEEE Trans. Geosci. Remote Sens.* **2017**, *55*, 3208–3222. [CrossRef]

32. Chen, Y.; Nasrabadi, N.M.; Tran, T.D. Sparse representation for target detection in hyperspectral imagery. *IEEE J. Sel. Top. Signal Process.* **2011**, *5*, 629–640. [CrossRef]

33. Zhang, Y.; Du, B.; Zhang, L.; Wang, S. A low-rank and sparse matrix decomposition-based mahalanobis distance method for hyperspectral anomaly detection. *IEEE Trans. Geosci. Remote Sens.* **2016**, *54*, 1376–1389. [CrossRef]

34. Lin, Z.; Liu, R.; Su, Z. Linearized alternating direction method with adaptive penalty for low rank representation. *Adv. Neural Inf. Process. Syst.* **2011**, 612–620.

35. Chen, S.; Yang, S.; Kalpakis, K.; Chang, C.I. Low-rank decomposition-based anomaly detection. *Proc. SPIE* **2013**, *8743*, 1–7.

36. Sun, W.; Liu, C.; Li, J.; Lai, Y.M.; Li, W. Low-rank and sparse matrix decomposition-based anomaly detection for hyperspectral imagery. *J. Appl. Remote Sens.* **2014**, *8*, 083641. [CrossRef]

37. Matteoli, S.; Diani, M.; Corsini, G. Impact of signal contamination on the adaptive detection performance of local hyperspectral anomalies. *IEEE Trans. Geosci. Remote Sens.* **2014**, *52*, 1948–1968. [CrossRef]

38. Liu, G.; Lin, Z.; Yu, Y. Robust subspace segmentation by low-rank representation. In Proceedings of the 27th International Conference on Machine Learning (ICML), Haifa, Israel, 21–24 June 2010; pp. 663–670.

39. Liu, G.; Lin, Z.; Yan, S.; Sun, J.; Yu, Y.; Ma, Y. Robust Recovery of Subspace Structures by Low-Rank Representation. *IEEE Trans. Pattern Anal. Mach. Intell.* **2013**, *35*, 171–184. [CrossRef] [PubMed]

40. Cai, J.F.; Candès, E.J.; Shen, Z. A Singular Value Thresholding Algorithm for Matrix Completion. *SIAM J. Optim.* **2010**, *20*, 1956–1982. [CrossRef]

41. BioucasDias; José, M.; Nascimento; José, M.P. Hyperspectral subspace identification. *IEEE Trans. Geosci. Remote Sens.* **2008**, *46*, 2435–2445. [CrossRef]

42. Mallat, S.G.; Zhang, Z. Matching pursuits with time-frequency dictionaries. *IEEE Trans. Signal Process.* **1993**, *41*, 3397–3415. [CrossRef]

43. Tropp, J.A.; Gilbert, A.C. Signal recovery from random measurements via orthogonal matching pursuit. *IEEE Trans. Inf. Theory* **2007**, *53*, 4655–4666. [CrossRef]

44. Zhu, L.; Wen, G. Hyperspectral Anomaly Detection via Background Estimation and Adaptive Weighted Sparse Representation. *Remote Sens.* **2018**, *10*, 272.

45. Kerekes, J. Receiver operating characteristic curve confidence intervals and regions. *IEEE Geosci. Remote Sens. Lett.* **2008**, *5*, 251–255. [CrossRef]

46. Fawcett, T. An introduction to ROC analysis. *Pattern Recogn. Lett.* **2006**, *27*, 861–874. [CrossRef]

47. Snyder, D.; Kerekes, J.; Hager, S. Target Detection Blind Test Dataset. Available online: http://dirsapps.cis.rit.edu/blindtest/ (accessed on 10 September 2018).

48. Stefanou, M.S.; Kerekes, J.P. A Method for Assessing Spectral Image Utility. *IEEE Trans. Geosci. Remote Sens.* **2009**, *47*, 1698–1706. [CrossRef]

49. Taghipour, A.; Ghassemian, H. Hyperspectral anomaly detection using attribute profiles. *IEEE Geosci. Remote Sens. Lett.* **2017**, *14*, 1136–1140. [CrossRef]

50. U.S. Army Corps of Engineers. Available online: http://www.tec.army.mil/Hypercurbe (accessed on 10 September 2018).

Article

Infrared Small Target Detection Based on Non-Convex Optimization with *Lp*-Norm Constraint

Tianfang Zhang [1], Hao Wu [1], Yuhan Liu [1], Lingbing Peng [1], Chunping Yang [1] and Zhenming Peng [1,2,*]

[1] School of Information and Communication Engineering, University of Electronic Science and Technology of China, Chengdu 610054, China; sparkcarleton@gmail.com (T.Z.); haowu_cn@163.com (H.W.); yuhanliu0211@outlook.com (Y.L.); lbpeng@163.com (L.P.); cpin2@163.com (C.Y.)

[2] Center for Information Geoscience, University of Electronic Science and Technology of China, Chengdu 611731, China

* Correspondence: zmpeng@uestc.edu.cn; Tel.: +86-130-7603-6761

Received: 31 January 2019; Accepted: 4 March 2019; Published: 7 March 2019

Abstract: The infrared search and track (IRST) system has been widely used, and the field of infrared small target detection has also received much attention. Based on this background, this paper proposes a novel infrared small target detection method based on non-convex optimization with *Lp*-norm constraint (NOLC). The NOLC method strengthens the sparse item constraint with *Lp*-norm while appropriately scaling the constraints on low-rank item, so the NP-hard problem is transformed into a non-convex optimization problem. First, the infrared image is converted into a patch image and is secondly solved by the alternating direction method of multipliers (ADMM). In this paper, an efficient solver is given by improving the convergence strategy. The experiment shows that NOLC can accurately detect the target and greatly suppress the background, and the advantages of the NOLC method in detection efficiency and computational efficiency are verified.

Keywords: low rank sparse decomposition; *Lp*-norm constraint; non-convex optimization; alternating direction method of multipliers; infrared small target detection

---

1. Introduction

In recent years, as an indispensable part of infrared search and track (IRST) system, infrared small target detection system is widely used in early warning systems, precision strike weapons and air defense systems [1–3]. On the one hand, since the imaging distance is usually several tens of hundreds of kilometers, the signal is attenuated by the atmosphere, and the target energy received by the IRST is weak. For the same reason, the target usually occupies a small area and lacks texture and structural information; on the other hand, the background and noise account for a large proportion, while the background is complex and constantly changing, resulting in a low signal-to-clutter ratio gain (SCR Gain) of the target in the image [4–7]. Therefore, infrared small target detection methods attract the attention of a large number of researchers [8–10].

At present, the mainstream infrared small target detection algorithm can be divided into two major categories: Track before detection (TBD) and detection before track (DBT). Among them, the TBD method jointly processes the information of multiple frames to track the infrared small target, and has higher requirements on computer performance, so the degree of attention is weak. The DBT method is to process the image in a single frame to get the target position, which is usually better in real time and has received more attention. Next, we will introduce the two methods and the research motivation of this article.

1.1. Track before Detection

Track before detection (TBD) methods use spatial and temporal information to estimate the target location by processing multiple adjacent frames. Traditional 3-D matched filtering [11], improved 3-D filtering [12] and Spatiotemporal multiscan adaptive matched filtering [13] are only for static backgrounds. However, the difference between the target and background in the infrared image tends to change rapidly, and the background is also complex and changeable. Therefore, the above methods are not effective.

Braganeto et al. used morphological connected operators to jointly consider target detection and tracking [14]; Dong et al. proposed a novel target detection method [15] by combining the difference of Gaussian (DOG), human visual system (HVS) and clustering methods; Li et al. proposed a biologically inspired multilevel approach for multiple moving target detection [16]; Li et al. proposed a spatio-temporal saliency approach [17]. However, since such methods usually require a large amount of computation and storage, and have high requirements for computer performance, TBD methods are not commonly used in practical applications.

1.2. Detection before Track

Detection before track (DBT) methods usually use the characteristics of small targets to process images on a single frame. The DBT methods can be roughly divided into three categories.

The background suppression based methods. This category of methods is based on the assumption of background consistency of infrared images, and usually adopts filters to suppress background and clutter. The Tophat method [18], Max-Mean and Max-Median method [19], facet model method [20,21] have been proposed and applied to the field of infrared small target detection. However, the assumptions and principles of the background suppression based methods are relatively simple, and the detection effect is not ideal.

The human visual system (HVS) based methods. Borji A [22] pointed out that the contrast between the target and background allows humans to observe small targets. Based on this point, Chen et al. [23] proposed the local contrast method (LCM). It derives the saliency map by sliding the window through each pixel to calculate the local contrast. Han J [24] increased the efficiency of the algorithm by increasing the sliding window step size and proposed improved local contrast method (ILCM). Deng H [25] proposed the weighted local difference measure (WLDM). Wei Y [26] proposed a multiscale patch-based contrast measure (MPCM) after analyzing the characteristics of bright and dark targets. Bai X [27] introduced the concept of derivative entropy into small target detection and proposed a derivative entropy-based contrast measure (DECM). Shi Y [28] proposed a high-boost-based multiscale local contrast measure (HB-MLCM). The prior knowledge of HVS based methods is simple, and usually the computational efficiency is relatively low, so the HVS based methods have been widely used. However, this category of method does not have an ideal facing complex background and noise, leading to low robustness.

The sparse and low-rank matrices recovery based methods. This category considers that the observed image is a linear combination of the target image, the background image, and the noise image, while assuming that the target image is sparse and the background image is low rank. Through the above process, a small target detection problem is transformed into an optimization problem, specifically the robust principal component analysis (RPCA) problem. Gao C [29] used the nuclear norm and the L1-norm as the characteristics of the optimal convex approximation of the rank function and the L0-norm and proposed infrared patch image (IPI) model. He et al. [30] proposed the low-rank representation (LRR) method. Wang C [31] proposed an adaptive target-background separation (T-BS) model. Dai Y [32] applied local steering kernel [33] to the penalty factor and proposed the weighted infrared patch image (WIPI) model. Dai Y [34] improved the way patch images are built, introduced the concept of a tensor [35,36] and proposed a reweighted infrared patch tensor (RIPT) model. Dai Y [37] relaxed the constraint of low-rank, added a non-negative prior, and proposed non-negative infrared patch image (NIPPS) model. Wang X [38] introduced total variation [39,40] to extract sharp edges

(TV-PCP) in the infrared image and obtained a purer target image. L Zhang [41] combined the l2,1 norm to describe the background and proposed a novel method based on non-convex rank approximation minimization joint l2,1 norm (NRAM). Since this category of method is assumed to be closer to the real situation, it will perform better than other categories, and with the continuous improvement of the solution algorithm, the convergence speed of such methods is also increasing.

1.3. Motivation

As can be seen from the above, the infrared small target detection methods can be described as a dazzling variety. Among them, the sparse and low-rank matrices recovery-based method has received much attention. However, since such methods usually use the L1-norm as an approximation of the L0-norm, the result may fall into the local minimum rather than the global minimum [42], which affects the constraints of the sparse item; consequently, the detection result is mixed with clutter, and the detection algorithm is poorly robust. Fortunately, there is still much room for improvement in the design of methods.

Previous work has demonstrated that the strategy of using Lp-norm regularization can greatly improve the ability of the algorithm to recover sparse signals compared to the L1-norm [43–46]. Besides, Lp minimization with $0 < p < 1$ recovers sparse signals from fewer linear measurements than does L1 minimization [43]. Another advantage of the Lp-norm is that when a sparse signal can be recovered, it often requires fewer iterations to converge the equation [44]. For RPCA problem, Chen X [47] theoretically established a lower bound of nonzero entries in solution of L2-Lp minimization. Furthermore, recent studies [48–51] have also given a solution to the RPCA problem of Lp-norm regularization. Although the optimization problem of the Lp-norm is non-convex, it has been studied before, and it has the advantages of being able to obtain a more sparse solution, fewer iterations to converge, and a theoretical basis for the L2-Lp minimization problem. The schatten q-norm can be understood as a sparse constraint on the singular value of the matrix, therefore obeying the above analysis.

Inspired by this, we aim to apply the constraints of the schatten q-norm and Lp-norm to the field of infrared small target detection and propose a novel infrared small target detection method based on non-convex optimization with Lp-norm constraint (NOLC). This method has the advantages of high detection accuracy, anti-noise, and fast convergence. Because of the excellent nature of the Lp-norm, the model is data-driven and can adapt to a variety of complex scenarios.

The main contributions of this article can be summarized as:

(1) Apply schatten q-norm and Lp-norm to the field of infrared small target detection, and propose NOLC method. This method transforms the NP-hard problem into a non-convex optimization problem, and it can restore sparser target images by enhancing constraints on a sparse item.

(2) An optimization solver is given to handle the non-convex optimization problem. This solver combines the ADMM method [52–54] to solve the problems. In order to speed up the convergence of this solver, an additional convergence condition is added to it. Similar optimization problems with this model can also use this solver.

(3) Through the specific experimental analysis, this paper gives the influence of different main parameters on the experimental results. Then, the set values of the key parameters are given. The experimental results for real infrared image sequences also verify the feasibility of this method.

The remaining part of this paper is organized as follows: Section 2 shows the principle of the NOLC method and solution of the non-convex optimization problem; Section 3 shows the experimental results, showing the effect of the method by analyzing the real infrared image sequences; The comparison between NOLC and other methods is given in the Section 4, highlighting the difference between this method and others; The conclusion is given in Section 5.

2. Methodology

This section will start with the basic schatten q-norm and Lp-norm, explain the application of these two norms in infrared small target detection, and propose a novel infrared small target detection method based on non-convex optimization with Lp-norm constraint (NOLC). Finally, a concrete solution method combining ADMM of this optimization is given.

2.1. Schatten q-norm and Lp-norm

Assume that matrix A has singular value decomposition $A = U * S * V^T$, where S denotes the singular value diagonal matrix. As we all know, the definition of the two norms of A is as Equations (1) and (2), where $\|A\|_{sq}$ represents schatten q-norm and $\|A\|_p$ represents Lp-norm.

$$\|A\|_{sq} = \left(\sum_{i=1}^{\min\{m,n\}} \sigma_i^q \right)^{1/q} , 0 < q < \infty \tag{1}$$

$$\|A\|_p = \left(\sum_{i=1}^{m} \sum_{j=1}^{n} |a_{ij}|^p \right)^{1/p} , 0 < p < \infty \tag{2}$$

where σ_i represents the ith singular value of matrix A, or can be expressed as the ith diagonal component of S; a_{ij} represents the pixel value of the ith row and the jth column of the matrix A. Since the matrix singular value is non-negative, the schatten q-norm of matrix A can be regarded as the Lp-norm of S. Therefore, we can understand the Schatten q-norm as a sparse constraint on singular values and it also obeys the following analysis of the Lp-norm.

For the optimization problem in Equation (3), geometrically, the constraint is a hyperplane and the Lp-norm is a ball blown from the origin point. As shown in Figure 1, when the blown ball is in contact with the hyperplane for the first time, the intersection is the optimal solution of problem (3).

$$\min_{X} \|X\|_p^p \ s.t. \ AX = b \tag{3}$$

Figure 1 shows the geometry of p when taking different values in 3D space. It can be seen that when p is greater than 1, the obtained optimal solution is not sparse, and when p is less than or equal to 1, the intersection point is on the coordinate axis, and two of the three elements are 0, so the optimal solution is sparse. Therefore, it can be geometrically stated that a sparse solution can be obtained when p is less than or equal to 1.

Broadly speaking, the values of p in the Equation (2) can range from 0 to positive infinity. But, in order to obtain the sparse solution, only the case where p is less than or equal to 1 is considered. In the special case where q and p equal to 0, the two norms can be expressed as Equations (4) and (5), where Equation (4) is a constraint on low rank and Equation (5) is a constraint on sparseness.

$$\|A\|_{s0} = \#(i) \ with \ \sigma_i \neq 0 = rank(A) \tag{4}$$

$$\|A\|_0 = \#(i) \ with \ a_i \neq 0 \tag{5}$$

However, these two functions above are non-convex and very difficult to solve, so they need to be approximated. Another special case is when q and p equal to 1, shown in Equations (6) and (7). These two norms are used in the reference [29] to approximate Equations (4) and (5) and constrain low rank and sparseness. Since Equations (6) and (7) are convex function, sub-problems with these two norm constraints can be easily solved.

$$\|A\|_{s1} = \sum_{i=1}^{\min\{m,n\}} \sigma_i = \|A\|_* \tag{6}$$

$$\|A\|_1 = \sum_{i=1}^{m} \sum_{j=1}^{n} |a_{ij}| \tag{7}$$

From the above analysis, the IPI model in reference [29] is a special case of the schatten q-norm and Lp-norm in this paper. It is worth mentioning that the strategy of using Lp-norm regularization can greatly improve the ability of the algorithm to recover sparse signals compared to the $L1$-norm [43–46]. Another advantage of the Lp-norm is that when a sparse signal can be recovered, it often requires fewer iterations to converge the equation. Based on this knowledge, we begin to introduce the model proposed in this paper.

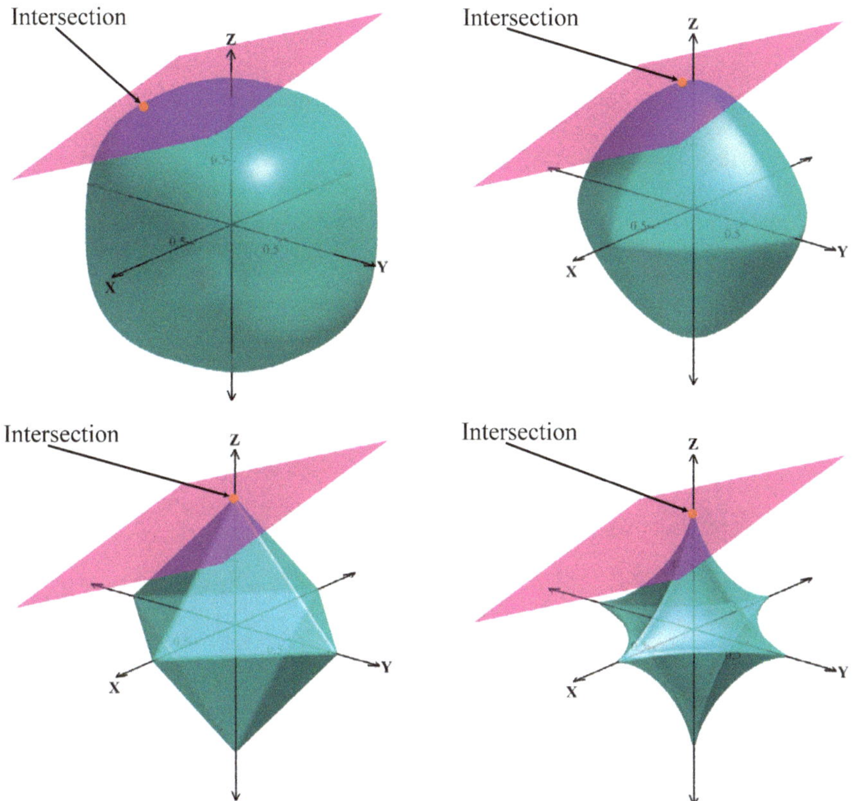

Figure 1. Geometry with different p values. From left top to right bottom, p equals to 2.8, 1.4, 1, 0.7, respectively.

2.2. The Proposed Method

In reference [10], when the noise can be approximated as additive, and the infrared small target image can be seen as a linear combination of target image, background image, and noise image. This assumption is also widely used in future models [30–34]. This model can be represented by the following equation.

$$f_D = f_B + f_T + f_N \tag{8}$$

where f_D denotes the infrared image; f_B, f_T and f_N represent the target image background image and noise image, respectively. Among them, because the target occupies a small area, it can be considered as a sparse matrix. The background contains many repetitive elements, so it can be considered as low rank matrix. The target image can be recovered by solving the model.

Then, by transforming the original image with the sliding window into patch image, the sparsity of the target and the low rank of the background are enhanced. The model is transformed into Equation (9):

$$D = B + T + N \tag{9}$$

where D, B, T and N denotes the patch images. Subsequently, we apply constraints on B and T using schatten q-norm and Lp-norm, respectively, and propose a method based on non-convex optimization with schatten q-norm and Lp-norm constraint (NOSLC). The objective function is expressed as follows.

$$\min_{B,T} \|B\|_{sq}^{q} + \lambda \|T\|_{p}^{p} \quad s.t. \|D - B - T\|_{F} \leq \delta \tag{10}$$

where λ is the penalty factor and δ denotes the noise level in the image; $\|\bullet\|_{F}$ denotes the Frobenius norm which is a special case of Lp-norm when p equals to 2.

As we mentioned earlier, the smaller the q and p values, the stronger the constraint on low rank and sparsity. However, we analyzed the real infrared image and found that the low rank property of the background patch image is not very strict compared to the sparsity of the target patch image.

Figure 2 shows the proportion of singular values greater than one and the proportion of target pixels for different infrared images. In the figure, six infrared images are analyzed, wherein the marked regions R1 to R3 are patch images with relatively large background changes, and are also regions that make the low rank property of background image less stringent. In the radar chart, the red dots indicate the proportion of singular values greater than 1, and the blue dots indicate the proportion of targets. It is obvious that the blue dots are squeezed together because their value is much smaller than that of the red dots. This image also shows that the sparsity of the target should be stricter than the low rank property of the background.

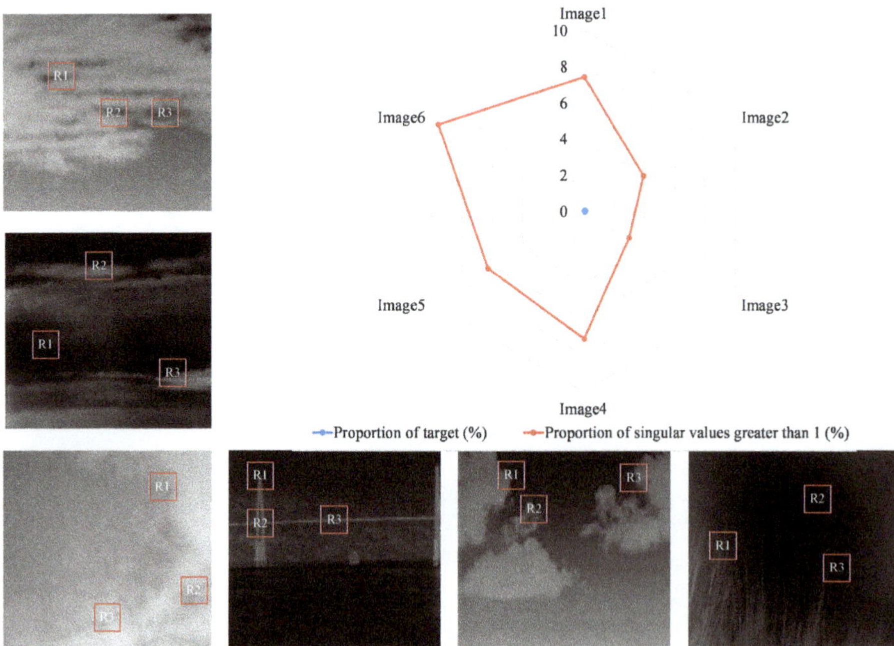

Figure 2. Illustration of low rank property and sparsity of infrared images.

Based on the above description, we know that it is unscientific to impose strong constraints on both the low rank property and sparsity, so we relax the constraint on the low rank property and let q equal to 1. Then, we propose a method based on non-convex optimization with Lp-norm constraint (NOLC). The objective function is shown in Equation (11).

$$\min_{B,T} \|B\|_* + \lambda\|T\|_p^p \qquad s.t. \|D - B - T\|_F \leq \delta \tag{11}$$

where $\|B\|_*$ denotes the nuclear norm of matrix B which is the special case of the schatten q-norm when q equals to 1. We have scaled down the constraints on low-rank property, so the model is not as sensitive as NOSLC to structural clutter in the background image which is inevitable. We qualitatively consider the two models presented above, and the NOLC method will achieve better results.

2.3. Solution of NOLC model

We obtained the physical model and objective function that we need to solve from the previous section. In this section we will give the solution to the NOLC model. Combined with the ADMM method, the Lagrange function of the objective function (11) is shown by Equation (12).

$$L(B,T,Y) = \|B\|_* + \lambda\|T\|_p^p + \langle Y, D - B - T \rangle + \frac{\rho}{2}\|D - B - T\|_F^2 \tag{12}$$

where $\langle \bullet, \bullet \rangle$ represents the inner product of two matrices, ρ is a penalty factor and Y is Lagrange multiplier matrix. Now we need to use an iterative method to minimize the Lagrange function. In this process, two sub-problems are solved. Next, we explain their solution method separately.

(a) The First Sub-Problem

The function is as follows:

$$\begin{aligned} B^{k+1} &= \underset{B}{\arg\min}\, L\left(B, T^k, Y^k\right) \\ &= \underset{B}{\arg\min}\, \|B\|_* + \frac{\rho}{2}\left\|B - \left(D + \rho^{-1}Y^k - T^k\right)\right\|_F^2 \end{aligned} \tag{13}$$

The above formula is a convex optimization problem and can be solved by the singular value shrinkage operator [54].

$$B^{k+1} = Q * S_{\rho^{-1}}\left[diag\left(\textstyle\sum\right)\right] * R^T \tag{14}$$

where $Q, \sum, R$ represents the singular value decomposition of matrix $D^k + \rho^{-1}Y^k - T^k$, that is to say $D^k + \rho^{-1}Y^k - T^k = Q * \sum * R^T$; $diag(\sum)$ denotes the diagonal elements of matrix $\sum$; $S_{\rho^{-1}}[\bullet]$ is the soft thresholding operator; its definition is given in the following formula.

$$S_\varepsilon[x] = \begin{cases} x - \varepsilon & if\ x > \varepsilon \\ x + \varepsilon & if\ x < -\varepsilon \\ 0 & others \end{cases} \tag{15}$$

(b) The Second Sub-Problem

For the second sub-problem, a non-convex optimization problem is involved.

$$\begin{aligned} T^{k+1} &= \underset{T}{\arg\min}\, L\left(B^{k+1}, T, Y^k\right) \\ &= \underset{T}{\arg\min}\, \lambda\|T\|_p^p + \frac{\rho}{2}\left\|T - \left(D + \rho^{-1}Y^k - B^{k+1}\right)\right\|_F^2 \end{aligned} \tag{16}$$

Since the elements in the matrix are linearly independent, this problem can be refined to each pixel to solve [50]. For each pixel, the optimization goal is:

$$x^* = \min_x \frac{1}{2}(x-a)^2 + \lambda|x|^p \tag{17}$$

Let the optimization function of each pixel be $g(x)$.

$$g(x) = \frac{1}{2}(x-a)^2 + \lambda|x|^p \tag{18}$$

In problem (17), we want to get the corresponding x value when $g(x)$ is the smallest. The curve of this function is shown in Figure 3. Obviously $g(x)$ is not a convex function, but the minimum point of $g(x)$ is easy to find.

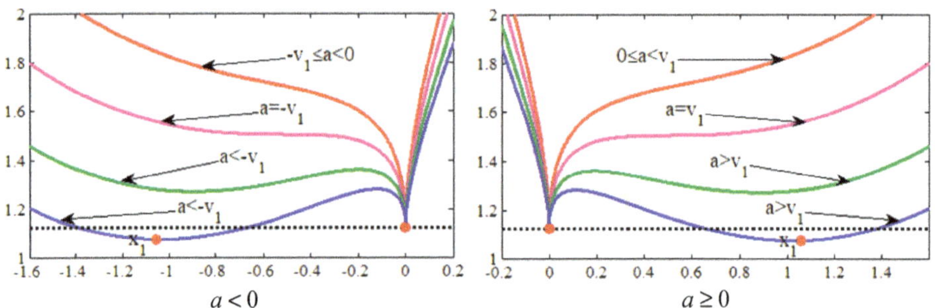

Figure 3. Illustration of the $g(x)$ curves for different a and λ.

By analyzing the first second and third derivative of $g(x)$, we can get the minimum point of $g(x)$, either 0 or x_1. We set two parameters $v = (\lambda p(1-p))^{\frac{1}{2-p}}$ and $v_1 = v + \lambda p|v|^{p-1}$. The solution to problem (17) is:

$$x^* = \begin{cases} 0 & a \le v_1 \\ \mathrm{argmin}_{x \in \{0,x_1\}} g(x) & a > v_1 \end{cases} \tag{19}$$

where x_1 is the solution of $g'(x) = 0$ in the case of $v < x < a$, and can be obtained by Newton iteration method. In the case where the initial value is set to v, it can be iteratively converged five times. The iterative formula of Newton method is as shown in Equation (20).

$$x_{n+1} = x_n - \frac{g'(x)}{g''(x)} \tag{20}$$

where $g'(x)$ and $g''(x)$ denote the first and second derivative of function $g(x)$. We define an operator $\mathcal{T}_\lambda[\bullet]$ to solve problem (17) in the matrix.

$$\mathcal{T}_{\lambda,p}[W] = \mathrm{argmin}_X \lambda\|X\|_p^p + \frac{1}{2}\|X - W\|_F^2 \tag{21}$$

By applying Equation (19) pixel by pixel, an optimal solution can be obtained. Therefore, it is obvious that problem (16) is solved by the definition of the operator $\mathcal{T}_\lambda[\bullet]$.

$$T^{k+1} = \mathcal{T}_{\rho^{-1}\lambda,p}\left[D + \rho^{-1}Y^k - B^{k+1}\right] \tag{22}$$

The specific process of solving the non-convex optimization problem (12) in combination with the ADMM is shown in Algorithm 1. So far, we have explained the definition and properties of schatten q-norm and Lp-norm, and have also described the principle and solution method of NOLC model in detail.

Algorithm 1 Solving the objective function of NOLC model

Input: Patch Image D, λ, p.
Output: Target patch image T and background patch image B.

1:    Initialization parameters: $B^0 = D$, $T^0 = 0$, $Y^0 = 0$, $\rho^0 = 1/(5 * std(D))$;

2:    **While not** converged **do**

3:    % *Update B^{k+1} via solving* $B^{k+1} = \underset{B}{\text{argmin}} L\left(B, T^k, Y^k\right)$

4:    $B^{k+1} = Q * S_{\rho^{-1}}[diag(\Sigma)] * R^T$;

5:    % *Update T^{k+1} via solving* $T^{k+1} = \underset{T}{\text{argmin}} L\left(B^{k+1}, T, Y^k\right)$

6:    $T^{k+1} = \mathcal{T}_{\rho^{-1}\lambda,p}\left[D + \rho^{-1}Y^k - B^{k+1}\right]$;

7:    % *Update Y^{k+1} and ρ^{k+1}*

8:    $Y^{k+1} = Y^k + \rho^k\left(D - B^{k+1} - T^{k+1}\right)$, $\rho^{k+1} = 1.5 * \rho^k$;

9:    % *Judge whether it has converged*

10:   *stopCriterion* $= \|D - B^{k+1} - T^{k+1}\|_F / \|D\|_F$

11:   *if stopCriterion* $< 10^{-7}$

12:   converged and stop iteration;

13:   *endif*

14:   **End while**

15:   **Return:** $B = B^{k+1}$, $T = T^{k+1}$.

2.4. Detection Procedure

Here are the specific implementation steps for the NOLC method proposed in this paper. Figure 4 also shows the detection steps.

(a) Traversing an infrared image $I(x, y)$ using a sliding window of length *len* and a *step* size into the patch image $D(x, y)$; the values of these two parameters will be discussed in detail in Section 3.

(b) Initialize some parameters: $lambda = L/\sqrt{\max(size(D))}$, the value of L and p will be discussed in Section 3; the recommended setting here is 1 and 0.4;

(c) Enter patch image $D(x, y)$ into Algorithm 1, and solve the target patch image $T(x, y)$ iteratively. It is worth mentioning that during the experiment we found that the non-zero elements in the target patch image no longer increase before the algorithm converges. In order to speed up the convergence of the algorithm, we set the non-zero element to no longer to increase as one of the conditions for the algorithm to stop iterating;

(d) Restore the target image $t(x, y)$ with the same sliding window as step (a);

(e) Threshold segmentation to the target image using the following formula, where *th* is the threshold for segmentation; μ and σ represents the respective mean and variance of the target image. Figure 5 shows the detection result of NOLC model.

$$th = \mu + k * \sigma \tag{23}$$

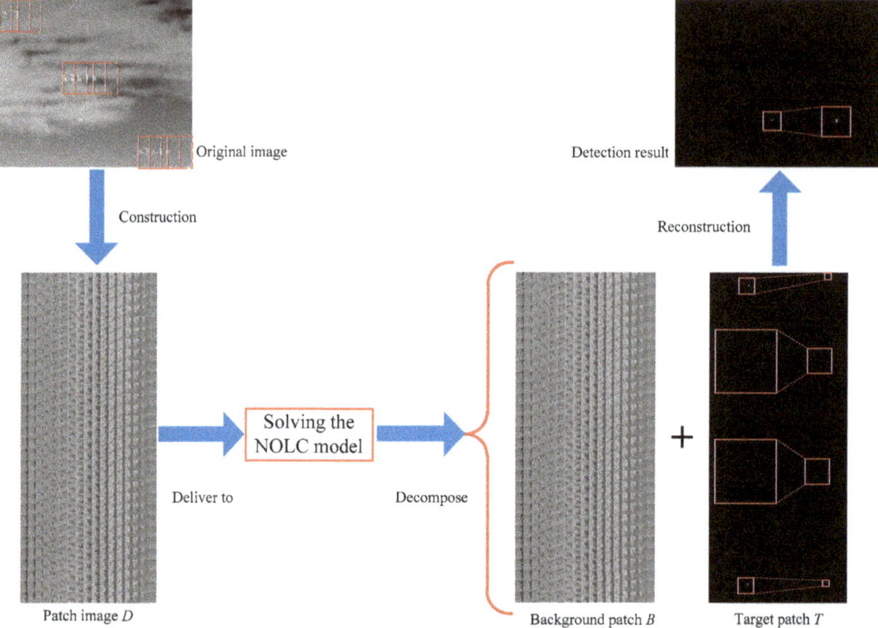

Figure 4. Detection flow of NOLC model.

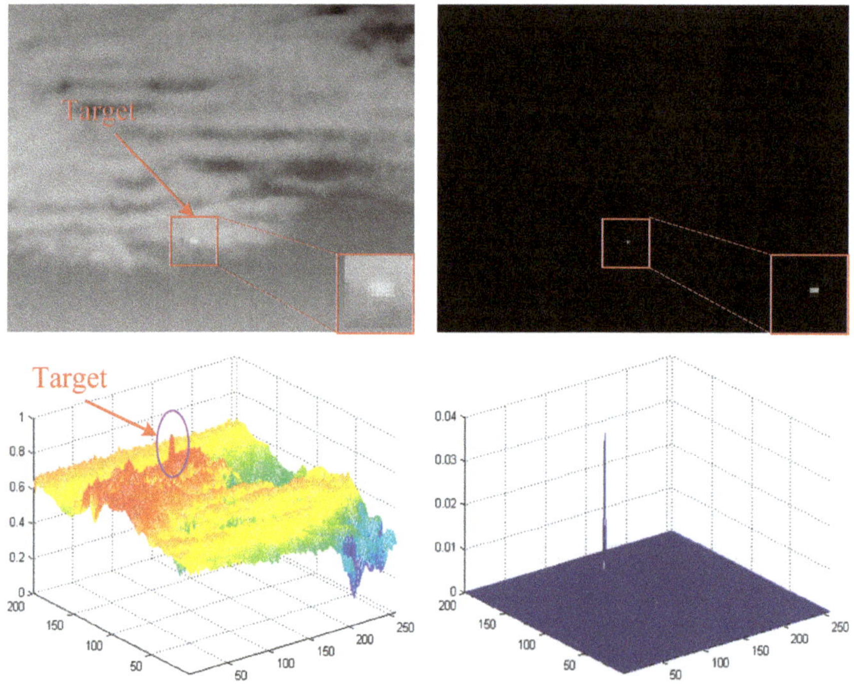

Figure 5. Infrared small target detection result of NOLC model.

3. Experiments

In this section we will introduce the evaluation indicators of this paper, discuss the impact of different parameter settings on the NOLC model, and compare NOLC with state-of-the-art, and finally, verify the validity of the NOLC model through the above steps.

3.1. Experimental Setting

This paper compares the NOLC model to the Tophat method (Tophat), Max Median method (MaxMedian), Local Contrast Method (LCM), Multiscale Patch-based Contrast Measure (MPCM), Infrared Patch Image (IPI) model and Reweighted Infrared Patch Tensor (RIPT) model, Non-Convex Rank Approximation Minimization (NRAM). The algorithm parameter settings used in this paper are given in Table 1. In addition, the specific information and image descriptions of the four sequences tested in this paper are summarized in Table 2. The software used in this article is MATLAB R2014a and the CPU is Core i5 7500, 3.4 GHz.

Table 1. Algorithm parameter setting.

Algorithm Name	Abbreviation	Parameter Setting
Tophat Method	Tophat	Structure shape: disk Structure size: 5×5
Max Median Method	MaxMedian	Support size: 5×5
Local Contrast Method	LCM	Window radius: 1,2,3,4
Multiscale Patch-based Contrast Measure	MPCM	Window radius: 1,2,3,4
Infrared Patch Image model	IPI	Patch size: 30×30 Slide step: 10, L: 1 Lambda: $L/\sqrt{\min(m,n)}$
Reweighted Infrared Patch Tensor model	RIPT	Patch size: 30×30 Slide step: 10, L: 1 Lambda: $L/\sqrt{\min(I,J,P)}$ h: 10, ε: 0.01
Non-Convex Rank Approximation Minimization	NRAM	Patch size: 30×30 Slide step: 10, L: 1 Lambda: $L/\sqrt{\min(m,n)}$
Non-Convex Optimization with Lp-Norm Constraint	NOLC	Patch size: 30×30 Slide step: 10 Lambda: $L/\sqrt{\max(size(D))}$ L: 1, p: 0.5

Table 2. Test sequence information.

Sequence	Resolution	Image Description
Seq1	256×200	The target is a long strip, the target is relatively pure, but there is a lot of horizontal cloud interference above the image.
Seq2	320×240	The target is close to a circle with noise interference around it, and there are a lot of irregular clouds at the edges of the image.
Seq3	256×172	The target is relatively bright, and it shuttles through the clouds. There are a lot of structural disturbances around it. At the same time, the target changes greatly in the field of view, and the background changes quickly.
Seq4	252×213	The target occupies a small number of pixels, and there is vertical clutter interference around the target, there is cluttered grass under the image, and the brightness around the target is uneven.

In order to objectively illustrate the effectiveness of the NOLC method, this paper uses quantitative evaluation indicators such as the receiver operating characteristic (ROC) curve, signal-to-clutter ratio gain (SCR Gain), background suppression factor (BSF) and iteration number.

(a)   ROC curve

The ROC curve is widely used in the evaluation of two-class models and also in the field of infrared small target detection. It can quantitatively describe the dynamic relationship between the true positive rate (TPR) and false positive rate (FPR), and give neutral and objective suggestions when evaluating algorithms. The abscissa of the ROC curve is TPR, which reflects the proportion of the target being correctly detected; the ordinate is FPR, which reflects the proportion of non-targets being misdetected as targets. Therefore, the closer the ROC curve is to the upper left corner, the better the algorithm works. When the ROC curve is applied to infrared small target detection, the abscissa and ordinate are defined as follows:

$$\text{FPR} = \frac{\%\text{number of pixels detected in background region}}{\%\text{real targets}} \tag{24}$$

$$\text{TPR} = \frac{\%\text{real targets detected}}{\%\text{real targets}} \tag{25}$$

Another key indicator of the ROC curve is the area under the curve (AUC). In general, the larger the AUC, the better the algorithm works.

(b)   SCR Gain and BSF

SCR Gain and BSF are indicators for measuring the degree of improvement of the target and the ability to suppress the background, respectively. Their definition of the target and background area is shown in Figure 6, defined as Equation (26).

$$\text{SCRG} = \frac{(S/C)_{out}}{(S/C)_{in}}, \text{BSF} = \frac{C_{in}}{C_{out}} \tag{26}$$

where S and C denote the signal (target region) amplitude and clutter (background region) standard deviation, respectively; *in* and *out* represent the original image and the detection result image. In the experiment, the values of a and b are 10 and 40, respectively. According to the definition, the larger the values of SCR Gain and BSF, the better the detection performance of the algorithm.

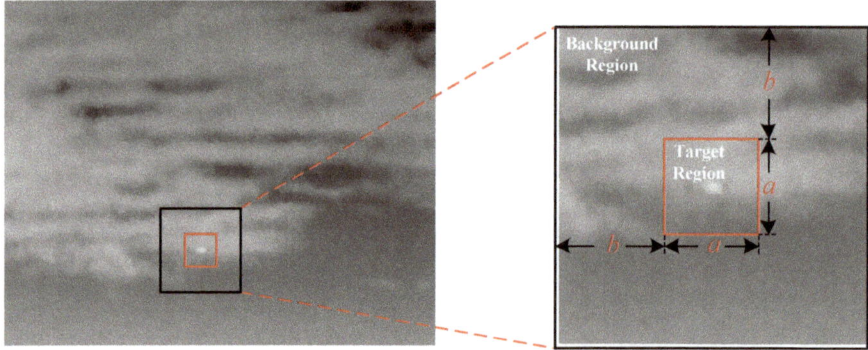

Figure 6. Neighborhood defined by SCR Gain and BSF.

(c)   Iteration number

All the sparse and low-rank matrices recovery based methods involve iterative solution, in which the iteration number of the algorithm directly affects the detection efficiency and running time. If a method has fewer iterations, it basically shows that the operation time is shorter, so the iteration number is also a key evaluation indicator.

3.2. Algorithm Validity

This section will prove the robustness of the NOLC model in various scenarios from the scene validity, and compare the proposed model with the IPI model and NOSLC model to prove the feasibility of NOLC.

(a)   Validity of Diverse Scene

The NOLC model strengthens the constraint on sparse items, and at the same time appropriately shrinks the constraints on low rank items, so it has a good detection effect. Figure 7 shows multiple original images, NOLC processing results, and their three-dimensional display.

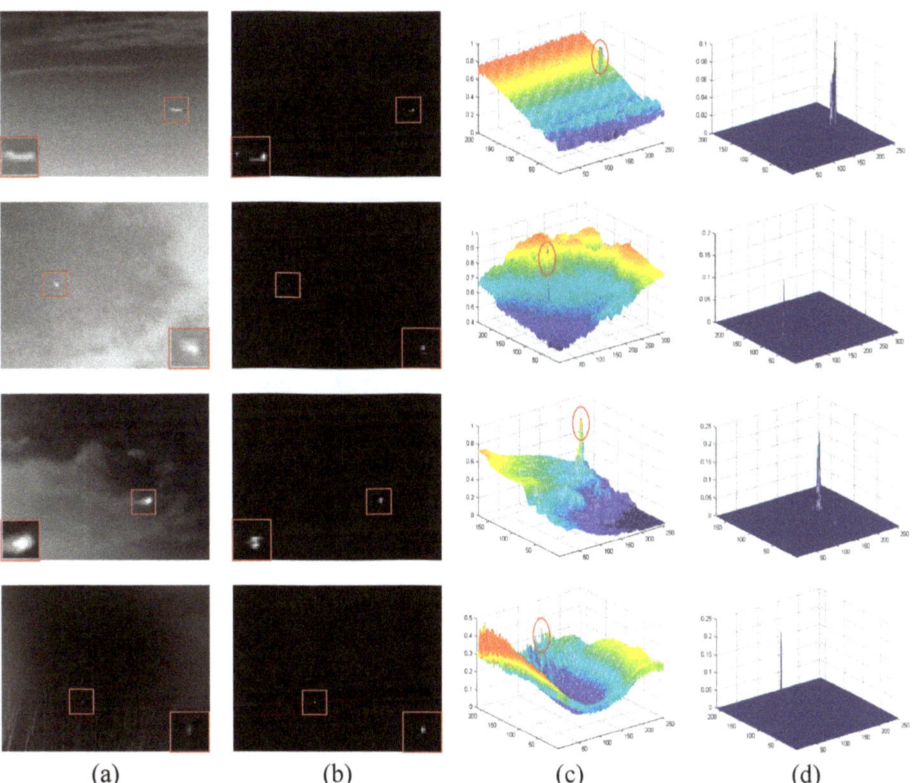

(a)   (b)   (c)   (d)

Figure 7. Display of the NOLC results of Seq1 to Seq4. (**a**) The original image; (**b**) the result of NOLC; (**c**) 3D display of (**a**); (**d**) 3D display of (**b**).

In order to better display the target information, the target region in the Figure 7 is enlarged and placed in the corner of the image. Each row in Figure 7 represents the presentation of the images in Sequence 1–4, with each column being the original image, NOLC processing results and their three-dimensional display. It can be seen from the Figure 7 that NOLC can accurately detect small targets regardless of whether the background is submerged in the clutter or the gray scale of the image is not uniform. As far as the result is concerned, the detected target image has only a corresponding target position, and the background region is suppressed to 0, so the effect of NOLC is remarkable.

(b)   Validity of the Proposed Algorithm

The previous section verifies that NOLC is effective. This section compares it with the IPI and NOSLC model mentioned in this paper, and the effectiveness of this method will be further confirmed. Figures 8 and 9 show the results of the three algorithms.

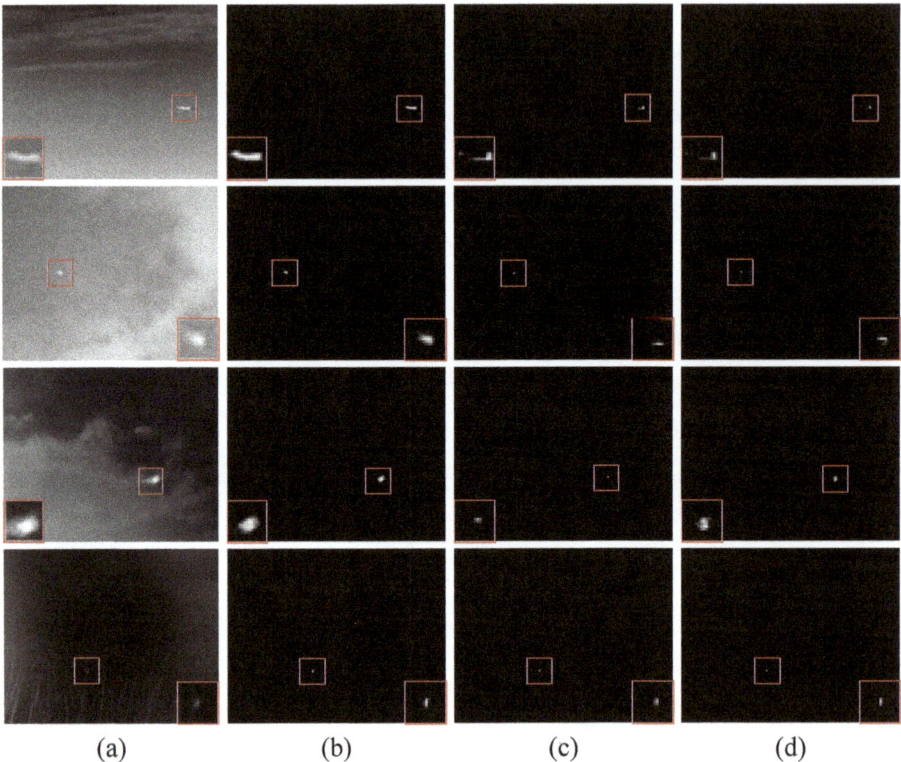

(a)   (b)   (c)   (d)

Figure 8. Comparison of IPI NOSLC and NOLC. (**a**) Original images; (**b**) IPI processing result; (**c**) NOSLC processing result; (**d**) NOLC processing result.

In Figure 8, from left to right are the original image, IPI, NOSLC and NOLC result image and from top to bottom are sequences 1–4. Similar to Figure 7, the target region is placed in the corner of the image in the processing result. For better illustration, Figure 9 shows a three-dimensional view of the corresponding position image of Figure 8. In the figure, the target position in the original image is circled in red, and the position of the clutter is circled with cyan in the 3D display of processing result. Since the clutter is relatively small in 3D display, it is not easy to visually see it, so it is marked in a cyan circle. It can be seen from the figure that all three methods can detect the target, but the IPI method contains much clutter when dealing with complex background images. The results of the NOSLC method are relatively low in terms of clutter, but because the constraints on the background low rank are stricter, the result is not very satisfactory. The NOLC method enhances the sparsity of the target and appropriately scales the background low rank property. The result is the lowest among the three methods in terms of clutter, and the background of most images is suppressed to 0. The test results of the sequence images are shown in Figure 10.

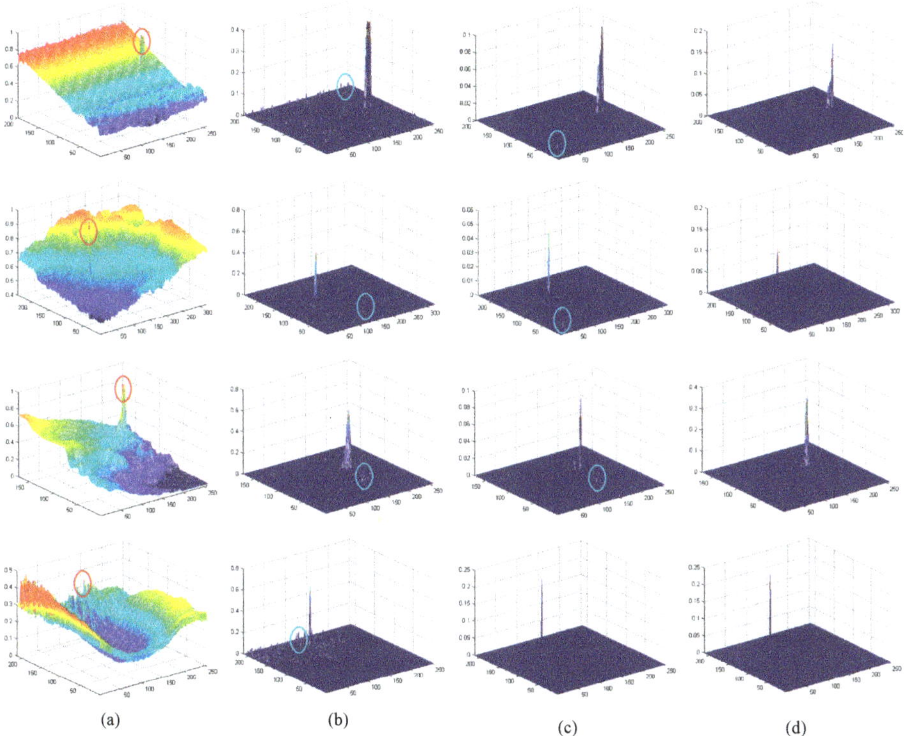

(a)   (b)   (c)   (d)

Figure 9. 3D display of Figure 8. (**a**) Original images; (**b**) IPI processing result; (**c**) NOSLC processing result; (**d**) NOLC processing result.

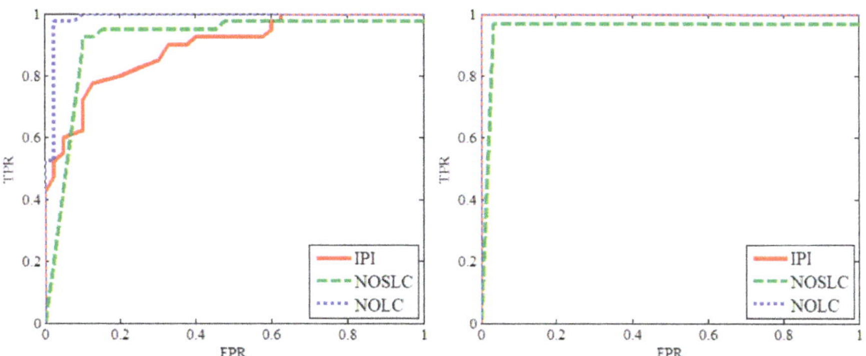

Figure 10. *Cont.*

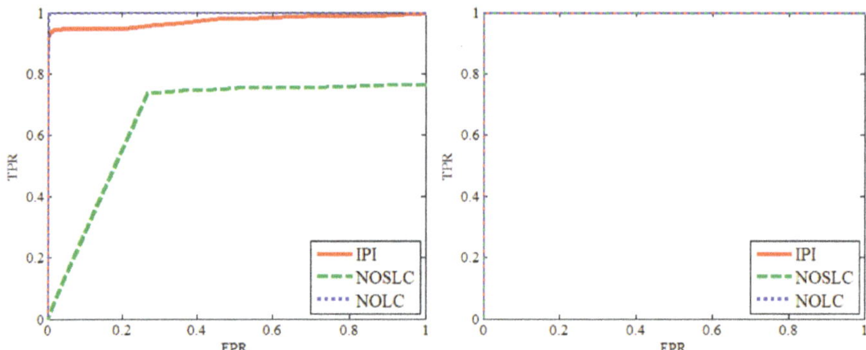

Figure 10. ROC curves for IPI, NOSLC and NOLC.

From the upper left corner to the lower right corner in Figure 10 are sequences 1–4. It can be seen that the three algorithms in Seq 4 have similar effects. The NOLC and IPI effects in Seq 2 are comparable and superior to NOSLC. In Seq 1, 3 the FPR of NOLC rises to 1 at the fastest, which is better than the other two. According to the above analysis, NOLC can not only accurately detect infrared small targets, but is also better than the IPI method and NOSLC method designed in this paper.

3.3. Parameter Analysis

In this section, we compare the four key parameters of NOLC to discuss the effect of parameter settings on the detection of NOLC method. The four parameters are the sliding window size *len*, the window sliding step size *step*, the *Lambda* initialization parameter L and the value of p in the Lp-norm. Figure 11 shows the ROC curve for comparison of these parameters.

The three columns from left to right in Figure 11 represent the ROC comparison chart of sequence 1–3, respectively. From top to bottom, the ROC curve comparison of the sliding window size *len*, sliding *step*, L and p parameters is shown. In the comparison experiment of the sliding window size *len*, we set the *len* values to 20, 30, 40, 50 and 60, and the remaining parameters are consistent. For qualitative considerations, if the *len* value is small, then the elements of each column in the patch image D will be relatively small, and the information contained will be less, the association between the columns will be missing, and the low rank and sparsity cannot be accurately guaranteed. On the contrary, if the *len* value is relatively large, it will not strictly conform to the constraint due to too many elements and redundant information. The first row of the ROC curve in Figure 11 also illustrates this. In the figure, when the *len* value is 30, a good ROC performance can be maintained in more sequences, and thus the *len* value can be taken as 30.

For the sliding window *step*, the *step* value is smaller, the window change is smaller each time, and the low rank property is stronger, but the small *step* greatly increases the block image matrix dimension and affects the algorithm detection efficiency. The second row in Figure 11 shows the ROC contrast image with *step* values of 6, 8, 10, 12, 14 when the remaining parameters are unchanged. In order to achieve a balance between algorithm efficiency and detection efficiency, the *step* value is recommended to be 10. In addition, the value of L also affects the detection effect. The third row in Figure 11 shows the ROC contrast image with different L values. It can be seen from the figure that the ROC curve performs best when the value of L is 1.

The fourth row of Figure 11 shows the ROC comparison of the last key parameter p. As mentioned in the second section, the smaller the p, the stronger the constraint on the low rank property and the efficiency of the algorithm is guaranteed. But when p is too small, the target cannot be detected. When the p value is increased, although the detection accuracy of the target can be ensured, it will increase the calculation time. Therefore, the choice of p value should be as small as possible. In combination with the ROC curve comparison in Figure 11, the p value is recommended to be 0.4.

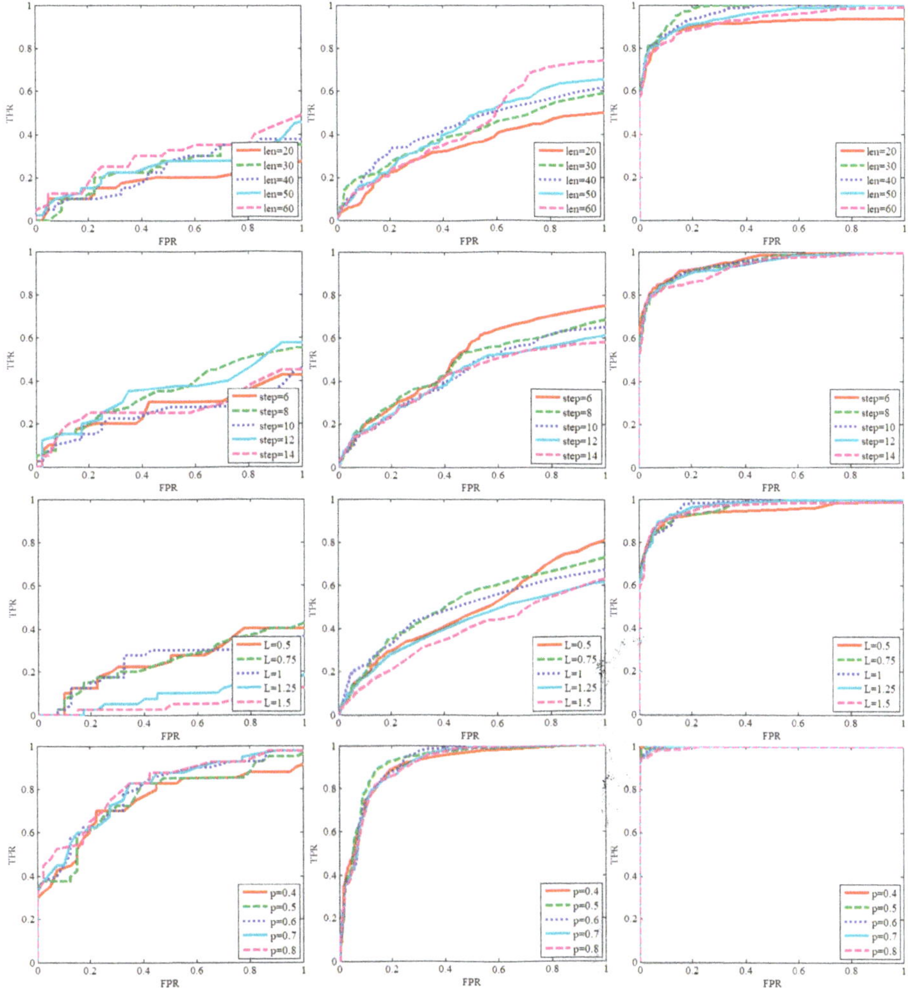

Figure 11. Parameter setting comparison.

3.4. Comparison to State-of-the-Art

The above sections verify the effectiveness of the diverse scene and the proposed method, and discuss the setting of key parameters. In this section, we compare the NOLC algorithm with other detection algorithms. The parameter settings of the seven contrasting algorithms and NOLC are presented in Table 1. We compared the NOLC model to the Tophat method (Tophat), Max Median method (MaxMedian), Local Contrast Method (LCM), Multiscale Patch-based Contrast Measure (MPCM), Infrared Patch Image (IPI) model, Reweighted Infrared Patch Tensor (RIPT) model and Non-Convex Rank Approximation Minimization (NRAM). The effect of all algorithms on a single frame image is shown in Figures 12 and 13.

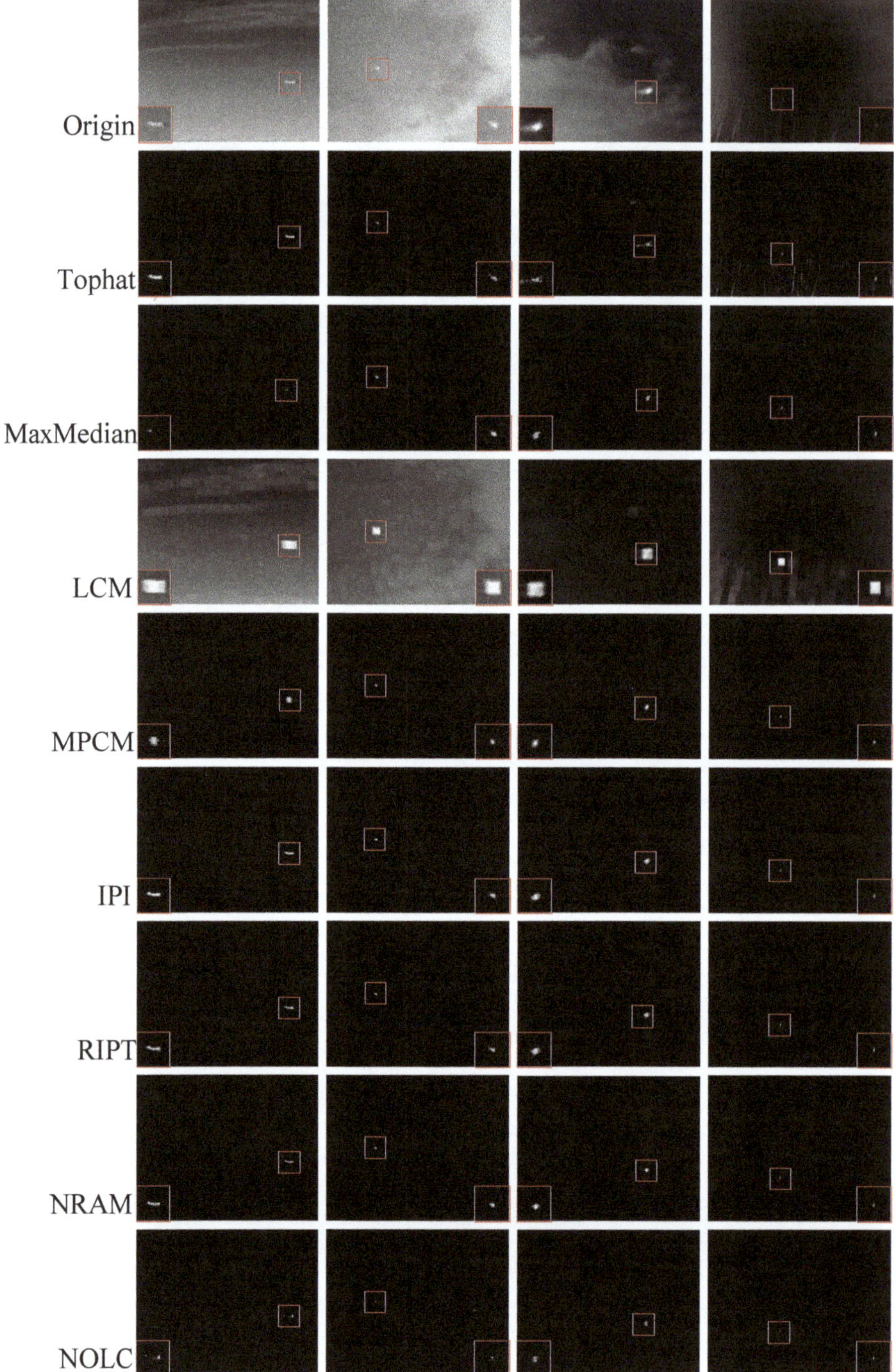

Figure 12. Performance of multiple methods.

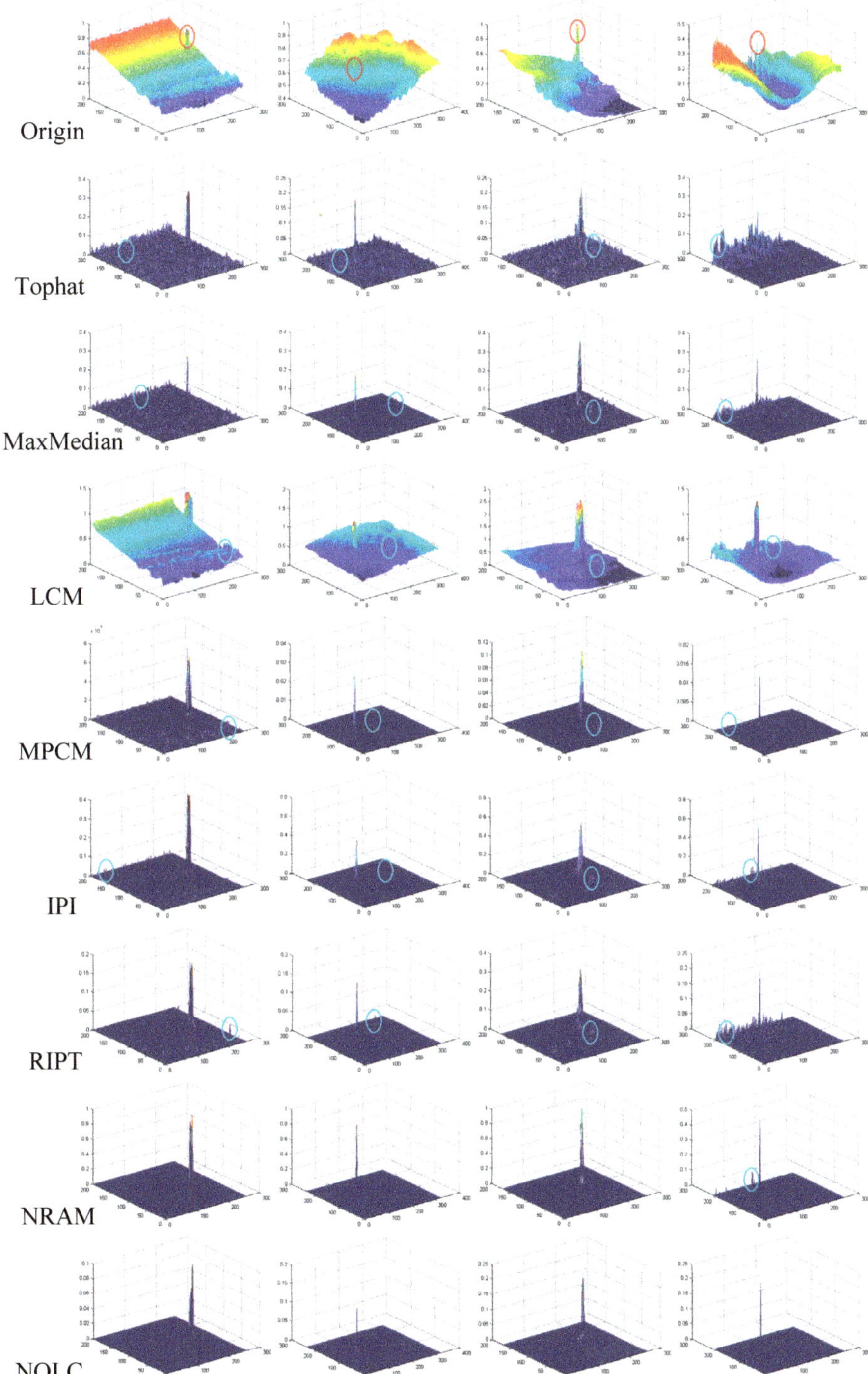

Figure 13. 3D display of original image and multiple method processing results.

Figure 12 shows the processing results of the original image and the above seven algorithms. For better display, the target region is framed in red and enlarged to the corner of the image. Figure 13 is a 3D representation of the corresponding position image of Figure 12. As in the previous section, the target position in the original image is circled in red, and the position of the clutter is circled in cyan in the 3D display of processing result.

In the above method, Tophat and MaxMedian are classic infrared small target detection methods. It can be seen that Tophat has a lot of clutter, while MaxMedian has less relative clutter but the target is greatly weakened. LCM and MPCM are typical methods based on the HVS method, but because the processing mechanism of LCM is relatively simple, the effect is not ideal. MPCM is able to accurately detect the target, but there is still a significant clutter in Sequence 1. IPI, RIPT, NRAM and NOLC are all sparse and low-rank matrices recovery based methods. Both IPI and RIPT can observe obvious clutter and have poor robustness. NRAM and NOLC can accurately detect the target while keeping the background basically suppressed to zero. However, as shown in the figure, the NRAM processing result has significant clutter.

To further demonstrate the superior performance of the NOLC method, we have experimented with four other complex scenarios. The processing results are shown in Figures 14 and 15 and the marking method is the same as above. The target position in the original image is circled in red, and the position of the clutter is circled in cyan in the 3D display of the processing result. The background of scene 1 is a large number of clouds, and the target occupies very few pixels and is disturbed by the clouds; scene 2 is a sea-sky background, in which there is sea level interference, and the bridge body appears as a structural disturbance in the picture, and the scene is very complicated; scene 3 is an air background, and irregular clouds appear on the edges. The image noise is relatively large, which also brings difficulty to the detection; the random noise in scene 4 is very strong, and there is a strong architectural disturbance in the lower left.

It can be seen from the experimental results that the background suppression based methods and the HVS based methods are very difficult to detect small targets in complex backgrounds. This is because the assumptions of the two methods are simple, and it is difficult to distinguish between clutter and target when encountering complex backgrounds. In contrast, because the assumptions of the sparse and low-rank matrices recovery based methods are supported by scientific physical models, they are superior in effect to other kinds of algorithms. However, there is still a lot of clutter in the processing results of IPI and RIPT. This is because the IPI model only limits the sparse item to a rough one, resulting in poor detection results. The RIPT method uses structural tensors to weight the sparse item, and the sparse constraints are still not strict, so the detection effect of RIPT is not ideal. As for the NRAM method, since the method only imposes constraints on the clutter, the contribution of this constraint to the detection effect is indirect, and the sparsity of the target is not strictly limited, so there is still clutter in the complex background. The NOLC method directly strengthens sparsely constrains and thus always finds sparse target locations in complex backgrounds, which explains why NOLC processing results have little clutter. This experiment also preliminarily illustrates the excellent robustness of NOLC.

The ROC comparison chart for the seven algorithms for the above four sequences is given by Figure 16 where the black line represents the curve of NOLC. From the top left to the bottom right, they represent sequences 1–4. As can be seen from the figure, the NOLC curve can always achieve a TPR of 1 when the FPR is relatively small, which means that the AUC of the NOLC is larger. To better compare the AUC of each of the curves in Figure 16, their specific values are listed in Table 3, where the maximum value of each sequence AUC is indicated in red and the second largest value is indicated in purple. From Table 3 we can quantitatively observe that the AUC of NOLC is the second largest in Sequence 2 and 3, and the rest are the largest. Therefore, it can be said that NOLC's performance in the sequence image test is remarkable.

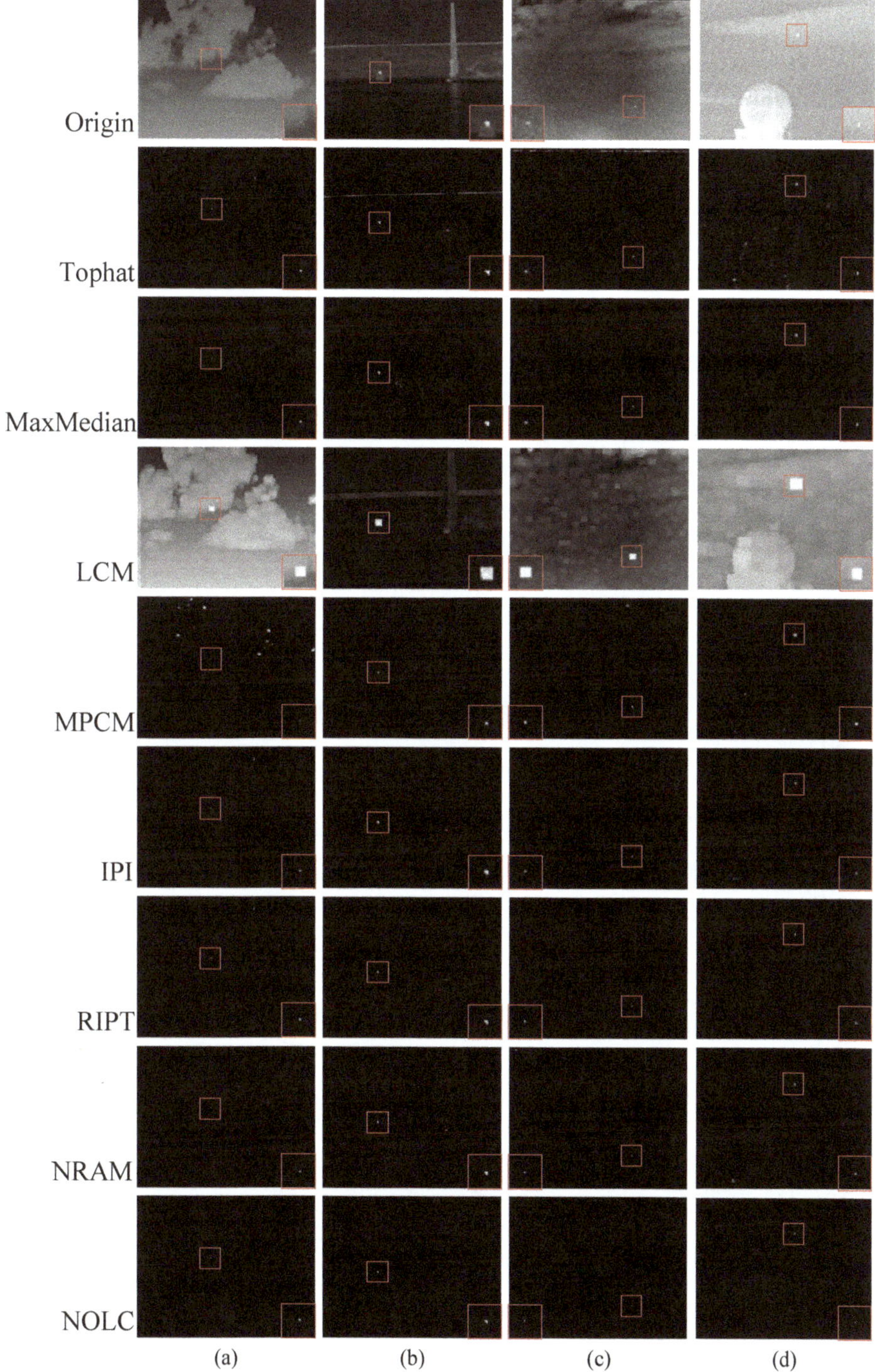

Figure 14. Comparison of four complex scenes. (a)–(d) are scenes 1–4.

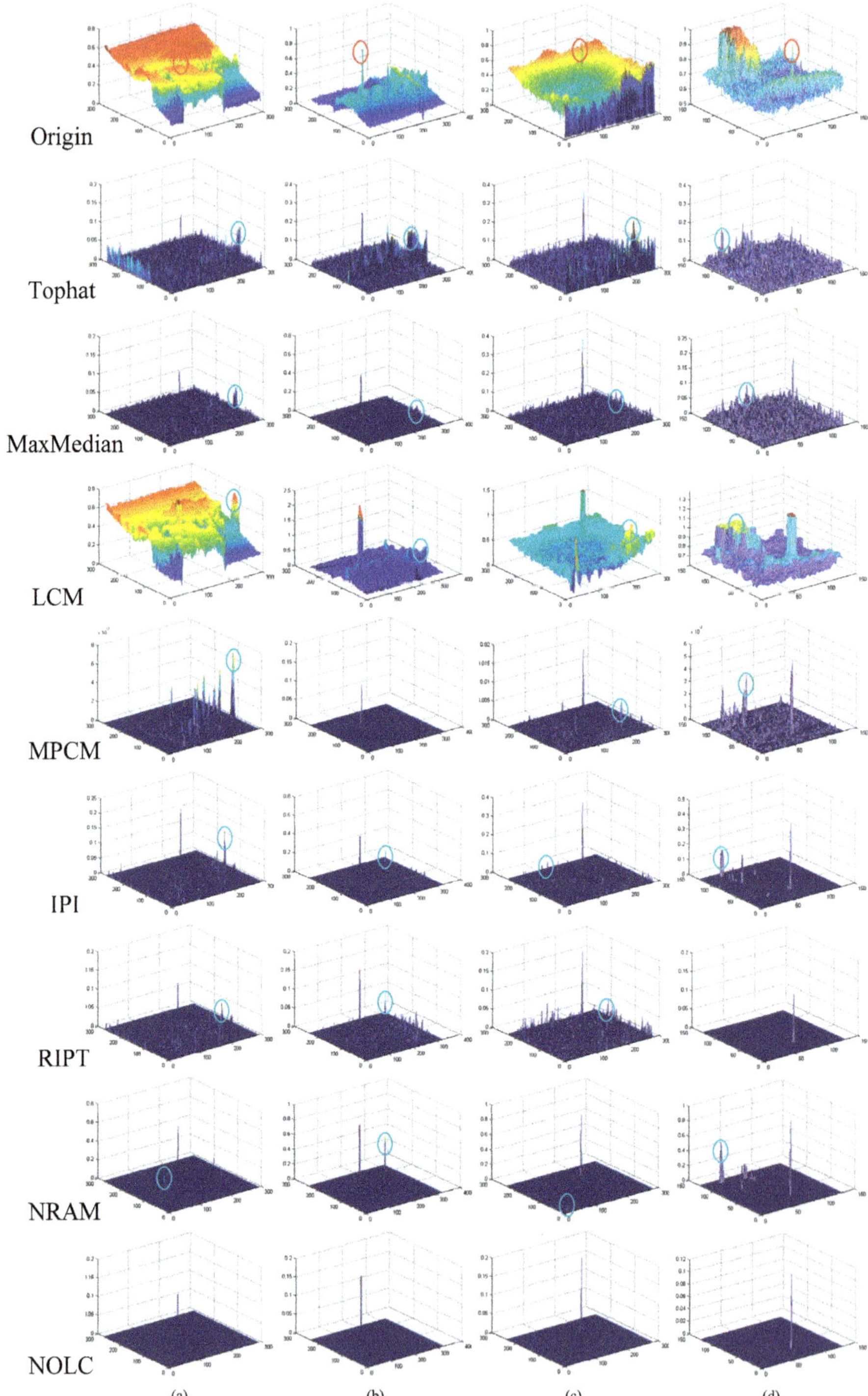

Figure 15. 3D display of Figure 14. (**a–d**) are scenes 1–4.

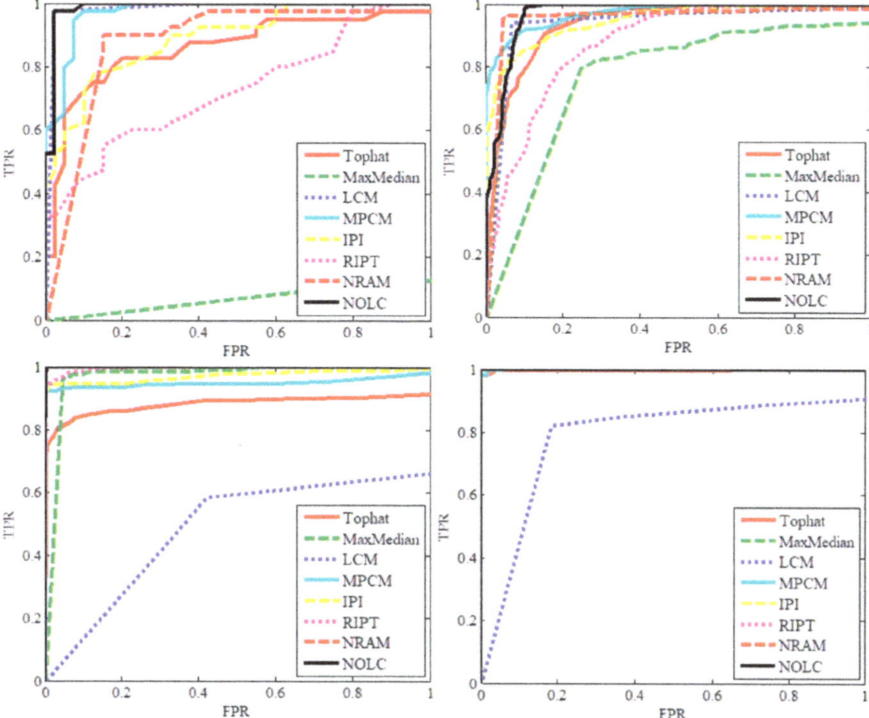

Figure 16. Seven algorithm comparison ROC curves.

Table 3. AUC of the ROC curves in Figure 14.

	Seq1	Seq2	Seq3	Seq4
Tophat	0.8591	0.9384	0.8814	0.9974
MaxMedian	0.0688	0.7698	0.9679	1.0000
LCM	0.9828	0.9369	0.4835	0.7855
MPCM	0.9731	0.9712	0.9494	0.9996
IPI	0.8900	0.9489	0.9721	1.0000
RIPT	0.7256	0.8781	0.9959	1.0000
NRAM	0.8769	0.9535	1.0000	1.0000
NOLC	0.9866	0.9657	0.9996	1.0000

Note: The maximum value of each sequence AUC is indicated in red and the second largest value is indicated in purple.

The test data for the other two key evaluation indicators, SCR Gain and BSF, are listed in Table 4. Similarly, the maximum value is indicated in red and the second largest is indicated in purple. You can see that the two classic methods do not perform very well. In the HVS based method, MPCM performs excellently with two maximum values and five second largest values. In the sparse and low-rank matrices recovery based methods, in addition to NOLC, the performance of RIPT is also excellent, with four maximum values and one second largest value. Overall, the comparison of the eight algorithms of NRAM and NOLC has the upper hand and has a maximum in each sequence. This shows that the two methods also do a better job of suppressing the background.

To further illustrate that the performance of the NOLC method is superior to the rest of the methods while verifying its robustness, we add normal noise with a mean of zero to the two sequences and compare it with IPI, RIPT and NRAM. The variance of the normal noise in Figure 17 is 0.04, 0.05 and 0.06. It can be seen that both IPI and RIPT are sensitive to noise, and the noise is very strong in the processing result. In the process of increasing the variance of the noise, the NRAM processing result is also mixed with a lot of clutter. While NOLC has always shown good performance, it can accurately detect the target and suppress the background very purely when the variance becomes larger. Figure 18 also illustrates the same fact in Seq 4. The above experiments show that the NOLC method is superior to other algorithms in terms of detection accuracy and algorithm robustness.

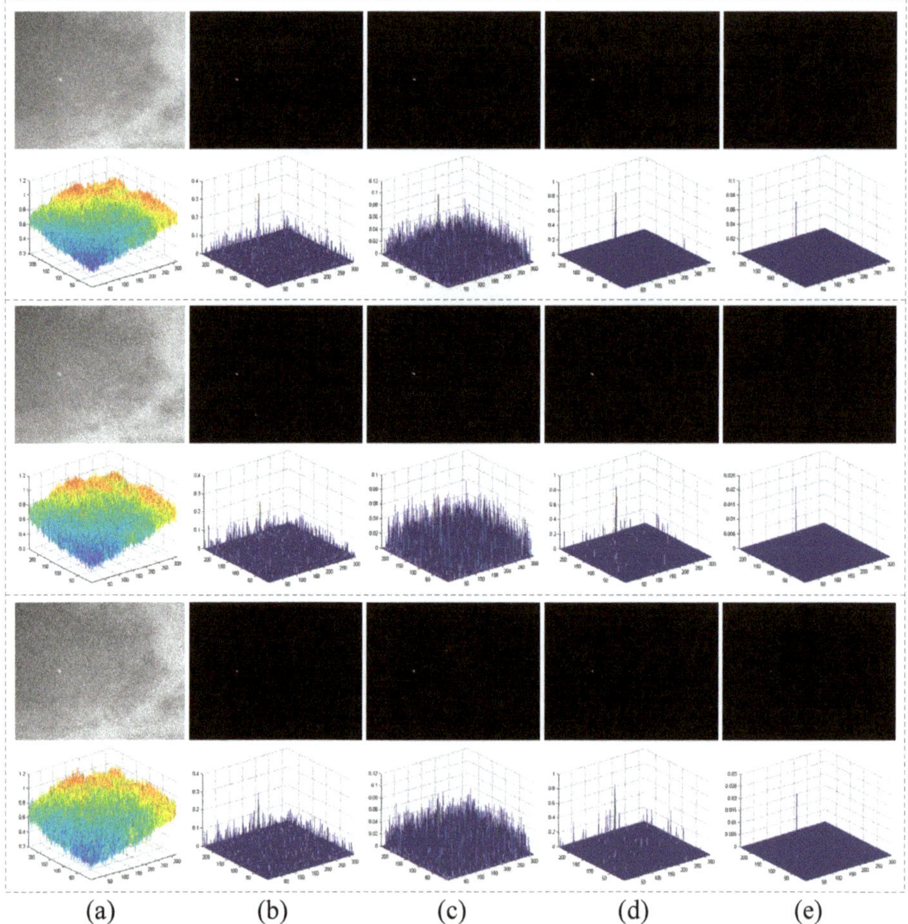

Figure 17. Comparison of processing results for Seq 2 noise images. The variance from top to bottom is 0.04, 0.05 and 0.06. (**a**) Noise image; (**b**) IPI processing result; (**c**) RIPT processing result; (**d**) NRAM processing result; (**e**) NOLC processing result.

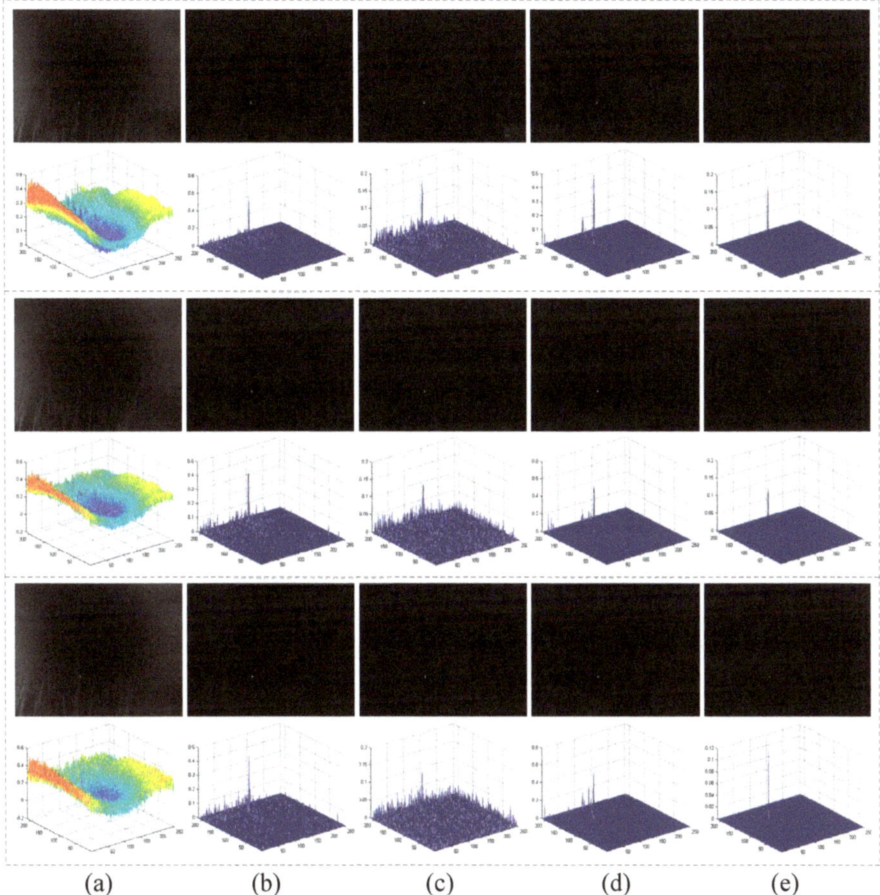

Figure 18. Comparison of processing results for Seq 4 noise images. The variance from top to bottom is 0.01, 0.02, 0.03. (**a**) Noise image; (**b**) IPI processing result; (**c**) RIPT processing result; (**d**) NRAM processing result; (**e**) NOLC processing result.

Table 4. Comparison of SCRG and BSF in various methods.

Evaluation Indicators	Seq1 SCR Gain	Seq1 BSF	Seq2 SCR Gain	Seq2 BSF	Seq3 SCR Gain	Seq3 BSF	Seq4 SCR Gain	Seq4 BSF
Tophat	10.78	5.196	3.936	2.764	11.68	14.62	2.451	1.864
MaxMedian	0.850	3.237	3.975	2.092	4.882	12.18	3.638	2.643
LCM	3.047	1.575	2.742	1.772	6.007	6.428	7.524	5.806
MPCM	102.2	111.3	193.8	138.9	Inf	Inf	129.1	116.9
IPI	121.3	53.50	48.06	25.19	218.5	190.6	22.10	12.80
RIPT	122.7	68.44	Inf	Inf	Inf	Inf	12.97	9.929
NRAM	Inf	Inf	Inf	Inf	Inf	Inf	Inf	Inf
NOLC	Inf	Inf	Inf	Inf	Inf	Inf	Inf	Inf

Note: The maximum value is indicated in red and the second largest is indicated in purple.

The last evaluation indicator is the iteration number. Since other methods do not involve iterative solution, four methods of IPI, RIPT, NRAM and NOLC are compared here. Figure 19 shows the iteration curves of the four methods in four sequences, where the algorithm name and the iteration number are given in the legend. It can be seen that since the IPI is solved by the accelerated proximal gradient (APG) method, the number of iterations is the highest, while RIPT and NRAM are solved by the faster ADMM method, and the iteration number is still higher than 10. The NOLC method not only uses the *Lp*-norm which can converge faster, but also improves the convergence judgment mechanism, so it can basically converge within 5 iterations. The convergence speed of NOLC is much better than the similar method, and it also has more advantages in running time.

In the experiments in this section, we show the detection effect of NOLC, and demonstrate the feasibility of NOLC by comparing IPI, NOSLC and NOLC. Then, in the aspects of the single frame effect, ROC curve, AUC, SCR Gain and BSF, the NOLC and other infrared small target detection methods are compared. It can be seen that the detection effect of NOLC has great advantages. The comparison of the image plus noise further illustrates the robustness of the NOLC method. Finally, the iteration number of NOLC and other sparse low-rank matrix reconstruction based methods are compared. The advantage of NOLC is explained again from the efficiency of the algorithm. All in all, NOLC is an excellent infrared small target detection method in terms of detection effect and running time.

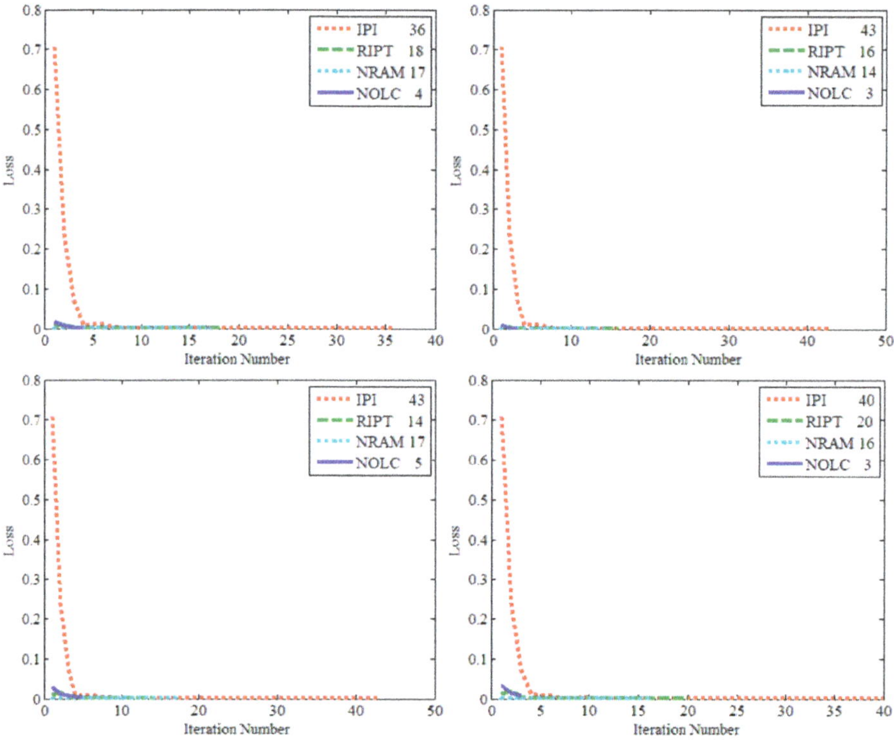

Figure 19. Iteration number comparison.

4. Discussion

The sparse and low-rank matrix recovery-based methods have been widely used by researchers, and a large number of methods are also applied to the field of infrared small target detection. However, starting from the IPI model, researchers often only pay attention to the use of additional constraint coefficients to improve the detection effect, while ignoring the difference in the sparse degree of low ranking items and sparse items in the infrared small target image. Experiments on six sets of actual data show that the sparsity degree difference between low-rank items and sparse terms is very large, even not within an order of magnitude, so it is unscientific to use only $L1$-norm constraints. Aiming at this property of infrared small target image, this paper uses the Lp-norm to constrain the sparse term and relaxes the constraint on the low ranking term, and the NOLC method is proposed.

Compared with other methods, the IPI model is the original method, and its principle and solution method are relatively simple. From the perspective of **background characteristics**, the RIPT model uses the local structure tensor as the penalty coefficient of the sparse term, in order to obtain a more accurate background. Because the results of local structure tensor are relatively rough and cannot be used as an ideal sparse penalty factor, RIPT does not work well in the face of complex backgrounds. The NRAM method is based on the **structural noise** and uses the $L21$-norm constraint. The $L21$-norm emphasizes that the row of the block image is sparse. To achieve this effect, the size of the structured noise must be smaller than the size of the sliding window to obtain the effect of row sparseness. However, for structured noise, its size often cannot meet the requirements (such as bridges and houses), which makes the NRAM using the $L21$-norm constrained noise term sometimes unconvincing. The NOLC method considers the difference between the target and the background sparsity from the perspective of the **target**, and directly uses the stricter Lp-norm to constrain the sparse item. This method can describe the target more directly and accurately than the IPI model and the NRAM method, can also obtain good detection effects under various complex backgrounds, and can always restore sparse targets in the noisy infrared small target image. The NOLC method improves the convergence strategy while utilizing the Lp-norm property, making the convergence speed better than other methods.

This article gives ample demonstration of the performance of NOLC through experiments. Firstly, the effect of the NOLC method in multiple scenarios is verified. Then, the key parameters in the method are analyzed and the values of the parameters are given. Then, compared with the existing methods, the results are also in line with the above analysis. NOLC is superior to other algorithms in detection accuracy, and can suppress most backgrounds to zero in terms of background suppression. Then, the noise infrared small target image is tested to verify the anti-noise ability of NOLC, and the robustness of the algorithm is further illustrated. Finally, comparing the iterative convergence speed of the four methods, NOLC also has obvious advantages.

In summary, the NOLC method has the advantages of high detection accuracy, anti-noise, fast convergence, etc. This method is not only a change of the metric, but an improvement of the performance brought by the improvement of the method. Recently, tensor-based infrared small target detection methods have also received extensive attention [34,55]. These method replace the matrix with tensor, and they can also provide good detection results.

5. Conclusions

In this paper, a novel infrared small target detection method based on non-convex optimization with Lp-norm constraint (NOLC) is proposed. The detection effect of the algorithm is enhanced by extending the original nuclear norm and $L1$-norm to the schatten q-norm and Lp-norm to strengthen the constraints on sparse items and appropriately scaling the constraints on low-rank items. At the same time, the NP-hard problem is transformed into a non-convex optimization problem. The NOLC model can not only accurately detect the target, but also greatly suppress the background area, achieving a good infrared small target detection effect. In the final part of the experiment, NOLC was compared with seven methods. It performed well on the ROC curve and also had very high SCR Gain and BSF. The comparison of the image plus noise further illustrates the robustness of the NOLC method. At the same time, it is also ahead of other algorithms in the number of iterations, which means that NOLC also leads in computing time.

Author Contributions: T.Z. proposed the original idea, performed the experiments and wrote the manuscript. H.W., Y.L., L.P. and C.Y. reviewed and edited the manuscript. Z.P. contributed to the direction, content, and revised the manuscript.

Remote Sens. **2019**, *11*, 559

Funding: This research was funded by the National Natural Science Foundation of China (61571096, 61775030), the Key Laboratory Fund of Beam Control, Chinese Academy of Sciences (2017LBC003) and Sichuan Science and Technology Program (2019YJ0167).

Conflicts of Interest: The authors declare no conflict of interest.

References

1. Shao, X.; Fan, H.; Lu, G.; Xu, J. An improved infrared dim and small target detection algorithm based on the contrast mechanism of human visual system. *Infrared Phys. Technol.* **2012**, *55*, 403–408. [CrossRef]
2. Chen, Y.; Xin, Y. An Efficient Infrared Small Target Detection Method Based on Visual Contrast Mechanism. *IEEE Geosci. Remote Sens. Lett.* **2016**, *13*, 962–966. [CrossRef]
3. Han, J.; Ma, Y.; Huang, J.; Mei, X.; Ma, J. An Infrared Small Target Detecting Algorithm Based on Human Visual System. *IEEE Geosci. Remote Sens. Lett.* **2016**, *13*, 452–456. [CrossRef]
4. Peng, Z.; Zhang, Q.; Wang, J.; Zhang, Q. Dim target detection based on nonlinear multi-feature fusion by Karhunen-Loeve transform. *Opt. Eng.* **2004**, *43*, 2954–2958.
5. Peng, Z.; Zhang, Q.; Guan, A. Extended target tracking using projection curves and matching pel count. *Opt. Eng.* **2007**, *46*, 066401.
6. Zheng, X.; Peng, Z.; Dai, J. Criterion to evaluate the quality of iInfrared target images based on scene features. *Elektronika ir Elektrotechnika* **2014**, *20*, 44–50. [CrossRef]
7. Fan, X.; Xu, Z.; Zhang, J.; Huang, Y.; Peng, Z. Dim small targets detection based on self-adaptive caliber temporal-spatial filtering. *Infrared Phys. Technol.* **2017**, *85*, 465–477. [CrossRef]
8. Wang, X.; Peng, Z.; Zhang, P.; He, Y. Infrared Small Target Detection via Nonnegativity-Constrained Variational Mode Decomposition. *IEEE Geosci. Remote Sens. Lett.* **2017**, *14*, 1700–1704. [CrossRef]
9. Wang, X.; Peng, Z.; Kong, D.; He, Y. Infrared Dim and Small Target Detection Based on Stable Multisubspace Learning in Heterogeneous Scene. *IEEE Trans. Geosci. Remote Sens.* **2017**, *99*, 1–13. [CrossRef]
10. Gu, Y.; Wang, C.; Liu, B.X.; Zhang, Y. A Kernel-Based Nonparametric Regression Method for Clutter Removal in Infrared Small-Target Detection Applications. *IEEE Geosci. Remote Sens. Lett.* **2010**, *7*, 469–473. [CrossRef]
11. Reed, I.S.; Gagliardi, R.M.; Stotts, L.B. Optical moving target detection with 3-D matched filtering. *IEEE Trans. Aerosp. Electron. Syst.* **2002**, *24*, 327–336. [CrossRef]
12. Li, M. Moving weak point target detection and estimation with three-dimensional double directional filter in IR cluttered background. *Opt. Eng.* **2005**, *44*, 107007. [CrossRef]
13. Modestino, J.W. Spatiotemporal multiscan adaptive matched filtering. In *Signal and Data Processing of Small Targets*; International Society for Optics and Photonics: Bellingham, WA, USA, 1995.
14. Braganeto, U.M.; Choudhury, M.; Goutsias, J.I. Automatic target detection and tracking in forward-looking infrared image sequences using morphological connected operators. *J. Electron. Imaging* **2004**, *13*, 802.
15. Dong, X.; Huang, X.; Zheng, Y.; Bai, S.; Xu, W. A novel infrared small moving target detection method based on tracking interest points under complicated background. *Infrared Phys. Technol.* **2014**, *65*, 36–42. [CrossRef]
16. Li, Y.; Tan, Y.; Li, H.; Li, T.; Tian, J. Biologically inspired multilevel approach for multiple moving targets detection from airborne forward-looking infrared sequences. *J. Opt. Soc. Am. A Opt. Image Sci. Vis.* **2014**, *31*, 734. [CrossRef] [PubMed]
17. Li, Y.; Zhang, Y.; Yu, J.G.; Tan, Y.; Tian, J.; Ma, J. A novel spatio-temporal saliency approach for robust dim moving target detection from airborne infrared image sequences. *Inf. Sci.* **2016**, *369*, 548–563. [CrossRef]
18. Tom, V.T.; Peli, T.; Leung, M.; Bondaryk, J.E. Morphology-based algorithm for point target detection in infrared backgrounds. In *Signal and Data Processing of Small Targets*; International Society for Optics and Photonics: Bellingham, WA, USA, 1993.
19. Venkateswarlu, R. Max-mean and max-median filters for detection of small targets. In *Signal and Data Processing of Small Targets*; International Society for Optics and Photonics: Bellingham, WA, USA, 1999; Volume 3809, pp. 74–83.
20. Wang, G.D.; Chen, C.Y.; Shen, X.B. Facet-based infrared small target detection method. *Electron. Lett.* **2005**, *41*, 1244–1246. [CrossRef]
21. Qi, S.; Xu, G.; Mou, Z.; Huang, D.; Zheng, X. A fast-saliency method for real-time infrared small target detection. *Infrared Phys. Technol.* **2016**, *77*, 440–450. [CrossRef]

22. Borji, A.; Itti, L. State-of-the-Art in Visual Attention Modeling. *IEEE Trans. Pattern Anal. Mach. Intell.* **2013**, *35*, 185–207. [CrossRef] [PubMed]

23. Chen, C.P.; Li, H.; Wei, Y.; Xia, T.; Tang, Y.Y. A Local Contrast Method for Small Infrared Target Detection. *IEEE Trans. Geosci. Remote Sens.* **2014**, *52*, 574–581. [CrossRef]

24. Han, J.; Ma, Y.; Zhou, B.; Fan, F.; Liang, K.; Fang, Y. A Robust Infrared Small Target Detection Algorithm Based on Human Visual System. *IEEE Geosci. Remote Sens. Lett.* **2014**, *11*, 2168–2172.

25. Deng, H.; Sun, X.; Liu, M.; Ye, C.; Zhou, X. Small Infrared Target Detection Based on Weighted Local Difference Measure. *IEEE Trans. Geosci. Remote Sens.* **2016**, *54*, 4204–4214. [CrossRef]

26. Wei, Y.; You, X.; Li, H. Multiscale patch-based contrast measure for small infrared target detection. *Pattern Recognit.* **2016**, *58*, 216–226. [CrossRef]

27. Bai, X.; Bi, Y. Derivative Entropy-Based Contrast Measure for Infrared Small-Target Detection. *IEEE Trans. Geosci. Remote Sens.* **2018**, *99*, 1–15. [CrossRef]

28. Shi, Y.; Wei, Y.; Yao, H.; Pan, D.; Xiao, G. High-Boost-Based Multiscale Local Contrast Measure for Infrared Small Target Detection. *IEEE Geosci. Remote Sens. Lett.* **2018**, *15*, 1–5. [CrossRef]

29. Gao, C.; Meng, D.; Yang, Y.; Wang, Y.; Zhou, X.; Hauptmann, A.G. Infrared Patch-Image Model for Small Target Detection in a Single Image. *IEEE Trans. Image Process.* **2013**, *22*, 4996–5009. [CrossRef] [PubMed]

30. He, Y.J.; Li, M.; Zhang, J.L.; An, Q. Small infrared target detection based on low-rank and sparse representation. *Infrared Phys. Technol.* **2015**, *68*, 98–109. [CrossRef]

31. Wang, C.; Qin, S. Adaptive detection method of infrared small target based on target-background separation via robust principal component analysis. *Infrared Phys. Technol.* **2015**, *69*, 123–135. [CrossRef]

32. Dai, Y.; Wu, Y.; Song, Y. Infrared small target and background separation via column-wise weighted robust principal component analysis. *Infrared Phys. Technol.* **2016**, *77*, 421–430. [CrossRef]

33. Takeda, H.; Farsiu, S.; Milanfar, P. Kernel Regression for Image Processing and Reconstruction. *IEEE Trans. Image Process.* **2007**, *16*, 349–366. [CrossRef] [PubMed]

34. Dai, Y.; Wu, Y. Reweighted Infrared Patch-Tensor Model with Both Nonlocal and Local Priors for Single-Frame Small Target Detection. *IEEE J. Sel. Top. Appl. Earth Observ. Remote Sens.* **2017**, *10*, 3752–3767. [CrossRef]

35. Goldfarb, D.; Qin, Z. Robust Low-rank Tensor Recovery: Models and Algorithms. *Siam J. Matrix Anal. Appl.* **2013**, *35*, 225–253. [CrossRef]

36. Wu, Z.; Wang, Q.; Jin, J.; Shen, Y. Structure tensor total variation-regularized weighted nuclear norm minimization for hyperspectral image mixed denoising. *Signal Process.* **2017**, *131*, 202–219. [CrossRef]

37. Dai, Y.; Wu, Y.; Song, Y.; Guo, J. Non-negative infrared patch-image model: robust target-background separation via partial sum minimization of singular values. *Infrared Phys. Tech.* **2017**, *81*, 182–194. [CrossRef]

38. Wang, X.; Peng, Z.; Kong, D.; Zhang, P.; He, Y. Infrared dim target detection based on total variation regularization and principal component pursuit. *Image Vis. Comput.* **2017**, *63*, 1–9. [CrossRef]

39. Kong, D.; Peng, Z.; Fan, H.; He, Y. Seismic random noise attenuation using directional total variation in shearlet domain. *J. Seismic Explor.* **2016**, *25*, 321–338.

40. Kong, D.; Peng, Z. Seismic random noise attenuation using shearlet and total generalized variation. *J. Geophys. Eng.* **2015**, *12*, 1024–1035. [CrossRef]

41. Zhang, L.; Peng, L.; Zhang, T.; Cao, S.; Peng, Z. Infrared small target detection via non-convex rank approximation minimization joint l2,1 norm. *Remote Sens.* **2018**, *10*, 1821. [CrossRef]

42. Liu, X.; Chen, Y.; Peng, Z.; Wu, J.; Wang, Z. Infrared image super-resolution reconstruction based on quaternion fractional order total variation with Lp quasinorm. *Appl. Sci.* **2018**, *8*, 1864. [CrossRef]

43. Chartrand, R. Exact Reconstruction of Sparse Signals via Nonconvex Minimization. *IEEE Signal Process. Lett.* **2007**, *14*, 707–710. [CrossRef]

44. Chartrand, R.; Staneva, V. Restricted isometry properties and nonconvex compressive sensing. *Inverse Probl.* **2010**, *24*, 657–682.

45. Chartrand, R.; Yin, W.T. Iteratively reweighted algorithms for compressive sensing. In Proceedings of the 2008 IEEE International Conference on Acoustics, Speech and Signal Processing, Las Vegas, NV, USA, 31 March–4 April 2008.

46. Elad, M. *Sparse and Redundant Representations*; Springer: New York, NY, USA, 2010; pp. 8–12.

47. Chen, X.; Xu, F.; Ye, Y. Lower bound theory of nonzero entries in solutions of l2-lp minimization. *SIAM J. Sci. Comput.* **2010**, *32*, 2832–2852. [CrossRef]

48. Marjanovic, G.; Solo, V. On lq Optimization and Matrix Completion. *IEEE Trans. Signal Process.* **2012**, *60*, 5714–5724. [CrossRef]

49. Kwak, N. Principal Component Analysis by Lp-Norm Maximization. *IEEE Trans. Cybern.* **2014**, *44*, 594–609. [CrossRef] [PubMed]

50. Nie, F.; Wang, H.; Huang, H.; Ding, C. Joint Schatten p-norm and lp-norm robust matrix completion for missing value recovery. *Knowl. Inf. Syst.* **2015**, *42*, 525–544. [CrossRef]

51. Xie, Y.; Gu, S.; Liu, Y.; Zuo, W.; Zhang, W.; Zhang, L. Weighted Schatten p-Norm Minimization for Image Denoising and Background Subtraction. *IEEE Trans. Image Process.* **2015**, *25*, 4842–4857. [CrossRef]

52. Lin, Z.; Chen, M.; Ma, Y. The Augmented Lagrange Multiplier Method for Exact Recovery of Corrupted Low-Rank Matrices. *arXiv* **2010**, arXiv:1009.5055.

53. Boyd, S.; Parikh, N.; Chu, E.; Peleato, B.; Eckstein, J. Distributed optimization and statistical learning via the alternating direction method of multipliers. *Found. Trends Mach. Learn.* **2011**, *3*, 1–122. [CrossRef]

54. Cai, J.F.; Candès, E.J.; Shen, Z. A Singular Value Thresholding Algorithm for Matrix Completion. *Siam J. Opt.* **2008**, *20*, 1956–1982. [CrossRef]

55. Zhang, L.; Peng, Z. Infrared Small Target Detection Based on Partial Sum of the Tensor Nuclear Norm. *Remote Sens.* **2019**, *11*, 382. [CrossRef]

Article

Infrared Small Target Detection Based on Partial Sum of the Tensor Nuclear Norm

Landan Zhang [1] and Zhenming Peng [1,2,*] 

[1] School of Information and Communication Engineering, University of Electronic Science and Technology of China, Chengdu 611731, China; zhanglandan@std.uestc.edu.cn
[2] Center for Information Geoscience, University of Electronic Science and Technology of China, Chengdu 611731, China
* Correspondence: zmpeng@uestc.edu.cn; Tel.: +86-13076036761

Received: 12 January 2019; Accepted: 11 February 2019; Published: 13 February 2019

Abstract: Excellent performance, real time and strong robustness are three vital requirements for infrared small target detection. Unfortunately, many current state-of-the-art methods merely achieve one of the expectations when coping with highly complex scenes. In fact, a common problem is that real-time processing and great detection ability are difficult to coordinate. Therefore, to address this issue, a robust infrared patch-tensor model for detecting an infrared small target is proposed in this paper. On the basis of infrared patch-tensor (IPT) model, a novel nonconvex low-rank constraint named partial sum of tensor nuclear norm (PSTNN) joint weighted l_1 norm was employed to efficiently suppress the background and preserve the target. Due to the deficiency of RIPT which would over-shrink the target with the possibility of disappearing, an improved local prior map simultaneously encoded with target-related and background-related information was introduced into the model. With the help of a reweighted scheme for enhancing the sparsity and high-efficiency version of tensor singular value decomposition (t-SVD), the total algorithm complexity and computation time can be reduced dramatically. Then, the decomposition of the target and background is transformed into a tensor robust principle component analysis problem (TRPCA), which can be efficiently solved by alternating direction method of multipliers (ADMM). A series of experiments substantiate the superiority of the proposed method beyond state-of-the-art baselines.

Keywords: infrared small target detection; local prior analysis; nonconvex tensor robust principle component analysis; partial sum of the tensor nuclear norm

---

1. Introduction

Infrared small target detection is of great importance in many military applications, such as early-warning systems, missile-tracking systems, and precision guided weapons. Unfortunately, infrared small target detection is still full of challenges, which is mainly related to the following. Firstly, because of the long imaging distance, small target is often spot-like, lacking texture and structural information; secondly, infrared imaging is also influenced by complex backgrounds, clutters, and atmospheric radiation, resulting in low signal-to-clutter (SCR) ratio in infrared images, and sometimes the target is even submerged by the background; thirdly, interferences such as artificial buildings, ships in the sea and birds in the sky also have a bad impact on detection ability. How to effectively suppress the background, improve the detection ability of the target, and reduce false alarms have always been difficult problems to solve.

In general, infrared small target detection methods can be divided into two categories: sequential-based and single-frame-based methods. Traditional sequential-based methods including pipeline filtering [1], 3D matched filtering [2], and multistage hypothesis testing [3] are applicable when the background is static and homogeneous, utilizing both spatial and temporal information to capture the target trajectory. However, in real applications, the movement between the target and imaging sensor is fast, coupled with various complex backgrounds, the performance of sequential-based methods degrades rapidly. Besides, those methods are unable to meet the real-time requirements due to the usage of multiple frames. Although there are still some studies on sequential-based methods [4,5], single-frame-based methods have attracted more research attention in recent years [6–8].

The prior information is the key to the success of single-frame-based methods, also in many other fields [9–11]. Up to now, the consistency of backgrounds [12–15], the saliency of targets [16–19], the sparsity of targets and the low rank of backgrounds [20–24] are the most used assumptions to detect infrared small targets in single image from different perspectives. The former two are local priors, whereas the latter two are nonlocal priors which are usually exploited simultaneously. Under simple scenes, the local priors are enough to distinguish target from background. Nevertheless, most real scenes are complex, which greatly limits the application of local priors. The nonlocal priors are more powerful and fit the real scenes well but still suffer from the sparse edges and noise. In fact, the combination of two types of prior information can improve the detection performance. Therefore, a suitable model for incorporating the local and nonlocal prior information plays a vital role in realizing high-efficiency detection methods.

1.1. Related Works on Single-Frame-Based Infrared Small Target Detection

According to the usage of prior information, the single-frame-based approaches can be mainly classified into two groups: filtering methods using local priors and optimizing methods using nonlocal priors. The first type of filtering methods exploits filters to estimate the background based on the prior information of background consistency. The target is enhanced by subtracting the predicted background from the original image. Conventional typical filters including Top-hat filter [12], two-dimensional least mean square (TDLMS) filter [15], and Max-mean filter [13] can catch the target easily under simple uniform scenes. Unfortunately, these filters cannot handle complex scenes full of edges and interferences well. In order to overcome this disadvantage, many improved filter were developed [25–28]. Another type of filtering methods highlights the small target based on the human visual system (HVS) via the calculation of saliency map. The contrast between target and its local neighborhood is a common measure to obtain the saliency map. Many HSV-based approaches such as Laplacian of Gaussian (LoG) filter [29], difference of Gaussian (DoG) filter [30], local contrast measure (LCM) [16], relative local contrast measure (RLCM) [19], multiscale patch-based contrast measure (MPCM) [31], weighted local difference measure (WLDM) [32], and multiscale gray and variance difference (MGVD) [33] measure were raised gradually. There are also methods to analyze visual saliency in the Fourier domain [34,35].

Unlike the filtering methods, optimizing methods employ the nonlocal self-correlation of infrared background and the sparsity of the target to reveal the data inner structure, which have been developed rapidly within the past decade. Assuming that the background comes from a single low-rank subspace, infrared patch image (IPI) model [20] regards the target as an outlier, so that the conventional target detection problem is converted to a robust principle component analysis (RPCA) [36] optimization problem. Compared with the traditional baselines, the detection ability has been significantly improved. Two obvious shortcomings of IPI are target over-shrinking and noise residuals mainly because of the low-rank regularization term which utilizes the nuclear norm. Subsequently, following this direction, more low-rank matrix recovery techniques were introduced into IPI model to get a better performance [21,37–39]. Considering that the original data are drawn from a union of low-rank subspaces, methods based on dictionary learning and sparse representation were proposed [24,40,41]. Unfortunately, either generating artificially or learning desired dictionaries to adapt to most scenarios is not easy but complex, especially when more dictionaries are needed. To dig out more useful information from the nonlocal configuration in patch space, Dai et al. [42] firstly generalized the IPI model to a novel infrared patch-tensor (IPT) model with the assumption that all the unfolding matrices are low rank, resulting in improved detection ability and reduction of computation time.

1.2. Motivation

For infrared small target detection, real-time processing and excellent performance are two fundamental expectations. However, one of the biggest problems of existing approaches is the imbalance between time and performance. Table 1 shows the computation time and performance of eight representative methods which concludes from our previous work [39]. Note that the time is obtained from processing an image of 256×200 pixels, and the full score of performance is five, the higher, the better. From Table 1, we can observe that the three filtering methods are fast but poor in performance, because of the simple assumptions regarding either the background or target. On the contrary, the six optimizing methods can obtain high-quality detection results but they are time consuming. The framework of optimization brings complex calculation and accurate detection results at the same time. How to simplify the calculation steps without destroying the detection performance is a crucial issue.

Table 1. The computation time and performance of eight representative methods.

	Tophat	LCM	MPCM	IPI	NIPPS	ReWIPI	SMSL	NRAM
Time (s)	0.022	0.074	0.089	11.907	7.486	15.469	1.245	3.378
Score	1	1	2	3	3.5	3	2.5	4

Experiments had shown the superiority of the RIPT model compared with state-of-the-art approaches (please see details in Ref. [42]). The intrinsic reasons lie in two aspects; for one thing, the novel patch-tensor model can extract more spatial correlations to reduce the interference, which is named the rare structure effect; for another, utilizing both local and nonlocal priors simultaneously increases the robustness of the RIPT upon various scenes and noise, as they are complementary when dealing with infrared small target detection. Nevertheless, the singleton model [43] used in RIPT may lead to a suboptimal value, since the sum of nuclear norms (SNN) [44] is not the convex envelope of the corresponding sum of ranks [45]. Furthermore, RIPT takes the difference of two eigenvalues derived from the structure tensor for the involvement of the local prior. The local structure weight map is illustrated in Figure 1, from which we can easily obtain the background edge information. An unfortunate fact worth mentioning is that the edge of the target is also highlighted. More specifically, it means the target would be over-shrunk, especially when the target lies upon boundaries such as those in Figure 1a, or there are no clear edges but the target is similar to that in Figure 1b. RIPT considers the background-related prior while ignoring the target-related prior since both of them can cause false alarms.

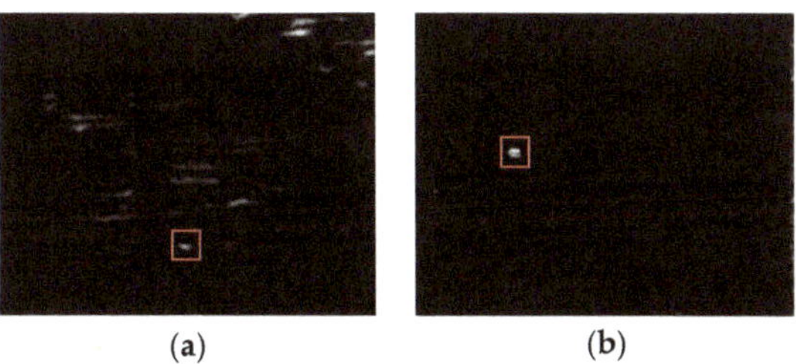

Figure 1. Illustration of the local structure weight map.

Inspired by the RIPT, the patch-tensor model can be exploited to seek out more intrinsic priors from a higher dimension. Another key factor is that RIPT with an additional stopping criterion is much faster than IPI. Hence, to alleviate the issue of imbalance and to overcome the two deficiencies of RIPT, this paper mainly makes three contributions.

- First, to avoid the problem of equal treatment on singular values and reduce some biases, we develop a nonconvex infrared small target detection model based on partial sum of tensor nuclear norm (PSTNN), which can approximate the tensor rank better, and convert the detection task into a problem of solving the tensor robust principle component analysis model.
- Second, by introducing the local prior which relates to background and target simultaneously as the local weight map, coupled with the reweighted scheme, thus the proposed model can preserve the target and suppress the background better, which assists us to complete the infrared small target detection task with good performance.
- Third, an efficient algorithm based on the alternating direction method of multipliers (ADMM) is designed for solving the proposed model accurately. Meanwhile, with the help of tensor singular value decomposition (t-SVD) and an extra stopping condition, the algorithm complexity and computation time are dramatically reduced, leading to a faster speed in comparison with similar state-of-the-art methods.

The rest of this paper is structured as follows. Some related notations and preliminaries about tensor and mathematical theorems are introduced in Section 2. In Section 3, the construction of local prior map and proposed model are described in detail, and the ADMM solver to the optimization problem is also provided. Extensive experiments on various scenes and sequences are conducted to verify the effectiveness of the proposed method in Section 4. Sections 5 and 6 present the discussion and conclusion of this paper, respectively.

2. Notations and Preliminaries

We first briefly introduce some necessary notions and preliminaries. In this paper, a tensor is denoted as $\mathcal{X}$, a matrix is denoted as X, a vector is denoted as x, and a scalar is denoted as x. A fiber is a vector obtained by fixing every index of $\mathcal{X}$ but one, a slice is a matrix obtained by fixing every index of $\mathcal{X}$ but two. For a three-order $\mathcal{X} \in \mathbb{R}^{n_1 \times n_2 \times n_3}$, its (i, j, k)-th entry is denoted as x_{ijk}, and we use $X_{i::}$, $X_{:i:}$, and $X_{::i}$ respectively representing the i-th horizontal, lateral and frontal slice. In most cases, the i-th frontal slice $X_{::i}$ is alternatively denoted as $X^{(i)}$. The mode-i unfolding of $\mathcal{X}$ denoted by $X_{(i)}$ is composed by taking the mode-i fibers as its columns, which is also known as matricization or flattening. We define the operator *unfold* that maps $\mathcal{X}$ to a matrix, namely, $X_{(i)} = unfold_i(\mathcal{X})$, and its inverse operator is *fold*. Besides, there are many acronyms used in this paper; we give a summary of these in Table 2 (excluding the acronyms of the comparison methods).

Table 2. Detailed parameter settings of the 10 tested methods.

Acronym	Full name
IPT [42]	Image Patch-Tensor
PSTNN [46]	Partial Sum of Tensor Nuclear Norm
t-SVD [47]	Tensor Singular Value Decomposition
RPCA [36]	Robust Principle Component Analysis
TRPCA [48]	Tensor Robust Principle Component Analysis
ADMM [49]	Alternating Direction Method of Multipliers
SNN [44]	Sum of Nuclear Norms
PSSV [50]	Partial Sum of Singular Values
PSVT [50]	Partial Singular Value Thresholding operator
TNN [51]	Tensor Nuclear Norm

2.1. Tensor Singular Value Decomposition

For a three-order $\mathcal{X} \in \mathbb{R}^{n_1 \times n_2 \times n_3}$, we denote $\overline{\mathcal{X}} \in \mathbb{C}^{n_1 \times n_2 \times n_3}$ as the result of DFT along its third dimension by using the matlab command *fft*, i.e., $\overline{\mathcal{X}} = fft(\mathcal{X}, [], 3)$. The inverse operator *ifft* computes $\mathcal{X}$ from $\overline{\mathcal{X}}$, i.e., $\mathcal{X} = ifft(\overline{\mathcal{X}}, [], 3)$.

Definition 1. *(tensor conjugate transpose)* [47] *The conjugate transpose of a tensor $\mathcal{X} \in \mathbb{R}^{n_1 \times n_2 \times n_3}$ is the tensor $\mathcal{X}^{\mathrm{T}} \in \mathbb{R}^{n_1 \times n_2 \times n_3}$ obtained by conjugate transposing each of the frontal slice and then reversing the order of transposed frontal slices 2 through n_3:*

$$\left(\mathcal{X}^{\mathrm{T}}\right)^{(1)} = \left(\mathcal{X}^{(1)}\right)^{\mathrm{T}} \quad and$$
$$\left(\mathcal{X}^{\mathrm{T}}\right)^{(i)} = \left(\mathcal{X}^{(n_3+2-i)}\right)^{\mathrm{T}}, \quad i = 2, \cdots, n_3 \tag{1}$$

Definition 2. *(identity tensor)* [47] *The identity tensor $\mathcal{I} \in \mathbb{R}^{n_1 \times n_2 \times n_3}$ is the tensor with its first frontal slice being the $n \times n$ identity matrix, and the other frontal slices being all zeros.*

Definition 3. *(orthogonal tensor)* [47] *A tensor $\mathcal{Q} \in \mathbb{R}^{n_1 \times n_2 \times n_3}$ is orthogonal if it satisfies*

$$\mathcal{Q}^{\mathrm{T}} * \mathcal{Q} = \mathcal{Q} * \mathcal{Q}^{\mathrm{T}} \tag{2}$$

Definition 4. *(f-diagonal tensor)* [47] *A tensor $\mathcal{X}$ is called f-diagonal if each frontal slice $X_{(i)}$ is a diagonal matrix.*

Theorem 1. *(t-SVD)* [47] *Let $\mathcal{X} \in \mathbb{R}^{n_1 \times n_2 \times n_3}$. Then it can be factorized as*

$$\mathcal{X} = \mathcal{U} * \mathcal{S} * \mathcal{V}^{\mathrm{T}} \tag{3}$$

where $\mathcal{U} \in \mathbb{R}^{n_1 \times n_1 \times n_3}$, $\mathcal{V} \in \mathbb{R}^{n_2 \times n_2 \times n_3}$ are orthogonal tensors, and $\mathcal{S} \in \mathbb{R}^{n_1 \times n_2 \times n_3}$ is an f-diagonal tensor.

The illustration of t-SVD decomposition of an $n_1 \times n_2 \times n_3$ tensor is in Figure 2. Note that t-SVD can be obtained via computing matrix SVDs in the Fourier domain. An efficient and fast way to compute t-SVD is shown in Algorithm 1 [52].

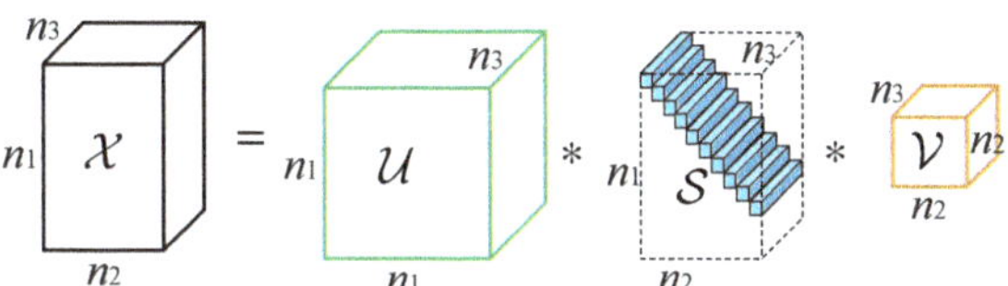

Figure 2. Illustration of tensor singular value decomposition.

Algorithm 1 T-SVD for three-order tensors

Input: $\mathcal{X} \in \mathbb{R}^{n_1 \times n_2 \times n_3}$
Output: T-SVD components $\mathcal{U}$, $\mathcal{S}$ and $\mathcal{V}$ of $\mathcal{X}$.
1. Compute $\overline{\mathcal{X}} = fft(\mathcal{X}, [], 3)$
2. Compute each frontal slice of $\overline{\mathcal{U}}, \overline{\mathcal{S}}$ and $\overline{\mathcal{V}}$ from $\overline{\mathcal{X}}$ by
    for $i = 1, \cdots, \lceil (n_3 + 1)/2 \rceil$ **do**
    $[\overline{U}^{(i)}, \overline{S}^{(i)}, \overline{V}^{(i)}] = \mathrm{SVD}(\overline{X}^{(i)})$;
    end for
    for $i = \lceil (n_3 + 1)/2 \rceil + 1, \cdots, n_3$ **do**
    $\overline{U}^{(i)} = \mathrm{conj}(\overline{U}^{(n_3-i+2)})$;
    $\overline{S}^{(i)} = \overline{S}^{(n_3-i+2)}$;
    $\overline{V}^{(i)} = conj(\overline{V}^{(n_3-i+2)})$;
    end for
3. **Compute** $\mathcal{U} = \mathrm{ifft}(\overline{\mathcal{U}}, [], 3), \mathcal{S} = \mathrm{ifft}(\overline{\mathcal{S}}, [], 3)$, and $\mathcal{V} = \mathrm{ifft}(\overline{\mathcal{V}}, [], 3)$.

2.2. Some Mathematical Preliminaries

Theorem 2. *(soft thresholding operator) [53] Let $\tau > 0$ and $X, Y \in \mathbb{R}^{n_1 \times n_2}$, define a l_1 norm minimization problem as*

$$\underset{X}{\operatorname{argmin}} \ \tau\|X\|_1 + \frac{1}{2}\|X - Y\|_F^2 \tag{4}$$

Then, Equation (4) could be solved by an elementwise soft thresholding operator defined as

$$S_\tau(x) = \operatorname{sign}(x) \times \max(|x| - \tau, 0) \tag{5}$$

Definition 5. *(partial sum of singular values, PSSV) [50] For a matrix $X \in \mathbb{R}^{n_1 \times n_2}$, the PSSV is defined as $\|X\|_{p=N} = \sum_{i=p+1}^{min(n_1,n_2)} \sigma_i(X)$, where $\sigma_i(X)(i = 1, \cdots, min(n_1, n_2))$ is the i-th largest singular value of X, and N is the preserved target rank.*

Theorem 3. *(partial singular value thresholding operator, PSVT) [50] Let $\tau > 0$, $l = min(n_1, n_2)$ and $X, Y \in \mathbb{R}^{n_1 \times n_2}$ which can be decomposed by SVD. Y can be considered as the sum of two matrices, $Y = Y_1 + Y_2 = U_{Y_1} D_{Y_1} V_{Y_1}^H + U_{Y_2} D_{Y_2} V_{Y_2}^H$, where U_{Y_1}, V_{Y_1} are the singular vector matrices corresponding to the N largest singular values, and U_{Y_2}, V_{Y_2} from the (N+1)-th to the last singular values. Define a complex minimization problem for PSSVas*

$$\underset{X}{\operatorname{argmin}} \ \tau\|X\|_{p=N} + \frac{\beta}{2}\|X - Y\|_F^2 \tag{6}$$

Then, the optimal solution of Equation (6) can be expressed by the PVST operator, which is defined as:

$$\begin{aligned} \mathcal{P}_{N,\tau}(Y) &= U_Y(D_{Y_1} + S_\tau[D_{Y_2}])V_Y^H \\ &= Y_1 + U_{Y_2} S_\tau[D_{Y_2}]V_{Y_2}^H \end{aligned} \tag{7}$$

where $\tau = \lambda/\beta$, $D_{Y_1} = diag(\sigma_1^Y, \cdots, \sigma_N^Y, 0, \cdots, 0)$, and $D_{Y_2} = diag(0, \cdots, 0, \sigma_{N+1}^Y, \cdots, \sigma_l^Y)$.

3. Proposed Method

Overall, an infrared image with small target can be described as follows [14]:

$$f_D = f_B + f_T + f_N \tag{8}$$

where f_D, f_B, f_T denotes the original image, background image, target image respectively, and f_N stands for the noise component. Depending on whether concentrating on merely the background, merely the target, or both of them leads to different methods to detect infrared small target. Unlike the general infrared image model, Gao et al. [20] generalized the traditional model into the IPI model, which can be formulated as

$$D = B + T + N \tag{9}$$

where D, B, T and N correspond to patch images of the original image, background image, target image and random noise, all of which are constructed by vectorizing the matrix within the sliding window. Since the infrared background is regarded as slowly transitional, that means that many local patches are approximately linearly correlated with each other. In other words, the configuration of nonlocal self-correlation leads to a low-rank background patch image. Besides, the small target only occupies a few pixels with respect to the whole image; thus the target patch image can be considered as a sparse matrix. Then, to separate the background and target is to solve an RPCA problem of recovering low-rank and sparse matrices. In terms of data dimensionality reduction and representation, the most popular method is PCA [54]. Recently, many other approaches spring up [36,55], and RPCA is an improvement of traditional PCA.

3.1. Infrared Patch-Tensor Model

To dig out more correlations among different patches, Dai et al. [42] proposed a novel target-background separation framework named the infrared patch-tensor model (IPT) based on a slightly different idea of construction. Transforming the original infrared image into a tensor is the first step. As indicated in Figure 3, without transforming each patch matrix into a vector, the original patch-tensor in IPT model is constructed by directly stacking the patches obtained via sliding a window from the top left to the bottom right over an image into a 3D cube. Hence, Equation (9) is transferred to the patch space:

$$\mathcal{D} = \mathcal{B} + \mathcal{T} + \mathcal{N} \tag{10}$$

where $\mathcal{D}, \mathcal{B}, \mathcal{T}, \mathcal{N} \in \mathbb{R}^{m \times n \times k}$ are the input patch-tensor, background patch-tensor, target patch-tensor, and noise patch-tensor, respectively. m and n are the patch height and width, and k is the patch number.

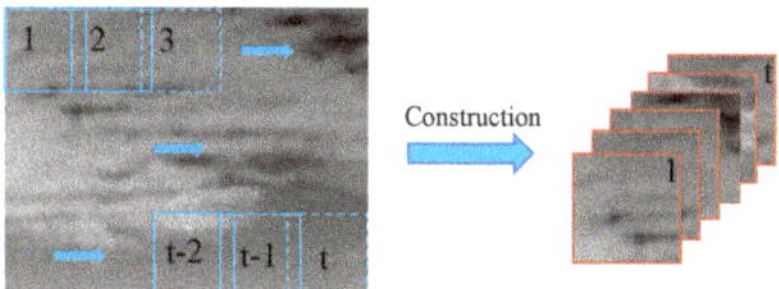

Figure 3. Illustration of tensor construction. The left is original image and the right is the constructed patch-tensor.

For a three-way tensor, we can get the mode-i ($1 \leq i \leq 3$) unfolding matrices by taking the corresponding fibers (i.e., columns, rows and tubes in tensor) as columns. Figure 4 illustrates the singular values of the mode-i ($1 \leq i \leq 3$) unfolding of the patch-tensor under typical scenes. Without any doubt, the curves of all the unfolding matrices changing sharply to zeroes demonstrate the low-rank property of the background patch-tensors along each mode. Particularly, the patch-image model could be seen as a special case of the patch-tensor model, as the patch-image is just the mode-3 flattening matrix of the corresponding patch-tensor. The IPT model not only generalizes the IPI model from matrix to tensor, but also encodes enough priors delivered by different flattening matrices with the spatial structure preserved. Therefore, we can impose a strong constraint on the unfolding matrices of background patch-tensor $\mathcal{B}$:

$$\mathrm{rank}(B_{(1)}) \leq r_1, \quad \mathrm{rank}(B_{(2)}) \leq r_2, \quad \mathrm{rank}(B_{(3)}) \leq r_3 \tag{11}$$

where r_1, r_2, and r_3 are nonnegative constants related to the complexity of the background image.

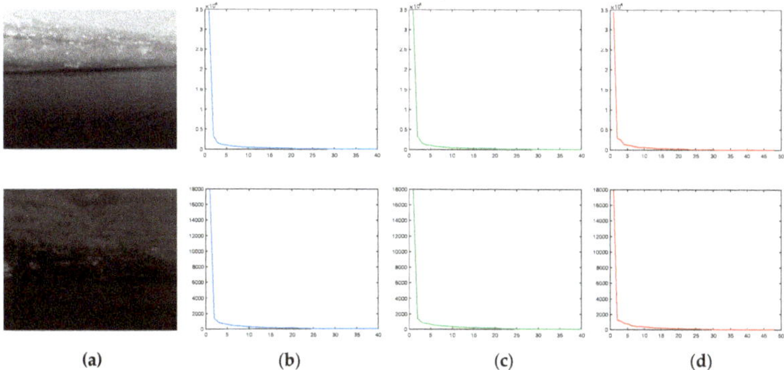

(a)   (b)   (c)   (d)

Figure 4. Illustration of the nonlocal self-correlation property of unfolding matrices. (a) Two representative scenes; (b)–(d) Singular values of mode-1, mode-2, and mode-3 unfolding matrices of the corresponding patch-tensors.

Obviously, the target patch-tensor $\mathcal{T}$ is actually a sparse tensor, which implies $\|\mathcal{T}\|_0 \leq k$, where k is a small integer that is totally determined by the size and the number of small targets. Assuming that the noise is additive white Gaussian noise and $\|\mathcal{N}\|_F \leq \delta$ for some $\delta > 0$, we have $\|\mathcal{D} - \mathcal{B} - \mathcal{T}\|_F \leq \delta$. Thus, we can obtain the following tensor robust principle component analysis (TRPCA) problem which attempts to separate the low-rank and sparse tensors:

$$\min_{\mathcal{B},\mathcal{T}} \ \text{rank}(\mathcal{B}) + \lambda\|\mathcal{T}\|_0$$
$$s.t. \ \mathcal{D} = \mathcal{B} + \mathcal{T} \tag{12}$$

where λ is a compromising parameter that controls the tradeoff between the target patch-tensor and the background patch-tensor, $\| \cdot \|_0$ denotes the l_0 norm, which counts the number of nonzero entries.

3.2. Local Prior Analysis

The grayscale-based measures that are used in most filtering methods are merely focusing on how to extract local prior such as local contrast [16,56,57], local entropy [58,59], and local difference [32,33,60]; nevertheless, this type of insufficient information is not enough to differentiate target and background. Conversely, optimizing methods with nonlocal property involved are more robust to complex scenes, but still suffer from background residuals in target components mainly because of the salient edges. Its intrinsic reason is because that the sparsity of the salient edges is similar to that of the targets. In fact, the stubborn edges can be easily identified by local prior, which means that the defects of optimizing methods can be alleviated via adding extra local prior. For this reason, the RIPT model employs structure tensor [61] to discriminate all of the image boundaries, since these boundaries tend to contaminate the sparse target matrix. The two highest eigenvalues λ_1 and λ_2 ($\lambda_1 \geq \lambda_2$) are applied to depict the local geometry structure. As the value of $\lambda_1 - \lambda_2$ highlights image boundaries clearly, the local structure weight patch-tensor used in the RIPT model is defined as:

$$\mathcal{W}_{LS} = \exp\left(h \cdot \frac{(\mathcal{L}_1 - \mathcal{L}_2) - d_{\min}}{d_{\max} - d_{\min}}\right) \tag{13}$$

where $\mathcal{L}_1$ and $\mathcal{L}_2$ are the corresponding patch-tensors of two obtained eigenvalue matrices, h is a weight-stretching parameter, $d_{\max}$ and $d_{\min}$ are the maximum and minimum of $\mathcal{L}_1 - \mathcal{L}_2$, respectively.

As analyzed in Section 1, the operator $\lambda_1 - \lambda_2$ that is utilized to calculate $\mathcal{W}_{LS}$ is completely poor at determining whether the edge components belong to the target or background. When serving as the local structure weight, such ambiguity causes the distortion of target shape, due to the similar weights between the background edge and the target edge. This situation becomes even worse with the increasing of h, as shown in Figure 5. We know that when locating at the corner region, $\lambda_1 \geq \lambda_2 \gg 0$; when locating at the edge region, $\lambda_1 \gg \lambda_2 \approx 0$; when locating at the flat region, $\lambda_1 \approx \lambda_2 \approx 0$. Hence, structure tensor tends to give lower values at corners even if some of them are part of the edges sometimes. As pointed out in [62], when the weight stretching parameter h decreases, the difference would be more significant, causing an increase in the false alarm rate. In summary, on one hand, to preserve the target and prevent it from being completely lost, a smaller h is needed; in contrast, to avoid the interference of residuals, a larger h is needed. This is contradictory and finding an appropriate value of h is difficult because the size of small target varies within a somewhat large range. Another disadvantage is that RIPT merely considers the background-edge-related prior while ignoring the target-related prior since both of them can cause false alarms.

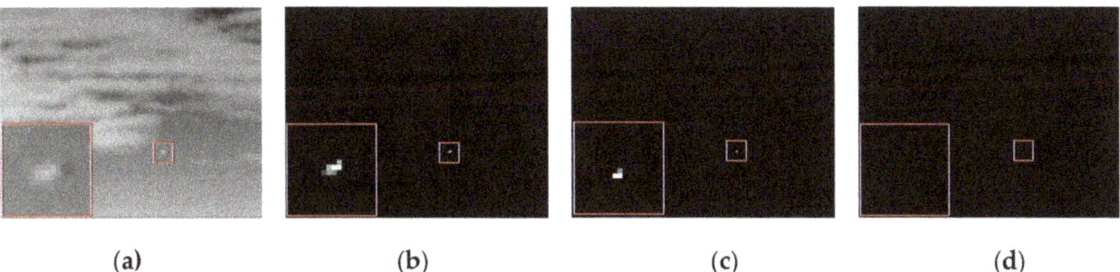

(a)   **(b)**   **(c)**   **(d)**

Figure 5. The phenomenon of target over-contraction with the increasing of h. (**a**) Original image; (**b**)–(**d**) The separated target image when h = 1,3,5, respectively.

Due to the objective existence of the target edge, it is hard to utilize operator $\lambda_1 - \lambda_2$ to only obtain the background prior. To alleviate the issue of target over-shrinking and corner disappearance, a new local structure descriptor related to the target prior without an additional stretching parameter was exploited. In [63], a "corner strength" function was computed to find the interest points:

$$w_{cs}(x,y) = \frac{\det(ST(x,y))}{\text{tr}(ST(x,y))} = \frac{\lambda_1 \lambda_2}{\lambda_1 + \lambda_2} \tag{14}$$

where (x,y) represents the pixel location, $ST(\cdot)$ denotes the structure tensor, $ST(x,y)$ is a matrix, $det(\cdot)$ and $tr(\cdot)$ are the determinant and trace of matrix respectively, and $w_{cs}(x,y)$ is the half of the harmonic mean of the eigenvalues (λ_1, λ_2). Figure 6 indicates the map of interest points of an infrared image (i.e., Figure 6c) compared with the local structure weight (i.e., Figure 6b), which demonstrates two underlying facts: (i) the target information is highlighted that fully complies with our expectation, and (ii) the corner regions that have been lost in the local structure weight map used in RIPT are identified. Furthermore, we replaced the subtraction operator as the maximum between two eigenvalues, namely:

$$w_{\text{m}}(x,y) = \max(\lambda_1, \lambda_2) \tag{15}$$

It should be noted that the same problems also exist in the maximum operator but not so badly.

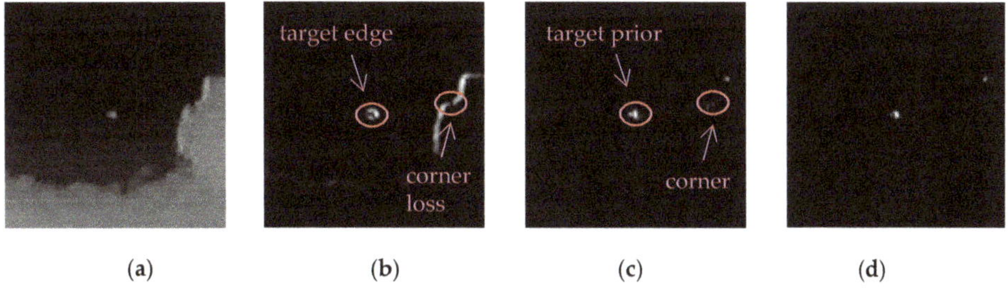

(a)   **(b)**   **(c)**   **(d)**

Figure 6. Comparison of different prior maps. (**a**) Original image; (**b**) The local structure weight map used in RIPT (calculated by Equation (13)); (**c**) The corner strength map (calculated by Equation (14)); (**d**) The prior weight map used in the proposed model (calculated by Equation (16)).

Thus, as shown in Figure 6d, the final version of prior weight map W_p is

$$W_p(x,y) = w_{cs}(x,y) \cdot w_{\text{m}}(x,y) = \max(\lambda_1, \lambda_2) \cdot \frac{\lambda_1 \lambda_2}{\lambda_1 + \lambda_2} \tag{16}$$

Then, the patch-tensor of the prior weight map with normalization is defined as:

$$\mathcal{W}_p = \frac{\mathcal{W}_p - w_{\min}}{w_{\max} - w_{\min}} \tag{17}$$

where $w_{\max}$ and $w_{\min}$ denote the maximum and minimum of W_p, respectively.

3.3. IPT Model Based on PSTNN

3.3.1. The Surrogate of Tensor Rank

Considering that the background changes slowly because of the high correlations among local and nonlocal patches, low rank is an intrinsic property of the infrared background. The straightforward measurement to access the low-rank characteristic of a tensor is the tensor rank. However, there is no direct way to extend the low-rankness from the matrices to tensors. More specially, due to the variety of tensor decomposition methods, the definition of tensor rank is not unique. The most popular definitions are CP rank [64] and Tucker rank [65]. Another difficulty lies in the tensor extension of RPCA (i.e., TRPCA) since the numerical algebra of tensors is fraught with hardness results [66]. How to choose a suitable tensor rank with a tight convex relaxation is of great importance.

In reweighted infrared patch-tensor (RIPT) model, the low-rank characteristic of the background patch-tensor is accessed via the sum of nuclear norms (SNN), which is based on the singleton model [43]. SNN, defined as $\sum_i \left\| X_{(i)} \right\|_*$, is used as a convex surrogate of $\sum_i rank(X_{(i)})$. A rational fact behind the regularizer SNN is that the nuclear norm is the tightest convex envelope to matrix rank within the unit ball of the spectral norm. Besides, instead of calculating the complex tensor nuclear norm, SNN calculates the simpler matrix nuclear norm. Nevertheless, SNN is not a tight convex relaxation of $\sum_i rank(X_{(i)})$ [45], which implies SNN has the limitation of obtaining suboptimal value. In other words, when served as a background constraint, SNN would produce false alarms.

Derived from t-SVD, the tensor nuclear norm (TNN) was proposed in [51] and successfully applied to image recovery which had shown its advancement compared to SNN. Generally, minimizing the TNN may cause some unavoidable biases [46]. Meanwhile, SNN and TNN treat each singular value equally which is irrational, since the larger singular values are generally associated with the image details; thus, they should be assigned smaller weights. To alleviate those phenomena, it's appropriate to adopt a nonconvex relaxation with unequal weights. In [46], Jiang et al. extended the partial sum of singular values (PSSV) [50] to the tensor version and presented the partial sum of the tensor nuclear norm (PSTNN) to replace the TNN as the nonconvex approximation of tensor csor $\mathcal{X} \in \mathbb{R}^{n_1 \times n_2 \times n_3}$ is defined as

$$\|\mathcal{X}\|_{\text{PSTNN}} = \sum_{i=1}^{n_3} \left\| \overline{\mathbf{X}}^{(i)} \right\|_{p=N} \tag{18}$$

where $\| \cdot \|_{p=N}$ denotes the PSSV. Since the infrared backgrounds could vary from simple to complex, it's better to employ an adaptively predicted rank constraint. On the contrary, considering that the small target only holds an extremely small part of the entire image, a simpler way to determine the parameter N is to set a fixed energy ratio without directly concentrating on the changeable backgrounds. To approximate the tensor rank with high accuracy, the PSTNN is a better candidate than SNN and TNN.

3.3.2. Model Construction

Likewise, we utilized the conventional way to relax the non-smooth and discrete l_0 norm. So the infrared small target detection model based on patch-tensors with the priors of target and background is as follows:

$$\min_{\mathcal{B},\mathcal{T}} \ \|\mathcal{B}\|_{\text{PSTNN}} + \lambda \|\mathcal{T} \odot \mathcal{W}_{rec}\|_1 \tag{19}$$

$$\text{s.t.} \ \mathcal{D} = \mathcal{B} + \mathcal{T}$$

where $\odot$ denotes the Hadamard product, $\mathcal{W}_{rec}$ is the tensor corresponding to elementwise reciprocals of the corresponding elements in $\mathcal{W}_p$, and $\| \cdot \|_1$ denotes the l_1 norm, which is the sum of the absolute values of all the elements.

In [67], Candès proposed a reweighted l_1 minimization to address the imbalance in which larger coefficients are penalized more heavily than smaller ones. Subsequently, the reweighted scheme achieved great success in many publications [68–70]. As indicated in Table 1, the computing time of optimizing methods is always a major concern. Therefore, to speed up the convergence rate, and reduce the time of the whole procedure, we adopted the reweighted scheme as well. The sparsity weight is defined as follows:

$$W_{sw}^{k+1} = \frac{c}{|\mathcal{T}^k| + \varepsilon} \tag{20}$$

where c is a nonnegative constant, $\varepsilon > 0$ is a small number to avoid division by zero, and $k+1$ denotes the $(k+1)$-th iteration. In some cases, c is fixed to 1 [42,62]. We combined the two weights to get a simplified form

$$W = W_{sw} \odot W_{rec} \tag{21}$$

Then, Equation (19) is rewritten as follows:

$$\min_{\mathcal{B},\mathcal{T}} \|\mathcal{B}\|_{\mathrm{PSTNN}} + \lambda\|\mathcal{T} \odot W\|_1 \tag{22}$$
$$\text{s.t. } \mathcal{D} = \mathcal{B} + \mathcal{T}$$

In addition, as the same as analyzed in [42], we observed that the number of nonzero entries in target patch-tensor stops changing after a few iterations, which is just a little proportion of the entire procedure if the stop condition is when the relative error is smaller (i.e., $\|\mathcal{B} + \mathcal{T} - \mathcal{D}\|_F^2 / \|\mathcal{D}\|_F^2$) than a given threshold. Hence, to better utilize this observation and alleviate the imbalance between computing time and performance, the algorithm stops the iterations once the number of nonzero entries ceases to decrease or the relative error is smaller than the given threshold.

3.3.3. Solution of the Proposed Model

The alternating direction method of multipliers (ADMM) [49] has a fast convergence rate and high accuracy. In this section, an ADMM-based solver is devised to solve Equation (22). The augmented Langrangian function of Equation (22) is defined as

$$L_\mu(\mathcal{B},\mathcal{T},W,\mathcal{Y}) = \|\mathcal{B}\|_{\mathrm{PSTNN}} + \lambda\|\mathcal{T} \odot W\|_1 + \langle \mathcal{Y}, \mathcal{B} + \mathcal{T} - \mathcal{D}\rangle + \frac{\mu}{2}\|\mathcal{B} + \mathcal{T} - \mathcal{D}\|_F^2 \tag{23}$$

where $\mathcal{Y}$ is the Lagrange multiplier, $\langle \cdot \rangle$ denotes the inner product of two tensors, $\| \cdot \|_F$ is the Frobenius norm, and $\mu > 0$ is a penalty factor.

Then, the problem $\mathrm{argmin}_{\mathcal{B},\mathcal{T},W,\mathcal{Y}} L_\mu(\mathcal{B},\mathcal{T},W,\mathcal{Y})$ in Equation (23) can be separated as several subproblems, and in the $(k+1)$-th step, $\mathcal{T}$ and $\mathcal{B}$ are updated as:

$$\mathcal{T}^{k+1} = \underset{\mathcal{T}}{\mathrm{argmin}}\lambda\|\mathcal{T} \odot W^k\|_1 + \frac{\mu^k}{2}\left\|\mathcal{B}^k + \mathcal{T} - \mathcal{D} + \frac{\mathcal{Y}^k}{\mu^k}\right\|_F^2 \tag{24}$$

$$\mathcal{B}^{k+1} = \underset{\mathcal{B}}{\mathrm{argmin}}\|\mathcal{B}\|_{\mathrm{PSTNN}} + \frac{\mu^k}{2}\left\|\mathcal{B} + \mathcal{T}^{k+1} - \mathcal{D} + \frac{\mathcal{Y}^k}{\mu^k}\right\|_F^2 \tag{25}$$

The subproblem (24) can be solved easily via *Theorem 2.3*:

$$\mathcal{T}^{k+1} = \mathcal{S}_{\frac{\lambda W^k}{\mu^k}}\left(\mathcal{D} - \mathcal{B}^k - \frac{\mathcal{Y}^k}{\mu^k}\right) \tag{26}$$

The subproblem (25) is calculated by *Theorem 2.2* utilizing Algorithm 1 in the Fourier domain, which is described in Algorithm 2 (please see Ref. [46] for details).

$\mathcal{Y}$ and μ update in the standard way:

$$\mathcal{Y}^{k+1} = \mathcal{Y}^k + \mu^k\left(\mathcal{D} - \mathcal{B}^{k+1} - \mathcal{T}^{k+1}\right) \tag{27}$$

$$\mu^{k+1} = \rho\mu^k \tag{28}$$

where $\rho > 1$. Finally, the whole process is described in Algorithm 3.

Algorithm 2 Solve Equation (25) using PSVT

Input: $\mathcal{A}^k = \mathcal{D} - \mathcal{T}^{k+1} - \frac{\mathcal{Y}^k}{\mu^k} \in \mathbb{R}^{n_1 \times n_2 \times n_3}$, λ, μ^k

1. Compute $\overline{\mathcal{A}}^k = fft(\mathcal{A}^k, [], 3)$
2. Compute each frontal slice of $\overline{\mathcal{B}}^{k+1}$ by
    for $i = 1, \cdots, \lceil (n_3 + 1)/2 \rceil$ **do**
    $\left(\overline{\mathcal{B}}^{k+1}\right)^{(i)} = \mathcal{P}_{N,\lambda/\mu^k}\left(\left(\overline{\mathcal{A}}^k\right)^{(i)}\right)$ (Operator $\mathcal{P}(\cdot)$ is defined in Equation (7));
    end for
    for $i = \lceil (n_3 + 1)/2 \rceil + 1, \cdots, n_3$ **do**
    $\left(\overline{\mathcal{B}}^{k+1}\right)^{(i)} = conj\left(\left(\overline{\mathcal{B}}^{k+1}\right)^{(n_3-i+2)}\right)$;
    end for
3. **Compute** $\mathcal{B}^{k+1} = ifft(\overline{\mathcal{B}}^{k+1}, [], 3)$

Algorithm 3 ADMM solver to the proposed model

Input: $\mathcal{D}$, $\mathcal{W}_p$, λ, μ^0, ε, N
Initialization: $\mathcal{B}^0 = \mathcal{T}^0 = \mathcal{Y}^0 = 0$, $\mathcal{W}_{sw} = 1$, $\mathcal{W}^0 = \mathcal{W}_{rec} \odot \mathcal{W}_{sw}$, $\mu^0 = 3 \times 10^{-3}$, $\rho = 1.1$, $c = 1$, $k = 0$
    while not converge **do**
    1. Fix the others and update $\mathcal{T}^{k+1}$ by Equation (26);
    2. Fix the others and update $\mathcal{B}^{k+1}$ by Algorithm 2;
    3. Fix the others and update $\mathcal{Y}^{k+1}$ by Equation (27);
    4. Fix the others and update $\mathcal{W}^{k+1}$ by
    $\mathcal{W}_{sw}^{k+1} = \frac{c}{|\mathcal{T}^k| + \varepsilon}$;
    $\mathcal{W}^{k+1} = \mathcal{W}_{rec} \odot \mathcal{W}_{sw}^{k+1}$;
    5. Update μ by Equation (28);
    6. Check the convergence conditions
    $\frac{\|\mathcal{B}^{k+1} + \mathcal{T}^{k+1} - \mathcal{D}\|_F^2}{\|\mathcal{D}\|_F^2} < \varepsilon$ or $\|\mathcal{T}^{k+1}\|_0 = \|\mathcal{T}^k\|_0$;
    7. Update k: $k = k+1$;
    end while
3. **Output:** $\mathcal{B}^k$, $\mathcal{T}^k$

3.4. The Whole Procedure of the Proposed Method

Figure 7 shows the whole procedure of the infrared small target detection method based on the proposed model, which can be described as follows:

(1). Local prior extraction. Given an infrared image, by calculating Equation (16), the prior weight map $\mathcal{W}_p$ related to the target and background information is obtained.

(2). Patch-tensor construction. By sliding a window of size $k \times k$ from top left to bottom right to transform the original infrared image $f_D \in \mathbb{R}^{m \times n}$ and the prior weight map $\mathcal{W}_p \in \mathbb{R}^{m \times n}$ into the original patch-tensor $\mathcal{D} \in \mathbb{R}^{k \times k \times t}$ and the prior weight patch-tensor $\mathcal{W}_p \in \mathbb{R}^{k \times k \times t}$ respectively, where t is the number of window sliding.

(3). Target-background separation. The input patch-tensor $\mathcal{D}$ is decomposed into a low-rank patch-tensor $\mathcal{B} \in \mathbb{R}^{k \times k \times t}$ and a sparse patch-tensor $\mathcal{T} \in \mathbb{R}^{k \times k \times t}$ via Algorithm 3.

(4). Image reconstruction and target detection. The target image $f_B \in \mathbb{R}^{m \times n}$ and background image $f_T \in \mathbb{R}^{m \times n}$ are reconstructed from the low-rank patch-tensor $\mathcal{B} \in \mathbb{R}^{k \times k \times t}$ and sparse patch-tensor $\mathcal{T} \in \mathbb{R}^{k \times k \times t}$, and the process of reconstruction is contrary to that of construction. Meanwhile, a one-dimensional median filter is exploited to determine the value of the position overlapped by several patches. Once the reconstruction is done, small targets are detected easily via adaptive threshold segmentation as in [20].

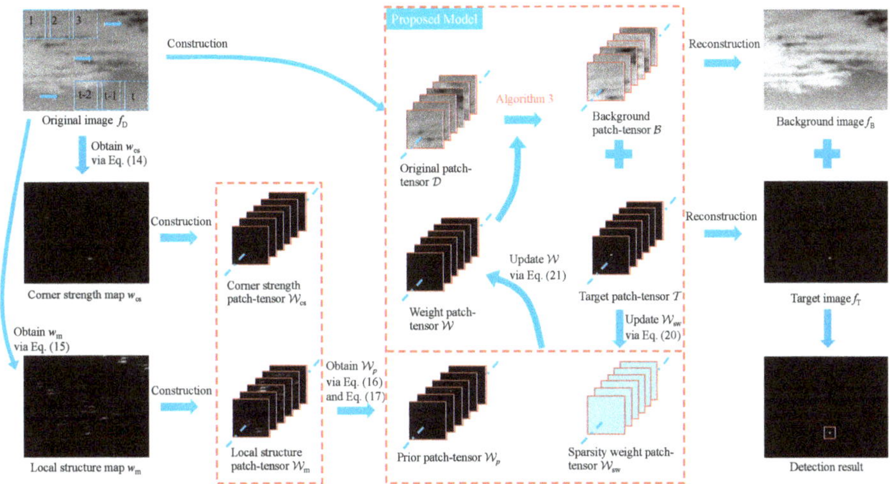

Figure 7. The overall procedure of the proposed model in this paper.

4. Experiment and Results

In this section, extensive experiments are conducted to verify the feasibility of the proposed model from different aspects including robustness against various scenes, robustness to noise, the ability of background suppression and target enhancement, target detection ability, and the computation time of the algorithm. To fully access the superiority of the proposed algorithm, nine state-of-the-art approaches are included for comparison.

4.1. Experimental Setup and Description

The diversity of scenes is one of the biggest challenges for detecting small targets embedded in infrared images. In order to validate the robustness of our approach to scenes, 24 infrared images with different varied scenes from uniform backgrounds with extremely dim targets to complex scenes with salient interferences and clutters were tested, which are displayed in Figure 8. All of the targets are marked with red (or green) square boxes. Moreover, for the sake of better observation and comparison, we had enlarged the target areas and then placed most of them in the lower left (right) corner of the image. Following this, six typical scenes were chosen from the 24 tested images to evaluate the performance of our method in the case of noise with different levels. Note that the added noise obeys the Gaussian distribution. Next, four sequences (Figure 8a–d) were used to quantify the detection ability of the proposed model. Finally, the algorithm complexity and computation time for different sizes are given. Nine methods including the Top-hat filter [12], Laplacian of Gaussian (LoG) filter [29], multiscale patch-based contrast measure (MPCM) [31], relative local contrast measure (RLCM) [19], infrared patch-image model (IPI) [20], nonnegative infrared patch-image model based on partial sum minimization of singular values (NIPPS) [21], reweighted IPI (ReWIPI) [38], nonconvex rank approximation minimization (NRAM) [39], and reweighted infrared patch-tensor model (RIPT) [42] were employed as the baselines. The same experiments were carried out with these baselines for all-round comparison. Given space limitations, only part of the experimental results are shown in this paper; the full extent can be found in the Appendices A and B. Table 3 summarizes the parameter settings of all the methods used in this paper. All of the optimizing methods, i.e., IPI, NIPPS, ReWIPI, NRAM, RIPT and the proposed method were solved via ADMM. In addition, all of the experiments were implemented with Matlab R2018a in Windows 7 based on Intel Celeron 2.90 GHz CPU with 4G of RAM.

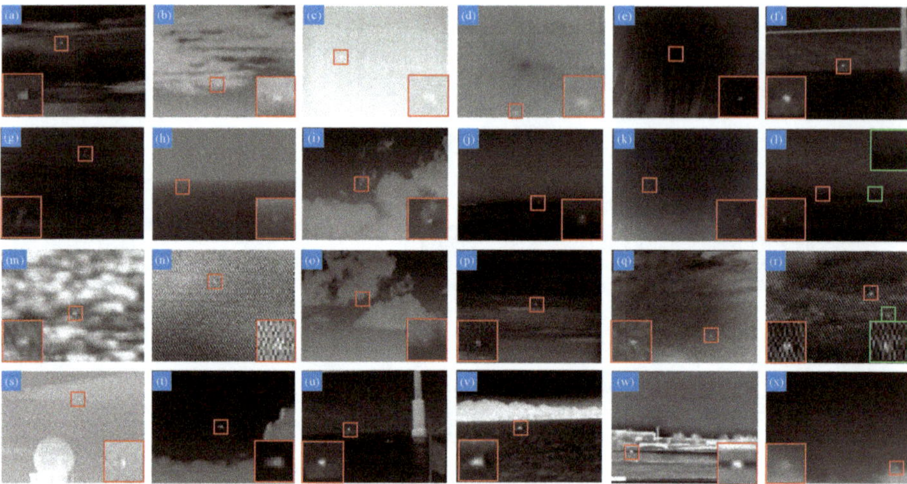

Figure 8. The 24 real scenes used in the experiments. For the sake of visualization, all of the images are changed to the same size.

Table 3. Detailed parameter settings of the 10 tested methods.

Method	Parameters
Top-hat [12]	Structure shape: disk, structure size: 3×3
LoG [29]	$\sigma = [0.50, 0.60, 0.72, 0.86, 1.03, 1.24, 1.49, 1.79, 2.14, 2.57, 3.09, 3.71]$
MPCM [31]	$N = 3, 5, 7, 9$, mean filter size : 3×3
RLCM [19]	$(K_1, K_2) = (2, 4), (5, 9)$, and $(9, 16)$
IPI [20]	Patch size : 50×50, sliding step : 10, $\lambda = 1/\sqrt{\min(m, n)}$, $\varepsilon = 10^{-7}$
NIPPS [21]	Patch size : 50×50, sliding step : 10, $\lambda = 2/\sqrt{\min(m, n)}$, $\varepsilon = 10^{-7}$
ReWIPI [38]	Patch size : 50×50, sliding step : 10, $\lambda = 2/\sqrt{\min(m, n)}$, $\varepsilon = 10^{-7}, \varepsilon_B = \varepsilon_T = 0.04$
NRAM [39]	Patch size : 50×50, sliding step : 10, $\lambda = 1/\sqrt{\min(m, n)}, \mu^0 = 3\sqrt{\min(m, n)}, \gamma = 0.002, C = \sqrt{\min(m, n)}/2.5, \varepsilon = 10^{-7}$
RIPT [42]	Patch size : 30×30, sliding step : 10, $\lambda = L/\sqrt{\min(m, n)}, L = 1, h = 1, \varepsilon = 10^{-7}$
Ours	Patch size : 40×40, sliding step : 40, $\lambda = 0.6/\sqrt{\max(n_1, n_2) * n_3}, \varepsilon = 10^{-7}$

4.2. Evaluation Metrics

In this subsection, for a comprehensive comparison with the aforementioned state-of-the-art approaches, several typical metrics, including the signal-to-clutter ratio gain (SCRG), the background suppression factor (BSF), and the receive operating characteristic (ROC) curve with the area under curve (AUC) were used, where the ROC curve shows the tradeoff between the detection probability P_d and false-alarm probability F_a. These metrics would reveal the ability of one method in target enhancement, background suppression, and target detection. The most widely used criterion SCRG is defined as

$$\mathrm{SCRG} = \frac{\mathrm{SCR_{out}}}{\mathrm{SCR_{in}}} \tag{29}$$

where subscripts out and in represent the original image and the obtained target image respectively, and SCR is a measurement of the difficulty of detecting a small target in an infrared image, whose definition is

$$\text{SCR} = \frac{|\mu_t - \mu_b|}{\sigma_b} \tag{30}$$

where μ_t is the average grayscale of the target area, μ_b and σ_b are the average pixel value and standard deviation of the surrounding local neighborhood region, respectively.

Another evaluation indicator is BSF, showing the background suppression quality of detection algorithms, which is defined as

$$\text{BSF} = \frac{\sigma_{in}}{\sigma_{out}} \tag{31}$$

where σ_{in} and σ_{out} stand for the standard deviation values before and after suppression in the local region. SCRG and BSF are calculated in the neighborhood region around the target, and Figure 9 shows the local region that is used in the experiment. Assuming that the target size is $a \times b$, then the local region size is $(a + 2d) \times (b + 2d)$; we set $d = 20$ in this paper.

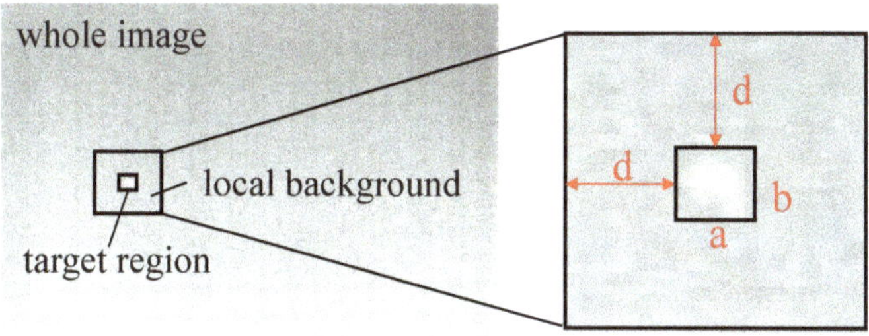

Figure 9. Local region of a small target in an infrared image.

In addition to the above two evaluation indicators, the detection probability P_d and false-alarm probability F_a is a pair of key metrics, which are defined as follows:

$$P_d = \frac{\text{number of true detections}}{\text{number of actual targets}} \tag{32}$$

$$F_a = \frac{\text{number of false detections}}{\text{number of images}} \tag{33}$$

The ROC curve is drawn according to P_d and F_a values, where F_a is abscissa and P_d is ordinate. The AUC is the area enclosed by the ROC curve and the coordinate axis. Except for ROC, for all the other metrics, the larger their value, the better the performance of the method.

4.3. Parameter Analysis

For the proposed model, there are several important parameters such as the patch size, the sliding step, the penalty factor μ, and the tradeoff constant λ that usually affect the robustness for different scenes. Hence, to obtain a better performance with real datasets, it is wise to choose proper parameters via experiments. The ROC curves on four real infrared sequences for different model parameters are given in Figure 10. Here, one point needs to be noted is that the performances obtained by tuning one of the parameters with the others fixed may not be globally optimal.

4.3.1. Patch Size

Patch size plays a vital role in determining not only the detection performance, but also the computation complexity of the algorithm. We hope for a larger patch size to make sure that the target is sparse enough due to the uncertainty of the target size; however, some noise with sparsity properties such as salient edges would also have a higher probability of being identified as target components, which degrades the separation results. On the other hand, a smaller patch size would lead to a smaller computational complexity in each inner loop with singular value decomposition (SVD), but the sparseness of the target is no longer so obvious. To figure out the influence of the patch size on Sequences 1–4, we varied the patch size from 20 to 60 with 10 intervals and the corresponding ROC curves are illustrated in the first row of Figure 10. By analyzing the ROC curves, we can conclude that the best performance is achieved when the patch size is set to 40 for all of the sequences. The worst performance is reached when the patch size is equal to 60 in most cases. This is because a too-large patch size would regard the salient non-target noise as the "true" target, also resulting in incorrect recovery, especially when the target is not so prominent. The performance of 20 depends on the target size as the target in Sequence 1 is very dim and small while it breaks down when dealing with larger targets in Sequences 3–4, which results from the lack of target sparsity. Another underlying fact is that our proposed model is a little sensitive to the patch size particularly when facing extreme complex scenes such as those in Sequence 1, the target of which is almost submerged. Therefore, we chose 40 as the best patch size utilized in the following experiments.

4.3.2. Sliding Step

Similar to the patch size, the sliding step has a direct impact on the construction of patch-tensor, which indirectly influences the computation time and detection performance simultaneously as well. The sliding step determines how many frontal slices we can obtain to compose the desired patch-tensor. Different from other similar models, we prefer a larger sliding step which results from the following reasons. (i) A smaller sliding step implies that there would be more frontal slices containing the target, leading to an insufficient sparseness of the target, and (ii) More frontal slices means an increased computation time of t-SVD in Algorithm 1, because more inner loops are needed to calculate the matrix SVD of each frontal slice. To investigate its actual influence, we show the effects of the sliding step in the second row of Figure 10 via varying it from 10 to 40 (based on the best value of the patch size) with five intervals. It can be observed that as the sliding step increases, the model works better. Ten is a commonly used value; however, it performs the worst. Furthermore, even if the sliding step changes slightly, this change has a great impact on the results, which means that the proposed model is very sensitive to this parameter. Hence, the best choice for the sliding step is 40.

4.3.3. Penalty Factor μ

μ controls the tradeoff between the low-rank background and sparse target, namely the PVST operator and soft-thresholding operator; thus, one has to choose μ carefully in order to ensure both optimality and a fast convergence rate. With a smaller μ, more details would be preserved in the background patch-tensor; nevertheless, the target may suffer from over-shrinking because its details are remained by the background. In contrast, a larger μ could protect the target, but might leave more non-target components in the target patch-tensor. To choose an appropriate value of μ for obtaining better detection ability and a lower false alarm ratio, we investigated the influence of penalty factor on Sequences 1-4 by changing μ from 1×10^{-3} to 9×10^{-3} with an interval of 0.002, as illustrated in the third row of Figure 10. From the results we can arrival at a conclusion that μ cannot be too large or too small, especially when $\mu = 1 \times 10^{-3}$; the target is totally lost in most cases. Therefore, 3×10^{-3} was used to get a better balance between the background patch-tensor and the target patch-tensor.

4.3.4. Compromising Parameter λ

λ is a compromising parameter that controls the tradeoff between the target patch-tensor and the background patch-tensor. Hence, it is of great importance to fine tuning λ. With reference to [48], we set λ as $L/\sqrt{\max(n_1, n_2) * n_3}$ and vary L from 0.2 to 1.4 instead of varying λ directly. We show the influence of λ on Sequences 1-4 in the fourth row of Figure 10. From the illustration, we can easily observe that when $L = 1.2$ and $L = 1.4$, the performance of the proposed method is always worst. That is because as λ increases, the target patch-tensor would be suppressed to keep the whole objective function at a minimum, and vice versa. In other words, on one hand, a larger λ leads to a cleaner target image, but the target would be over-shrunk; on the other hand, a smaller λ can keep the target complete, but background residuals would be kept too. How to find the balance is a serious task. The experimental results shows that the performance is relative well when $L = 0.6$. Then, $\lambda = 0.6/\sqrt{\max(n_1, n_2) * n_3}$ was used at the end.

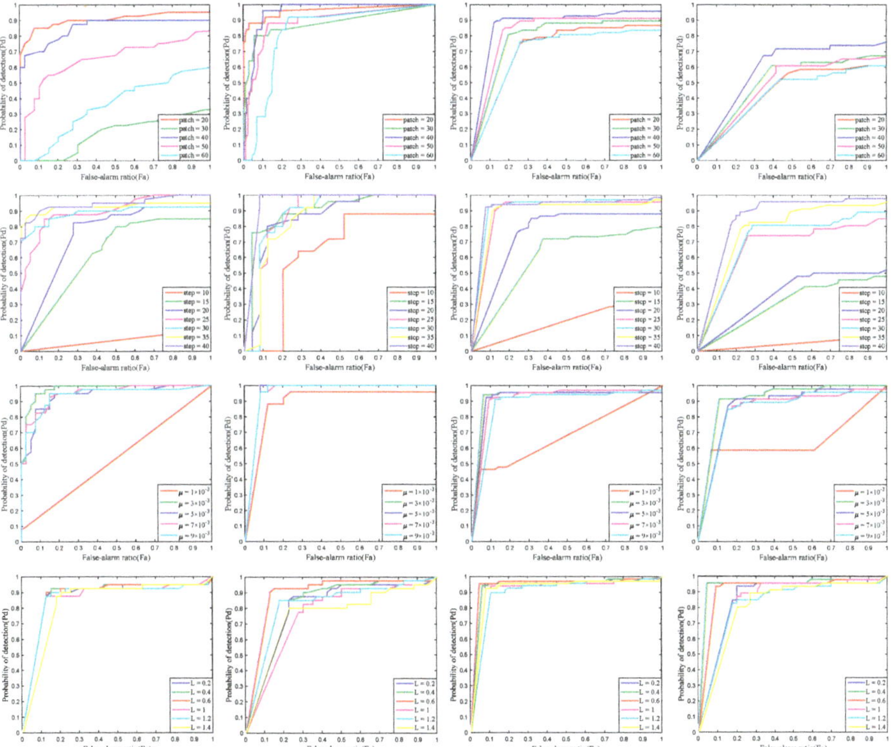

Figure 10. Detection performances under different parameters. Rows 1: ROC curves with respect to different patch sizes, Rows 2: ROC curves with respect to different sliding steps, Rows 3: ROC curves with respect to different penalty factors, Rows 4: ROC curves with respect to different compromising parameter.

4.4. Qualitative Evaluation

In this subsection, the proposed method is compared with nine state-of-the-art methods from qualitative aspects, i.e., robustness to different scenes and Gaussian noise, which reflects the ability of target enhancement and background suppression of each approach. Note that due to the large number of images, the results of all the methods except the proposed model and RIPT model are in the Appendices A and B.

4.4.1. Robustness to Different Scenes

One major challenge of infrared small target detection lies in its variety, which has two-fold meanings. Firstly, infrared scenes are diverse, such as sky background with thick clouds such as those in Figure 8b, a sea background with buildings and moving ships such as those in Figure 8f, a messy background with lots of salient interferences such as those in Figure 8w, etc. Secondly, the size of the small target is not fixed, but varies within a large range. For instance, as shown in Figure 8o, the target embedded in the cloud layer can be viewed as a point target, while the target in Figure 8t is much bigger than the aforementioned one. Therefore, a useful way to verify whether a detection method is good or not is to test its robustness against different scenes containing different target sizes. The separated target images obtained from the proposed model under 24 different scenes are displayed in Figure 11, from which we can observe that the backgrounds are totally wiped out, remaining merely the desired targets. Meanwhile, the shape of the targets has also been basically preserved.

Figure 12 indicates the results processed by the RIPT model; as analyzed in Section 3.2, it is easy to observe that the RIPT model is suitable for dealing with a spot-like target, but when it comes to a non-spot-like target, the issue of over-shrinking happens, which results from the local structure weight treating the target edge and background edge equally, as shown in Figure 12c,t,u. In addition, the suboptimality of SNN brings about remaining residuals (noise) in target images such as those in Figure 12a,n. One more point worth mentioning is that the RIPT model may suffer from totally losing the target when the background and target are both dim, such as in Figure 12d,h. The results of handling the remaining methods with various scenes are displayed in Figures A1–A8 in Appendix A, from which it is clear that they all lack robustness. Hence, compared with these baselines, it's fair to say that the proposed method shows advancement in dealing with different scenes and targets simultaneously.

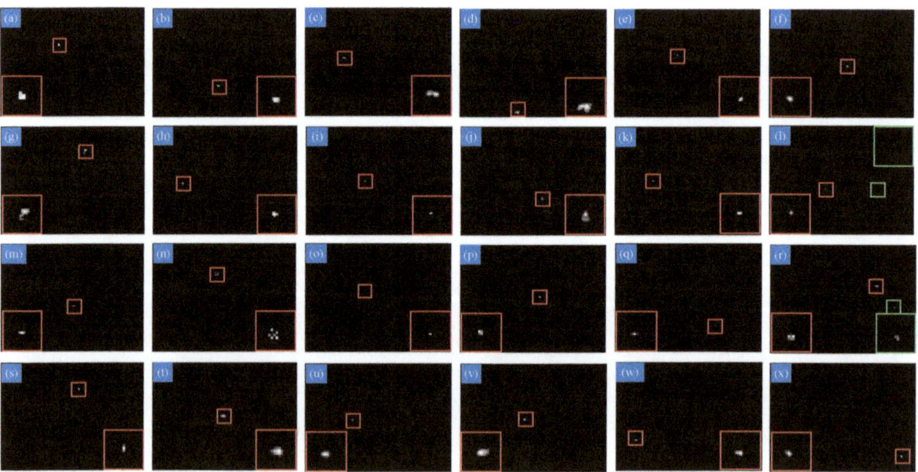

Figure 11. The separated target images of the proposed model under 24 scenes.

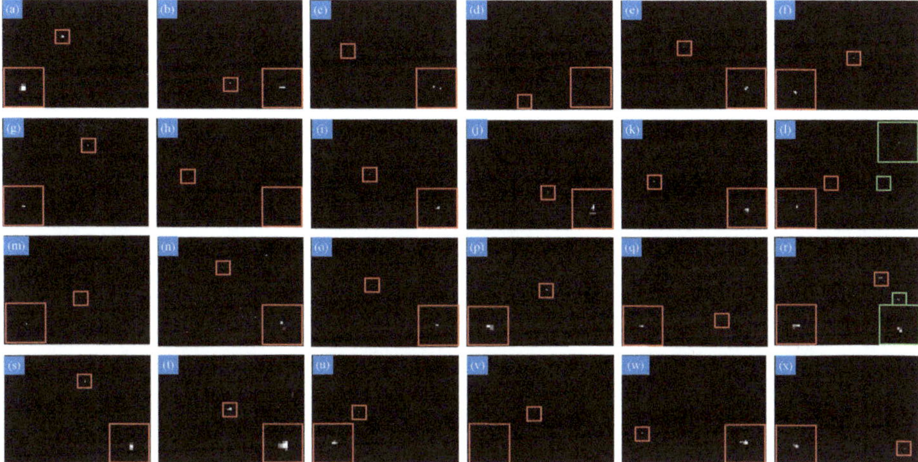

Figure 12. The separated target images of the RIPT model under 24 scenes.

4.4.2. Robustness to Noise

In addition to various scenes, noise is also a key factor that affects the detection results. In Figure 13, we further evaluated the proposed model in terms of noise with different levels under six scenes selected from Figure 8. Gaussian noise with a mean of zero was imposed to the images in the first row and third row of Figure 13, respectively. When the standard deviation is 10, the proposed method performs relatively well regarding background suppression and target enhancement, as well as preserving the shape of the target. When the standard deviation increases to 20, the proposed method still accurately locates the targets and wipes out the backgrounds in Figure 13s,u,x. Unfortunately, in Figure 13t,v,w, the detected results deviate from the real targets regardless of shape or size. This is acceptable considering the noise is so dense that the target can hardly be detected. We can also conclude that as long as the target in the contaminative image is still relative salient such as in Figure 13a–f, the proposed method can work.

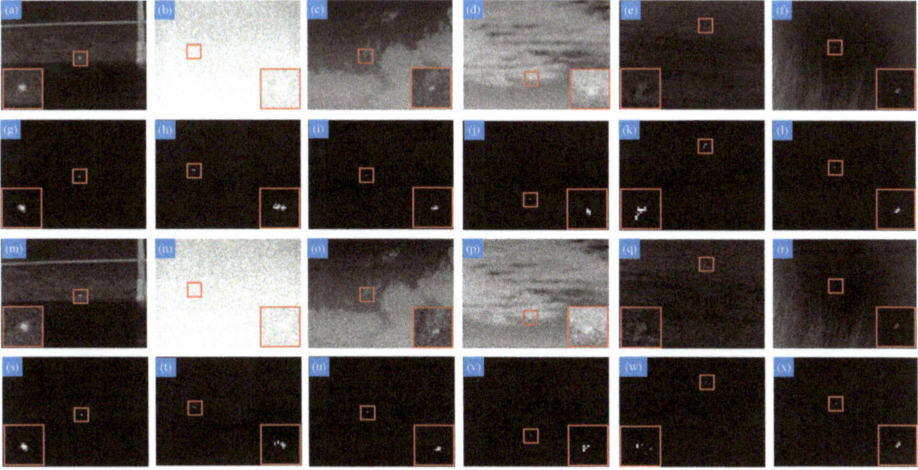

Figure 13. The first and third row are infrared images with additive white Gaussian noise with standard deviations of 10 and 20, and the second and fourth rows are the corresponding detection results by the proposed method.

We show in Figure 14 the performance of the RIPT model dealing with different levels of noise. As can be seen from the figure, the target is more likely to be lost. Furthermore, although the target is still salient within a noise-containing background (Figure 14a,m for instance), the target recovered via RIPT is only spot-like, which demonstrates its weakness in handling slightly larger targets one more time. The results of the remaining optimizing methods facing noise are displayed in Figures A9–A12 in Appendix B. We can easily observe that they all have unsatisfactory performances, especially when the standard deviation is 20.

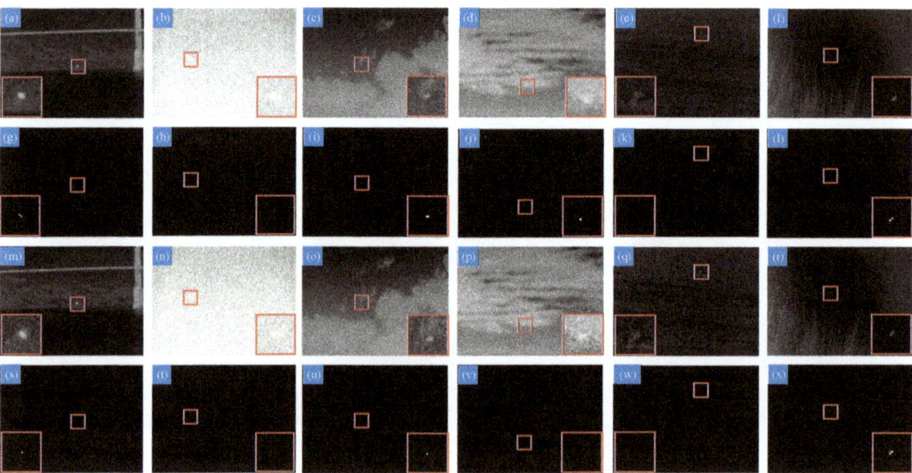

Figure 14. The first and third row are infrared images with additive white Gaussian noise with standard deviations of 10 and 20, and the second and fourth rows are the corresponding detection results by the RIPT method.

4.4.3. Visual Comparison with Baselines

To further visually compare the performance of all the competing methods, the results obtained by all the tested methods on Sequences 1–4 are displayed in Figures 15–18, and the detailed descriptions of four sequences are shown in Table 4. Note that for the convenience of observation, the contrast of the results obtained by Top-hat, LoG and RLCM is adjusted. For conventional Top-hat transformation, it can highlight the target to a certain extent in Figures 15a, 16a, 17a and 18a; however, it is extremely sensitive to noise and clutters, which would produce many false alarms. The intrinsic reason is mainly relevant to the usage of the fixed structural element without considering the surrounding neighborhood. Besides, the fixed structural element with a fixed shape is difficult to perfectly match all the targets. LoG, MPCM and RLCM are all HVS-based approaches but the performance of LoG is much worse than the latter two. We can obviously see that LoG is also vulnerable to edges and noise which results from the calculation of Gaussian scale space and its second derivative, making the target and edges both enhanced, especially in the case of complex background such as those in Figures 15b and 16b. The main difference between MPCM and RLCM is the definition of local contrast measure, leading to distinguishing a detection ability. For MPCM, its local contrast measure is defined based on the difference between the current patch and its adjacent background patches; while for RLCM, the local contrast is associated with the mean grayscale value of each cell. Their improvement is apparent when facing uniform scenes, and the RLCM is slightly better than the MPCM from Figures 17c and 18d. Nevertheless, just as the results in Figures 15 and 16, the phenomenon of enhancing non-target pixels still exists, which is caused by the inaccuracy of the local dissimilarity measure; in some cases, they are even brighter than the real target.

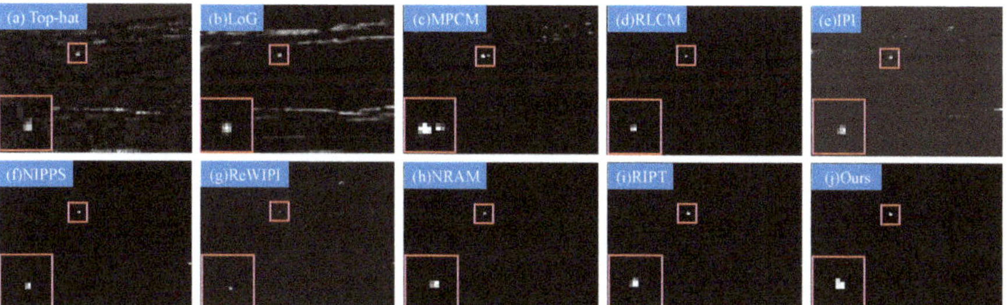

Figure 15. Results of the different approaches to Sequence 1.

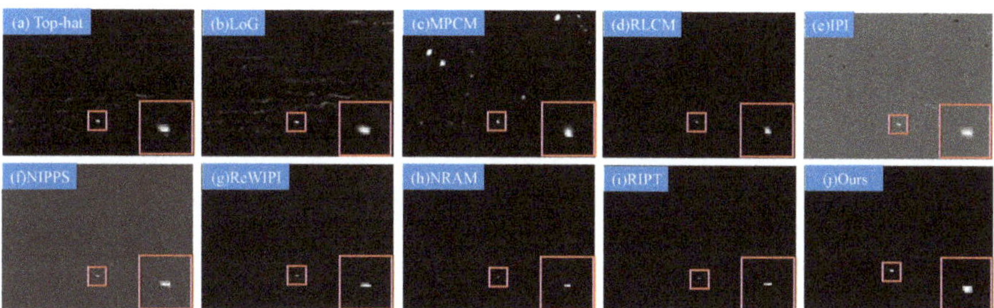

Figure 16. Results of the different approaches to Sequence 2.

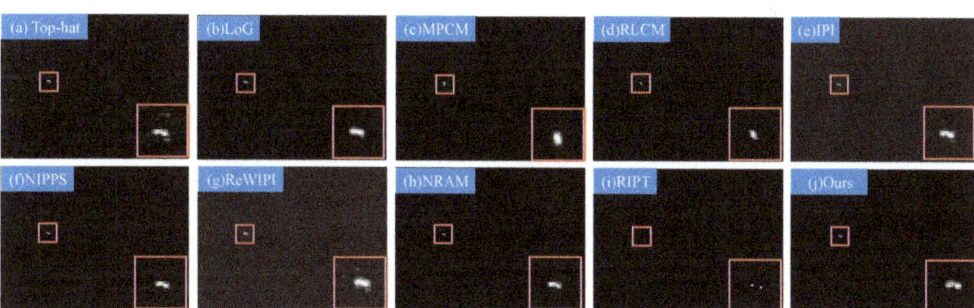

Figure 17. Results of the different approaches to Sequence 3.

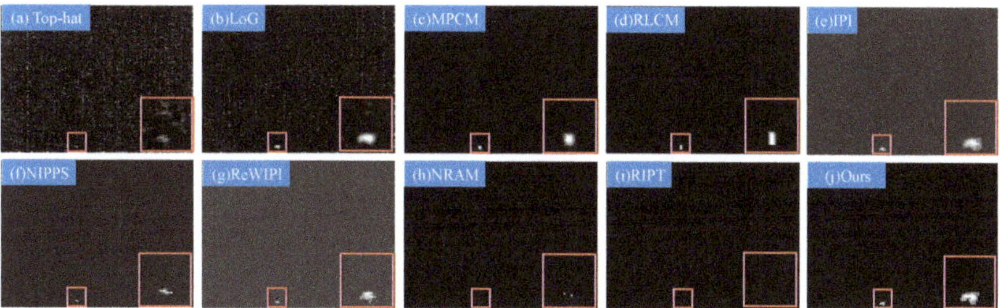

Figure 18. Results of the different approaches to Sequence 4.

Table 4. Detailed descriptions of four real sequences.

	Frame Number	Size	Background Description	Target Description
Sequence 1	52	128×128	Sky scene with banded cloud	Tiny and dim
Sequence 2	30	256×200	Heavy banded cloud and floccus	Small, size varies a lot
Sequence 3	67	320×240	Very bright, heavy noise	Moves fast with changing shape, brightness
Sequence 4	46	320×240	Very blurry with black holes in the middle	Keeps moving in the sequence and changing shape and brightness

Generally speaking, the rest of the optimizing methods show superiority in both target enhancement and background suppression. From the figures, there's no doubt that IPI suffers from residuals in the recovered target image, because the matrix nuclear norm treats all the singular values equally, which usually leads to suboptimal solutions. Via minimizing the partial sum of singular values, NIPPS achieves a better performance than IPI. However, as observed in Figures 15f and 18f, either a complex scene including highlight interferences and intensive noise or a particularly dim scene is still a challenge for NIPPS. To overcome the deficiencies of initial IPI, the ReWIPI adopts weighted technology to restore the background and target simultaneously. We can see from the results that the ReWIPI lacks robustness to different scenarios although it does well in Figure 16g. NRAM provides a tighter surrogate of rank with nonconvex rank approximation involved, which implies that the separated background image would be more accurate so that the problem of residuals could be solved. NRAM reaches the desired results except for the last sequence, from which the target is almost disappeared.

Unlike these matrix-level methods, RIPT directly stacks the patches into a tensor named patch-tensor without vectorizing each patch into a vector, which successfully converts a low-rank matrix recovery problem into a tensor recovery problem. As an extension of the IPI model, RIPT accurately captures the low-rank property of the matrix that is obtained by unfolding the patch-tensor along each mode, and thus achieving better detection performance. However, there are two issues for which the RIPT model has not been resolved: namely, salient noise such as that in Figure 15i, and target distortion with the possibility of completely loss such as that in Figure 18i. The proposed method shows superior performance not only in the preservation of the target but also in the suppression of backgroundcompared with the above baselines, especially in Figures 15 and 18. Basically, all the methods are not performing well, except for ours. Furthermore, the computation time of the proposed method is less than that of the similar optimizing methods, which will be discussed later.

4.5. Quantitative Evaluation

Apart from visually validating the robustness of our method through single-frame images with different backgrounds and different noise levels, in this subsection, the detection performance of our model and other baselines was further measured via quantitative evaluation indicators including the signal-to-clutter ratio gain (SCRG), background suppression factor (BSF), and ROC curves on four real sequences. Table 5 lists the experimental results for all 10 tested approaches for Sequences 1–4. It should note that **inf** (i.e., infinity) represents the background is completely wiped out in the local region. Since NRAM and RIPT are not able to detect the target in Sequence 4 in some cases, when calculating SCRG and BSF for the last sequence, we don't take these two methods into account. It can be clearly seen that the proposed method achieves the highest values in terms of SCRG and BSF in all of the datasets, showing great advantages in background suppression and target enhancement. On the other hand, RIPT gets the second highest scores sometimes in terms of the two metrics, which suggests that the tensor model can indeed seek more spatial information to improve the robustness. Filtering methods get very low scores in comparison with optimizing methods, resulting from the simple assumption based on background homogeneity or target saliency.

Table 5. SCRG and BSF values of the ten methods.

Method	Sequence 1		Sequence 2		Sequence 3		Sequence 4	
	SCRG	BSF	SCRG	BSF	SCRG	BSF	SCRG	BSF
Top-hat	1.04	1.99	9.56	1.90	0.36	0.22	0.58	12.46
LoG	8.25	1.88	7.33	1.30	1.30	0.30	2.28	7.86
MPCM	9.77	23.72	14.4	4.1	8.72	7.90	20.38	14.54
RLCM	28.97	35.14	30.63	62.99	2.05	1.82	2.22	16.25
IPI	106.12	140.32	43.33	16.73	8.05	1.88	5.36	2.66
NIPPS	456.15	544.79	180.08	118.16	43.2	35032728557	13.71	24.33
ReWIPI	242.14	641.92	302.55	153.61	5.10	1.35	5.42	4.16
NRAM	1004.48	677.2	687.02	178.69	109.83	inf	—	—
RIPT	523.44	222.97	690.32	276.02	46.8	inf	—	—
Ours	1059.58	1229.65	697.77	315.87	147.67	inf	46.34	60.21

NOTES: Underline with bold represents the highest value and underline represents the second highest value.

To further demonstrate the advantage of the proposed method, ROC curves corresponding to the four sequences that reflect overall detection ability of one method are plotted in Figure 19, and the AUC values are also listed in Table 6. A higher AUC value means that an algorithm has better performance. The performance of RLCM fluctuates greatly; for Sequences 1 and 2, RLCM works very well, but fails dealing with other sequences. The reason comes down to the local contrast measure utilized by RLCM, which merely relates to the mean grayscale of each cell, being extremely unsuitable for handling the low-contrast background embedded within a blurred target. Another interesting thing is that the AUC values of RIPT are only at a medium level, which is due to the problem of the excessive shrinkage of a slightly larger target, resulting in a relatively low detection probability. The ROC curves of IPI and ReWIPI obtained from handling Sequence 1 confirm that they are not enough to cope with complex scenes full of salient edges and clutters. In general, the proposed method always gets the highest detection probability with respect to the same false-alarm ratio, indicating that the proposed model outperforms other state-of-the-art methods in target detection performance.

Table 6. Area under curve (AUC) values of the 10 methods.

	Top-hat	LoG	MPCM	RLCM	IPI	NIPPS	ReWIPI	NRAM	RIPT	Ours
Sequence 1	0.311	0.861	0.613	0.986	0.387	0.829	0.173	1	0.987	1
Sequence 2	0.743	0.932	0.863	0.900	0.938	0.933	0.957	0.967	0.928	0.990
Sequence 3	0.604	0.927	0.930	0.181	0.938	0.856	0.849	0.944	0.606	0.945
Sequence 4	0.340	0.347	0.877	0.021	0.862	0.917	0.925	0.707	0.503	0.933

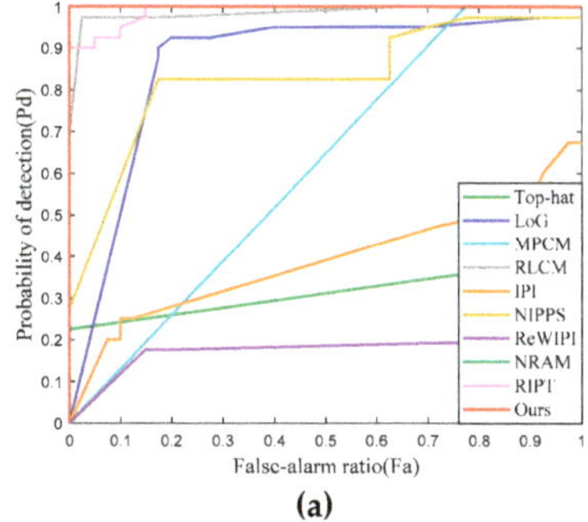

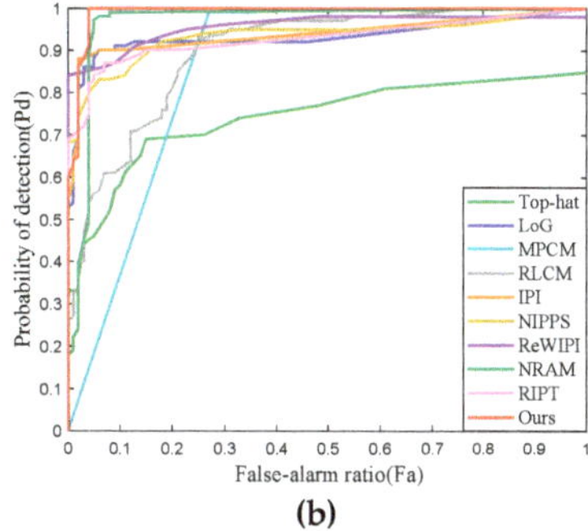

Figure 19. *Cont.*

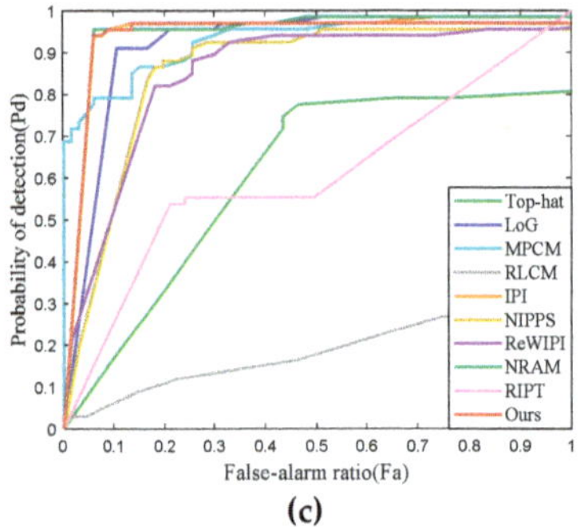 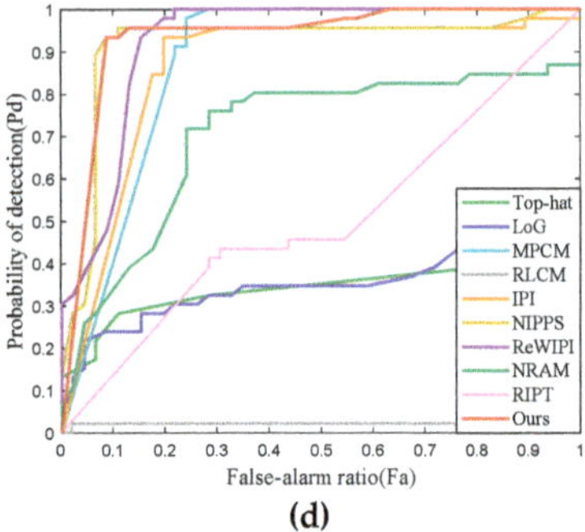

Figure 19. ROC curves of detection results of four real sequences. (**a**) Sequence 1; (**b**) Sequence 2; (**c**) Sequence 3; (**d**) Sequence 4.

4.6. Algorithm Complexity and Computational Time

In addition to high accuracy, real-time performance is also a basic requirement in infrared small target detection. However, it's hard to balance good detection ability and real-time performance. For filtering-based methods, simple assumptions coupled with simple calculations are fast, but not effective. For the optimizing-based methods, one major challenge is that the algorithms are time-consuming because of their high complexity, which is mainly associated with SVD. Therefore, the computational efficiency of different methods is discussed in this part. Suppose that the size of the original infrared image is $M \times N$, in which m and n are the columns and rows of the patch-image, and the size of the patch-tensor is $n_1 \times n_2 \times n_3$. The computation cost of Top-hat is $O(I^2 \log I^2 MN)$, where I denotes the size of the structural element. Due to the computation complexity of Gaussian filtering being $O(M^2 N^2)$, considering the use of k different scales, the final cost of LoG is $O(kM^2 N^2)$. For MPCM and RLCM, the major time-consuming part is calculating the saliency map pixel by pixel. The computation of MPCM and RLCM for a specific pixel needs an $O(l^2)$ cost, where $l(l = 1, 2, \cdots, L)$ is the processing window scale. Further, the total cost of them is $O(L^3 MN)$. Furthermore, for low-rank matrix-based methods, the computation complexity is mainly derived from the matrix SVD, which has a computational complexity of $O(mn^2)$. RIPT needs to calculate the SVD of an unfolding matrix along each mode with the sizes of $n_1 \times (n_2 n_3)$, $n_2 \times (n_1 n_3)$ and $n_3 \times (n_1 n_2)$, respectively. Therefore, the cost of the RIPT model is $O(n_1 n_2 n_3 (n_1 n_2 + n_2 n_3 + n_1 n_3))$. For the proposed model, the dominant factor of the complexity cost is calculating the SVD and FFT in Algorithm 2. Considering that merely the frontal slice with the size of $n_1 \times n_2$ is utilized to calculate FFT, the final computation cost of proposed model is $O(n_1 n_2 n_3 log(n_1 n_2) + n_1 n_2^2 \lceil (n_3 + 1)/2 \rceil)$, which shows a great reduction compared with RIPT. Note that because of the introduction of the new iteration stop condition, RIPT and our method would actually be faster.

Table 7 summarizes the algorithm complexity of all the methods, and lists their average computing time for Sequences 1–4. We can observe that all the optimizing methods based on the matrix level are extremely sensitive to changes in image size. In other words, as the size increases, the computation time increases dramatically, which is a big drawback of these methods. In contrast, the tensor-based approaches improve significantly, and this gap is more pronounced as the size increases. Among all the low-rank optimizing methods, the proposed method costs the least time. Although it still slightly slower than the filtering methods, considering the excellent performance, this is undoubtedly acceptable.

Table 7. Comparison of computational complexity and average computing time (in seconds) of the 10 methods.

Methods	Complexity	Sequence 1	Sequence 2	Sequence 3	Sequence 4
Top-hat	$O(I^2 \log I^2 MN)$	0.015	0.015	0.018	0.016
LoG	$O(kM^2N^2)$	0.019	0.035	0.048	0.046
MPCM	$O(L^3MN)$	0.038	0.074	0.097	0.096
RLCM	$O(L^3MN)$	0.895	2.941	4.414	4.385
IPI	$O(mn^2)$	0.327	8.717	22.063	21.941
NIPPS	$O(mn^2)$	0.321	7.561	19.182	18.097
ReWIPI	$O(mn^2)$	1.030	14.978	39.612	41.559
NRAM	$O(mn^2)$	0.494	3.661	8.357	8.341
RIPT	$O(n_1 n_2 n_3 (n_1 n_2 + n_2 n_3 + n_1 n_3))$	0.211	1.079	1.279	3.217
Ours	$O(n_1 n_2 n_3 \log(n_1 n_2) + n_1 n_2^2 \lceil (n_3 + 1)/2 \rceil)$	0.081	0.136	0.127	0.217

5. Discussion

Even though many scholars are working in the field of infrared small target detection, there is still room for improvement in this field. Based on simple assumptions, filtering methods enable real-time detection whereas they cannot work well under complex scenes. Exploiting the nonlocal self-correlation property of infrared backgrounds and the sparsity of targets, optimizing methods show a strong detection ability and robustness in comparison with filtering methods, but they are time-consuming. The cornerstone of early optimizing methods is the construction of an infrared patch-image (IPI), which completely destroys the original structural information. To utilize more spatial prior, an infrared patch-tensor (IPT) model was proposed, introducing the tensor recovery technology into this filed.

By employing the IPT model with involving target-related and background-related priors, the proposed method fully considers the nonlocal configuration and local structure of infrared images, showing great performance not only in target enhancement but also in background suppression. Moreover, with the help of an extra stopping condition and reweighted scheme, the complexity of the ADMM solver for the proposed method is dramatically reduced, which is indicated in Table 6. Hence, we meet the requirement of alleviating the issue of imbalance between the computation time and detection performance.

Series experiments including robustness to various scenes, robustness to noise, target enhancement, background suppression, detection ability, and computation time were carried out to compare the proposed method and other baselines. The experimental results demonstrated that the proposed method outperforms the nine representative state-of-the-art methods, including Top-hat, LoG, MPCM, RLCM, IPI, NIPPS, ReWIPI, NRAM, and RIPT.

6. Conclusions

To cope with the issue of imbalance between the detection performance and computation time of current methods and further improve the robustness to noise and various scenes, a robust infrared patch-tensor model based on partial sum of tensor nuclear norm was proposed in this paper. Furthermore, the local prior which relates to the background and target simultaneously was introduced into the model as an effective means of suppressing edge residuals. Then, the traditional infrared small target detection task is transformed into a problem of solving the nonconvex tensor robust principal component analysis model. By incorporating a reweighted scheme with an accelerated version of t-SVD, an efficient algorithm based on ADMM was designed to solve this new model. Extensive experiments illustrated that the proposed method outperforms the state-of-the-art methods both in background suppression and target enhancement, achieving strong robustness and a great improvement in time reduction.

There are still some issues worth considering. For example, although we utilize the energy ratio to estimate the preserved target rank, finding a better way of determining it is still needed.

Author Contributions: L.Z. proposed the original idea, performed the experiments and wrote the manuscript. Z.P. contributed to the direction, content, and revised the manuscript.

Funding: This work was funded by National Natural Science Foundation of China (61571096 and 61775030), the Key Laboratory Fund of Beam Control, Chinese Academy of Sciences (2017LBC003), and Sichuan Science and Technology Program (19YYJC0019).

Acknowledgments: The authors would thank the published code of Gao's model and Dai's model for comparison. The same appreciation goes to the Laboratory of Imaging Detection and Intelligent Perception (IDIP) at the University of Electronic Science and Technology of China.

Conflicts of Interest: The authors declare no conflict of interest.

Appendix A

These are the obtained results of the other eight approaches that are not shown in the main body of this paper.

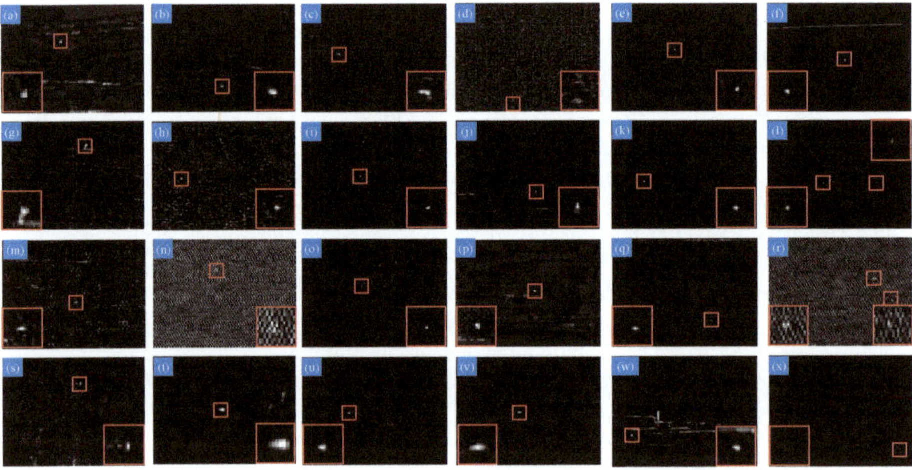

Figure A1. The separated target images of Top-hat under 24 scenes.

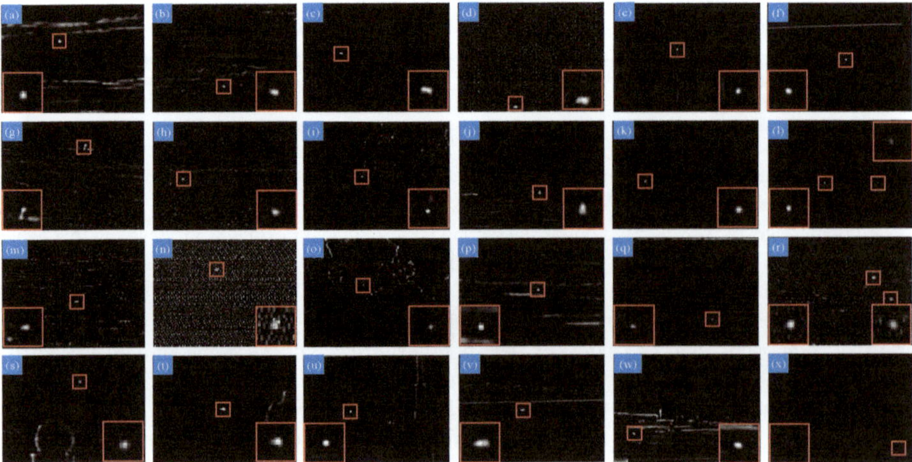

Figure A2. The separated target images of LoG under 24 scenes.

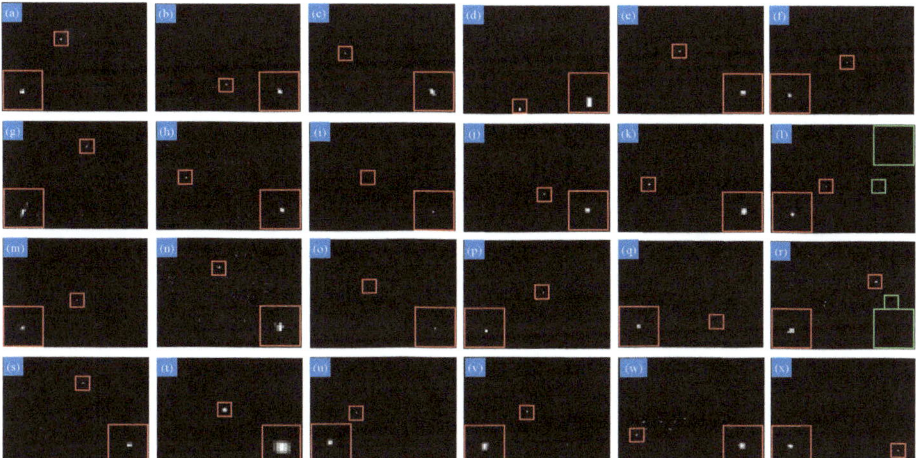

Figure A3. The separated target images of MPCM under 24 scenes.

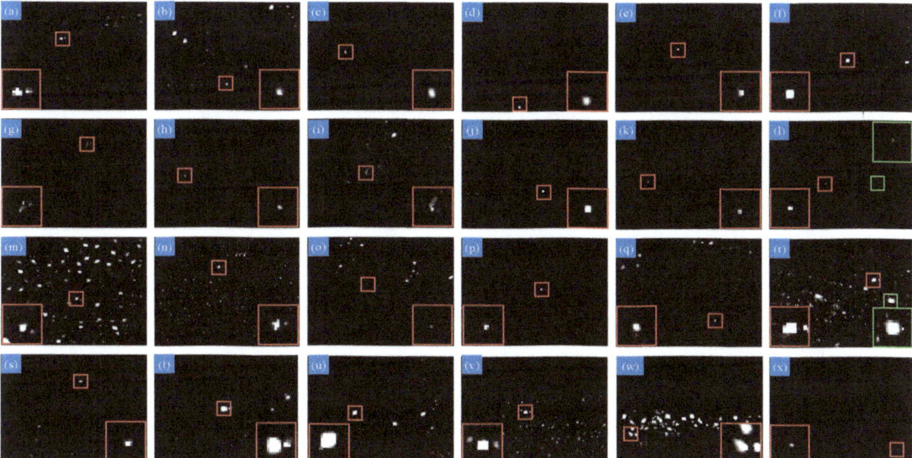

Figure A4. The separated target images of RLCM under 24 scenes.

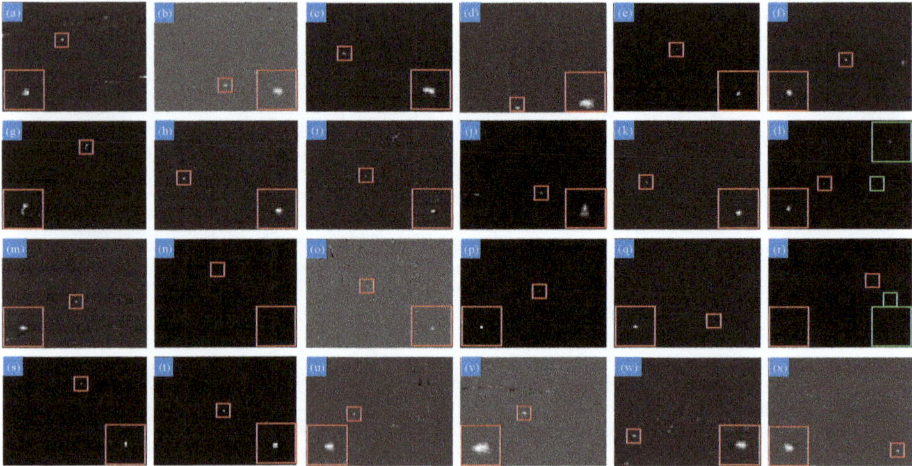

Figure A5. The separated target images of IPI under 24 scenes.

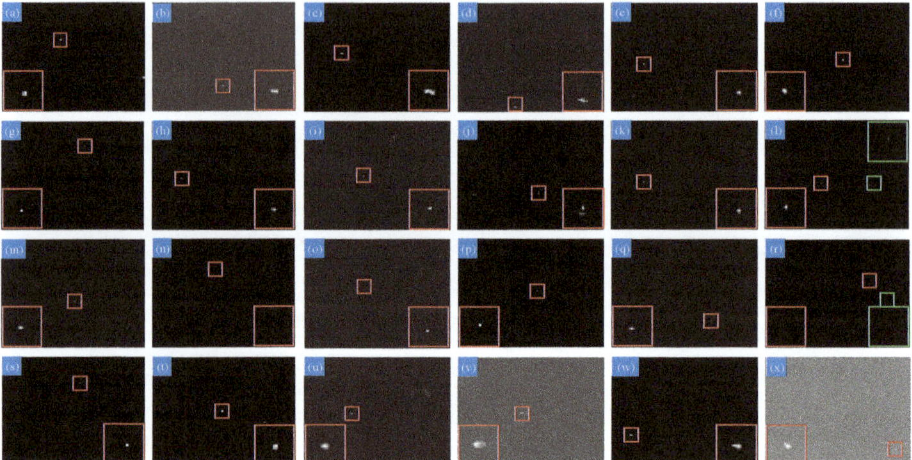

Figure A6. The separated target images of NIPPS under 24 scenes.

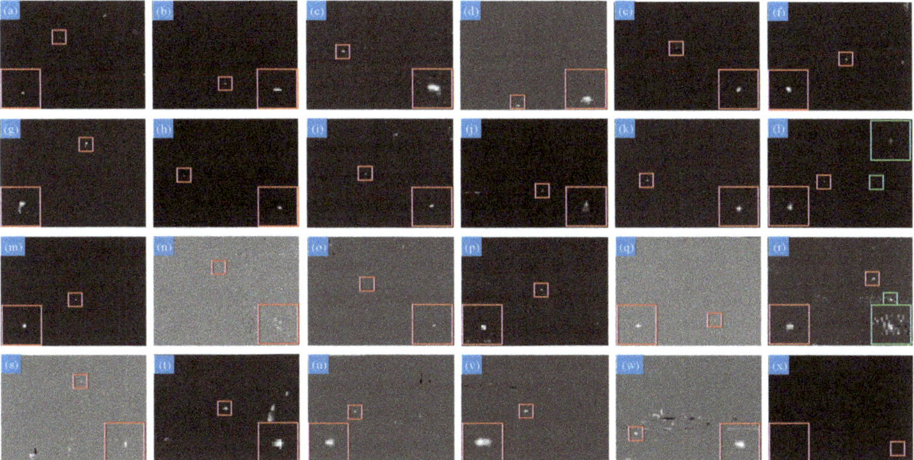

Figure A7. The separated target images of ReWIPI under 24 scenes.

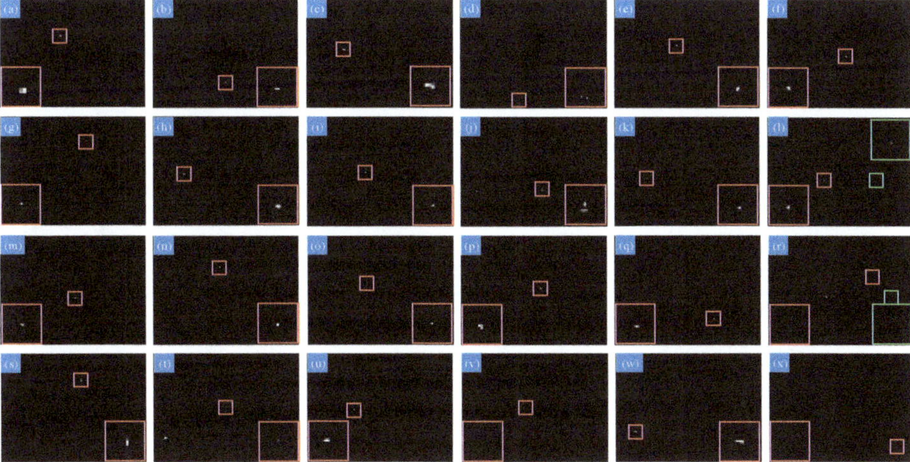

Figure A8. The separated target images of NRAM under 24 scenes.

Appendix B

These are the obtained results of the four optimizing approaches (i.e., IPI, NIPPS, ReWIPI, and NRAM) that are not shown in the main body of this paper. Note that the performances of filtering methods even without noise are not satisfactory; therefore, we didn't take them into account in this part.

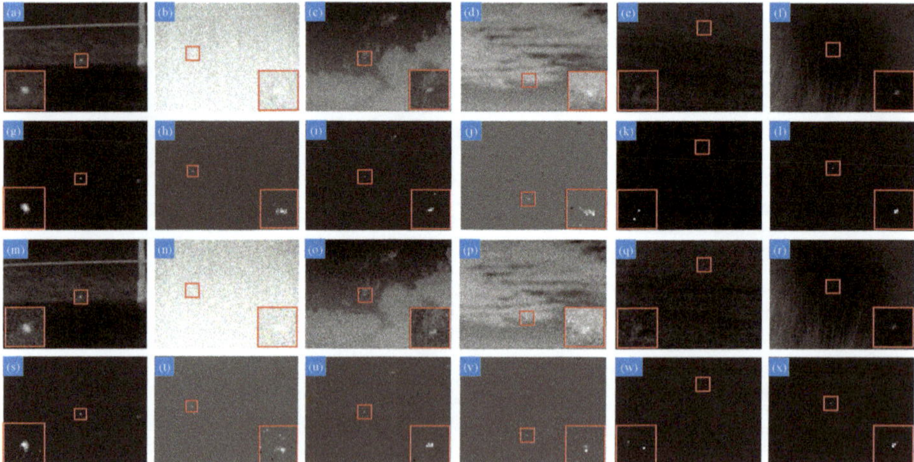

Figure A9. The first and third row show infrared images with additive white Gaussian noise with standard deviation of 10 and 20, and the second and fourth rows are the corresponding detection results by IPI.

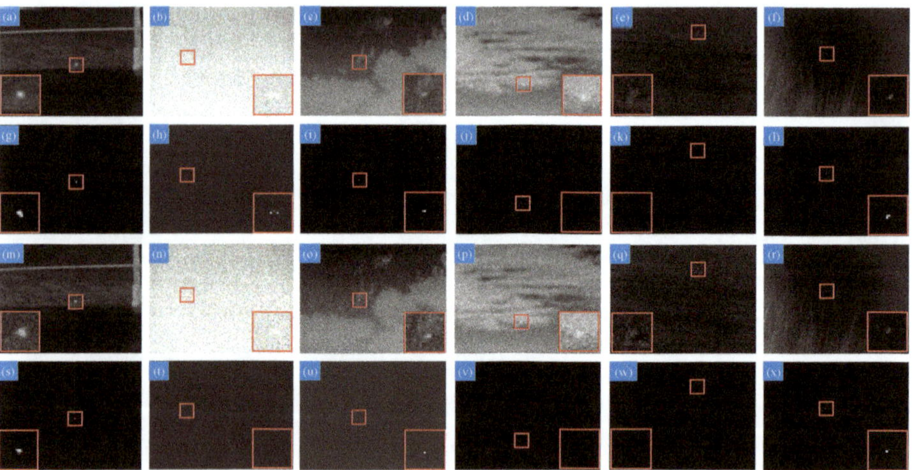

Figure A10. The first and third row show infrared images with additive white Gaussian noise with standard deviation of 10 and 20, and the second and fourth rows are the corresponding detection results by NIPPS.

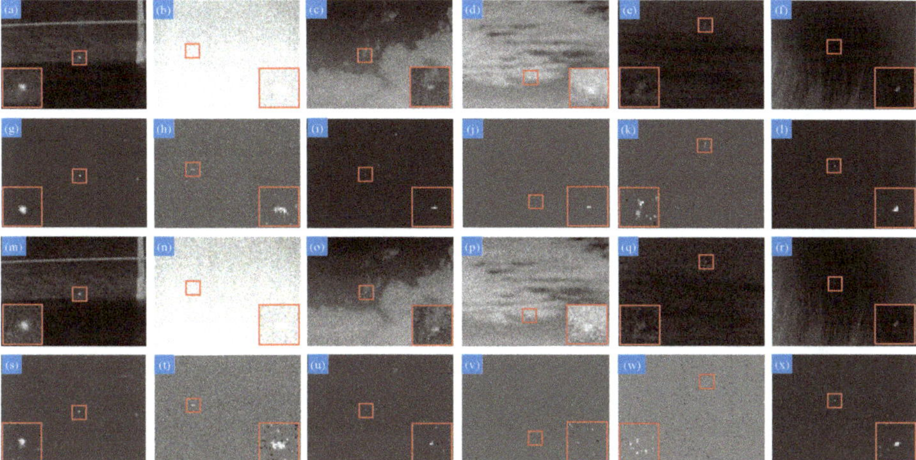

Figure A11. The first and third row show infrared images with additive white Gaussian noise with standard deviation of 10 and 20, and the second and fourth rows are the corresponding detection results by ReWIPI.

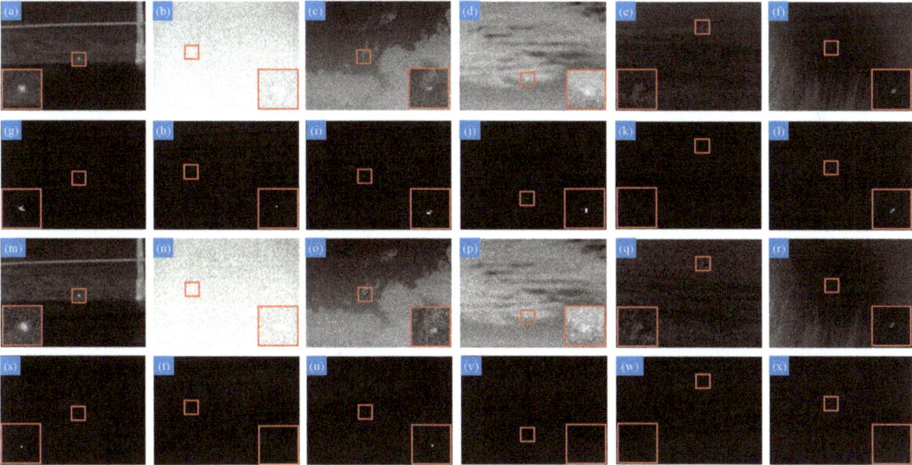

Figure A12. The first and third row show infrared images with additive white Gaussian noise with standard deviation of 10 and 20, and the second and fourth rows are the corresponding detection results by NRAM.

References

1. Wang, B.; Xu, W.H.; Zhao, M.; Wu, H.D. Antivibration pipeline-filtering algorithm for maritime small target detection. *Opt. Eng.* **2014**, *53*. [CrossRef]
2. Reed, I.S.; Gagliardi, R.M.; Stotts, L.B. Optical moving target detection with 3-D matched filtering. *IEEE Trans Aerosp Electron. Syst.* **1988**, *24*, 327–336. [CrossRef]
3. Blostein, S.D.; Richardson, H.S. A sequential detection approach to target tracking. *IEEE Trans. Aerosp. Electron. Syst.* **1994**, *30*, 197–212. [CrossRef]
4. Fan, X.S.; Xu, Z.Y.; Zhang, J.L.; Huang, Y.M.; Peng, Z.M. Infrared Dim and Small Targets Detection Method Based on Local Energy Center of Sequential Image. *Math. Probl. Eng.* **2017**, *2017*. [CrossRef]

5. Peng, Z.M.; Zhang, Q.H.; Wang, J.R.; Zhang, Q.P. Dim target detection based on nonlinear multifeature fusion by Karhunen-Loeve transform. *Opt. Eng.* **2004**, *43*, 2954–2958. [CrossRef]

6. Liu, D.P.; Cao, L.; Li, Z.Z.; Liu, T.M.; Che, P. Infrared Small Target Detection Based on Flux Density and Direction Diversity in Gradient Vector Field. *IEEE J. Sel. Top. Appl. Earth Obs. Remote Sens.* **2018**, *11*, 2528–2554. [CrossRef]

7. Gao, C.Q.; Wang, L.; Xiao, Y.X.; Zhao, Q.; Meng, D.Y. Infrared small-dim target detection based on Markov random field guided noise modeling. *Pattern Recognit.* **2018**, *76*, 463–475. [CrossRef]

8. Fan, X.S.; Xu, Z.Y.; Zhang, J.L.; Huang, Y.M.; Peng, Z.M. Dim small targets detection based on self-adaptive caliber temporal-spatial filtering. *Infrared Phys. Technol.* **2017**, *85*, 465–477. [CrossRef]

9. Wang, Q.; Gao, J.Y.; Yuan, Y. Embedding Structured Contour and Location Prior in Siamesed Fully Convolutional Networks for Road Detection. *IEEE Trans. Intell. Transp. Syst.* **2018**, *19*, 230–241. [CrossRef]

10. He, K.M.; Sun, J.; Tang, X.O. Single Image Haze Removal Using Dark Channel Prior. *IEEE Trans. Pattern Anal. Mach. Intell.* **2011**, *33*, 2341–2353. [CrossRef]

11. Peng, X.; Feng, J.S.; Xiao, S.J.; Yau, W.Y.; Zhou, J.T.; Yang, S.F. Structured AutoEncoders for Subspace Clustering. *IEEE Trans. Image Process.* **2018**, *27*, 5076–5086. [CrossRef] [PubMed]

12. Tom, V.T.; Peli, T.; Leung, M.; Bondaryk, J.E. Morphology-based algorithm for point target detection in infrared backgrounds. In Proceedings of the Signal and Data Processing of Small Targets 1993, Orlando, FL, USA, 12–14 April 1993; pp. 2–12.

13. Deshpande, S.D.; Er, M.H.; Venkateswarlu, R.; Chan, P. Max-mean and max-median filters for detection of small targets. In Proceedings of the Signal and Data Processing of Small Targets 1999, Denver, CO, USA; pp. 74–84.

14. Gu, Y.F.; Wang, C.; Liu, B.X.; Zhang, Y. A Kernel-Based Nonparametric Regression Method for Clutter Removal in Infrared Small-Target Detection Applications. *IEEE Geosci. Remote Sens. Lett.* **2010**, *7*, 469–473. [CrossRef]

15. Hadhoud, M.M.; Thomas, D.W. The two-dimensional adaptive LMS (TDLMS) algorithm. *IEEE Trans. Circuits Syst.* **1988**, *35*, 485–494. [CrossRef]

16. Chen, C.L.P.; Li, H.; Wei, Y.T.; Xia, T.; Tang, Y.Y. A Local Contrast Method for Small Infrared Target Detection. *IEEE Trans. Geosci. Remote Sens.* **2014**, *52*, 574–581. [CrossRef]

17. Han, J.H.; Ma, Y.; Zhou, B.; Fan, F.; Liang, K.; Fang, Y. A Robust Infrared Small Target Detection Algorithm Based on Human Visual System. *IEEE Geosci. Remote Sens. Lett.* **2014**, *11*, 2168–2172. [CrossRef]

18. Dong, X.B.; Huang, X.S.; Zheng, Y.B.; Shen, L.R.; Bai, S.J. Infrared dim and small target detecting and tracking method inspired by Human Visual System. *Infrared Phys. Technol.* **2014**, *62*, 100–109. [CrossRef]

19. Han, J.H.; Liang, K.; Zhou, B.; Zhu, X.Y.; Zhao, J.; Zhao, L.L. Infrared Small Target Detection Utilizing the Multiscale Relative Local Contrast Measure. *IEEE Geosci. Remote Sens. Lett.* **2018**, *15*, 612–616. [CrossRef]

20. Gao, C.Q.; Meng, D.Y.; Yang, Y.; Wang, Y.T.; Zhou, X.F.; Hauptmann, A.G. Infrared Patch-Image Model for Small Target Detection in a Single Image. *IEEE Trans. Image Process.* **2013**, *22*, 4996–5009. [CrossRef] [PubMed]

21. Dai, Y.M.; Wu, Y.Q.; Song, Y.; Guo, J. Non-negative infrared patch-image model: Robust target-background separation via partial sum minimization of singular values. *Infrared Phys. Technol.* **2017**, *81*, 182–194. [CrossRef]

22. Wang, X.Y.; Peng, Z.M.; Zhang, P.; He, Y.M. Infrared Small Target Detection via Nonnegativity-Constrained Variational Mode Decomposition. *IEEE Geosci. Remote Sens. Lett.* **2017**, *14*, 1700–1704. [CrossRef]

23. Wang, X.Y.; Peng, Z.M.; Kong, D.H.; Zhang, P.; He, Y.M. Infrared dim target detection based on total variation regularization and principal component pursuit. *Image Vis. Comput.* **2017**, *63*, 1–9. [CrossRef]

24. He, Y.J.; Li, M.; Zhang, J.L.; An, Q. Small infrared target detection based on low-rank and sparse representation. *Infrared Phys. Technol.* **2015**, *68*, 98–109. [CrossRef]

25. Bai, X.Z.; Zhou, F.Z.; Xie, Y.C.; Jin, T. Modified Top-hat transformation based on contour structuring element to detect infrared small target. In Proceedings of the 3rd IEEE Conference on Industrial Electronics and Applications, Singapore, 3–5 June 2008.

26. Bai, X.Z.; Zhou, F.G. Analysis of new top-hat transformation and the application for infrared dim small target detection. *Pattern Recognit.* **2010**, *43*, 2145–2156. [CrossRef]

27. Bae, T.W.; Zhang, F.; Kweon, I.S. Edge directional 2D LMS filter for infrared small target detection. *Infrared Phys. Technol.* **2012**, *55*, 137–145. [CrossRef]

28. Cao, Y.; Liu, R.M.; Yang, J. Small target detection using two-dimensional least mean square (TDLMS) filter based on neighborhood analysis. *Int. J. Infrared Millimeter Waves* **2008**, *29*, 188–200. [CrossRef]
29. Kim, S.; Lee, J. Scale invariant small target detection by optimizing signal-to-clutter ratio in heterogeneous background for infrared search and track. *Pattern Recognit.* **2012**, *45*, 393–406. [CrossRef]
30. Wang, X.; Lv, G.F.; Xu, L.Z. Infrared dim target detection based on visual attention. *Infrared Phys. Technol.* **2012**, *55*, 513–521. [CrossRef]
31. Wei, Y.T.; You, X.G.; Li, H. Multiscale patch-based contrast measure for small infrared target detection. *Pattern Recognit.* **2016**, *58*, 216–226. [CrossRef]
32. Deng, H.; Sun, X.P.; Liu, M.L.; Ye, C.H.; Zhou, X. Small Infrared Target Detection Based on Weighted Local Difference Measure. *IEEE Trans. Geosci. Remote Sens.* **2016**, *54*, 4204–4214. [CrossRef]
33. Gao, J.; Guo, Y.; Lin, Z.; An, W.; Li, J. Robust Infrared Small Target Detection Using Multiscale Gray and Variance Difference Measures. *IEEE J. Sel. Top. Appl. Earth Obs. Remote Sens.* **2018**. [CrossRef]
34. Li, J.; Duan, L.Y.; Chen, X.W.; Huang, T.J.; Tian, Y.H. Finding the Secret of Image Saliency in the Frequency Domain. *IEEE Trans. Pattern Anal. Mach. Intell.* **2015**, *37*, 2428–2440. [CrossRef] [PubMed]
35. Tang, W.; Zheng, Y.B.; Lu, R.T.; Huang, X.S. A Novel Infrared Dim Small Target Detection Algorithm based on Frequency Domain Saliency. In Proceedings of the 2016 IEEE Advanced Information Management, Communicates, Electronic and Automation Control Conference (IMCEC 2016), Xi'an, China, 3–5 October 2016; pp. 1053–1057.
36. Lin, Z.; Chen, M.; Ma, Y. The Augmented Lagrange Multiplier Method for Exact Recovery of Corrupted Low-Rank Matrices. *arXiv*, 2010; arXiv:1009.5055v3.
37. Dai, Y.M.; Wu, Y.Q.; Song, Y. Infrared small target and background separation via column-wise weighted robust principal component analysis. *Infrared Phys. Technol.* **2016**, *77*, 421–430. [CrossRef]
38. Guo, J.; Wu, Y.Q.; Dai, Y.M. Small target detection based on reweighted infrared patch-image model. *IET Image Process.* **2018**, *12*, 70–79. [CrossRef]
39. Zhang, L.; Peng, L.; Zhang, T.; Cao, S.; Peng, Z. Infrared Small Target Detection via Non-Convex Rank Approximation Minimization Joint l2, 1 Norm. *Remote Sens.* **2018**, *10*, 1821. [CrossRef]
40. Liu, D.P.; Li, Z.Z.; Liu, B.; Chen, W.H.; Liu, T.M.; Cao, L. Infrared small target detection in heavy sky scene clutter based on sparse representation. *Infrared Phys. Technol.* **2017**, *85*, 13–31. [CrossRef]
41. Wang, X.Y.; Peng, Z.M.; Kong, D.H.; He, Y.M. Infrared Dim and Small Target Detection Based on Stable Multisubspace Learning in Heterogeneous Scene. *IEEE Trans. Geosci. Remote Sens.* **2017**, *55*, 5481–5493. [CrossRef]
42. Dai, Y.M.; Wu, Y.Q. Reweighted Infrared Patch-Tensor Model With Both Nonlocal and Local Priors for Single-Frame Small Target Detection. *IEEE J. Sel. Top. Appl. Earth Obs. Remote Sens.* **2017**, *10*, 3752–3767. [CrossRef]
43. Goldfarb, D.; Qin, Z.W. Robust Low-Rank Tensor Recovery: Models and Algorithms. *Siam J. Matrix Anal. A* **2014**, *35*, 225–253. [CrossRef]
44. Liu, J.; Musialski, P.; Wonka, P.; Ye, J.P. Tensor Completion for Estimating Missing Values in Visual Data. *IEEE Trans. Pattern Anal. Mach. Intell.* **2013**, *35*, 208–220. [CrossRef]
45. Romera-Paredes, B.; Pontil, M. A New Convex Relaxation for Tensor Completion. In Proceedings of the Advances in Neural Information Processing Systems, Harrahs and Harveys, Lake Tahoe, 5–10 December 2013; pp. 2967–2975.
46. Jiang, T.X.; Huang, T.Z.; Zhao, X.L.; Deng, L.J. A novel nonconvex approach to recover the low-tubal-rank tensor data: when t-SVD meets PSSV. *arXiv*, 2017; arXiv:1712.05870.
47. Kilmer, M.E.; Martin, C.D. Factorization strategies for third-order tensors. *Linear Algebra Appl* **2011**, *435*, 641–658. [CrossRef]
48. Lu, C.Y.; Feng, J.S.; Chen, Y.D.; Liu, W.; Lin, Z.C.; Yan, S.C. Tensor Robust Principal Component Analysis: Exact Recovery of Corrupted Low-Rank Tensors via Convex Optimization. In Proceedings of the 2016 IEEE Conference on Computer Vision and Pattern Recognition (CVPR), Las Vegas, NV, USA, 27–30 June 2016; pp. 5249–5257. [CrossRef]
49. Boyd, S.; Parikh, N.; Chu, E.; Peleato, B.; Eckstein, J. Distributed optimization and statistical learning via the alternating direction method of multipliers. *Found. Trends®Mach. Learn.* **2011**, *3*, 1–122. [CrossRef]

50. Oh, T.H.; Tai, Y.W.; Bazin, J.C.; Kim, H.; Kweon, I.S. Partial Sum Minimization of Singular Values in Robust PCA: Algorithm and Applications. *IEEE Trans. Pattern Anal. Mach. Intell.* **2016**, *38*, 744–758. [CrossRef] [PubMed]

51. Zhang, Z.M.; Ely, G.; Aeron, S.; Hao, N.; Kilmer, M. Novel methods for multilinear data completion and de-noising based on tensor-SVD. *IEEE Conf. Comput. Vis. Pattern Recognit. (CVPR)* **2014**, 3842–3849. [CrossRef]

52. Lu, C.; Feng, J.; Chen, Y.; Liu, W.; Lin, Z.; Yan, S. Tensor Robust Principal Component Analysis with A New Tensor Nuclear Norm. *arXiv*, 2018; arXiv:1804.03728.

53. Hale, E.T.; Yin, W.T.; Zhang, Y. FIXED-POINT CONTINUATION FOR l(1)-MINIMIZATION: METHODOLOGY AND CONVERGENCE. *SIAM J. Optim.* **2008**, *19*, 1107–1130. [CrossRef]

54. Turk, M.; Pentland, A. Eigenfaces for Recognition. *J. Cogn. Neurosci.* **1991**, *3*, 71–86. [CrossRef] [PubMed]

55. Huang, Z.; Zhu, H.; Zhou, J.T.; Peng, X. Multiple Marginal Fisher Analysis. *IEEE Trans. Ind. Electron.* **2018**, 99. [CrossRef]

56. Chen, Y.W.; Song, B.; Wang, D.J.; Guo, L.H. An effective infrared small target detection method based on the human visual attention. *Infrared Phys. Technol.* **2018**, *95*, 128–135. [CrossRef]

57. Liu, J.; He, Z.; Chen, Z.; Shao, L. Tiny and Dim Infrared Target Detection Based on Weighted Local Contrast. *IEEE Geosci. Remote Sens. Lett.* **2018**, *15*, 1780–1784. [CrossRef]

58. Deng, H.; Sun, X.P.; Liu, M.L.; Ye, C.H.; Zhou, X. Infrared Small-Target Detection Using Multiscale Gray Difference Weighted Image Entropy. *IEEE Trans. Aerosp. Electron. Syst.* **2016**, *52*, 60–72. [CrossRef]

59. Bai, X.Z.; Bi, Y.G. Derivative Entropy-Based Contrast Measure for Infrared Small-Target Detection. *IEEE Trans. Geosci. Remote Sens.* **2018**, *56*, 2452–2466. [CrossRef]

60. Guo, Y.; Lin, Z.; An, W. Infrared Small Target Detection Using Multiscale Gray and Variance Difference. Proceedings of Chinese Conference on Pattern Recognition and Computer Vision (PRCV), Guangzhou, China, 19 November 2018; pp. 53–64.

61. Bigun, J.; Granlund, G.H.; Wiklund, J. Multidimensional Orientation Estimation with Applications To Texture Analysis and Optical-Flow. *IEEE Trans. Pattern Anal. Mach. Intell.* **1991**, *13*, 775–790. [CrossRef]

62. Wang, H.; Yang, F.; Zhang, C.; Ren, M. Infrared Small Target Detection Based on Patch Image Model with Local and Global Analysis. *Int. J. Image Graph.* **2018**, *18*, 1850002. [CrossRef]

63. Brown, M.; Szeliski, R.; Winder, S. Multi-image matching using multi-scale oriented patches. *IEEE Comput. Soc. Conf.* **2005**, *1*, 510–517.

64. Carroll, J.D.; Chang, J.J. Analysis of individual differences in multidimensional scaling via an N-way generalization of "Eckart-Young" decomposition. *Psychometrika* **1970**, *35*, 283–319. [CrossRef]

65. Tucker, L.R. Some mathematical notes on three-mode factor analysis. *Psychometrika* **1966**, *31*, 279–311. [CrossRef] [PubMed]

66. Hillar, C.J.; Lim, L.H. Most Tensor Problems Are NP-Hard. *J. ACM* **2013**, *60*. [CrossRef]

67. Candes, E.J.; Wakin, M.B.; Boyd, S.P. Enhancing Sparsity by Reweighted l(1) Minimization. *J. Fourier Anal. Appl.* **2008**, *14*, 877–905. [CrossRef]

68. Gu, S.H.; Xie, Q.; Meng, D.Y.; Zuo, W.M.; Feng, X.C.; Zhang, L. Weighted Nuclear Norm Minimization and Its Applications to Low Level Vision. *Int. J. Comput. Vis.* **2017**, *121*, 183–208. [CrossRef]

69. Lu, C.Y.; Tang, J.H.; Yan, S.C.; Lin, Z.C. Nonconvex Nonsmooth Low Rank Minimization via Iteratively Reweighted Nuclear Norm. *IEEE Trans. Image Process.* **2016**, *25*, 829–839. [CrossRef]

70. Peng, Y.G.; Suo, J.L.; Dai, Q.H.; Xu, W.L. Reweighted Low-Rank Matrix Recovery and its Application in Image Restoration. *IEEE Trans. Cybern.* **2014**, *44*, 2418–2430. [CrossRef]

Article

Infrared Small-Faint Target Detection Using Non-i.i.d. Mixture of Gaussians and Flux Density

Yang Sun, Jungang Yang *, Miao Li and Wei An

The College of Electronic Science and Technology, National University of Defense Technology, Changsha 410073, China; sunyang_kd@163.com (Y.S.); lm8866@nudt.edu.cn (M.L.); anwei@nudt.edu.cn (W.A.)
* Correspondence:yangjungang@nudt.edu.cn; Tel.: +86-1857-061-6407

Received: 25 October 2019; Accepted: 25 November 2019; Published: 28 November 2019

Abstract: The robustness of infrared small-faint target detection methods to noisy situations has been a challenging and meaningful research spot. The targets are usually spatially small due to the far observation distance. Considering the underlying assumption of noise distribution in the existing methods is impractical; a state-of-the-art method has been developed to dig out valuable information in the temporal domain and separate small-faint targets from background noise. However, there are still two drawbacks: (1) The mixture of Gaussians (MoG) model assumes that noise of different frames satisfies independent and identical distribution (i.i.d.); (2) the assumption of Markov random field (MRF) would fail in more complex noise scenarios. In real scenarios, the noise is actually more complicated than the MoG model. To address this problem, a method using the non-i.i.d. mixture of Gaussians (NMoG) with modified flux density (MFD) is proposed in this paper. We firstly construct a novel data structure containing spatial and temporal information with an infrared image sequence. Then, we use an NMoG model to describe the noise, which can be separated with the background via the variational Bayes algorithm. Finally, we can select the component containing true targets through the obvious difference of target and noise in an MFD maple. Extensive experiments demonstrate that the proposed method performs better in complicated noisy scenarios than the competitive approaches.

Keywords: infrared small-faint target detection; non-independent and identical distribution (non-i.i.d.) mixture of Gaussians; flux density; variational Bayesian

---

1. Introduction

Distant and faint target detection is of great importance to infrared systems, as anti-missile techniques and early-warning systems. Due to the unique characteristic of these military tasks, the targets need to be detected accurately as early as possible in the infrared search and track systems to provide ample time for deployment and striking back. However, the target usually occupies only a few pixels and lacks texture information due to the very far observation distance. The backgrounds are very complex, including sky background and sea–sky background, which means the acquired infrared images are usually contaminated by a clutter background and a varying noise. The contrast between targets, background and the varying noise might be very poor. The low signal-to-clutter ratio (SCR) and signal-to-noise ratio (SNR) make the infrared targets very faint. Therefore, robust infrared small and faint target detection technique remains a valuable research hotspot [1–3].

To achieve a satisfying target detection performance, many approaches have been proposed for different scenarios, including two types: Track-before-detection (TBD) approaches [4,5] and detection-before-track (DBT) approaches [6–8]. TBD approaches have good detection performance for targets with continuous track motion, such as 3D matched filters [9] and its improved versions [10,11]. DBT approaches focus on suppressing the clutter background while enhancing the target in single frame, and are more efficient than TBD approaches. TBD approaches are widely used in practical engineering. At present, the common types of DBT methods are filtering, human visual system (HVS) and multi-feature based approaches. Filtering methods analyze spatial continuity of an input image, and the target is modeled as a break point, such as max-median filter [12], top-hat filter [13] and 2D least mean square (TDLMS) filter [14]. HVS based methods [15–17] assume that there is a significant contrast between background and target regions. Multi-feature based methods [18–20] represent the target characteristics and background region with features used to train the classifiers.

Moreover, the low-rank and sparse component recovery based approach, as a subdiscipline of the low-rank representation (LRR) [21], has become very popular in recent years. In this approach, the background regions are assumed to change gradually, and a special low-rank data structure can be constructed with the original images, such as a 2D matrix and a 3D tensor. With the recovery of the low-rank background, the dim target can be separated from the original image. Gao proposed an infrared image-patch (IPI) model [22], which constructs a low-rank matrix by sliding window. The IPI model uses vanilla nuclear norm minimization (NNM) [23] and l_1 [24] to regularize the background and the target, respectively. The performance of NNM in a low-rank component estimation problem would degrade in a noisy scenario. The solution for this problem is to replace NNM with a more suitable regularizer. Thus, Dai proposed a weighted IPI approach [25] and a non-negative IPI approach [26], and Guo proposed a reweighted WIPI model (ReWIPI) based on weighted nuclear norm minimization (WNNM) [27]. In the view of the dimension of data, Dai proposed a reweighted infrared patch-tensor (RIPT) method to generalize the low-rank matrix to low-rank tensor for mining more spatial information [28]. However, the RIPT method unfolds the background patch tensor as three matrices and regularizes it via the sum of nuclear norms (SNN) [29], which is suboptimal and inefficient. To remedy this issue, Sun proposed a weighted tensor nuclear norm with IPT (WNRIPT) method [30].

However, most of the existing low-rank component recovery based approaches [22,25–28,30] only use the Frobenius loss term [31] to constrain the noise, which models the noise as an independent and identically distributed (i.i.d.) Gaussian distribution. In practical applications, the infrared images usually include complex instrumental noise that degrades the performance for target detection. The complex noise degrades the performance of the target detection severely. A robust method, capable of distinguishing different kinds of noise, is needed.

To this end, a state-of-the-art method [32] digs out valuable information in time domain and uses a mixture of Gaussian (MoG) noise models [33] to model the target component and noise component together. The MoG model characterizes each pixel in the image and updates the mixed Gaussian model after the new image is acquired. It matches each pixel in the current image with the MoG model, and the matched pixels are classified into background regions [34,35]. Finally, the Markov random field (MRF) method [34] is used to detect the target. However, the noise distribution in different frames is modeled as i.i.d. MoG distributions substantially in [32], which is not suitable for complex noisy scenarios. In addition, the MRF model does not provide a robust noise estimate in complex scenarios, since its performance is based on the assumption that the noise component does not arise in the neighborhood region of the targets. However, the noise permeates through the whole image, including the target.

We propose a small and faint target detection approach based on a non-i.i.d. MoG (NMoG) model [36] and modified flux density (MFD) maple [37]. The noise distributions in different frames (sequences of images) is assumed to follow non-i.i.d. for improving the robustness in real scenarios. The target is considered as a kind of noise extracted from the background via an NMoG low-rank matrix factorization (NMoG-LRMF) model, solved by a variational Bayes (VB) algorithm. In a second step, the MFD maple [37] method is used to distinguish the true target from the noise, accounting for the fact that target flux density differs from the noise in infrared gradient vector field.

This paper is organized as follows. The proposed method is described in Section 2. Section 3 provides the experimental results to validate the effectiveness of the proposed method. Finally, we conclude our work in Section 4.

2. The Proposed Model

2.1. Spatio-Temporal Patch Model

Given an infrared image sequence, we can get a 3D cube patch tensor by storing each frame into its slice. We vectorize each slice and get a 2D matrix. The procedure is given in Figure 1. Note that it is possible to reconstruct the image sequence from the processed 2D matrix via inverse operation. Assume an infrared image sequence $f_1, f_2, \cdots, f_P \in \mathbb{R}^{m \times n}$ transformed into a matrix F with size of $N \times P$, where $N = m \times n$ and P denote the spatial and temporal dimensions. We divide F into background component B and noise E, described as:

$$F = B + E, \tag{1}$$

and the small-faint target component T is considered as a sparse noise component in E [32].

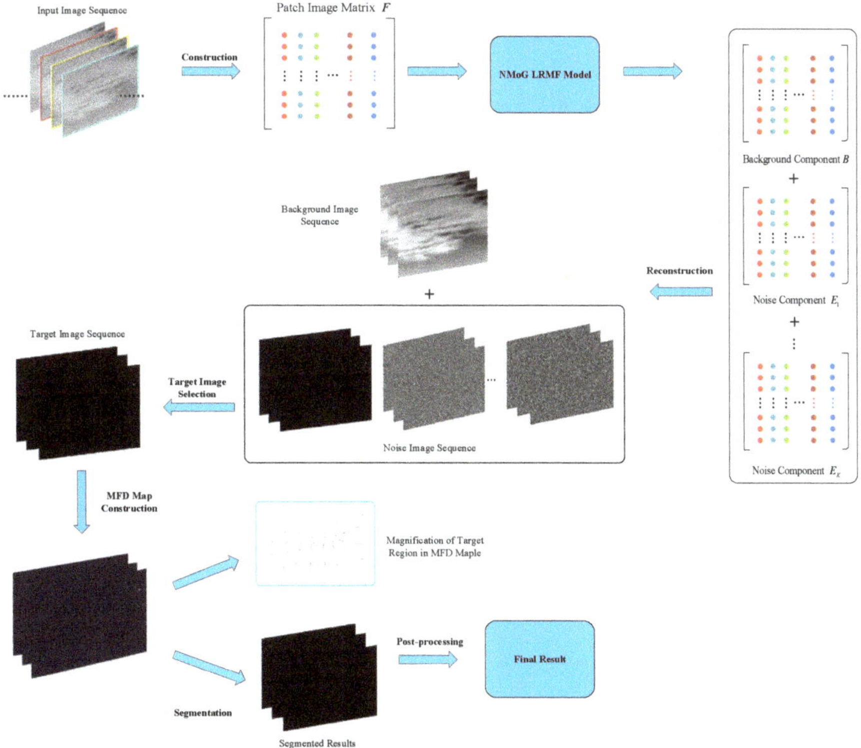

Figure 1. The framework of the proposed method.

2.2. Background Component

In low-rank recovery based methods, background regions are assumed to vary slowly, and there are a lot of repeated regions. The low-rank matrix B [32] is modeled as follows:

$$B = UV^T = \sum_{l=1}^{R} u_{.l} v_{.l}^T, \tag{2}$$

where $U \in \mathbb{R}^{N \times R}$ and $V \in \mathbb{R}^{P \times R}$, and their l-th columns are represented as $u_{.l}$ and $v_{.l}$. R is the initial rank of B. The intrinsic low-rank nature of B is guaranteed by assuming $u_{.l}$ and $v_{.l}$ generated according to a Gaussian distribution:

$$\mathbf{u}_{.l} \sim N\left(\mathbf{u}_{.l} \middle| 0, \gamma_l^{-1} I_N\right), \mathbf{v}_{.l} \sim N\left(\mathbf{v}_{.l} \middle| 0, \gamma_l^{-1} I_P\right), \tag{3}$$

where I_N (I_P) is the $N \times N$ ($P \times P$) identity matrix. γ_l denotes the precision of $u_{.l}$ and $v_{.l}$ that satisfies:

$$\gamma_l \sim \text{Gam}\left(\gamma_l \middle| \xi_0, \delta_0\right), \tag{4}$$

where $\text{Gam}\left(\gamma_l \middle| \xi_0, \delta_0\right)$ represents a gamma distribution, and ξ_0, δ_0 are scales. The low-rank component can be estimated accurately by this model [38].

2.3. Noise Component

In [32], the noise of different frames are assumed to be i.i.d., which is not practical in real and complex scenarios. Thus, we use the NMoG model [36] to model the noise distributions in different frames, namely noise distribution of images in different frames are nonidentical. The ij-th element of the noise E can be divided into K components as below:

$$e_{ij} \sim \sum_{k=1}^{K} \pi_{jk} N\left(e_{ij} \middle| \mu_{jk}, \tau_{jk}^{-1}\right), \tag{5}$$

where π_{jk} denotes the mixing proportion that is non-negative, and $\sum_{k=1}^{K} \pi_{jk} = 1$. μ_{jk} and τ_{jk} denote mean and precision, respectively. Instead of setting the MoG parameters, i.e., π_{jk}, μ_{jk} and τ_{jk}, as unchanging value for k-th Gaussian component, we vary them in different frames. Equation (5) can be equivalently expressed as a two-level generative model by introducing the indicator variables z_{ij}. z_{ij} is the hidden variable generated from Multinomial distribution with parameter π_j:

$$e_{ij} \sim \prod_{k=1}^{K} N\left(e_{ij} \middle| \mu_{jk}, \tau_{jk}^{-1}\right)^{z_{ijk}}$$
$$z_{ij} \sim \text{Multinominal}\left(z_{ij} \middle| \pi_j\right) \tag{6}$$

where $z_{ij} = (z_{ij1}, \ldots, z_{ijK}) \in \{0,1\}^K$, $\sum_{k=1}^{K} z_{ijk} = 1$. Multinomial( ) represents the multinomial Dirichlet distribution. The conjugate priors of μ_{jk}, τ_{jk} and the mixing proportions $\pi_j = [\pi_{j1}, \ldots, \pi_{jK}]$ are also defined for completing the Bayesian model:

$$\mu_{jk}, \tau_{jk} \sim N\left(\mu_{jk} \middle| m_0, \left(\beta_0 \tau_{jk}\right)^{-1}\right) \text{Gam}\left(\tau_{jk} \middle| c_0, d\right)$$
$$d \sim \text{Gam}\left(d \middle| \eta_0, \lambda_0\right) \tag{7}$$
$$\pi_j \sim \text{Dir}\left(\pi_j \middle| \alpha_0\right)$$

where β_0, m_0, c_0, d are the hyper-parameters, and d satisfies Gam distribution with hyper-parameters η_0, λ_0. Dir(.) is a Dirichlet distribution parameterized by $\alpha_0 = (\alpha_{01}, \ldots, \alpha_{0K})$. Then, the noise component can be modeled by Equations (6) and (7).

Combining Equations (2)–(7) together, Bayes' theorem is used to estimate from F the values of all parameters:

$$p\left(U, V, \mathcal{Z}, \mu, \tau, \pi, \gamma, d \mid F\right) \tag{8}$$

where $\mathcal{Z} = \{z_{ij}\}_{N \times P}, \mu = \{\mu_{jk}\}_{B \times K}, \tau = \{\tau_{jk}\}_{B \times K}, \pi = (\pi_1, \ldots, \pi_P)$ and $\gamma = (\gamma_1, \ldots, \gamma_R)$.

2.4. Variational Inference

In this section, the posterior of parametric model Equation (8) is inferred by the VB approach [39]. VB obtains the objective parameters x finding the minimum Kullback–Leibler (KL) divergence between the approximated distribution $q(x)$ and the actual distribution $p(x \mid D)$ with the known observation D, which can be formulated as below:

$$q^*(x) = \min_{q \in \Omega} KL\left(q(x) \,\|\, p(x \mid D)\right), \tag{9}$$

where Ω is the constrained probability densities for obtaining the feasible solution. We can factorize $q(\theta)$ as $q(\theta) = \prod_i q_i(\theta_i)$ by mean field theory, and the posterior distribution Equation (8) can be approximated with the following form:

$$p\left(U, V, \mathcal{Z}, \mu, \pi, \tau, \gamma, d\right) = \prod_i q\left(u_{i\cdot}\right) \prod_j q\left(u_{j\cdot}\right)$$
$$\prod_{ij} q\left(z_{ij}\right) \times \prod_j q\left(\mu_j, \tau_j\right) q\left(\pi_j\right) \prod_l q\left(\gamma_l\right) q\left(d\right). \tag{10}$$

2.4.1. Estimation of Noise Component

For the noise component in the j-th frame, we need to estimate four parameters, $\mu_j, \tau_j, \mathcal{Z}$ and π_j. Firstly we update μ_j and τ_j in the following way:

$$q^*\left(\mu_j, \tau_j\right) = \prod_k N\left(\mu_{jk} \,\middle|\, m_{jk}, \frac{1}{\beta_{jk} \tau_{jk}}\right) \mathrm{Gam}\left(\tau_{jk} \,\middle|\, c_{jk}, d_{jk}\right), \tag{11}$$

where

$$m_{jk} = \frac{1}{\beta_{jk}} \left\{ m_0 \beta_0 + \sum_i \left\langle z_{ijk} \right\rangle \left(f_{ij} - \left\langle u_{i\cdot} \right\rangle \left\langle v_{j\cdot} \right\rangle^T \right) \right\},$$

$$\beta_{jk} = \beta_0 + \sum_i \left\langle z_{ijk} \right\rangle, c_{jk} = c_0 + \frac{1}{2} \sum_i \left\langle z_{ijk} \right\rangle,$$

$$d_{jk} = \langle d \rangle + \frac{1}{2} \left\{ \sum_i \left\langle z_{ijk} \right\rangle \left\langle \left(f_{ij} - u_{i\cdot} v_{j\cdot}^T \right)^2 \right\rangle + \beta_0 m_0^2 \right. \tag{12}$$

$$\left. - \frac{1}{\beta_{jk}} \left(\sum_i \left\langle z_{ijk} \right\rangle \left(f_{ij} - \left\langle u_{i\cdot} \right\rangle \left\langle v_{j\cdot} \right\rangle^T \right) + \beta_0 m_0 \right)^2 \right\},$$

where f_{ij} denotes the ij-th element of the F. The variables z_{ij} can be derived in closed form as below:

$$q\left(z_{ij}\right) = \prod_k r_{ijk}^{z_{ijk}} \tag{13}$$

where

$$r_{ijk} = \frac{\rho_{ijk}}{\sum_k \rho_{ijk}}, \tag{14}$$

$$\ln\rho_{ijk} = \left\langle \ln\pi_{jk} \right\rangle - \frac{1}{2}\ln 2\pi + \frac{1}{2}\left\langle \ln\tau_{jk} \right\rangle$$
$$- \frac{1}{2}\left\langle \tau_{jk}\left(f_{ij} - \mu_{jk} - \mu_{i\cdot}v_{j\cdot}^{T}\right)^{2} \right\rangle. \tag{15}$$

Finally, we update π_j by:

$$q\left(\pi_j\right) = \prod_k \pi_{jk}^{\alpha_{jk}-1}, \tag{16}$$

where $\alpha_{jk} = \alpha_0 + \sum_i \left\langle z_{ijk} \right\rangle$, and the hyper-parameter d is updated by the following equation:

$$q\left(d\right) = \mathrm{Gam}\left(d\,|\,\eta,\lambda\right), \tag{17}$$

where $\eta = \eta_0 + c_0 KP$ and $\lambda = \lambda_0 + \sum_{j,k}\left\langle \tau_{jk} \right\rangle$.

2.4.2. Estimation of Background Component

For the background component, we need to estimate three parameters, including U, V and γ. $u_{i\cdot}$ $(i = 1, \ldots, N)$ can be estimated as follows:

$$q\left(u_{i\cdot}\right) = N\left(u_{i\cdot}\,|\,\mu_{u_{i\cdot}}, \Sigma_{u_{i\cdot}}\right), \tag{18}$$

where

$$\mu_{u_{i\cdot}} = \left\{ \sum_{j,k}\left\langle z_{ijk} \right\rangle \left\langle \tau_{jk} \right\rangle \left(f_{ij} - \left\langle \mu_{jk} \right\rangle\right)\left\langle v_{j\cdot} \right\rangle \right\}\Sigma_{u_{i\cdot}},$$
$$\Sigma_{u_{i\cdot}} = \left\{ \sum_{j,k}\left\langle z_{ijk} \right\rangle \left\langle \tau_{jk} \right\rangle \left\langle v_{j\cdot}^{T}v_{j\cdot} \right\rangle + \left\langle \Gamma \right\rangle \right\}^{-1}.$$

Similarly, $v_{j\cdot}$ $(j = 1, \ldots, P)$ is estimated by:

$$q\left(v_{j\cdot}\right) = N\left(v_{j\cdot}\,\Big|\,\mu_{v_{j\cdot}}, \Sigma_{v_{j\cdot}}\right), \tag{19}$$

where

$$\mu_{v_{j\cdot}} = \left\{ \sum_{j,k}\left\langle z_{ijk} \right\rangle \left\langle \tau_{jk} \right\rangle \left(f_{ij} - \left\langle \mu_{jk} \right\rangle\right)\left\langle u_{j\cdot} \right\rangle \right\}\Sigma_{v_{j\cdot}},$$
$$\Sigma_{v_{j\cdot}} = \left\{ \sum_{j,k}\left\langle z_{ijk} \right\rangle \left\langle \tau_{jk} \right\rangle \left\langle u_{j\cdot}^{T}u_{j\cdot} \right\rangle + \left\langle \Gamma \right\rangle \right\}^{-1}.$$

$\Gamma = \mathrm{diag}\left(\left\langle \gamma \right\rangle\right)$, γ_l is a decisive factor for guaranteeing low-rank property of B by removing the corresponding rows when its value is very large [38], which can be estimated by:

$$q\left(\gamma_l\right) = \mathrm{Gam}\left(\gamma_r\,|\,\xi_l,\delta_l\right), \tag{20}$$

where

$$\xi_l = \xi_0 + \tfrac{1}{2}\left(m + n\right),$$
$$\delta_l = \delta_0 + \tfrac{1}{2}\sum_i \left\langle u_{il}^2 \right\rangle + \tfrac{1}{2}\sum_j \left\langle v_{jl}^2 \right\rangle.$$

In the following experiments, we set $m_0 = 0$, and $\alpha_0, \beta_0, c_0, d_0, \eta_0, \lambda_0, \xi_0, \delta_0$ are initialized with 10^{-6} [36].

2.5. Target Extraction

In this section, we firstly select the noise component containing small-faint target. Then, the MFD method [37] is used to extract the target from the noise.

2.5.1. Selecting Noise Component Containing Target

We obtain the noise component E separating it from the background component, and we can divide it into K components $E^1, \ldots, E^K$ according to the maximum probability criteria [32]:

$$e^m_{ij} = \begin{cases} e_{ij}, & \text{if } m = \arg \max_{k=1,\ldots,K} \left(r_{ijk} \right) \\ 0, & \text{else} \end{cases}.$$
(21)

The K components are restored to sequences $\bar{E}^1, \ldots, \bar{E}^K$ by the aforementioned method in Section 2.1. Note that the intensity of the true target is quite different from the noise. Instead of using variance guided method in [32], we calculate the difference between the minimum and maximum of intensity and select the largest one $\bar{E}^i$ as the component containing target, which can be described as follows:

$$i = \arg \max_{k=1,\ldots,K} \left(\max \left(\bar{E}^k \right) - \min \left(\bar{E}^k \right) \right)$$
(22)

The following experimental results demonstrate that this method is effective.

2.5.2. Extracting Target by MFD

Figure 2 gives the results of a representative infrared noisy image using NMoG method with $K = 3$, and subfigure (c) is the slice containing the true target. It is observed from Figure 2c that the restored slice containing true target is still contaminated by pixel noise. Thus, we use the MFD method [37] to wipe out the noise and enhance the target.

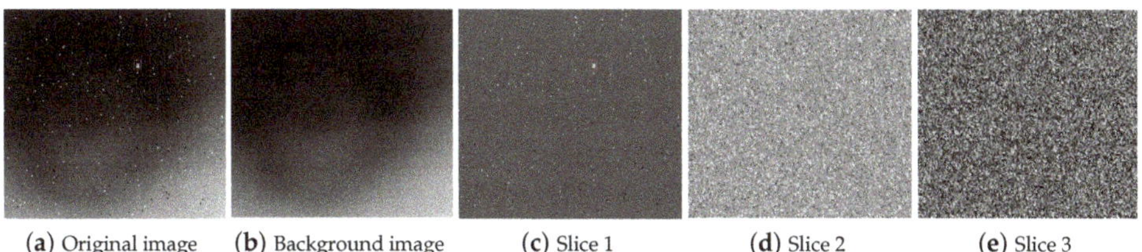

(**a**) Original image   (**b**) Background image   (**c**) Slice 1   (**d**) Slice 2   (**e**) Slice 3

Figure 2. The results of NMoG method with $K = 3$.

The noise component E containing the target is firstly transformed into a gradient vector field by:

$$I(x,y) = \begin{bmatrix} e'_x(x,y) \\ e'_y(x,y) \end{bmatrix}$$
$$e'_x(x,y) = \frac{e(x+1,y) - e(x-1,y)}{2}$$
$$e'_y(x,y) = \frac{e(x,y+1) - e(x,y-1)}{2}$$
(23)

where $e(x,y)$ denotes the value of E at location (x,y), $e'_x(x,y)$ and $e'_y(x,y)$ are the gradient value in the x-direction and y-direction.

From Figure 3b,d, it can be observed that both the true target and bright noise residuals are a sink in gradient vector field. But the gradient vectors of noise pixel focus on 4 directions, and MFD method can compute the flux density of each pixel after removing its four largest gradient vectors, which is defined as follows [37]:

$$\mathrm{MFD}_s(x,y) = \sum_{(x',y') \in O'(x,y,s)} \frac{I(x',y') \cdot \vec{n}_o(x'-x, y'-y)}{8s - 4}$$
(24)

where MFD_s is s-scale MFD, s denotes the scale variable, O' denotes the subset of O, which removes four pixels containing the four largest gradient vectors. Note that the number of pixels on the curve is $8s - 4$. O represents the neighborhood region as:

$$O\left(x,y,s\right) = \left\{\left(x',y'\right) \middle| \max\left(\left|x'-x\right|,\left|y'-y\right|\right) = s\right\} \tag{25}$$

and the norm vector on the boundary point $\vec{n}_o\left(x,y\right)$ is defined as follows:

$$\vec{n}_o\left(x,y\right) = \begin{bmatrix} n_{ox}\left(x,y\right) \\ n_{oy}\left(x,y\right) \end{bmatrix}$$
$$n_{ox}\left(x,y\right) = \begin{cases} -1, & x = k \\ 1, & x = -k \\ 0, & \text{else} \end{cases} \tag{26}$$
$$n_{oy}\left(x,y\right) = \begin{cases} -1, & y = k \\ 1, & y = -k \\ 0, & \text{else} \end{cases}$$

where $n_{ox}\left(x,y\right)$ and $n_{oy}\left(x,y\right)$ are the value in the x-direction and y-direction.

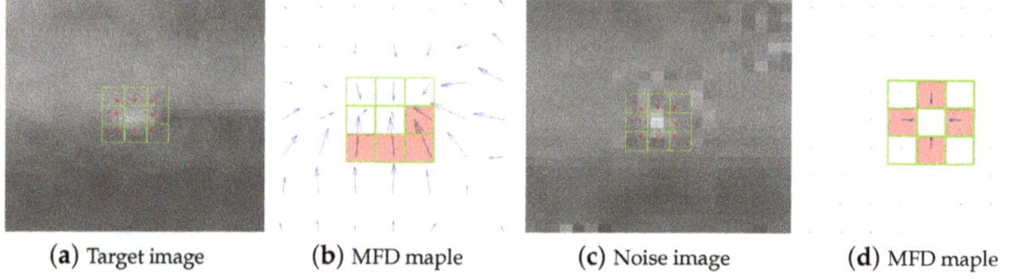

(**a**) Target image   (**b**) MFD maple   (**c**) Noise image   (**d**) MFD maple

Figure 3. Modified flux density (MFD) maple. (**a,c**) Boundaries and norm vectors for flux density calculation of the target and noise pixel. (**b,d**) The corresponding flux density of (**a,c**); the details are (**b**) scale = 1, flux density = 24.38; (**d**) scale = 1, flux density = −0.29.

Figure 3 shows that noisy pixels are wiped out according to their MFD value. This is because the MFD value of the noisy pixels is much smaller than that of the real target, which is usually a negative element. Following this property, the corresponding noise pixels are wiped out in the target image. Thus, we obtain an initial result by the following equation:

$$T\left(x,y\right) = \bar{E}^i\left(x,y\right) * \text{MFD}_s\left(x,y\right)_+ \tag{27}$$

where $T\left(x,y\right)$ denotes the initial target image, $\text{MFD}_s\left(x,y\right)_+$ is the result by setting the positive elements and negative elements in the original MFD maple to 1 and 0, respectively. Finally, we use an adaptive threshold to further separate the target [22], which is described as below:

$$T = \mu + k\sigma \tag{28}$$

where μ and σ are the mean value and standard deviation of the small target image. k is a empirical value, and we set it as 0.05 in our experiment. The framework of our method is shown in Figure 1, and the detection procedure is given in Algorithm 1.

Algorithm 1

Input: Infrared image sequence $f_1, f_2, \cdots, f_P \in \mathbb{R}^{m \times n}$.
Initialize: Set parameters $(m_0, \beta_0, c_0, d_0, \eta_0, \lambda_0) = 10^{-6}$ in noise prior. Low-rank background component $\mathbf{U}_0, \mathbf{V}_0$ and α_0 parameters in the model prior (ξ_0, δ_0), scale parameter $s = 1$ in MFD method, iteration number $t = 1$.
Step 1: Construct the spatio-temporal patch image $\mathbf{F}$ with the input infrared image sequence using the method in Section 2.1.
Step 2: Build NMoG noise model under the Bayesian framework by Equations (2) and (5).
Step 3: **While** not converged do:
1. Update approximate posterior of noise component $\mathcal{Z}^t, \pi^t$ by Equations (13)–(16), μ^t, τ^t by Equations (11) and (12) and d^t by Equation (17), respectively.
2. Update approximate posterior of background component U, V by by Equations (18) and (19).
3. Update approximate posterior of parameters in noise component γ^t by Equation (20).
4. Update $t = t + 1$.
end While
Step 4: Noise component E by $E = F - UV^t$. Decompose E into K components by Equation (21), and reconstruct noise components into the corresponding image sequences by method in Section 2.1.
Step 5: Select the true target images by Equation (22).
Step 6: Calculate the original MFD map of the target images by Equations (23) and (24).
Step 7: Obtain the separated target images by using both MFD maple and adaptive threshold, which can be computed by Equation (27).
Output: Separated target image sequence.

---

3. Experiments

To validate the effectiveness of the proposed approach, extensive experiments are performed on simulated and real infrared image sequences in this section.

3.1. Metrics and Comparative Methods

In this paper, we use the receiver operating characteristic (ROC) to show the relationship between the detection probability P_d and false alarm rate F_a, and they are described as below [22,25–28,32]:

$$P_d = \frac{\text{number of true detections}}{\text{number of actual targets}} \tag{29}$$

$$F_a = \frac{\text{number of false detections}}{\text{number of images}} \tag{30}$$

In addition, the local signal-to-noise ratio gain (LSNRG), background suppression factor (BSF) , signal to clutter ratio gain (SCRG) and contrast gain (CG) metrics are also used in our work, and the detailed definitions can be found in [28,32]. We also introduce a local background region for computing the LSNRG and SCRG [22], which is displayed in Figure 4. The width of neighboring region d is set as 20 here.

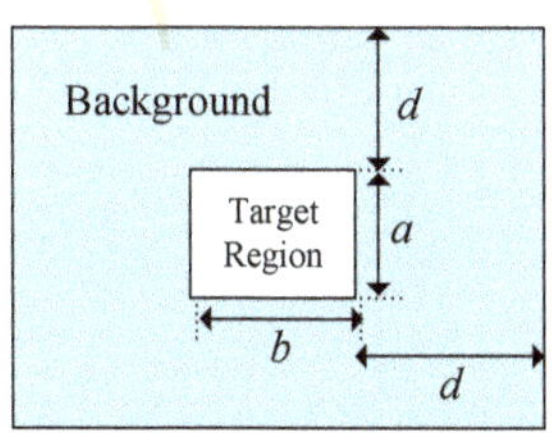

Figure 4. The neighboring background region.

Nonetheless, the accuracy of the low-rank background estimation is also an important metric, since less estimation error means better preservation of strong edges in the background component. Thus, we use another metric, namely accuracy of background recovery (ABR), which is defined as:

$$\text{ABR} = \frac{\|\boldsymbol{B}_{\text{out}}\|_F}{\|\boldsymbol{B}_{\text{in}}\|_F} \tag{31}$$

where $\boldsymbol{B}_{\text{in}}$ and $\boldsymbol{B}_{\text{out}}$ are the background before and after processing.

The five baseline methods for comparison including two classical filtering methods, i.e., top-hat [13] and max-median filtering [12], and three low-rank matrix analysis methods IPI [22] and RIPT [28] (using spatial information) and the MRF-MoG [32] (using spatio-temporal information and assuming i.i.d. MoG noise) method. Table 1 gives the detailed parameter settings, where n_1, n_2, n_3 denotes the dimensions of the infrared patch tensor [28].

Table 1. Parameter setting of methods.

Methods	Acronyms	Parameter Settings
max-median filter	max-median	Support size: 5×5
top-hat method	top-hat	Structure shape: Square, structure size: 3×3
Infrared Patch-Image Mode	IPI	Patch size: 50×50, sliding step: 10, $\lambda = \dfrac{L}{\sqrt{\min(n_1,n_2,n_3)}}$, $L = 1, \varepsilon = 1e - 7$
Reweighted Infrared Patch-Tensor Model	RIPT	Patch size: 50×50, sliding step: 10, $\lambda = \dfrac{L}{\sqrt{\min(n_1,n_2,n_3)}}$, $L = 1, h = 10, \varepsilon = 1e - 7$
Mixture of Gaussians with Markov random field	MoG with MRF	Noise component number: $K = 3$
Mixture of Non-i.i.d. Gaussians with Modified Flux Density	NMoG with MFD	Noise component number: $K = 3$

3.2. Simulated and Real Datasets

The noise of real infrared images usually includes five typical types: Gaussian noise, Poisson noise, impulse noise, dead pixels or lines, and salt and pepper noise. To validate the effectiveness of the proposed approach in complex noisy situations, five consecutive real infrared image sequences are used as original images to add the mixture of the above five types of noises, and these original images are approximately noise-free. Additive white Gaussian noise with two SNR value are added to each frame of five sequences, and the SNR are in the range of [10, 15] dB and [15, 20] dB, respectively. The location of pixels corrupted by different noises are chosen randomly. We choose forty frames of Sequences 1–4 to add with various types of noise and different intensity. Finally, we add the mixture of noise to each frame in sequence 5. The details are described in Table 2, and their representative frames are displayed in the first column of Figure 9.

Table 2. Characteristics of noisy infrared sequences.

Sequence	Number of Frames	Image Resolution (pixels)	Noise Characteristics	Background Characteristics	SCR	$\overline{\text{SCR}}$
1	135	220×140	Gaussian + Deadline Noise	Sea-Sky Clutters	$0.25 \sim 10.11$	3.49
2	108	280×228	Gaussian + Salt and Pepper Noise	Heavy cloud-sky clutters	$0.13 \sim 8.24$	2.71
3	114	250×200	Gaussian + Poisson Noise	Heavy cloud-sky clutters	$0.11 \sim 3.30$	1.33
4	123	281×240	Gaussian + Impulse Noise	Heavy cloud-sky clutters	$0.02 \sim 4.32$	1.90
5	102	200×150	Mixture Noise	Heavy cloud-sky clutters	$0.05 \sim 10.24$	3.09

SCR is defined as follows [40]:

$$\text{SCR} = \frac{|\mu_t - \mu_b|}{\sigma_b} \tag{32}$$

where μ_t is the average pixel value of the target region, μ_b and σ_b denote the average pixel value and the standard deviation of the neighborhood region. Based on definition of SCR, the average SCR value of targets is used to characterize the noisy sequence, which is defined as follows [22]:

$$\overline{\text{SCR}} = \frac{1}{N} \sum_{i=1}^{N} \text{SCR}_i \tag{33}$$

where N denotes the number of targets and SCR_i denotes ith target.

Then we also carry out comparison experiments with three real infrared image sequences contaminated by heavy noise.

3.3. Effect of Component Number

Here, we vary K from 2 to 7 for analyzing the influence of noise component parameter K on the performance of the proposed model. For quantitative analysis, the experiments have fixed false-alarm rates (F_a) by changing the segmentation thresholds on Sequences 1–5, which are given in Table 3. The bold format number indicates the highest score. Besides, we also display the ROC curves in Figure 5. We can observe from the result that there is no significant difference in performance when K is larger than 2. From Figure 5a,d, it can be seen that F_a of $K = 2$ are higher than that of other K values, this is because the target component might contain the sparse noise, which could not be wiped out by the threshold. However, it is also improper to set K too large. From Figure 5a,c–e, the probability of detection is decreasing as K becoming larger when $K \geq 4$ due to the true targets might lose in the separated target component. In addition, considering the computation complexity is increasing with larger K, K is set as 3 in experiments.

Table 3. The detection performance of the proposed method with different K values.

Metric	K	Sequence 1	Sequence 2	Sequence 3	Sequence 4	Sequence 5
F_a		$F_a = 0.01/\text{image}$	$F_a = 0.1/\text{image}$	$F_a = 0.5/\text{image}$	$F_a = 2/\text{image}$	$F_a = 0.25/\text{image}$
P_d	2	0.98	0.90	0.90	0.90	0.85
	3	**1.00**	**0.90**	**0.94**	**0.93**	**0.87**
	4	0.96	0.90	0.87	0.85	0.86
	5	0.96	0.90	0.86	0.84	0.80
	6	0.96	0.89	0.84	0.84	0.81
	7	0.96	0.89	0.85	0.82	0.80

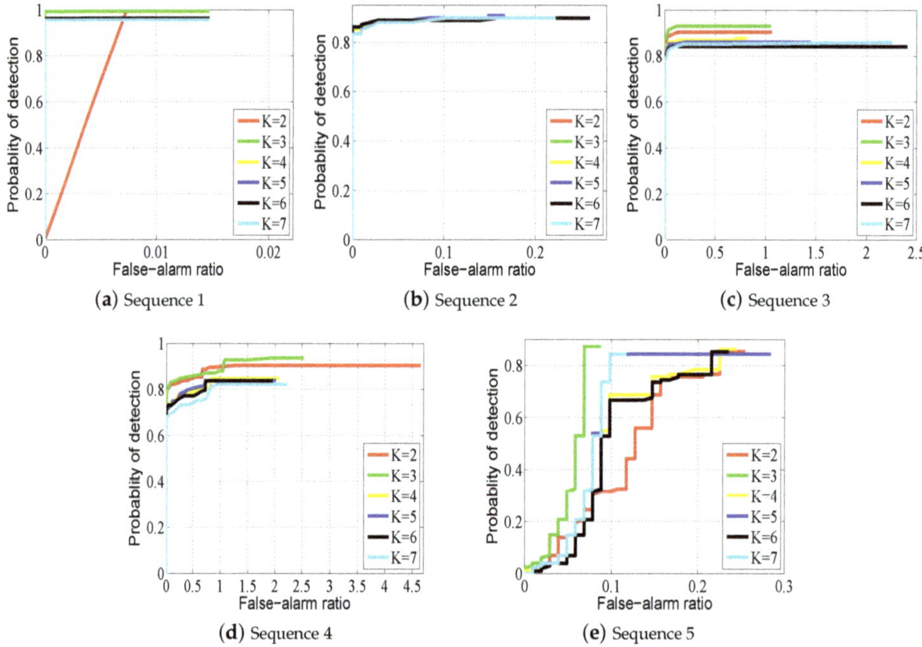

Figure 5. The receiver operating characteristic (ROC) curves of different values for the parameter *K* on Sequences 1–5.

3.4. Effect of MFD

To demonstrate the superiority of the MFD method over other methods, we perform comparative experiments on a representative image of simulated Sequence 5, including the MRF [32] and the ablated version (NMoG without MFD). From Figure 6, we can observe that the MFD method can effectively wipe out the bright noise, while the other two methods lose the true target and have many residual noise pixels, and these residuals could cause a high false alarm ratio.

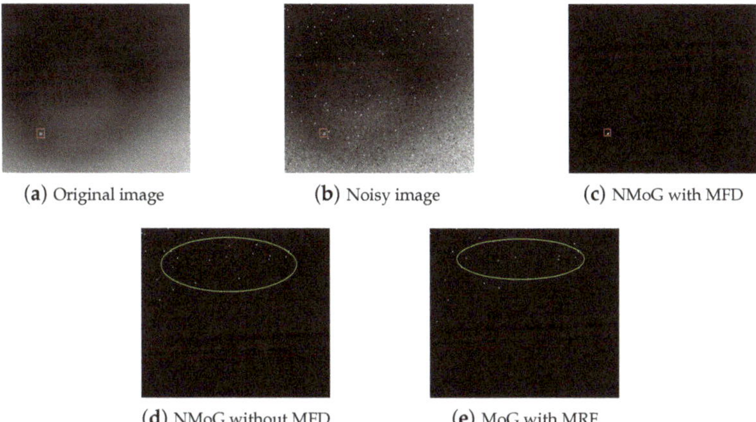

(**a**) Original image   (**b**) Noisy image   (**c**) NMoG with MFD

(**d**) NMoG without MFD   (**e**) MoG with MRF

Figure 6. The results of different methods on a representative image of Sequence 5. (**a–e**) are the original image, the noisy image, the results of the NMoG (non-independent and identical distribution (i.i.d.) mixture of Gaussians (MoG)) with MFD, NMoG without MFD and MoG with Markov random field (MRF), respectively. The red rectangles denote the targets and the green ellipses are representative examples of noise.

3.5. Performance of Multiple Targets Scene

Considering the number of targets may change in different scenes, such as antimissile systems, we test the effectiveness of the proposed method in multi-target scenarios (the number of the targets is 3). The method of embedding a synthetic target into the images can be found in [22]. The representative images and the corresponding results are displayed in the first row and second row of Figure 7. All the targets are detected successfully by the proposed method.

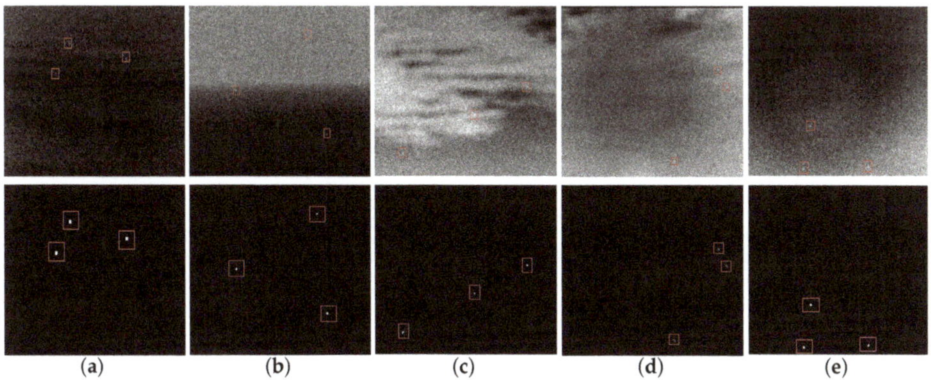

(**a**)   (**b**)   (**c**)   (**d**)   (**e**)

Figure 7. Multiple target scenes. The first and second row of (**a–e**) are five original images and corresponding results processed by the proposed method, respectively.

3.6. Comparisons to Baseline Methods

3.6.1. Experiments on Simulated Data

In this experiment, we focus on analyzing and comparing the performance of different approaches on real infrared images with synthetic noise. To illustrate the difference between the original images and noisy images, we display the gray histograms of five representative frames in Figure 8. The representative images are chosen from one image of the corresponding 40 noisy images of Sequences 1–4 and from one image of Sequence 5 randomly. It can be observed from Figure 8 that the distributions of original and noisy images are quite different. Figure 9 shows the corresponding target images of different approaches. We can observe that both max-median filter and top-hat filter can not suppress the noise pixels clearly, and these residuals would increase F_a. Besides, top-hat filter loses the target in Sequences 2 and 5. The performances of both max-median filter and top-hat filter are limited by the filtering size required to be fixed as an input parameter without any knowledge of the target size. Their performances degrade heavily when the filter size deviates from the target size.

From the comparison between the results of filtering based approaches and low-rank based approaches, we conclude that the latter can achieve better performance than the former ones. All the targets can be detected by IPI method, but many noise pixels are also retained due to the deficiency effects [28], especially for Sequences 2, 4 and 5. This phenomenon demonstrate that the IPI method is quite sensitive to salt and pepper noise and impulse noise. The RIPT approach has better background suppression ability than IPI approach, but we can find that it is also sensitive to salt and pepper noise from the corresponding results of Sequences 2 and 5. Moreover, the RIPT method fails in Sequence 3. MoG-MRF only detects the true targets of Sequence 1 and 4, the unsatisfying performance of MoG-MRF is because the i.i.d. MoG assumption is not suitable to the case when the noise distribution between different frames is nonidentical. Besides, the segmentation performance of MRF would degrade when the noise pixel is adjacent to true targets in complex noisy cases. From the last column of Figure 9, it can be observed that all targets are detected correctly by the proposed model while noise pixels and clutters being suppressed completely.

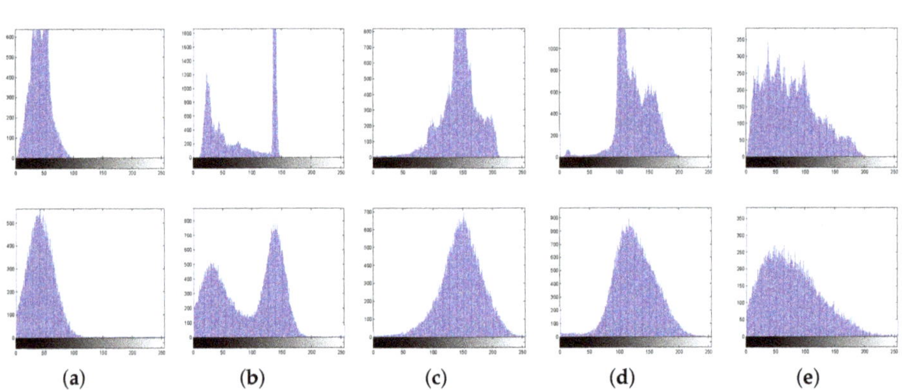

(a)   (b)   (c)   (d)   (e)

Figure 8. The histograms of the representative frames in original and noisy Sequences 1–5. The first row of (**a–e**) are the histograms of five original infrared images for experiments. The second row of (**a–e**) are corresponding histograms of noisy infrared images.

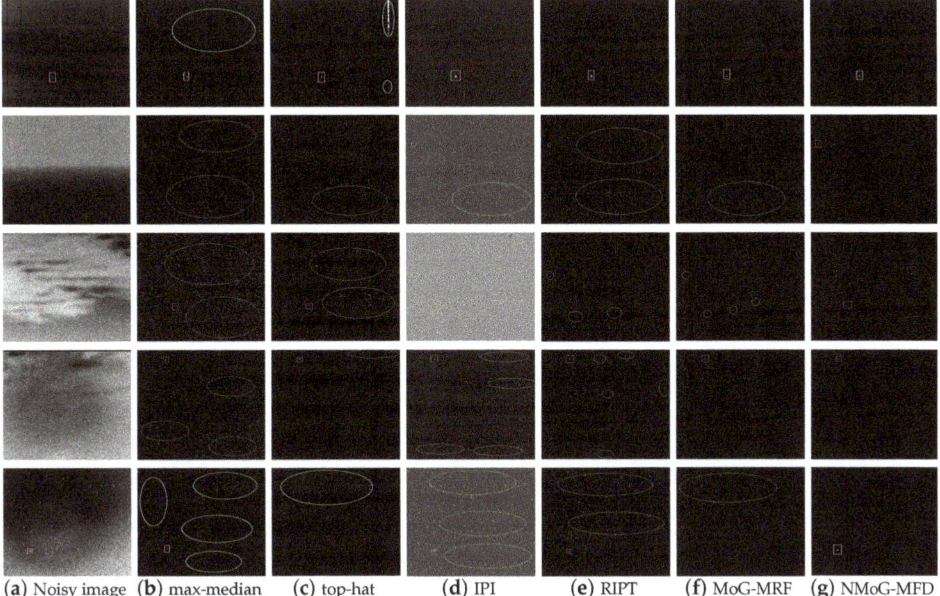

(a) Noisy image   (b) max-median   (c) top-hat   (d) IPI   (e) RIPT   (f) MoG-MRF   (g) NMoG-MFD

Figure 9. Separated target images of the representative frames in Sequences 1–5 by different methods. Row 1: 31-st frame of Sequence 1. Row 2: 3-rd frame of Sequence 2. Row 3: 41-st frame of Sequence 3. Row 4: 4-th frame of Sequence 4.Row 5: 22-nd frame of Sequence 5. (**a**) columns are the noisy images, respectively. (**b–g**) columns are the separated target images of (**b**) MaxMedian, (**c**) TopHat, (**d**) IPI, (**e**) RIPT (**f**) MoG with MRF and (**g**) NMoG with MFD methods.

In addition, we also use five metrics to analyze the performance of different approaches quantitatively. The LSNRG, BSF and SCRG values of different approaches for the representative images are given in Tables 4 and 5. The Inf means that the standard deviation of neighboring background is zero, and the high scores in the above three metrics only reflect the good suppression performance in a local region. Note that the values of low-rank based methods in the above three metrics are usually Inf, as the results of RIPT method, MoG-MRF method and the proposed method on Sequences 1 and 4. Considering the above phenomenon, the average CG and ABR values of all images are also computed for further comparison [32], as listed in Table 6. For quantitative analysis, the experiments have fixed false-alarm rates (F_a) by changing the segmentation thresholds on Sequences 1–5, which are given in Table 7. In conclusion, the proposed approach achieves the best performance. In conclusion, the proposed approach achieves the best performance.

Moreover, we show the ROC curves of different approaches in Figure 10. From the result, we can see that the F_a of max-median on Sequences 2 and 5 are very high. The performance of the proposed approach is superior to other approaches on Sequences 1–3 and 5, which achieves the highest P_d with very low F_a, this is because the noise pixels and background residuals are suppressed thoroughly by the proposed method. As for Sequence 4, IPI achieves higher P_d than that of the proposed method when $F_a \leq 1.1$. However, the proposed method can achieve higher probability of detection when $F_a > 1.1$. The ROC curves of IPI and RIPT on Sequences 2 and 5 demonstrate that they are sensitive to salt and pepper noise, and the performance of MoG with MRF method is not satisfying due to the identical noise distribution assumption fails in complex noise case.

Table 4. Quantitative evaluation of different methods for the representative images of Sequences 1–3.

	31st Frame of Sequence 1			3rd Frame of Sequence 2			41st Frame of Sequence 3		
Method	**LSNRG**	**BSF**	**SCRG**	**LSNRG**	**BSF**	**SCRG**	**LSNRG**	**BSF**	**SCRG**
Max-Meidan	1.6209	14.8329	2.7474	0.4829	7.0193	3.6076	0.4640	14.4141	0.0792
top-hat	Inf	Inf	Inf	**Miss**	**Miss**	**Miss**	1.0311	2.3991	0.2823
IPI	4.1182	1.5731	6.207	0.976	1.5658	0.6981	0.9828	1.3869	0.2679
RIPT	Inf	Inf	Inf	0.6207	6.4585	13.3034	0	6.0941	32.9413
MoG-MRF	Inf	Inf	Inf	**Miss**	**Miss**	**Miss**	0	6.8651	53.2848
NMoG-MFD	Inf	Inf	Inf	Inf	Inf	Inf	Inf	Inf	Inf

Table 5. Quantitative evaluation of different methods for the representative images of Sequences 4 and 5.

	4th Frame of Sequence 4			22nd Frame of Sequence 5		
Method	**LSNRG**	**BSF**	**SCRG**	**LSNRG**	**BSF**	**SCRG**
Max-Meidan	1.2486	9.5636	3.2359	0.2618	2.4965	0.366
top-hat	Inf	Inf	Inf	**Miss**	**Miss**	**Miss**
IPI	2.1088	3.8868	4.8508	0.9329	1.3979	0.4839
RIPT	Inf	Inf	Inf	0.6674	2.901	3.6575
MoG-MRF	Inf	Inf	Inf	**Miss**	**Miss**	**Miss**
NMoG-MFD	Inf	Inf	Inf	Inf	Inf	Inf

Table 6. The evaluation results of average contrast gain (CG) and accuracy of background recovery (ABR) values of different methods for all image sequences.

	Sequence 1		Sequence 2		Sequence 3		Sequence 4		Sequence 5	
Method	**CG**	**ABR**	**CG**	**ABR**	**CG**	**ABR**	**CG**	**ABR**	**CG**	**ABR**
Max-Meidan	2.3312	0.9221	1.2661	0.9457	2.4625	0.9519	1.7005	0.868	1.4464	0.8981
top-hat	3.8321	0.9066	5.0661	0.9286	5.6079	0.9398	4.1474	0.9289	3.1628	0.9192
IPI	2.6185	0.8601	1.5188	0.8374	1.7097	0.9067	2.5819	0.9447	1.8794	0.8861
RIPT	3.1073	0.9179	3.7042	0.9303	6.1015	0.9423	3.0008	0.9321	2.1123	0.8993
MoG-MRF	4.7533	0.9327	5.2441	0.9700	6.3158	0.9508	4.0663	0.9584	3.8108	0.9435
NMoG-MFD	**4.7798**	**0.9801**	**5.6895**	**0.9841**	**8.1627**	**0.9837**	**5.1757**	**0.9849**	**3.8432**	**0.9825**

Table 7. The detection performance of different methods on Sequences 1–5.

Metric	**Methods**	**Sequence 1**	**Sequence 2**	**Sequence 3**	**Sequence 4**	**Sequence 5**
F_a		$F_a = 1/\text{image}$	$F_a = 15/\text{image}$	$F_a = 2.5/\text{image}$	$F_a = 2/\text{image}$	$F_a = 11/\text{image}$
P_d	max-median	0.84	0	0.49	0.30	0
	Top-hat	0.46	0.28	0.41	0.22	0.26
	IPI	0.91	0.05	0.93	0.90	0.27
	RIPT	0.91	0.17	0.93	0.91	0.54
	MoG-MRF	0.90	0.52	0.88	0.74	0.75
	NMoG-MFD	**1.00**	**0.90**	**0.94**	**0.93**	**0.87**

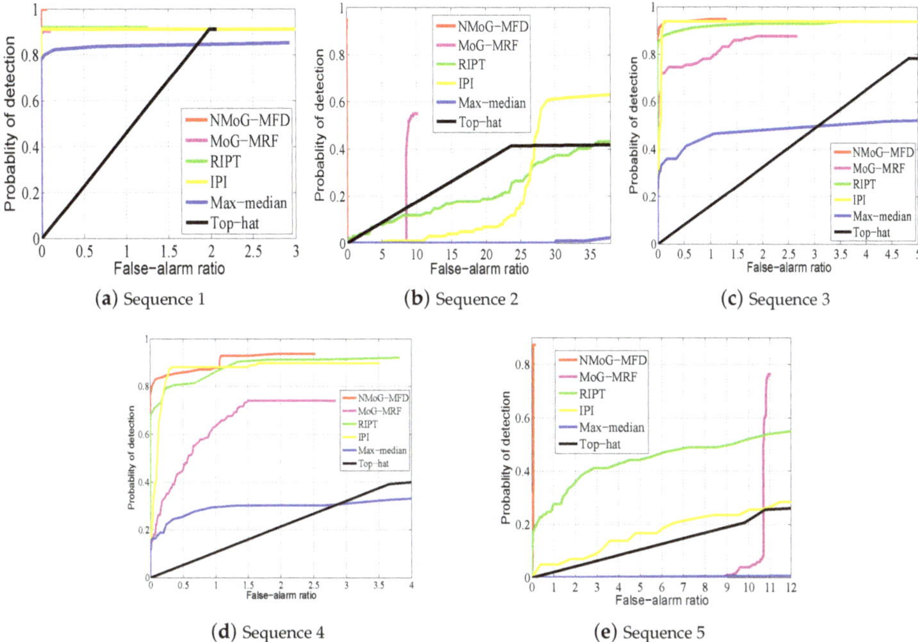

Figure 10. The ROC curves of different methods on Sequences 1–5. (**a**) result of Sequence 1, (**b**) result of Sequence 2, (**c**) result of Sequence 3, (**d**) result of Sequence 4, (**e**) result of Sequence 5.

3.6.2. Experiments on Real Data

We also carry out additional experiments on three real and noisy infrared image sequences, namely, Sequences 6–8. Briefly, we use the most important metric, i.e., the ROC curves of 6 tested method on real image sequences, to compare their performance, which are shown in Figure 11. In addition, Table 8 shows the quantitative analysis, and the proposed approach achieves the highest P_d with the same F_a. The results demonstrate the superiority of the proposed approach on target detection, background clutter and noise suppression ability over other competitive methods, because the NMoG model and MFD maple improve the robustness of the proposed approach to different kinds of noise.

Table 8. The detection performance of different methods on Sequences 6–8.

Metric	Methods	Sequence 6	Sequence 7	Sequence 8
F_a		$F_a = 2/\text{image}$	$F_a = 2/\text{image}$	$F_a = 2/\text{image}$
	max-median	0	0	0
	Top-hat	0.11	0.14	0.51
	IPI	0.86	0.33	0.04
P_d	RIPT	0.34	0.28	0.32
	MoG-MRF	0.63	0.14	0.85
	NMoG-MFD	**0.89**	**0.90**	**1.00**

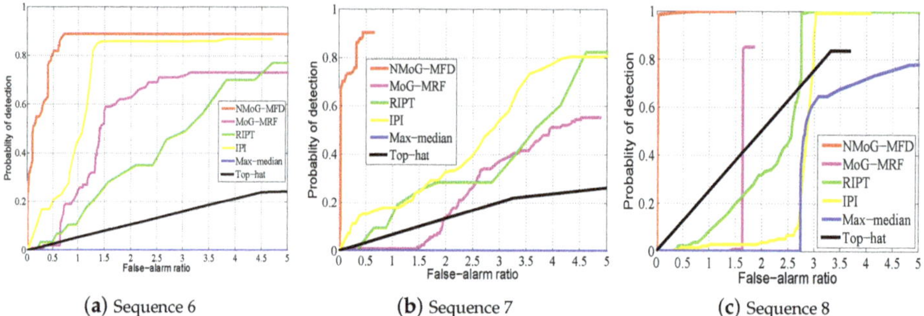

(a) Sequence 6   **(b)** Sequence 7   **(c)** Sequence 8

Figure 11. ROC curves of different methods on real noisy Sequences 6–8. (**a**) result of Sequence 6, (**b**) result of Sequence 7, (**c**) result of Sequence 8.

3.7. Complexity Analysis

Here, we analyze and compare the complexity of different approaches, which are listed in Table 9. (m, n) and L denote the image size and the structure element, respectively. (n_1, n_2, n_3) represent the dimensions of the tensor in RIPT model, and the details can be found in [30]. As for the proposed method, let $\boldsymbol{F} \in \mathbb{R}^{N \times P}$, we firstly need to infer the parameters in NMoG model, and its complexity is $\mathcal{O}\left((N+P) R^3 + KNPR\right)$ in each iteration. For computing MFD maple of an image with size of $m \times n$, the whole computational cost is $\mathcal{O}\left(mn(2s+1)^2\right)$. For target segmentation, the cost of this step is $\mathcal{O}(mn)$. Thus, the entire computation cost of the proposed method is $\mathcal{O}\left(t\left((N+P) R^3 + kNPR\right) + mn(2s+1)^2 + mn\right)$, where t is the iteration number. The MoG with MRF method uses median operation to reconstruct image sequences, and its complexity is $\mathcal{O}(mnw)$, where w denotes the number of overlapped pixels during the transformation from the spatio-temporal patch image to the reconstruction image [32]. In addition, we compare the computational time of different approaches on whole Sequence 6. It can be observed from the result that MOG with MRF method is the slowest while the top-hat filter is the fastest. The processing time of the RIPT approach is shorter than the IPI approach and max-median filter. The proposed approach is slower than RIPT method and two filtering methods, but the superiority of its performance over other baseline methods can make up for this deficiency.

Table 9. Algorithm complexity and computational time comparisons of different methods.

Method	Complexity	Times(s)
max-median	$\mathcal{O}(mnL^2)$	392.997661
top-hat	$\mathcal{O}(mnL^2 logL)$	2.639046
IPI	$\mathcal{O}(mn^2)$	682.764355
RIPT	$\mathcal{O}\left(tn_1 n_2 n_3 \left(n_1 n_2 + n_2 n_3 + n_1 n_3\right)\right)$	224.866089
MoG-MRF	$\mathcal{O}\left(t\left((N+P) R^3 + kNPR\right) + mn(w+1)\right)$	3002.7214
NMoG-MFD	$\mathcal{O}\left(t\left((N+P) R^3 + kNPR\right) + mn(2s+1)^2 + mn\right)$	482.9220

4. Conclusions

In this paper, we propose a novel infrared small and faint target detection approach based on NMoG and MFD models for complex and noisy scenarios. The proposed model can finely accord with the noise characteristics embedded in real infrared image sequences by using the NMoG model. We model the recovery of a low-rank background component and noise component as an LRMF model, which can be solved by the VB algorithm. Finally, the target can be extracted correctly from the noise by using MFD maple. Experimental results show that the proposed approach performs better than other competitive approaches, since it is more robust to complex noisy scenarios in real application.

Remote Sens. **2019**, *11*, 2831

Author Contributions: Conceptualization, Y.S. and W.A.; methodology, Y.S.; software, Y.S.; validation, Y.S., J.Y. and M.L.; formal analysis, Y.S.; investigation, Y.S.; resources, Y.S.; data curation, Y.S.; writing—original draft preparation, Y.S.; writing—review and editing, M.L.; visualization, Y.S.; supervision, W.A.; project administration, W.A.; funding acquisition, J.Y.

Funding: This work was supported in part by the National Natural Science Foundation of China under Grant (No. 61605242).

Conflicts of Interest: The authors declare no conflict of interest. The funders had no role in the design of the study; in the collection, analyses, or interpretation of data; in the writing of the manuscript, or in the decision to publish the results.

References

1. Chen, Y.; Xin, Y. An Efficient Infrared Small Target Detection Method Based on Visual Contrast Mechanism. *IEEE Geosci. Remote Sens. Lett.* **2016**, *13*, 962–966. [CrossRef]
2. Bai, X.; Chen, Z.; Zhang, Y.; Liu, Z.; Lu, Y. Infrared Ship Target Segmentation Based on Spatial Information Improved FCM. *IEEE Trans. Cybern.* **2016**, *46*, 3259–3271. [CrossRef] [PubMed]
3. Deng, H.; Sun, X.; Liu, M.; Ye, C.; Zhou, X. Entropy-based window selection for detecting dim and small infrared targets. *Pattern Recognit.* **2017**, *61*, 66–77. [CrossRef]
4. Chen, H.; Bar-Shalom, Y.; Pattipati, K.R.; Kirubarajan, T. MDL approach for multiple low observable track initiation. *IEEE Trans. Aerosp. Electron. Syst.* **2002**, *39*, 862–882. [CrossRef]
5. Grossi, E.; Lops, M. Sequential Along-Track Integration for Early Detection of Moving Targets. *IEEE Trans. Signal Process.* **2008**, *56*, 3969–3982. [CrossRef]
6. Li, Y.; Liang, S.; Bai, B.; Feng, D. Detecting and tracking dim small targets in infrared image sequences under complex backgrounds. *Multimed. Tools Appl.* **2014**, *71*, 1179–1199. [CrossRef]
7. Liu, Z.; Zhou, F.; Chen, X.; Bai, X.; Sun, C. Iterative infrared ship target segmentation based on multiple features. *Pattern Recognit.* **2014**, *47*, 2839–2852. [CrossRef]
8. Han, J.; Ma, Y.; Huang, J.; Mei, X.; Ma, J. An Infrared Small Target Detecting Algorithm Based on Human Visual System. *IEEE Geosci. Remote Sens. Lett.* **2016**, *13*, 452–456. [CrossRef]
9. Reed, I.S.; Gagliardi, R.M.; Stotts, L.B. Optical moving target detection with 3D matched filtering. *IEEE Trans. Aerosp. Electron. Syst.* **2002**, *24*, 327–336. [CrossRef]
10. Yang, W.; Sun, X. Moving weak point target detection and estimation with three-dimensional double directional filter in IR cluttered background. *Opt. Eng.* **2005**, *44*, 107007.
11. Liu, X.; Zuo, Z. A Dim Small Infrared Moving Target Detection Algorithm Based on Improved Three-Dimensional Directional Filtering. In *Advances in Image and Graphics Technologies*; Springer: Berlin/Heidelberg, Germany, 2013; Volume 363, pp. 102–108.
12. Venkateswarlu, R. Max-mean and max-median filters for detection of small targets. *Proc. SPIE-Int. Soc. Opt. Eng.* **1999**, *3809*, 74–83.
13. Fortin, R. Detection of dim targets in digital infrared imagery by morphological image processing. *Opt. Eng.* **1996**, *35*, 1886–1893.
14. Hadhoud, M.M.; Thomas, D.W. The two-dimensional adaptive LMS (TDLMS) algorithm. *Circuits Syst. IEEE Trans.* **1988**, *35*, 485–494. [CrossRef]
15. Kim, S.; Yang, Y.; Lee, J.; Park, Y. Small Target Detection Utilizing Robust Methods of the Human Visual System for IRST. *J. Infrared Millim. Terahertz Waves* **2009**, *30*, 994–1011. [CrossRef]
16. Chen, C.L.P.; Li, H.; Wei, Y.; Xia, T.; Tang, Y.Y. A Local Contrast Method for Small Infrared Target Detection. *IEEE Trans. Geosci. Remote Sens.* **2013**, *52*, 574–581. [CrossRef]
17. Han, J.; Ma, Y.; Zhou, B.; Fan, F.; Liang, K.; Fang, Y. A Robust Infrared Small Target Detection Algorithm Based on Human Visual System. *IEEE Geosci. Remote Sens. Lett.* **2014**, *11*, 2168–2172.
18. Tun, H.U.; Zhao, J.J.; Yuan, C.; Wang, F.L.; Jie, Y. Infrared Small Target Detection Based on Saliency and Principle Component Analysis. *J. Infrared Millim. Waves* **2010**, *29*, 303–306.
19. Wang, C.; Qin, S. Adaptive detection method of infrared small target based on target-background separation via robust principal component analysis. *Infrared Phys. Technol.* **2015**, *69*, 123–135. [CrossRef]
20. Bi, Y.; Bai, X.; Jin, T.; Guo, S. Multiple Feature Analysis for Infrared Small Target Detection. *IEEE Geosci. Remote Sens. Lett.* **2017**, *14*, 1333–1337. [CrossRef]

21. Guangcan, L.; Zhouchen, L.; Shuicheng, Y.; Ju, S.; Yong, Y.; Yi, M. Robust recovery of subspace structures by low-rank representation. *IEEE Trans. Pattern Anal. Mach. Intell.* **2013**, *35*, 171–184.

22. Gao, C.; Meng, D.; Yang, Y.; Wang, Y.; Zhou, X.; Hauptmann, A.G. Infrared Patch-Image Model for Small Target Detection in a Single Image. *IEEE Trans. Image Process.* **2013**, *22*, 4996–5009. [CrossRef] [PubMed]

23. Candès, E.J.; Tao, T. The Power of Convex Relaxation: Near-optimal Matrix Completion. *IEEE Trans. Inf. Theor.* **2010**, *56*, 2053–2080. [CrossRef]

24. Mohimani, H.; Babaie-Zadeh, M.; Jutten, C. A fast approach for overcomplete sparse decomposition based on smoothed L0 norm. *IEEE Trans. Signal Process.* **2008**, *57*, 289–301. [CrossRef]

25. Dai, Y.; Wu, Y.; Song, Y. Infrared small target and background separation via column-wise weighted robust principal component analysis. *Infrared Phys. Technol.* **2016**, *77*, 421–430. [CrossRef]

26. Dai, Y.; Wu, Y.; Song, Y.; Guo, J. Non-negative infrared patch-image model: Robust target-background separation via partial sum minimization of singular values. *Infrared Phys. Technol.* **2017**, *81*, 182–194. [CrossRef]

27. Guo, J.; Wu, Y.; Dai, Y. Small target detection based on reweighted infrared patch-image model. *IET Image Process.* **2018**, *12*, 70–79. [CrossRef]

28. Dai, Y.; Wu, Y. Reweighted Infrared Patch-Tensor Model With Both Nonlocal and Local Priors for Single-Frame Small Target Detection. *IEEE J. Sel. Top. Appl. Earth Obs. Remote Sens.* **2017**, *10*, 3752–3767. [CrossRef]

29. Liu, J.; Musialski, P.; Wonka, P.; Ye, J. Tensor completion for estimating missing values in visual data. *IEEE Trans. Pattern Anal. Mach. Intell.* **2013**, *35*, 208–220. [CrossRef]

30. Sun, Y.; Yang, J.; Long, Y.; Shang, Z.; An, W. Infrared Patch-Tensor Model With Weighted Tensor Nuclear Norm for Small Target Detection in a Single Frame. *IEEE Access* **2018**, *6*, 76140–76152. [CrossRef]

31. Shabalin, A.A.; Nobel, A.B. Reconstruction of a low-rank matrix in the presence of Gaussian noise. *J. Multivar. Anal.* **2013**, *118*, 67–76. [CrossRef]

32. Gao, C.; Wang, L.; Xiao, Y.; Zhao, Q.; Meng, D. Infrared Small-dim Target Detection Based on Markov Random Field Guided Noise Modeling. *Pattern Recognit.* **2017**, *76*, 463–475. [CrossRef]

33. Meng, D.; Torre, F.D.L. Robust Matrix Factorization with Unknown Noise. In Proceedings of the 2013 IEEE International Conference on Computer Vision, Sydney, Australia, 1–8 December 2013; pp. 1337–1344.

34. Cao, X.; Zhao, Q.; Meng, D.; Chen, Y.; Xu, Z. Robust Low-Rank Matrix Factorization Under General Mixture Noise Distributions. *IEEE Trans Image Process* **2016**, *25*, 4677–4690. [CrossRef] [PubMed]

35. Zhao, Q.; Meng, D.; Xu, Z.; Zuo, W.; Zhang, L. Robust principal component analysis with complex noise. In Proceedings of the 31st International Conference on Machine Learning (ICML 2014), Beijing, China, 21–26 June 2014; pp. 55–63.

36. Chen, Y.; Cao, X.; Zhao, Q.; Meng, D.; Xu, Z. Denoising Hyperspectral Image With Non-i.i.d. Noise Structure. *IEEE Trans. Cybern.* **2017**, *48*, 1054–1066. [CrossRef] [PubMed]

37. Liu, D.; Cao, L.; Li, Z.; Liu, T.; Che, P. Infrared Small Target Detection Based on Flux Density and Direction Diversity in Gradient Vector Field. *IEEE J. Sel. Top. Appl. Earth Obs. Remote. Sens.* **2018**, *11*, 2528–2554. [CrossRef]

38. Babacan, S.D.; Luessi, M.; Molina, R.; Katsaggelos, A.K. Sparse Bayesian Methods for Low-Rank Matrix Estimation. *IEEE Trans. Signal Process.* **2012**, *60*, 3964–3977. [CrossRef]

39. Bishop, C.M.; Nasrabadi, N.M. *Pattern Recognition and Machine Learning*; Springer: New York, NY, USA, 2006; p. 049901.

40. Gao, C.; Zhang, T.; Li, Q. Small infrared target detection using sparse ring representation. *IEEE Aerosp. Electron. Syst. Mag.* **2012**, *27*, 21–30.

 *remote sensing*

Article

Mask Sparse Representation Based on Semantic Features for Thermal Infrared Target Tracking

Meihui Li [1,2], **Lingbing Peng** [1], **Yingpin Chen** [3], **Suqi Huang** [1,2], **Feiyi Qin** [1,2] **and Zhenming Peng** [1,2,*]

[1] School of Information and Communication Engineering, University of Electronic Science and Technology of China, Chengdu 611731, China
[2] The Laboratory of Imaging Detection and Intelligent Perception, University of Electronic Science and Technology of China, Chengdu 610054, China
[3] School of Physics and Information, Minnan Normal University, Zhangzhou 363000, China
* Correspondence: zmpeng@uestc.edu.cn; Tel.: +86-1307-603-6761

Received: 11 July 2019; Accepted: 20 August 2019; Published: 21 August 2019

Abstract: Thermal infrared (TIR) target tracking is a challenging task as it entails learning an effective model to identify the target in the situation of poor target visibility and clutter background. The sparse representation, as a typical appearance modeling approach, has been successfully exploited in the TIR target tracking. However, the discriminative information of the target and its surrounding background is usually neglected in the sparse coding process. To address this issue, we propose a mask sparse representation (MaskSR) model ,which combines sparse coding together with high-level semantic features for TIR target tracking. We first obtain the pixel-wise labeling results of the target and its surrounding background in the last frame, and then use such results to train target-specific deep networks using a supervised manner. According to the output features of the deep networks, the high-level pixel-wise discriminative map of the target area is obtained. We introduce the binarized discriminative map as a mask template to the sparse representation and develop a novel algorithm to collaboratively represent the reliable target part and unreliable target part partitioned with the mask template, which explicitly indicates different discriminant capabilities by label 1 and 0. The proposed MaskSR model controls the superiority of the reliable target part in the reconstruction process via a weighted scheme. We solve this multi-parameter constrained problem by a customized alternating direction method of multipliers (ADMM) method. This model is applied to achieve TIR target tracking in the particle filter framework. To improve the sampling effectiveness and decrease the computation cost at the same time, a discriminative particle selection strategy based on kernelized correlation filter is proposed to replace the previous random sampling for searching useful candidates. Our proposed tracking method was tested on the VOT-TIR2016 benchmark. The experiment results show that the proposed method has a significant superiority compared with various state-of-the-art methods in TIR target tracking.

Keywords: thermal infrared target tracking; semantic features; mask sparse representation; particle filter framework; ADMM

---

1. Introduction

With the improvement of the imaging quality and resolution of thermal cameras, thermal infrared (TIR) target tracking has begun to attract many researchers' attention in recent years. Compared with visual target tracking, TIR target tracking is capable of working in total darkness and is less susceptible to changes in external environment, such as lighting and shadows. Thus, it is important for both military and civil use [1,2]. However, there are some adverse factors that could influence the accuracy and robustness of the TIR target tracking. Firstly, the TIR images have the characteristics of low-contrast, low signal-to-noise ratio, low signal-to-clutter ratio and lack of color information [3,4], which cause a lot of difficulty in distinguishing the moving target from the background. Secondly, the deformation and scale change of the moving target also bring great challenges to the tracking task.

To handle these difficulties, several TIR tracking methods have been proposed, which can be categorized into discriminative tracking methods [5–11] and generative tracking methods [12–18]. Discriminative approaches formulate tracking as a classification task, which aims to find the target area whose features are most discriminative to the background. By comparison, generative approaches focus more on building an appearance model to describe the target. Accordingly, the final tracking result is determined by finding the candidate area with the maximum likelihood score. Sparse representation has drawn much attention in the generative tracking branch due to its good adaption to target appearance changes [13,14,17]. In the sparse representation-based method, the target templates are linearly combined to describe candidate images, while the negative templates are used to handle target partial occlusion, deformation, etc.

First, sparse representation-based tracking methods adopt a global model to describe the target, which is susceptible to target local appearance changes [17,19,20]. Afterwards, some local sparse models [21–23] are proposed successively, in which each target is divided into several rectangular image blocks by a sliding window. These local blocks are treated equally in the sparse coding process, regardless of the diverse discriminant capabilities of different object local parts. However, as shown in Figure 1, the human body wrapped by the yellow line is much easier to distinguish compared with the remaining area in the red bounding box, which is also annotated as the tracking target but actually belongs to the background. Current local sparse representation-based trackers neglect this problem and are prone to tracking drift when there are too many non-distinguishable pixels in some of the local patches.

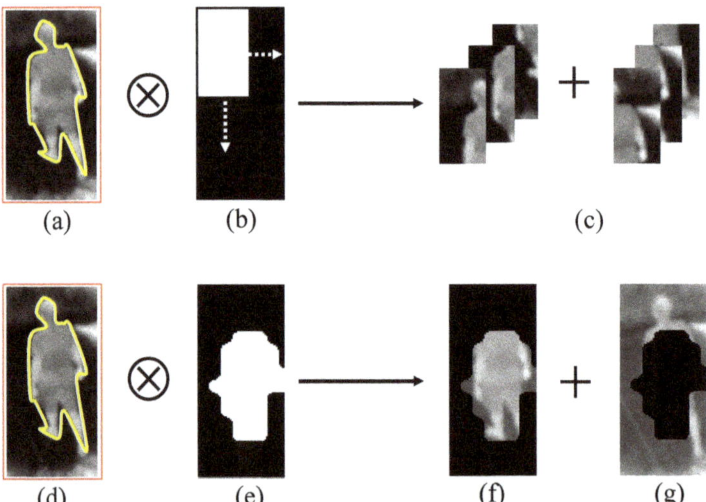

Figure 1. Comparison of the target partition using sliding window and semantic mask template. The upper part of the illustration shows the target partition approach using sliding window, and the lower part shows the target partition approach using semantic mask template: (**a**) tracking target area; (**b**) sliding window; (**c**) target local parts; (**d**) tracking target area; (**e**) semantic mask template; (**f**) reliable target partition; and (**g**) unreliable target partition.

This observation motivates an approach that can adaptively extract distinguishable/reliable pixels from the whole target area, and then use the reliable target part to refine the reconstruction output of the unreliable target part. Considering the benefit of strong discriminative ability of the deep convolutional neural networks (DCNN) [7,8,10,11,24], we propose a supervised learning manner to extract high-level semantic features of the target area. Based on the convolutional neural networks pre-trained for image classification, DCNN can learn information of salient objects at any position of the input image. In [25], a soft-mask module is added to an optical flow estimation network, which aims to mask out parts with consistency motions. The mask filters are trained by fixing the pre-trained weights. In this paper, we propose to add a channel selection layer after convolutional layers, which is more specific to the tracking task. With the pixel-wise labeling results of the target and its surrounding background in the last frame, the output channels are sorted and filtered to obtain target-specific features from DCNN.

The binarized semantic features are introduced as the mask template to extract reliable pixels with powerful discriminative capability, as shown in the lower part of Figure 1. In the proposed MaskSR model, the reliable target part (with label 1) and the unreliable target part (with label 0) correspond to their respective dictionary sets. For each candidate image, the MaskSR model enables representing its two local parts collaboratively by adding l_1 regularization to the difference between the sparse coefficients of the reliable part and unreliable part, aiming to preserve the category consistency of the same candidate area. On the other hand, the fidelity term of the reliable target part is assigned to a larger weight to ensure its superiority to the unreliable part in sparse coding. Therefore, our model fully considers the reliability of different target parts in distinguishing the target from the background. The multi-parameter problem is solved by a customized alternating direction method of multipliers (ADMM). The proposed mask sparse representation model is applied to achieve TIR object tracking under the particle filter framework. In the conventional particle filter method, the target motion parameters should be set in advance to perform Gaussian random sampling on the next frame. Moreover, to ensure efficient calculation, the number of particles cannot be too large, which makes it uncertain whether the scattered random particles cover the real target region. To solve the above two problems, we improve the random particle sampling strategy to discriminative particle selection, which is achieved by the kernel correlation filter method. Experiments on VOT-TIR2016 benchmark show that the developed method is effective for TIR object tracking.

In summary, the contributions of this paper include the following three points:

- To improve the ability of distinguishing the target from the clutter background, we propose a mask sparse representation method for target appearance modeling. In this model, the distinguishable and reliable pixels of the target are identified and are utilized to refine the reconstruction output of the unreliable target part.
- With the pixel-wise labeling results of the target and its surrounding background in the last frame, we develop a supervised manner to learn a high-level pixel-wise discriminative map of the target area. The binarized discrimination map is introduced in the MaskSR model to indicate discrimination capabilities of different object parts.
- The proposed MaskSR model is introduced in an improved particle filter framework to achieve TIR target tracking. We achieved state-of-the-art performance on VOT-TIR2016 benchmark, in terms of both robustness and accuracy evaluations.

The rest of this paper is organized as follows. In Section 2, some works that are closely related to ours are introduced. In Section 3, we present the details of our tracking framework. Section 4 shows the experiment results of the proposed tracker and the comparison results to other state-of-the-art tracking methods. Section 5 is the conclusion of the whole paper.

2. Related Work

Our work is focus on the formulation of the target appearance model and candidate searching strategy. Thus, we first review some TIR tracking methods based on deep learning and sparse representation. Then, the development of particle filter framework for object tracking is discussed afterwards.

2.1. Deep Learning-Based TIR Tracking Method

Deep convolutional neural networks (CNN) have made great progress in the visual classification task. However, there are some limitations for the usage of CNN in the TIR object tracking, which is mainly caused by the lack of labeled infrared image data and the unfitness of the location estimation task compared with label prediction. Many methods have been developed to address these two problems recently. In [11], an image-to-image transition model is employed to generate synthetic TIR data, on which they can train end-to-end optimal features for TIR tracking. By comparison, most existing methods directly adopt a pre-trained network on visual image set and transfer it to the TIR data. For example, in [8,26], a pre-trained Siamese network is utilized as a similarity function to evaluate the similarity between the initial target and candidates. To improve the accuracy of location estimation, some spatial related methods have been proposed [7,8,10] recently. The presented spatial-aware Siamese network in [8] combines spatial and semantic features of TIR object together to enhance the discriminative ability of the coalesced hierarchical feature. In [7], features are extracted from multiple convolutional layers and are used to construct multiple weak trackers to give response maps of the target's location. The evaluation result in [27] has shown that the learned infrared features perform favorably against the hand-crafted features (HOG and Gist) in the correlation filter-based tracking framework.

2.2. Sparse Representation-Based TIR Tracking Method

From the presence of the l_1 tracker, the sparse representation model has been widely applied in object tracking, including the field of TIR object tracking. In [28], a discriminative sparse representation model is presented for infrared dim moving target tracking, in which the dictionary is composed of a target dictionary and a background dictionary. A sparsity-based discriminative classifier is proposed in [9] to evaluate the confidence of different target templates, of which the best template is used for calculating the convolution score of the candidate images. To explore the underlying relationship of multiple candidates, a low-rank sparse learning method is proposed in [13] that describes corruptions adaptively by finding the maximum-likelihood estimation solution of the residuals. Later, a multi-task Laplacian sparse representation is proposed in [1] to refine the sparse coefficients by deploying the similarity of each candidate pair. Due to the low-rank property of the infrared background, some decomposition-based methods have been proposed for TIR object tracking. A block-wise sparse representation-based tracker is proposed in [29], in which the infrared image is divided into overlapped blocks. These blocks are further decomposed into low-rank target components and sparse occlusion components with adaptive weighting parameters of different parts. A total variation term is further added to constrain the occlusion matrix in [18] to prevent the noise pixel from being separated into the occlusion term. Apart from the pure TIR object tracking, some methods integrate the RGB information of the corresponding visual data with the thermal information to achieve RGB-T object tracking [16,30–33]. In these methods, the joint sparse representation model is employed to ensure multiple modalities in appearance representation.

2.3. Particle Filter for Tracking

Particle filter framework models object tracking as a state estimation process, which is implemented by a Bayesian inference filter with Monte Carlo simulation. The dynamics between the states in two adjacent frames is usually modeled by a Brownian motion. In most tracking methods [19,28,34], the state parameters are predicted independently by a Gaussian distribution. However, in these methods, many particles are needed to cover the states of the real target. In [15,35,36], the result of the saliency extraction is utilized as a prior knowledge of the transition probability model to limit the particle sampling process, which can improve the efficiency of particle sampling significantly. In [37], an improved particle filter framework is proposed to enhance the mean state estimation and resampling procedures, in which the number of high-weighted particles are determined adaptively by applying the k-means clustering over all particles' weights. In [38], a multi-task correlation particle filter (MCPF) is proposed for object tracking, which can cover object state space well with a few particles. In this method, each particle corresponds to an image region enclosed by a bounding box instead of a single target state. The above-mentioned methods employ the particle filter approach to estimate the target space with affine space. In [39], Li et al. directly used it to infer whether the reliable patches are on the tracked object. In contrast to the traditional particle filters, they do not need to remove and resample particles at each frame. Instead, the posterior of each reliable patch can be employed to estimate the scale and position of the tracked target through a Hong Voting-like scheme.

3. Proposed Approach

In this section, we first introduce the method of building the target appearance model for TIR images, which is composed of two individual components, the target mask generation part in Section 3.1 and the mask sparse representation part in Sections 3.2 and 3.3. Then, the proposed appearance model is applied to an improved particle filter framework with discriminative particle selection to achieve TIR object tracking, which is illustrated in Section 3.4. The algorithm overview and update strategy are shown in Section 3.5.

Besides, we use a uniform rule to define the notations in the following context. Capital letters are used to define matrices, bold lowercase letters are used to define vectors, and ordinary lowercase letters are used to define scalars.

3.1. Target Mask Generation

The network structure of the VGG-Net19 has received considerable attention in many CNN based trackers [7,24,40]. In this work, we adopt the popular VGG-Net19 pre-trained on the ImageNet dataset and transfer the first four convolutional layers of it to extract features of the TIR images. To obtain the high-level semantic attributes specific to the target area, we propose to add a channel selection layer after the layer of conv 4-4 to account for the channel entry with target area enhancement. This process is shown in Figure 2.

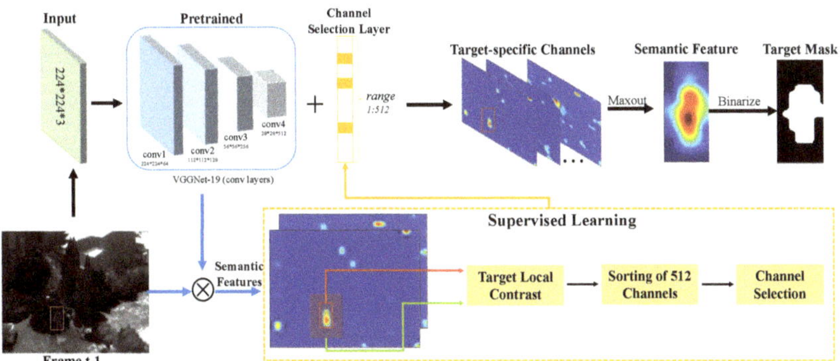

Figure 2. Illustration of generating binary mask template of the target based on CNN features.

In the online training stage, our goal is to use the given target and background classification labels to obtain high-level feature channels specific to the target area. The feature maps are firstly resized to the same size as the input image. Then, we use the local contrast value to evaluate the saliency of the target area in the feature maps. Denote $T_{x,y} \in R^{w \times h}$ as the target area, where (x,y) and (w,h) represent the target center position and target size, respectively, which are calibrated in the last frame. Its surrounding background is denoted as $B_{x,y} \in R^{w(1+s) \times h(1+s)}$, which is centered on (x,y) and is s

times larger than the target size. The average gray values of the target and its surrounding background are defined as follows:

$$t_{x,y} = \frac{1}{n_T} \sum T_{x,y} \tag{1}$$

$$b_{x,y} = \frac{1}{n_B} \left(\sum B_{x,y} - \sum T_{x,y} \right) \tag{2}$$

where n_T and n_B denote the target pixel number and background pixel number, respectively. The contrast value c^j on the jth channel is defined as follows:

$$c^j = \frac{t^j_{x,y}}{b^j_{x,y}} \tag{3}$$

where $t^j_{x,y}$ and $b^j_{x,y}$ are the target area and background area extracted from the jth channel. After the local contrast values of all L channels are sorted, the indicating values of the first few channels are set to 1 and others are set to 0, which forms the channel selection layer. In this way, channels corresponding to larger local contrast are output as target-specific feature maps, while other entries are removed. Assuming that each feature map models a single part or multiple parts of the target, we adopt a maxout operation to extract useful target information among the output channels. The obtained feature map is further binarized to form a binary mask template of the target $m \in R^d$, where d is the dimension of the target.

3.2. Mask Sparse Representation Model

By adding the binary mask template m to the input infrared image, the tracking object is divided into two partitions. Pixels corresponding to label 1 definitely belong to the reliable target part, while pixels corresponding to label 0 are denoted as the unreliable target part. Let $Y = \{y_1, y_2, \ldots y_n\} \in R^{d \times n}$ denote the candidate target set, where d and n represent the dimension of the target and the number of candidates, respectively. Let $D = [D_{pos}, D_{neg}]$ denote the dictionary base, which is composed of a positive dictionary set $D_{pos} = \{d_1, d_2, \ldots d_p\}$ and a negative dictionary set $D_{neg} = \{d_{p+1}, d_{p+2}, \ldots d_{p+q}\}$. Thus, the reliable candidate partition is denoted as $T_r = \{m \otimes y_1, m \otimes y_2, \ldots m \otimes y_n\}$, the unreliable candidate partition is denoted as $T_{r'} = \{(1 - m) \otimes y_1, (1 - m) \otimes y_2, \ldots (1 - m) \otimes y_n\}$, the reliable dictionary partition is denoted as $D_r = \{m \otimes d_1, m \otimes d_2, \ldots m \otimes d_{p+q}\}$, and the unreliable dictionary partition is denoted as $D_{r'} = \{(1 - m) \otimes d_1, (1 - m) \otimes d_2, \ldots (1 - m) \otimes d_{p+q}\}$. We use the reliable dictionary partition as the basis to reconstruct the reliable candidate partition. Meanwhile, the unreliable dictionary partition is utilized as the basis to reconstruct the unreliable candidate partition. The mask sparse representation model is shown as follows:

$$\arg\min_{x_r, x_{r'}} \frac{w}{2} \|D_r x_r - y_r\|_2^2 + \frac{1}{2} \|D_{r'} x_{r'} - y_{r'}\|_2^2 + \lambda_1 \|x_r\|_1 + \lambda_2 \|x_{r'}\|_1 + \lambda_3 \|x_r - x_{r'}\|_1 \tag{4}$$

where x_r and $x_{r'}$ are the sparse coefficient vectors corresponding to representation of the reliable target part and the unreliable target part, respectively. w is the reliable weight, which is a constant larger than 1. λ_1, λ_2 and λ_3 are balance parameters.

The first and second terms of Equation (4) represent the reconstruction error of the reliable target part and the unreliable target part, respectively. According to Section 3.1, the reliable part is the target area corresponding to more salient semantic features, which means this part has better discriminative ability on distinguishing the target from its surrounding background compared with the unreliable part. Therefore, a larger weight is assigned to the first penalty function to ensure a higher reconstruction accuracy of the reliable target part. When w is set to 1, these two terms can be combined together, and the mask sparse representation model is equal to the traditional sparse representation model.

For the representation of a single candidate, the obtained non-zero coefficients of the reliable part and the unreliable part may correspond to different dictionary subsets, which will cause ambiguity on deciding which category the candidate area belongs to. To solve this problem, a constraint term $\|x_r - x_{r'}\|_1$ is added to the mask sparse representation model. The difference between the coefficients x_r and $x_{r'}$ is induced to be sparse by an l_1 norm, which aims to encourage one candidate target to share the same template basis d across different target partitions.

3.3. Optimization Approach

The objective function defined in Equation (4) is a convex problem which includes two variables x_r and $x_{r'}$ to be solved. We adopt the alternating direction method of multipliers (ADMM) to optimize one variable by fixing another one. More in detail, we first solve over $x_r^{(k+1)}$ given $\left(x_{r'}^{(k)}, z_1^{(k)}, z_3^{(k)}, u_1^{(k)}, u_3^{(k)}\right)$, and then for $x_{r'}^{(k+1)}$ given $\left(x_r^{(k+1)}, z_2^{(k)}, z_3^{(k)}, u_2^{(k)}, u_3^{(k)}\right)$. The algorithm flow of ADMM is summarized in Algorithm 1. See Appendix A for formula derivation.

Algorithm 1 Optimization approach for solving the proposed mask sparse representation model via ADMM

Input: dictionary D_r and $D_{r'}$, candidate y_r and $y_{r'}$, reliable weight w, regularized parameters λ_1, λ_2 and λ_3, penalty parameters ρ_1, ρ_2 and ρ_3, relaxation parameters α, iteration number MAX_ITER
Initialize: $x_{r'}^{(k)} = z_1^{(k)} = z_2^{(k)} = z_3^{(k)} = u_1^{(k)} = u_2^{(k)} = u_3^{(k)} = 0 \in R^{(p+q)\times 1}$
while not converged **do**

Step 1: update variable $x_r^{(k+1)}$: $x_r^{(k+1)} = \arg\min_{x_r} L_{\rho_1,\rho_3}\left(x_r; x_{r'}^{(k)}, z_1^{(k)}, z_3^{(k)}, u_1^{(k)}, u_3^{(k)}\right)$

Step 2: update variable $x_{r'}^{(k+1)}$: $x_{r'}^{(k+1)} = \arg\min_{x_{r'}} L_{\rho_2,\rho_3}\left(x_{r'}; x_r^{(k+1)}, z_2^{(k)}, z_3^{(k)}, u_2^{(k)}, u_3^{(k)}\right)$

Step 3: update auxiliary variables $z_1^{(k+1)}$, $z_2^{(k+1)}$ and $z_3^{(k+1)}$:

$$z_1^{(k+1)} = \arg\min_{z_1} L_{\rho_1}\left(z_1; x_r^{(k+1)}, u_1^{(k)}\right)$$
$$z_2^{(k+1)} = \arg\min_{z_2} L_{\rho_2}\left(z_2; x_{r'}^{(k+1)}, u_2^{(k)}\right)$$
$$z_3^{(k+1)} = \arg\min_{z_3} L_{\rho_3}\left(z_3; x_r^{(k+1)}, x_{r'}^{(k+1)}, u_3^{(k)}\right)$$

Step 4: update dual variables $u_1^{(k+1)}$, $u_2^{(k+1)}$, $u_3^{(k+1)}$:

$$u_1^{(k+1)} \Big/ \rho_1 = u_1^{(k)} \Big/ \rho_1 + \left(x_r^{(k+1)} - z_1^{(k+1)}\right)$$
$$u_2^{(k+1)} \Big/ \rho_2 = u_2^{(k)} \Big/ \rho_2 + \left(x_{r'}^{(k+1)} - z_2^{(k+1)}\right)$$
$$u_3^{(k+1)} \Big/ \rho_3 = u_3^{(k)} \Big/ \rho_3 + \left(x_r^{(k+1)} - x_{r'}^{(k+1)} - z_3^{(k+1)}\right)$$

end while
Output: sparse coefficient vectors $x_r^{(k+1)}$, $x_{r'}^{(k+1)}$

3.4. Particle Filter Framework with Discriminative Particle Selection

In the particle filter-based tracking method, the posterior distribution of the target state Z_t at time t is approximated by a finite set of particles $I^{1:t}$ via the Bayesian inference:

$$p\left(Z^t \Big| I^{1:t}\right) \propto p\left(I^t \Big| Z^t\right) \int p\left(Z^t \Big| Z^{t-1}\right) p\left(Z^{t-1} \Big| I^{1:t-1}\right) dZ^{t-1} \tag{5}$$

where $p\left(Z^t \Big| Z^{t-1}\right)$ represents the state transition model and $p\left(I^t \Big| Z^t\right)$ is the observation model. The optimal target state for time t is obtained from the maximal estimation of $p\left(Z^t \Big| I^{1:t}\right)$. Thus, the construction of these two models formulate the core problem of object tracking.

In our tracking approach, the mask sparse representation method is employed as the observation model, where reconstruction errors generated from two target partitions are adopted to calculate the likelihood probability of candidate samples:

$$p\left(I^t\,\middle|\,Z^t\right) = exp\left(-\frac{\|D_r\,(:,1:p)\,x_r\,(1:p) - y_r\|_2^2 + \|D_{r'}\,(:,1:p)\,x_{r'}\,(1:p) - y_{r'}\|_2^2}{\sigma^2}\right) \tag{6}$$

From Equation (6), we can see that the efficiency of the likelihood estimation is determined by the number of particles at time t. In the traditional particle filter framework, the state parameters of Z^t are generally denoted as $(x, y, s, \theta, \alpha, \phi)$, which represent displacement in x-axis, displacement in y-axis, scale, rotation, aspect ratio and skew angle, respectively [19]. In the conventional particle filter method, the state transition parameters between two frames are modeled by Gaussian distribution, with every state parameter being treated independently with each other:

$$p\left(Z^t\,\middle|\,Z^{t-1}\right) = N\left(Z^t; Z^{t-1}, \Phi\right) \tag{7}$$

where $\Phi = \left(\sigma_x, \sigma_y, \sigma_s, \sigma_\theta, \sigma_\alpha, \sigma_\phi\right)$ represents the affine variance. To ensure that the real target state is covered in the state transition process, many particles are needed, which will increase the computation cost of solving the mask sparse model. The visualization of the random particle sampling modeled by Gaussian distribution is shown in Figure 3a. To address this contradictory issue, we propose a discriminative particle selection method to construct the state model more effectively.

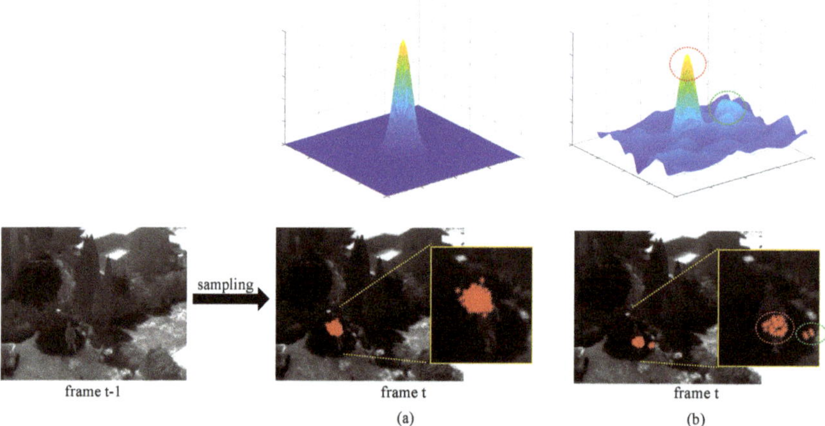

Figure 3. Visualization of particle distribution: (**a**) 300 particles are sampled, which are modeled by the Gaussian distribution; snf (**b**) 50 discriminative particles are drawn according to the peak values of the response map obtained from the correlation filter.

We note that the output of the correlation filter [41] can provide a rough prediction of the existence of the tracking object. On the other hand, the training of the correlation filter is very efficient, which can achieve millisecond order of magnitude. As shown in Figure 3b, the positions of the peak values appearing on the response map are selected as latent target states, to which the target areas correspond are further modeled by the mask sparse representation method. In the simple scenario, there is a single peak in the response map, which is the position of the target. In complex scenarios, multiple peaks appear in the response map, as shown in Figure 3. These local peaks have potential discriminative ability for the target and are selected to form the candidate set. After obtaining the placement state of the target, a scale filter is applied to obtain the optimal target scale, the details of which are described in [42].

3.5. Algorithm Overview and Update Strategy

The algorithm flow of our proposed tracking approach is shown in Algorithm 2. The method of obtaining the target mask has been described in Section 3.1. Detailed theory of the correlation filter and the scale filter can be found in [41,42]. Steps 1–5 of the tracking implementation process are described in Sections 3.2 and 3.3. In this subsection, we first introduce the details on how to construct and update dictionary for target representation, and then present the update criteria for Steps 7–9.

Algorithm 2 The proposed approach for TIR object tracking

Input:

image sequence $\left\{ f_1, f_2, \ldots f_{frame\_end} \right\}$
target position in the first frame s_1
target deep features in the first frame $feature_1$

Initialize:
construct object dictionary D
obtain target mask m
correlation filter
scale filter

for $f = 2$ to $frame\_end$ **do**

    1. generate discriminative particles with correlation filter
    2. construct the mask sparse representation model according to Eq (4)
    3. compute the likelihood value of each particle (candidate) by Eq (14)
    4. obtain the optimal target position
    5. compute the optimal scale factor by scale filter
    6. update object dictionary D
    7. update target mask m
    8. update correlation filter
    9. update scale filter
end for

Output: target states: $s_2 : s_{frame\_end}$

In this work, positive and negative dictionaries are constructed separately. The target state in the first frame is initialized by the ground truth data. Firstly, we adopt the areas surrounding the real target position as positive templates, and areas far away from the real target position as negative templates. Then, the eigenbasis vectors extracted from the positive template set are employed as the positive dictionary basis, which aims to preserve the information different observations have in common. The negative templates are directly utilized as the negative dictionary basis. Both the positive dictionary and the negative dictionary need to be updated in the tracking process to adapt to target appearance changes, as well as scene variations. For the positive dictionary, the target templates need to be updated frequently due to the inevitable appearance changes caused by target motion. However, if we update the templates too frequently, wrong tracking results may be introduced into the template set and cause tracking drift. Thus, we employ the cumulative probability-based method [21] to update the earlier accurate tracking results at a slow pace and update the newly entrant templates at a fast pace. The update probabilities for templates from older to newer ones are generated as:

$$L_p = \left\{ 0, \frac{1}{2^n - 1}, \frac{3}{2^n - 1}, \ldots 1 \right\} \tag{8}$$

The template to be replaced is determined by which interval the random number $r \in [0,1]$ lies in. The new positive dictionary is formulated by adding p to the end of the old dictionary:

$$q = \arg\min_{q} \frac{1}{2} \left\| p - \begin{bmatrix} U & D_{neg} \end{bmatrix} \begin{bmatrix} q \\ e \end{bmatrix} \right\|_2^2 + \lambda \left\| \begin{bmatrix} q \\ e \end{bmatrix} \right\|_1 \tag{9}$$

where U represents the eigenbasis vectors and p is the new observation. The new entrant q is the target area removing noises and occlusion.

We propose a relatively strict criterion to update the negative dictionary with a slow pace to avoid bringing the target into it. The likelihood probability of the optimal observation in the second frame is denoted as a reference value $conf_{ref}$. When the maximum likelihood probability in the current frame exceeds $th \times conf_{ref}$, the current tracking result is regarded as a reliable new target. Then, the background areas extracted from this frame are used to form the new negative dictionary. Otherwise, the negative dictionary remains unchanged.

When the target result is considered to be reliable, the target mask, correlation filter and scale filter are updated with a fixed learning rate. Equation (10) takes the update for target mask as an example.

$$m = (1 - \gamma)\, m_{old} + \gamma m_{new} \tag{10}$$

4. Experiments

We first set the experiment environment in Section 4.1, including the parameters of our tracking approach and the testing dataset. The evaluation metrics for method comparison are introduced in Section 4.2. The parameter setting for optimization is discussed in Section 4.3. The quantitative and qualitative comparisons of our tracker with other state-of-the-art methods are given in Sections 4.4 and 4.5, respectively.

4.1. Experiment Setup

The corresponding parameters of our tracker are given as follows. In the candidate searching stage, we crop a searching area which is 1.5 times larger than the size of the target in the last frame. The regularization parameter of the KCF tracker is set to 10^{-4}. Fifty discriminative particles are drawn according to the peak values of the correlation filter response map. In the mask sparse representation stage, the infrared images are input into the VGG-Net19 pre-trained on the ImageNet dataset to extract deep features. Ten channels are selected from the convolution layer conv 4-4 as the output of target specific feature maps. The weight of the fidelity term for the reliable target part is set to 1.5. The regularization parameters of the MaskSR model λ_1, λ_2 and λ_3 are set to 0.01, 0.01 and 0.005, respectively. In the optimization stage, the penalty parameters ρ_1, ρ_2 and ρ_3 are set to 1. For the scale searching, we use the same parameters as DSST method [42], which includes 17 scales with a scale factor of 1.02. The learning rates of the correlation filter and scale filter are set to 0.01 and 0.1, respectively. The update rate of the binary mask is set to 0.01. We conducted the simulation experiments of our proposed method in Matlab 2017b combined with the Matconvnet toolbox. The proposed method ran at 1.2 fps averagely on a laptop with an Intel i7-6700HQ CPU at 2.60 GHz and 16.0 GB RAM.

We carried out the comparison experiment on the VOT-TIR2016 benchmark. This dataset includes 25 TIR sequences, with the minimum length of 92 frames and the maximum length of 1420 frames. The tracking objects include pedestrian, vehicle and animal with five challenging attributes annotated on each frame: camera motion, dynamics change, motion change, occlusion and size change.

4.2. Evaluation Metrics

The benchmark for VOT-TIR2016 has a re-start scheme, which means when the tracking fails, the tracker will be re-initialized after five frames. Accordingly, two performance measures, accuracy (A) and robustness (R), are used as evaluation metrics [43]. The accuracy is calculated by the overlap rate between the predicted bounding box and the ground truth during successful tracking period. The robustness measures the likelihood that the tracker will not fail in S frames, which is based on the number of tracking failures in a new sequence. It is calculated by:

$$R_0 = \sum_{j=0}^{Q} F^j$$
$$R = e^{-S\frac{R_0}{Q}}$$

(11)

where Q represents the sequence length on each attributes and F^j is the failure number. Another measure called expected average overlap (EAO) is used to combine A and R together. To calculate this measure, the tracker is only initialized at the beginning of the sequence. When it drifts off the target, the remaining overlap rate is set to 0. Thus, the average overlap is computed by:

$$\Phi_{N_s} = \frac{1}{N_s} \sum_{i=1}^{N_s} \Phi_i$$

(12)

where Φ_i is the per-frame overlap including the zero overlaps after failure. The EAO measure Φ is calculated over an interval $[N_{lo}, N_{hi}]$ as follows. The interval is provided by the benchmark.

$$\Phi = \frac{1}{N_{hi} - N_{lo}} \sum_{N_s = N_{lo}:N_{hi}} \Phi_{N_s}$$

(13)

4.3. Parameter Analysis

Several parameters play important roles in solving the MaskSR model. In this section, we set two comparison experiments to discuss the effect of the penalty parameter ρ and the regularization parameter λ_3 on the convergence of ADMM.

(1) Effect of ρ_1, ρ_2 and ρ_3

The penalty parameter ρ is usually set to 1 in the standard ADMM algorithm. To test the effect of different ρ on the convergence speed, we conducted several numerical examples. The convergence of ADMM was evaluated by the primal residuals $\left\|r^{(k+1)}\right\|_2$ and dual residuals $\left\|s^{(k+1)}\right\|_2$, which are denoted by:

$$r^{(k+1)} = x^{(k+1)} - z^{(k+1)}$$
$$s^{(k+1)} = z^{(k+1)} - z^{(k)}$$

(14)

Figure 4a shows the dual residuals and primal residuals when $\rho_1 = 0.8, 1.0, 1.2$, respectively. Similarly, Figure 4b,c shows the convergence performance with different ρ_2 and ρ_3. We can see that, with the increase of ρ, the convergence speed of dual residuals decreases; however, the convergence speed of primal residuals improves. Thus, we define $\rho = 1$ to balance the convergence performance of these two characters.

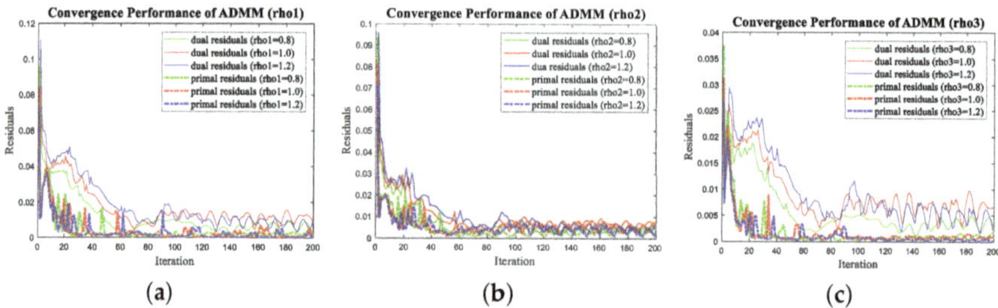

Figure 4. Convergence of primal residuals and dual residuals with different penalty parameters: (**a**) testing on penalty ρ_1; (**b**) testing on penalty ρ_2; and (**c**) testing on penalty ρ_3.

(2) Effect of λ_3

The parameter λ_3 influences the sparseness degree of $x_r - x_r'$. A larger λ_3 can lead to a better performance on refining the representation result of the unreliable target part. However, when λ_3 is set too large, the optimization process cannot converge. As shown in Figure 5, when λ_3 is set to 0.01, which is equal to the value of λ_1 and λ_2, both the dual residual plot (Figure 5a) and the primal residual plot (Figure 5b) diverge. Thus, we set λ_3 to 0.005 to guarantee the convergence of the optimization process.

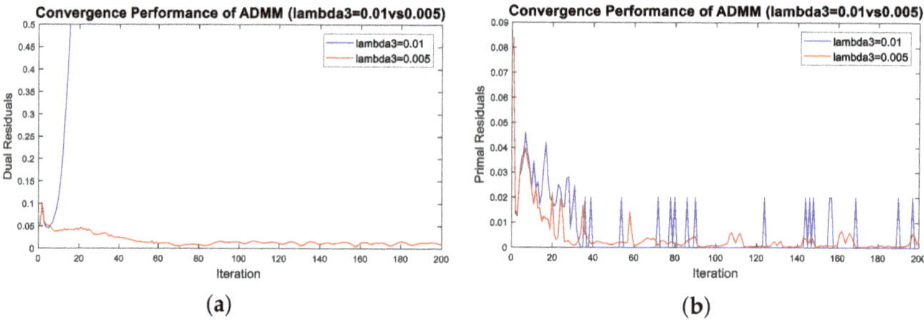

Figure 5. Convergence of primal residuals and dual residuals with different regularization parameters: (**a**) primal residuals of $\lambda_3 = 0.01$ vs. $\lambda_3 = 0.005$; and (**b**) dual residuals of $\lambda_3 = 0.01$ vs. $\lambda_3 = 0.005$.

4.4. Quantitative Comparison

We compared our tracker with other 19 state-of-the-art trackers on VOT-TIR2016 in the quantitative comparison experiment: two convolutional neural network based trackers, deepMKCF [44] and MDNet_NoTrain [43]; six discriminative correlation filter-based trackers, DSST [42], MvCFT [45], NSAMF [46], SKCF [47], SRDCF [48] and Staple+ [43]; seven part-based trackers, BDF [49], BST [43], DPCF [50], DPT [51], FCT [43], GGT2 [52] and LT_FLO [43]; one mean-shift based tracker, PKLTF [49]; one tracking-by-detection tracker, DAT [43]; and two fusion based trackers, LOFT_Lite [43] and MAD [43]. We removed the SRDCFir tracker [43] because it uses motion threshold to focus more on the performance evaluation of the target appearance model of different trackers.

There are three types of AR raw plot and AR rank plot in Figure 6. The mean AR raw plot and mean AR rank plot were obtained by the average values and averages ranks of seven attributes (including six challenging attributes and one empty tag). The weighted mean AR raw plot and weighted mean AR rank plot take the sequence length of each attribute into account. The pooled plots gather all frames and compute values and ranks on a single combined sequence. In all three rank plots, the proposed method achieves the best robustness, which means our tracker has the least failure probability on sequences with 100 frames. In the accuracy evaluation, the proposed tracker is not as good as the MDNet_NoTrain tracker, deepMKCF tracker, Staple+ and DSST tracker according to the pooled measurement. However, the accuracy difference between these trackers is very slight. On the other hand, the low failure number of our tracker will also influence the average value of the overlap rate. Thus, we further show the EAO comparison of 20 trackers in Figures 7 and 8, which show the proposed tracker gives the best overall performance in the TIR object tracking.

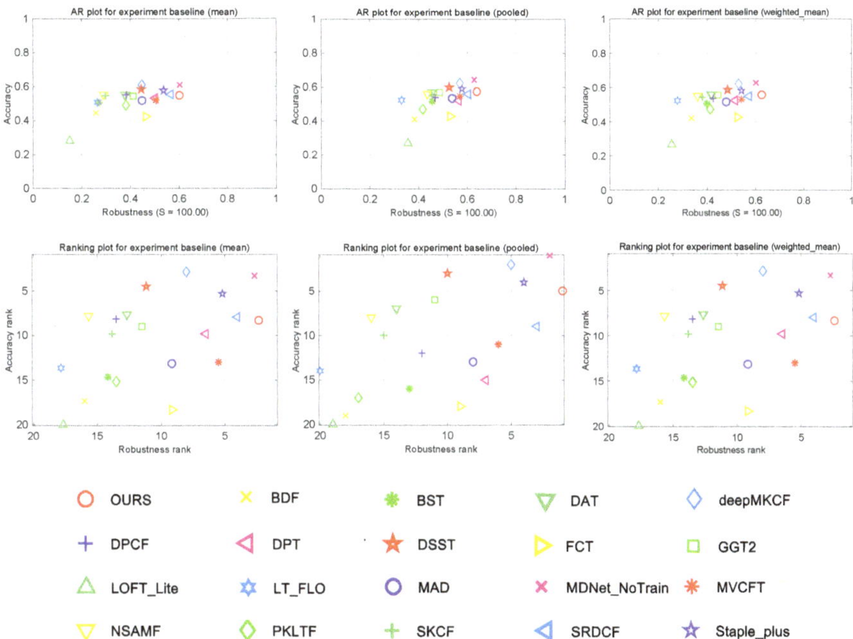

Figure 6. The overall AR raw plots and the AR rank plots of the 20 compared trackers on VOT-TIR2016.

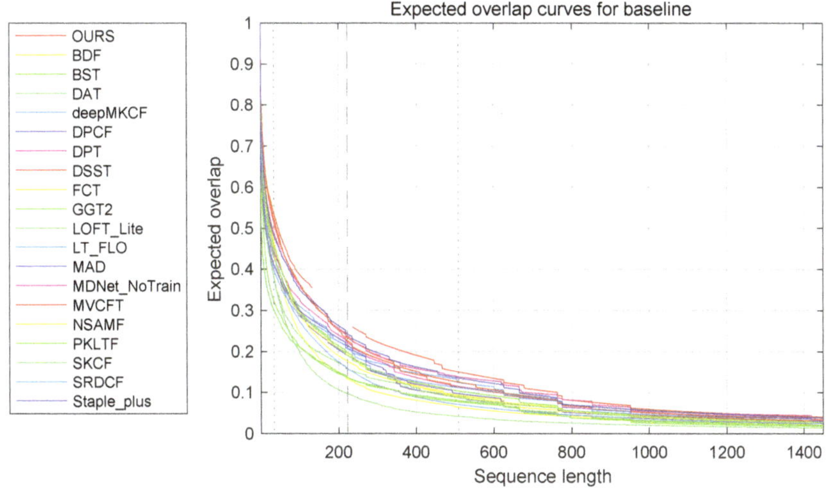

Figure 7. Expected overlap curves of the 20 compared trackers on VOT-TIR2016.

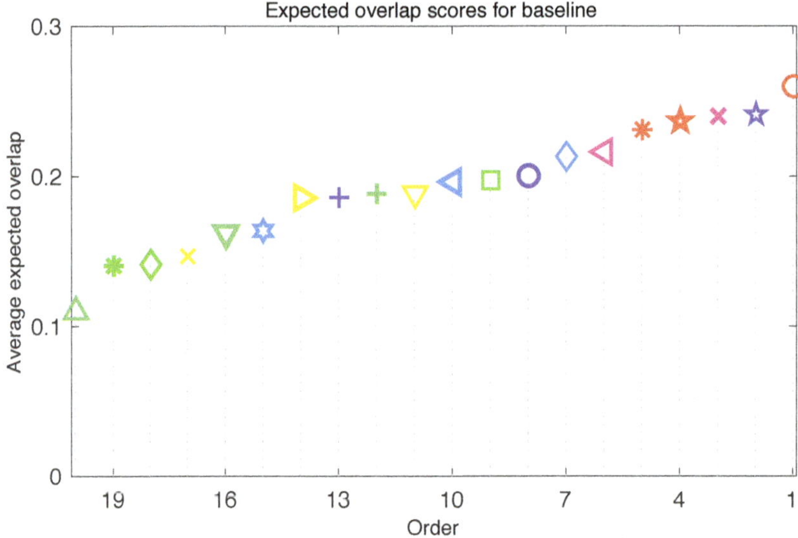

Figure 8. Expected overlap scores of 20 compared trackers on VOT-TIR2016 (see Figure 6 for legend).

To illustrate the tracking performance of trackers on different challenging scenarios, we show the accuracy ranking plot and robustness ranking plot with respect to six visual attributes in Figure 9: camera motion, dynamics change, empty tag, motion change, occlusion and size change. In the robustness evaluation, our tracker ranks first in the situation of camera motion, dynamics change, size change and empty. In the two other situations of occlusion and motion change, our tracker ranks fourth and sixth, , respectively. The MDNet_NoTrain tracker and SRDCF tracker achieve the best performance in the occlusion and motion change scenarios, respectively. According to the accuracy ranking, our tracker achieves better performance in the situation of size change, motion change and empty. By comparison, two CNN based trackers, the MDNet tracker and deepMKCF tracker, locate the target more accurately in the tracking process. As shown in Table 1, the accuracy of the MDNet_NoTrain tracker is 1.8% and 9.7% higher than the proposed tracker in the situation of empty and size change, respectively. However, the robustness of the proposed tracker is 1.5% and 4.5% higher than the MDNet_NoTrain tracker, respectively. Similarly, the accuracy of the deepMKCF tracker is 20.5% and 17.2% higher than the proposed tracker, while the robustness of the proposed tracker is 19.6% and 211% higher than the deepMKCF tracker, respectively. Generally speaking, the correlation filter based trackers and CNN based trackers have better performance on the TIR object tracking.

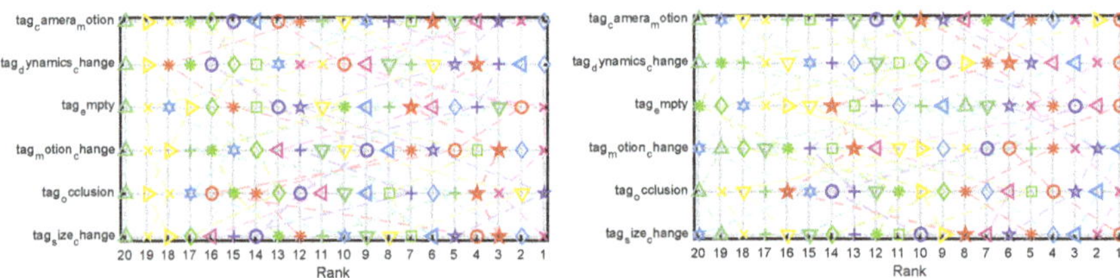

Figure 9. Accuracy ranking and robustness ranking of 20 trackers on six different attributes (see Figure 6 for legend).

Table 1. Quantitative results of expected average overlap (EAO), Accuracy (A) and Robustness (R) of the eight best trackers. The best, second best and the third best trackers in different situations are marked by */**/***, respectively.

Measurements		Staple+	MDNet_N	DSST	MVCFT	DPT	deepMKCF	MAD	Ours
ALL	EAO	0.241 **	0.240 ***	0.237	0.231	0.216	0.213	0.200	**0.260** *
Camera Motion	A	0.584 ***	0.611 **	0.559	0.520	0.561	**0.623** *	0.494	0.517
	R	0.517**	0.496***	0.410	0.465	0.418	0.490	0.382	**0.586***
Dynamics Change	A	0.568 ***	0.518	0.574 **	0.467	0.523	**0.612** *	0.483	0.522
	R	0.389 ***	0.532**	0.322	0.322	0.389 ***	0.182	0.266	**0.576** *
Empty	A	0.544	**0.624** *	0.579	0.522	0.585	0.589 ***	0.542	0.613 **
	R	0.460	0.473	0.404	0.480 *	0.480 *	0.422	0.480 *	**0.480** *
Motion Change	A	0.514	**0.613** *	0.551 ***	0.509	0.474	0.592 **	0.490	0.521
	R	**0.867** *	0.848 **	0.684	0.789 ***	0.684	0.717	0.752	0.752
Occlusion	A	**0.658** *	0.627 **	0.625 ***	0.562	0.573	0.607	0.570	0.520
	R	0.591 **	**0.664** *	0.349	0.468	0.496	0.468	0.392	0.557 ***
Size Change	A	0.595	**0.654** *	0.612 ***	0.544	0.474	0.643 **	0.520	0.596
	R	0.627	0.682 **	0.607	0.627	0.607	0.637 ***	0.560	**0.713** *

4.5. Qualitative Comparison

To display the tracking results more intuitively, we give a qualitative comparison for eight trackers with better EAO ranks in the quantitative experiment, which is shown in Figure 10. Due to the re-start scheme in the VOT-TIR2016 benchmark, there is no sense in displaying the predicted bounding box for the sequence frames after re-initialization. Thus, when a tracker drifts off the target, the later tracker results are placed on top left corner of the images without re-initialization. Six representative sequences are selected in the qualitative experiment: "boat2", "crouching", "quadrocopter", "car2", "garden" and "excavator". Generally speaking, the proposed method has a better performance than the seven other trackers. In Figure 10a ("boat2"); the predicted bounding boxes of the SRDCF and MvCFT tracker are far larger than the real target size. In the sequence "crouching" shown in Figure 10b, four trackers, namely Staple+, SRDCF, DPT, and deepMKCF, fail to locate the target when the target is occluded by another person. Targets in other two sequences, "car2" and "garden", also suffer from severe occlusion; only the proposed method locates the target correctly among the eight trackers. For the sequence "quadrocopter" shown in Figure 10c, the appearance change of the target is slight, however the background around target has a dramatic variation in the tracking process. The proposed method uses the binary mask to extract reliable target part, which can improve the tracking performance in the situation of background clutter significantly. The target in Figure 10f is almost submerged in the background. Only the MDNet_NoTrain and the proposed trackers track the target successfully.

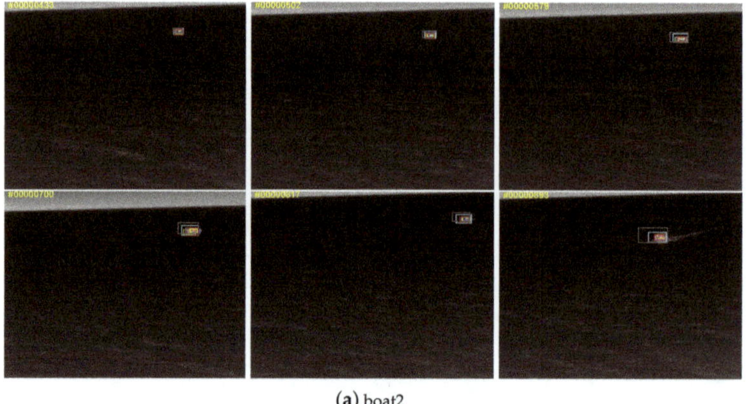

(**a**) boat2

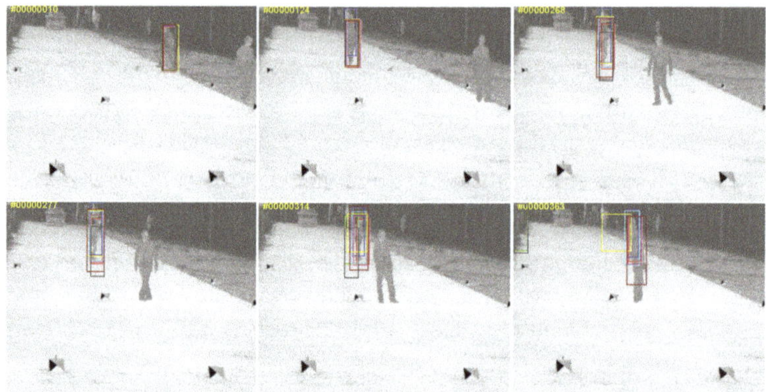

(**b**) crouching

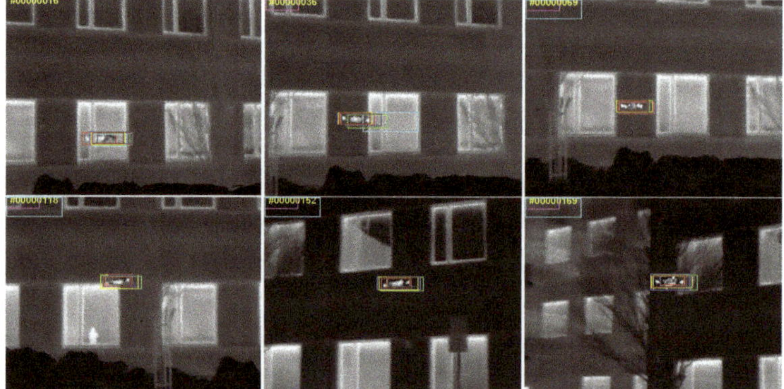

(**c**) quadrocopter

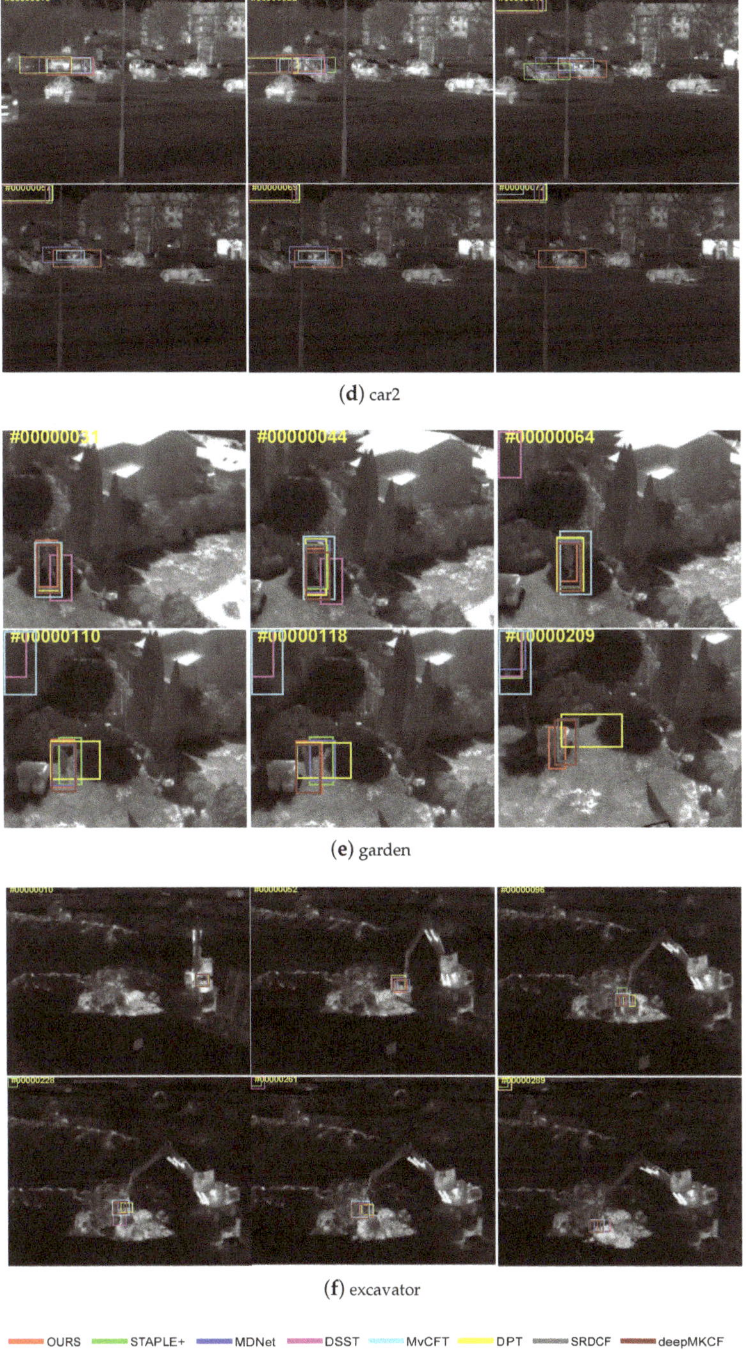

(**d**) car2

(**e**) garden

(**f**) excavator

Figure 10. Visualized tracking results of several state-of-the-art trackers on representative sequences.

5. Conclusions

In this paper, we propose a MaskSR-based appearance model to achieve TIR target tracking in an improved particle filter framework. This model considers different discriminant capabilities of different target parts at a pixel level, which can enhance the importance of the distinguishable target pixels in the reconstruction process while weakening the diverse effect of target appearance changes and background clutters. Moreover, to improve the tracking efficiency, a discriminative particle selection strategy is proposed to replace the previous random sampling strategy, which can greatly reduce the number of represented particles and improve the tracking accuracy simultaneously. The proposed method was evaluated on the VOT-TIR2016 benchmark with a re-initialized scheme when tracking fails. The experiment results of accuracy, robustness and expected average overlap show that the proposed tracker is superior to 19 other state-of-the-art trackers for TIR object tracking. Future improvement can be made by applying a regression-based strategy to train the channel selection layer and using a more accurate segmentation method to divide the target.

Considering applying the proposed method to real applications, future improvement can be made by redesigning the program using C or C++, which are advantageous for running speed and are more convenient to be transplanted to the hardware platform. On the other hand, the improvement of sensors on imaging quality will significantly improve the accuracy and robustness of the proposed tracking in the real application.

Author Contributions: M.L. conceptualized and performed the algorithm, analyzed the experiment data and wrote the paper; Z.P. is the research supervisor; L.P. and Y.C. helped modify the language; and S.H. and F.Q. provided technical assistance to the research. The manuscript was discussed by all co-authors.

Funding: This research was funded by National Natural Science Foundation of China (61571096 and 61775030), the Key Laboratory Fund of Beam Control, Chinese Academy of Science (2017LBC003), Sichuan Science and Technology Program (2019YJ0167) and Minnan Normal University Teaching Reform (JG201918).

Conflicts of Interest: The authors declare no conflict of interest.

Appendix A

The ADMM algorithm is designed to solve equality constrained problems. Thus, we rewrite Equation (4) in the following form by introducing auxiliary variables z_1, z_2 and z_3:

$$\min \frac{w}{2}\|D_r x_r - y_r\|_2^2 + \frac{1}{2}\|D_{r'} x_{r'} - y_{r'}\|_2^2 + \lambda_1\|z_1\|_1 + \lambda_2\|z_2\|_1 + \lambda_3\|z_3\|_1$$
$$s.t. \begin{cases} x_r - z_1 = 0 \\ x_{r'} - z_2 = 0 \\ x_r - x_{r'} - z_3 = 0 \end{cases} \tag{A1}$$

The augmented Lagrangian expression of Equation (A1) is formulated as:

$$\begin{aligned} L_{\rho_1,\rho_2,\rho_3}\left(x_r, x_{r'}, z_1, z_2, z_3, u_1, u_2, u_3\right) &= \frac{w}{2}\|D_r x_r - y_r\|_2^2 + \frac{1}{2}\|D_{r'} x_{r'} - y_{r'}\|_2^2 \\ &+ \lambda_1\|z_1\|_1 + \lambda_2\|z_2\|_1 + \lambda_3\|z_3\|_1 + \langle u_1, x_r - z_1 \rangle + \frac{\rho_1}{2}\|x_r - z_1\|_2^2 + \langle u_2, x_{r'} - z_2 \rangle \\ &+ \frac{\rho_2}{2}\|x_{r'} - z_2\|_2^2 + \langle u_3, x_r - x_{r'} - z_3 \rangle + \frac{\rho_3}{2}\|x_r - x_{r'} - z_3\|_2^2 \end{aligned} \tag{A2}$$

For Steps 1 and 2 in Algorithm 1, the solution for these two sub-problems can be easily derived as:

$$\begin{aligned} x_r^{(k+1)} &= \arg\min_{x_r} \frac{w}{2}\|D_r x_r - y_r\|_2^2 + \frac{\rho_1}{2}\left\|x_r - z_1^{(k)} + u_1^{(k)}\big/\rho_1\right\|_2^2 + \frac{\rho_3}{2}\left\|x_r - x_{r'}^{(k)} - z_3^{(k)} + u_3^{(k)}\big/\rho_3\right\|_2^2 \\ &= \left(wD_r'D_r + \rho_1 I + \rho_2 I\right)^{-1}\left(wD_r'y_r + \rho_1\left(z_1^{(k)} - \frac{u_1^{(k)}}{\rho_1}\right)\right) + \rho_3\left(x_{r'}^{(k)} + z_3^{(k)} - \frac{u_3^{(k)}}{\rho_3}\right) \end{aligned} \tag{A3}$$

$$x_{r'}^{(k+1)} = \arg\min_{x_r'} \tfrac{1}{2} \left\| D_{r'} x_{r'} - y_{r'} \right\|_2^2 + \tfrac{\rho_2}{2} \left\| x_{r'} - z_2^{(k)} + u_2^{(k)} \middle/ \rho_2 \right\|_2^2 + \tfrac{\rho_3}{2} \left\| x_{r'} - x_r^{(k+1)} + z_3^{(k)} - u_3^{(k)} \middle/ \rho_3 \right\|_2^2$$
$$= \left(D_{r'}{}' D_{r'} + \rho_2 I + \rho_3 I \right)^{-1} \left(w D_{r'}{}' y_{r'} + \rho_2 \left(z_2^{(k)} - \frac{u_2^{(k)}}{\rho_2} \right) \right) + \rho_3 \left(x_r^{(k+1)} - z_3^{(k)} + \frac{u_3^{(k)}}{\rho_3} \right) \tag{A4}$$

Obviously, $\left(w D_{r}{}' D_r + \rho_1 I + \rho_2 I \right)^{-1}$ and $\left(D_{r'}{}' D_{r'} + \rho_2 I + \rho_3 I \right)^{-1}$ can be pre-calculated because they are not included in the iteration process. The computation cost of solving this sub-problem is $O\left((p+q) \times d \right)$.

For Step 3, due to the presence of the non-derivate function $\|z_i\|_1$ in the optimization problem, we need to introduce the soft-threshold operator to solve these sub-problems. This operator is defined as follows:

$$S_{\lambda/\rho}(x) = sign(x) \max \left\{ |x| - \frac{\lambda}{\rho}, 0 \right\} \tag{A5}$$

where x is a scalar, representing the elements in a vector. Thus, the solution of Step 3 is:

$$z_1^{(k+1)} = S_{\lambda_1/\rho_1} \left(x_r^{(k+1)} - \frac{u_1(x)}{\rho_1} \right) \tag{A6}$$

Similarly,

$$z_2^{(k+1)} = S_{\lambda_2/\rho_2} \left(x_{r'}^{(k+1)} - \frac{u_2(x)}{\rho_2} \right) \tag{A7}$$

$$z_3^{(k+1)} = S_{\lambda_3/\rho_3} \left(x_{r'}^{(k+1)} - x_r^{(k+1)} - \frac{u_3(x)}{\rho_3} \right) \tag{A8}$$

The computation cost of this sub-problem is $O(p+q)$.

References

1. Li, C.; Sun, X.; Wang, X.; Zhang, L.; Tang, J. Grayscale-Thermal Object Tracking via Multitask Laplacian Sparse Representation. *IEEE Trans. Syst. Man Cybern. Syst.* **2017**, *47*, 673–681. [CrossRef]
2. Zhang, L.; Peng, L.; Zhang, T.; Cao, S.; Peng, Z. Infrared Small Target Detection via Non-Convex Rank Approximation Minimization Joint l2,1 Norm. *Remote Sens.* **2018**, *10*, 1821. [CrossRef]
3. Zhang, L.; Peng, Z. Infrared Small Target Detection Based on Partial Sum of the Tensor Nuclear Norm. *Remote Sens.* **2019**, *11*, 382. [CrossRef]
4. Zhang, T.; Wu, H.; Liu, Y.; Peng, L.; Yang, C.; Peng, Z. Infrared Small Target Detection Based on Non-Convex Optimization with Lp-Norm Constraint. *Remote Sens.* **2019**, *11*, 559. [CrossRef]
5. Yu, X.; Yu, Q.; Shang, Y.; Zhang, H. Dense structural learning for infrared object tracking at 200+ Frames per Second. *Pattern Recognit. Lett.* **2017**, *100*, 152–159. [CrossRef]
6. Berg, A.; Ahlberg, J.; Felsberg, M. Channel coded distribution field tracking for thermal infrared imagery. In Proceedings of the IEEE Conference on Computer Vision and Pattern Recognition Workshops (CVPRW), Las Vegas, NV, USA, 26 June–1 July 2016; pp. 9–17.
7. Liu, Q.; Lu, X.; He, Z.; Zhang, C.; Chen, W.S. Deep convolutional neural networks for thermal infrared object tracking. *Knowl. Based Syst.* **2017**, *134*, 189–198. [CrossRef]
8. Li, X.; Liu, Q.; Fan, N.; He, Z.; Wang, H. Hierarchical spatial-aware Siamese network for thermal infrared object tracking. *Knowl. Based Syst.* **2019**, *166*, 71–81. [CrossRef]
9. Qian, K.; Zhou, H.; Wang, B.; Song, S.; Zhao, D. Infrared dim moving target tracking via sparsity-based discriminative classifier and convolutional network. *Infrared Phys. Technol.* **2017**, *86*, 103–115. [CrossRef]
10. Zulkifley, M.A.; Trigoni, N. Multiple-Model Fully Convolutional Neural Networks for Single Object Tracking on Thermal Infrared Video. *IEEE Access* **2018**, *6*, 42790–42799. [CrossRef]
11. Zhang, L.; Gonzalez-Garcia, A.; Weijer, J.V.d.; Danelljan, M.; Khan, F.S. Synthetic Data Generation for End-to-End Thermal Infrared Tracking. *IEEE Trans. Image Process.* **2019**, *28*, 1837–1850. [CrossRef] [PubMed]

12. Shi, Z.; Wei, C.; Fu, P.; Jiang, S. A Parallel Search Strategy Based on Sparse Representation for Infrared Target Tracking. *Algorithms* **2015**, *8*, 529–540. [CrossRef]

13. He, Y.; Li, M.; Zhang, J.; Yao, J. Infrared Target Tracking Based on Robust Low-Rank Sparse Learning. *IEEE Geosci. Remote Sens. Lett.* **2016**, *13*, 232–236. [CrossRef]

14. Gao, S.J.; Jhang, S.T. Infrared Target Tracking Using Multi-Feature Joint Sparse Representation. In Proceedings of the International Conference on Research in Adaptive and Convergent Systems, Odense, Denmark, 11–14 October 2016; pp. 40–45. [CrossRef]

15. Zhang, X.; Ren, K.; Wan, M.; Gu, G.; Chen, Q. Infrared small target tracking based on sample constrained particle filtering and sparse representation. *Infrared Phys. Technol.* **2017**, *87*, 72–82. [CrossRef]

16. Lan, X.; Ye, M.; Zhang, S.; Zhou, H.; Yuen, P.C. Modality-correlation-aware sparse representation for RGB-infrared object tracking. *Pattern Recognit. Lett.* **2018**, in press. [CrossRef]

17. Li, Y.; Li, P.; Shen, Q. Real-time infrared target tracking based on l1 minimization and compressive features. *Appl. Opt.* **2014**, *53*, 6518–6526. [CrossRef]

18. Wan, M.; Gu, G.; Qian, W.; Ren, K.; Chen, Q.; Zhang, H.; Maldague, X. Total Variation Regularization Term-Based Low-Rank and Sparse Matrix Representation Model for Infrared Moving Target Tracking. *Remote Sens.* **2018**, *10*, 510. [CrossRef]

19. Bao, C.; Wu, Y.; Ling, H.; Ji, H. Real time robust l1 tracker using accelerated proximal gradient approach. In proceedings of the IEEE Conference on Computer Vision and Pattern Recognition (CVPR), Providence, RI, USA, 16–21 June 2012; pp. 1830–1837.

20. Zhang, T.; Ghanem, B.; Liu, S.; Ahuja, N. Robust visual tracking via multi-task sparse learning. In Proceedings of the IEEE Conference on Computer Vision and Pattern Recognition (CVPR), Providence, RI, USA, 16–21 June 2012; pp. 2042–2049. [CrossRef]

21. Jia, X.; Lu, H.; Yang, M. Visual tracking via adaptive structural local sparse appearance model. In Proceedings of the IEEE Conference on Computer Vision and Pattern Recognition (CVPR), Providence, RI, USA, 16–21 June 2012; pp. 1822–1829. [CrossRef]

22. Li, Z.; Zhang, J.; Zhang, K.; Li, Z. Visual Tracking With Weighted Adaptive Local Sparse Appearance Model via Spatio-Temporal Context Learning. *IEEE Trans. Image Process.* **2018**, *27*, 4478–4489. [CrossRef]

23. Zhang, T.; Xu, C.; Yang, M. Robust Structural Sparse Tracking. *IEEE Trans. Pattern Anal. Mach. Intell.* **2019**, *41*, 473–486. [CrossRef]

24. Ma, C.; Huang, J.B.; Yang, X.; Yang, M.H. Hierarchical convolutional features for visual tracking. In Proceedings of the IEEE international conference on computer vision (ICCV), Santiago, Chile, 7–13 December 2015; pp. 3074–3082,

25. Zhang, X.; Ma, D.; Ouyang, X.; Jiang, S.and Gan, L.; Agam, G. Layered optical flow estimation using a deep neural network with a soft mask. In Proceedings of the 27th International Joint Conference on Artificial Intelligence (IJCAI), Morgan Kaufmann, Stockholm, Sweden, 13–19 July 2018; pp. 1170–1176.

26. Liu, Q.; Yuan, D.; He, Z. Thermal infrared object tracking via Siamese convolutional neural networks. In Proceedings of the International Conference on Security, Pattern Analysis, and Cybernetics (SPAC), Shenzhen, China, 15–17 December 2017; pp. 1–6.

27. Gundogdu, E.; Koc, A.; Solmaz, B.; Hammoud, R.I.; Aydin Alatan, A. Evaluation of feature channels for correlation-filter-based visual object tracking in infrared spectrum. In Proceedings of the IEEE Conference on Computer Vision and Pattern recognition Workshops (CVPRW), Las Vegas, NV, USA, 26 June–1 July 2016; pp. 24–32.

28. Li, Z.; Li, J.; Ge, F.; Shao, W.; Liu, B.; Jin, G. Dim moving target tracking algorithm based on particle discriminative sparse representation. *Infrared Phys. Technol.* **2016**, *75*, 100–106. [CrossRef]

29. Li, M.; Lin, Z.; Long, Y.; An, W.; Zhou, Y. Joint detection and tracking of size-varying infrared targets based on block-wise sparse decomposition. *Infrared Phys. Technol.* **2016**, *76*, 131–138. [CrossRef]

30. Li, C.; Zhao, N.; Lu, Y.; Zhu, C.; Tang, J. Weighted Sparse Representation Regularized Graph Learning for RGB-T Object Tracking. In Proceedings of the 25th ACM International Conference on Multimedia, New York, NY, USA, 23–27 October 2017; pp. 1856–1864. [CrossRef]

31. Lan, X.; Ye, M.; Shao, R.; Zhong, B.; Jain, D.K.; Zhou, H. Online Non-negative Multi-modality Feature Template Learning for RGB-assisted Infrared Tracking. *IEEE Access* **2019**, *7*, 67761–67771. [CrossRef]

32. Lan, X.; Ye, M.; Shao, R.; Zhong, B.; Yuen, P.C.; Zhou, H. Learning Modality-Consistency Feature Templates: A Robust RGB-Infrared Tracking System. *IEEE Trans. Ind. Electron.* **2019**, *66*, 9887–9897. [CrossRef]
33. Li, C.; Cheng, H.; Hu, S.; Liu, X.; Tang, J.; Lin, L. Learning Collaborative Sparse Representation for Grayscale-Thermal Tracking. *IEEE Trans. Image Process.* **2016**, *25*, 5743–5756. [CrossRef] [PubMed]
34. Li, Y.; Zhu, J.; Hoi, S.C. Real-Time Part-Based Visual Tracking via Adaptive Correlation Filters. In Proceedings of the IEEE Conference on Computer Vision and Pattern Recognition (CVPR), Boston, MA, USA, 7–12 June 2015; pp. 4902–4912.
35. Wang, F.; Zhen, Y.; Zhong, B.; Ji, R. Robust infrared target tracking based on particle filter with embedded saliency detection. *Inf. Sci.* **2015**, *301*, 215–226. [CrossRef]
36. Shi, Z.; Wei, C.; Li, J.; Fu, P.; Jiang, S. Hierarchical search strategy in particle filter framework to track infrared target. *Neural Comput. Appl.* **2018**, *29*, 469–481. [CrossRef]
37. Chiranjeevi, P.; Sengupta, S. Rough-Set-Theoretic Fuzzy Cues-Based Object Tracking Under Improved Particle Filter Framework. *IEEE Trans. Fuzzy Syst.* **2016**, *24*, 695–707. [CrossRef]
38. Zhang, T.; Xu, C.; Yang, M. Learning Multi-Task Correlation Particle Filters for Visual Tracking. *IEEE Trans. Pattern Anal. Mach. Intell.* **2019**, *41*, 365–378. [CrossRef]
39. Li, Y.; Zhu, J.; Hoi, S.C. Reliable patch trackers: Robust visual tracking by exploiting reliable patches. In Proceedings of the IEEE Conference on Computer Vision and Pattern Recognition (CVPR), Boston, MA, USA, 7–12 June 2015; pp. 353–361.
40. Qi, Y.; Zhang, S.; Qin, L.; Yao, H.; Huang, Q.; Lim, J.; Yang, M.H. Hedged deep tracking. In Proceedings of the IEEE Conference on Computer Vision and Pattern Recognition (CVPR) , Las Vegas, NV, USA, 27 June 2016; pp. 4303–4311.
41. Henriques, J.F.; Caseiro, R.; Martins, P.; Batista, J. High-Speed Tracking with Kernelized Correlation Filters. *IEEE Trans. Pattern Anal. Mach. Intell.* **2015**, *37*, 583–596. [CrossRef] [PubMed]
42. Danelljan, M.; Häger, G.; Khan, F.S.; Felsberg, M. Discriminative Scale Space Tracking. *IEEE Trans. Pattern Anal. Mach. Intell.* **2017**, *39*, 1561–1575. [CrossRef]
43. Felsberg, M.; Kristan, M.; Matas, J.; Leonardis, A.; Pflugfelder, R.; Häger, G.; Berg, A.; Eldesokey, A.; Ahlberg, J.; Čehovin, L. The Thermal Infrared Visual Object Tracking VOT-TIR2016 Challenge Results. In Proceedings of the International Conference on Computer Vision (ECCV), Amsterdam, The Netherlands, 8–16 October 2016; pp: 824–849.
44. Tang, M.; Feng, J. Multi-kernel correlation filter for visual tracking. In Proceedings of the IEEE International Conference on Computer Vision (ICCV), Santiago, Chile, 7–13 December 2015; pp. 3038–3046.
45. Li, X.; Liu, Q.; He, Z.; Wang, H.; Zhang, C.; Chen, W.S. A multi-view model for visual tracking via correlation filters. *Knowl. Based Syst.* **2016**, *113*, 88–99. [CrossRef]
46. Possegger, H.; Mauthner, T.; Bischof, H. In defense of color-based model-free tracking. In Proceedings of the IEEE Conference on Computer Vision and Pattern Recognition (CVPR), Boston, MA, USA, 7–12 June 2015; pp. 2113–2120.
47. Montero, A.S.; Lang, J.; Laganiere, R. Scalable kernel correlation filter with sparse feature integration. In proceedings of the IEEE International Conference on Computer Vision Workshop (ICCVW), Santiago, Chile, 7–13 December 2015; pp. 587–594.
48. Danelljan, M.; Hager, G.; Shahbaz Khan, F.; Felsberg, M. Learning Spatially Regularized Correlation Filters for Visual Tracking. In Proceedings of the IEEE International Conference on Computer Vision (ICCV), Santiago, Chile, 7–13 December 2015; pp. 4310–4318.
49. Felsberg, M.; Berg, A.; Hager, G.; Ahlberg, J.; Kristan, M.; Matas, J.; Leonardis, A.; Cehovin, L.; Fernandez, G.; Vojír, T.; et al. The thermal infrared visual object tracking VOT-TIR2015 challenge results. In Proceedings of the IEEE International Conference on Computer Vision Workshops (ICCVW), Santiago, Chile, 7–13 December 2015; pp. 76–88.
50. Akin, O.; Erdem, E.; Erdem, A.; Mikolajczyk, K. Deformable part-based tracking by coupled global and local correlation filters. *J. Vis. Commun. Image Represent.* **2016**, *38*, 763–774. [CrossRef]

51. Lukežič, A.; Zajc, L.Č.; Kristan, M. Deformable parts correlation filters for robust visual tracking. *IEEE Trans. Cybern.* **2017**, *48*, 1849–1861. [CrossRef] [PubMed]
52. Du, D.; Qi, H.; Wen, L.; Tian, Q.; Huang, Q.; Lyu, S. Geometric Hypergraph Learning for Visual Tracking. *IEEE Trans. Cybern.* **2017**, *47*, 4182–4195. [CrossRef] [PubMed]

Article

Infrared Optical Observability of an Earth Entry Orbital Test Vehicle Using Ground-Based Remote Sensors

Qinglin Niu [1,2], Xiaying Meng [1,2], Zhihong He [2] and Shikui Dong [1,2,*]

[1] Key Laboratory of Aerospace Thermophysics of Ministry of Industry and Information Technology, Harbin Institute of Technology, 92 West Dazhi Street, Harbin 150001, China; qinglinniu@163.com (Q.N.); meng_xiaying@163.com (X.M.)

[2] School of Energy Science and Engineering, Harbin Institute of Technology, 92 West Dazhi Street, Harbin 150001, China; zhihong_he@hit.edu.cn

* Correspondence: dongsk@hit.edu.cn

Received: 9 September 2019; Accepted: 13 October 2019; Published: 16 October 2019

Abstract: Optical design parameters for a ground-based infrared sensor rely strongly on the target's optical radiation properties. Infrared (IR) optical observability and imaging simulations of an Earth entry vehicle were evaluated using a comprehensive numerical model. Based on a ground-based IR detection system, this model considered many physical mechanisms including thermochemical nonequilibrium reacting flow, radiative properties, optical propagation, detection range, atmospheric transmittance, and imaging processes. An orbital test vehicle (OTV) was selected as the research object for analysis of its observability using a ground-based infrared system. IR radiance contours, maximum detecting range (MDR), and thermal infrared (TIR) pixel arrangement were modeled. The results show that the distribution of IR radiance is strongly dependent on the angle of observation and the spectral band. Several special phenomena, including a strong receiving region (SRR), a characteristic attitude, a blind zone, and an equivalent zone, are all found in the varying altitude MDR distributions of mid-wavelength infrared (MWIR) and long-wavelength infrared (LWIR) irradiances. In addition, the possible increase in detectivity can greatly improve the MDR at high altitudes, especially for the backward and forward views. The difference in the peak radiance of the LWIR images is within one order of magnitude, but the difference in that of the MWIR images varies greatly. Analyses and results indicate that this model can provide guidance in the design of remote ground-based detection systems.

Keywords: ground-based detection; infrared imaging; observability; detecting distance; earth entry vehicle

---

1. Introduction

The use of ground-based remote sensing detectors is becoming an important method of accessing information on trajectories, positions, and flight conditions in the growing field of space technology. Recently, a very promising type of orbital test vehicle (OTV) came to the attention of many space agencies [1]. It is believed to be a candidate for the next generation of space planes and can be reused repeatedly due to low launch costs and high-speed maneuverability. A typical representative of this type of OTV is the X-37B spaceplane [2]. The aircraft can maintain operations in space for several months at a time, like a satellite, and can then return to the Earth's atmosphere on its own. During the entry phase, it is essential to track the vehicle's trajectory and flight behavior. Up to now, thermal infrared (TIR) remote sensing technology is widely used for monitoring the background environment and aerial targets [3,4]. However, numerical studies of this technique are rare due to the attendant complexity of the physical processes.

The study of the observability of aircraft based on the TIR effect is a thermal–optical problem. During the entry phase, the air around the aircraft undergoes strong compression along with high frictional forces acting on the aircraft body, resulting in hot reaction air flows. In the high-temperature flow field, many chemical reactions occur including dissociation, ionization, and recombination [5,6]. Under these conditions, air components, consisting of atoms, ions, and molecules, radiate strong optical radiation [7]. For the gas molecules, the process of vibrational transition produces infrared radiation. In addition, the aero-heating effect is also serious and causes a rapid increase in the temperature of the aircraft's surface, from which strong radiation can also be emitted. Infrared radiation from gases (including air dissociation products and trace air components) in the flow field and the surface is partially absorbed by the surrounding gases in the transmission process. The absorption process has strong spectral band selectivity and can be divided into two regions: (a) self-radiation emitted from the surface and hot gases in the high-temperature region, and (b) the atmospheric transmission effect in the low-temperature region. The infrared radiation of the target and the radiation noise of the environmental background are received by the optical sensor using Earth's atmospheric attenuation. The radiation is converted into electrical signals and then recognized by the infrared (IR) detector.

Lots of investigations on target detection were conducted for analysis of the aircraft IR signature. Mahulikar et al. [8–10] took a low-altitude fighter as a research object to analyze the role of atmosphere in IR signature, and the relationship between IR signature level and target susceptibility. Pan et al. [11] predicted the IR radiation and stealth characteristics for the cabin of a supersonic aircraft. Huang and Ji [12] investigated the effect of environmental radiation on the long-wave IR signature of a cruise aircraft. In these studies, they focused mainly on the surface emission and the exhaust plume under a low-temperature low-altitude condition. However, a comprehensive model that can be used for analysis of the observability of hypersonic vehicles considering the high-temperature gas effect is still rarely reported.

In the context of multi-mode detection requirements, ground-based remote detection saw much development [13–15]. Some relevant observations of hypersonic aircraft were conducted with the aid of TIR emissions. These experiments focused mainly on two aspects: TIR imaging of the space transportation system (STS) and radiative heating of sample return capsules (SRC). For instance, NASA carried out a series of hypersonic thermodynamic IR measurements (HYTHIRM) that relied on aerial and ground-based infrared imaging systems [15,16]. The infrared images were used to determine the surface temperature distribution on the viewable windward surface of the shuttle orbiter. These observations of the SRC reentering the Earth's atmosphere [17–19] were mainly concerned with the near-infrared band, with the aim of verifying the radiation excitation mechanisms and flow structures. However, those observations did not provide evaluation models and did not report on the maximum detection range (MDR).

The MDR is an important parameter in the design of optical instruments, which reflects the performance of the detection system. In most cases, it is appropriate that the target is treated as a point source when the aircraft's irradiance image only fills one or a few pixels of the sensor [20]. Prior literature [13,20,21] indicated that the MDR of an infrared imaging system is a function of factors such as background environment, target radiation characteristics, atmospheric transmittance, and the system threshold signal-to-noise ratio (SNR). Recently, Zhao et al. [21] proposed a spectral bisection method for calculating the operating distance of IR systems based on the MODTRAN (moderate spectral resolution atmospheric transmittance algorithm and computer model) program. Ren et al. [22] suggested a new formula for calculating the atmospheric transmittance based on the LOWTRAN (low-resolution atmospheric transmission) database. Huang et al. [20] reported a photoelectric detection method based on a long-wavelength infrared (LWIR, 8–14 μm) fisheye imaging system. In these literature sources, the target was specified as a uniform low-temperature gray body without gas emissions. However, such a treatment is overly simplistic. In fact, the surface temperature of a hypersonic vehicle may reach

2000 K with a non-uniform distribution [23]. The TIR emission can also be radiated from gases in the shock layer and wake flows [24]. This means that the temperature of both the aircraft's surfaces and the reacting flows may influence the evaluation of infrared optical observability.

Up to now, it remains a challenge to establish models to investigate the detectability and imaging of a hypersonic vehicle based on its detailed radiative properties. To obtain the irradiance received by a detector, lots of parameters should be calculated such as reacting flows, surface temperature, species concentrations, absorption coefficients, reconstructed nodes, and optical path. These require knowledge of fluid mechanics, spectroscopy, thermochemistry, and optics.

In this study, a comprehensive physical model was proposed to simulate the MDR and the TIR image of an Earth entry OTV. Firstly, the hot reacting flows and surface temperatures were simulated using a thermochemical gas-solid interaction computational fluid dynamics (CFD) solver. In addition, the optical radiative properties of radiating species were evaluated in thermal equilibrium and nonequilibrium. Then, TIR radiance characteristics were computed by solving the radiative transfer equation (RTE) in a fluid inclusion. Furthermore, using the concept of the point source, the MDR was simulated in different bands, trajectory points, and observation angles. Finally, the effects of sensor detectivity on the MDR and the TIR images in the aperture of the detector were discussed and analyzed.

2. Description of Physical Processes in Ground-Based Observation

An Earth entry OTV, with similar geometry to the X-37B [2], was used in this study. After entering the atmosphere, the aircraft flies in a typical flight path, and its velocity as function of altitude for the X-37B was reported in Reference [25], as shown in the lower right of Figure 1. During the entry phase, the air around the aircraft is heated to an extremely high temperature. Under this condition, there are two strong TIR radiation sources: (1) hot air components and gaseous products from dissociation, ionization, and recombination chemical reactions, and (2) glowing aircraft surfaces. These TIR radiations can be received by an IR detection system after being attenuated by passing the Earth's atmosphere.

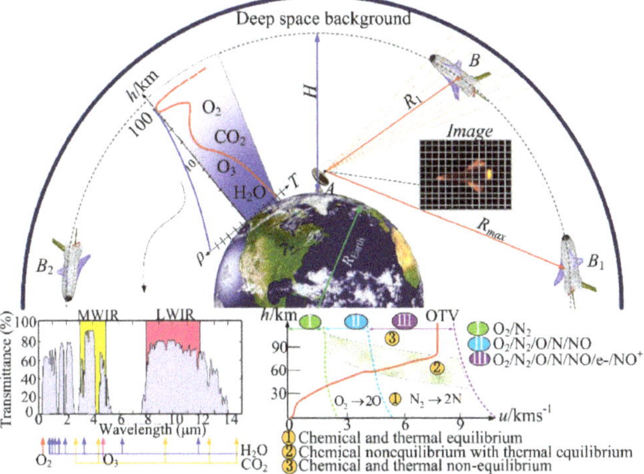

Figure 1. Diagrammatic sketch of orbital test vehicle (OTV) thermal infrared (TIR) observation. Atmospheric transmittance (gray) within the wavelengths of 0.4–14 μm at a low altitude is shown in the lower left corner. The flight path of the OTV and altitude-varying thermal and chemical properties are shown in the lower right corner. The variations of density and temperature with altitude are normalized using the corresponding above sea level (ASL) conditions.

For the air around the OTV, the flows are hypersonic and go through chemical nonequilibrium and thermal nonequilibrium conditions due to the drastic environmental changes. Figure 1 shows three typical thermal–chemical regions [26]. For the vehicle surface, the wall temperature depends on the aero-heating, structure heat conduction, radiation, etc. Due to the presence of atmospheric windows as illustrated in the lower left of Figure 1, the spectral bands of interest are generally medium-wavelength infrared (MWIR), with a wavelength of 3–5 µm, and LWIR [27]. At a flight altitude H above sea level (ASL), the ground-based infrared system A can receive the TIR radiation from aircraft B or B_1.

During the observation process, changes in the aspect angle between the detector and the aircraft may exert an arbitrary effect on observability. Considering the Earth's radius R_{earth} = 6371 km, there is an MDR above the horizon R_{max}, as shown in Figure 1. Below this MDR, the TIR intensity and distribution of the target can be imaged in the aperture of the infrared system. This study focuses mainly on the MDR and the TIR imaging during OTV entry.

3. Computational Methods

3.1. Description of CFD Solver

For hypersonic flows above 40 km, the time scale of the chemical and the internal energy exchange processes is comparable with the characteristic time of flows [5]. The internal energy exchange should be described through multiple temperatures. Recently, our research group carried out a series of simulations of hypersonic reacting flows using an in-house code [5,6]. In the code, a two-temperature CFD solver is available for predictions of thermal–chemical nonequilibrium flows. On assuming continuous flows are valid, three-dimensional Reynolds-averaged Navier–Stokes (N–S) equations are solved with a structured implicit scheme with the finite volume method (FVM). The viscous and inviscid fluxes are computed using a central difference and Roe's averaging scheme [28], respectively. Yee's symmetric total variation diminishing (STVD) limiter [29] is employed for accurate predictions of the shock layer. The two-equation shear stress transport (SST) with compressible correction is used for the flow simulations in the supersonic–hypersonic regime.

3.2. Optical Radiation and Transfer Models

3.2.1. Optical Radiative Properties of High-Temperature Gases

Studies [30,31] demonstrated that gaseous molecules of NO, CO_2, and H_2O are the main radiating components of air. Among these species, CO_2 and H_2O belong to the set of trace components and have a low number density. For instance, the volume fraction of CO_2 at ground level is approximately 3.628×10^{-4}, which is two orders of magnitude lower than that of H_2O. In hypersonic flows, their number densities are associated with the degree of compression of the flow field. For NO, it is the product of the combination reaction $O + N \rightarrow NO$ in air, and its formation is related to the dissociation reactions $N_2 \rightarrow 2N$ and $O_2 \rightarrow 2O$. Generally, high-altitude hypersonic flows are in local thermodynamic nonequilibrium (non-LTE) [32]. In this case, the optical radiation properties of radiating species should be evaluated under non-LTE conditions.

Currently, the new total internal partition sums (TIPS) routine [33] can be used for partition function calculations for some components, including CO_2 at temperatures below 5000 K and H_2O at temperatures below 6000 K. Based on the known partition function, the spectral lines of the corresponding molecules can be calculated with the aid of the high-temperature database HITEMP [34] (only the spectral lines in the standard conditions are provided). The TIPS routine provides an applicable range for NO at temperatures under 3500 K. The application is limited for high-temperature flows. Thus, a partition function suitable for high temperatures should be used. According to one of the basic principles of quantum mechanics, the reduced partition function of NO can be determined using Equation (1), neglecting the interaction between rotational and vibrational states [34].

$$Q(T) = \sum_{\substack{vibrational \\ states}} d_{vib} e^{-hc\frac{G_{vib}}{kT_{ev}}} \sum_{\substack{rotational \\ states}} d_{rot} e^{-hc\frac{F_{rot}}{kT}}, \tag{1}$$

where d_{vib} and d_{rot} are the degeneracy factors for states, and G_{vib} and F_{rot} are the term values of vibrational and rotational states, respectively. These parameters can be imported from Reference [35].

Relying on the partition function $Q(T)$, the high-temperature line intensity at a given wavenumber η can be calculated using the correction formula below.

$$S_\eta(T) = S_\eta(T_{ref}) \frac{Q(T_{ref})}{Q(T)} \frac{e^{\frac{-E_l}{k_B T}}}{e^{\frac{-E_l}{k_B T_{ref}}}} \frac{1 - e^{\frac{-hc\eta}{k_B T}}}{1 - e^{\frac{-hc\eta}{k_B T_{ref}}}}, \tag{2}$$

where $S(T_{ref})$ is the line intensity under standard conditions; h, c, and k_B are the Planck constant, the speed of light, and the Boltzmann constant, respectively. E_l stands for the energy of the lower state. The absorption coefficient of each species within the specified wavenumber and temperature intervals can then be calculated using the line-by-line (LBL) method [36].

$$\kappa(\eta - \eta_0) = S_\eta(T)\Phi(\eta - \eta_0)N, \tag{3}$$

where N is the number density of species, and Φ is the line shape function, for which the Voigt line profile [37] is often recommended. Finally, the total absorption coefficient of the mixture can be computed on the assumption that the absorption coefficient is cumulative for each species.

3.2.2. Optical Radiative Properties of High-Temperature Surfaces

For the surface, the radiation intensity is determined using the temperature and emissivity, along with the radiative properties of the surface element calculated in accordance with Planck's radiation law for a gray body [36].

$$I_{\lambda,sur} = \frac{C_1}{\pi} \frac{\varepsilon(\lambda)}{\lambda^5 \left(e^{\frac{C_2}{\lambda T}} - 1\right)}, \tag{4}$$

where C_1 and C_2 are the first and second radiation constants, respectively; ε is the emissivity, $I_{\lambda,sur}$ is the radiance for a thermal source of the surface element, and λ is the wavelength which can be converted to the wavenumber η. The surface emission requires coupling with the gas radiation along the optical path of light propagation.

High-temperature gas radiation differs from the gray-body radiation characteristics of the surface. Its self-emission and self-absorption properties need to be taken into consideration. The total radiation spectral intensity can be calculated using discrete path intervals. Specifically, it can be described using the RTE [36].

$$\frac{\partial I_\lambda(s, \vec{s})}{\partial s} = \kappa_\lambda(s)\left(I_{b,\lambda}(\vec{s}) - I_\lambda(s, \vec{s})\right), \tag{5}$$

where λ indicates the wavelength, and I_λ is the local spectral intensity. $I_{b,\lambda}$ is the Planck blackbody function, whereas s and $\vec{s}$ represent the position and the optical path vector, respectively.

Methods commonly used to solve the RTE include the line-of-sight (LOS), ray tracing (RT), and Monte Carlo (MC) methods. Under the condition of an absence of scattering particles, the LOS method is equivalent to the other two. Thus, the LOS method was applied in this study due to a compromise between computational cost and accuracy. The LOS method was introduced in our previous studies [5,38]. LOS starts with a surface element, and the surface emission $I_{\lambda,sur}$ can be treated as the initial value of the RTE. According to the RTE, the spectral intensity of the target can be calculated by summing the radiance from each path interval as follows:

$$I_{\eta,tar} = \sum_{j=1}^{M} I^0_{\eta m,j-0.5}\left(\kappa_{\eta m,j\to M} - \kappa_{\eta m,j-1\to M}\right)\Delta L_n,\tag{6}$$

where M is the number of segments in optical path, and $I_{\eta,tar}$ is the radiance at the wavenumber η in a inclusion with a cutoff value equal to the ambient condition.

3.3. Infrared Optical Observability of Ground-Based Sensors

3.3.1. Detection Range Model

For the optical detection system, the spectrum intensity that arrives at the detector is given by

$$P_\lambda = \tau_0(\lambda)\frac{A_0}{R^2}\tau_a(\lambda, R)A_t\left(I_{\lambda,tar} - I_{\lambda,bg}\right),\tag{7}$$

where R is the distance between the target and the detector, $\tau(\lambda,R)$ is the atmospheric transmittance with a distance of R, A_t is the effective radiation area of the target surface, $\tau_0(\lambda)$ stands for the spectral transmittance of the optical system, A_0 is the pupil area of the objective lens system, and $I_{\lambda,bg}$ denotes the background radiance received by detector.

The optical radiant power must be converted into a signal voltage, which is integrated within the wavelengths of λ_l–λ_u and has the following form [21]:

$$\Delta V_s = \frac{V_n}{\sqrt{A_d g \Delta f}} \cdot \int_{\lambda_l}^{\lambda_u} D^*(\lambda)P_\lambda d\lambda,\tag{8}$$

where $D^*(\lambda)$ is the normalized system detectivity, Δf is the frequency bandwidth of the detector circuitry, A_d is the pixel area of the detector, g is the photoconductive gain, and λ_u and λ_l stand for the upper and lower limits of wavelengths for the band of interest.

According to the above equations, the detection distance of the optical system with respect to a point target can be written as

$$R = \left[\frac{A_0 \int_{\lambda_l}^{\lambda_u} \tau_0(\lambda)D^*(\lambda)\tau_a(\lambda, R)A_t\left(I_{\lambda,tar} - I_{\lambda,bg}\right)d\lambda}{(A_d \cdot \Delta f)^{\frac{1}{2}}\frac{\Delta V_s}{V_n}}\right]^{\frac{1}{2}}\tag{9}$$

In Equation (9), $\Delta V_s/V_n$ is the SNR of the system. Based on the noise equivalent flux density (NEFD) [39], which is defined as the incoming TIR power per unit area at the aperture, the sensor parameters (A_0, D^*, A_d, and Δf) can be integrated into an evaluation parameter. In this study, the background is the deep space. The basic value of the NEFD is 10^{-12} W/cm^2, and the SNR is specified as 5. According to these threshold values, R is the longest detecting range, namely, the MDR.

3.3.2. Atmospheric Transmittance and Radiance

In addition to the main components of nitrogen and oxygen, the atmosphere has a variety of trace components that possess properties of radiation emission and absorption in the corresponding spectral bands. The atmospheric transmittance is a complex parameter due to the selective absorption of atmospheric molecules and the change in atmospheric density with altitude. Therefore, the TIR radiation emitted from the high-temperature fluid inclusion can be absorbed partially by these components, which means that the atmospheric transmittance and self-emission need to be calculated. In the atmospheric environment, the spectral radiation and transmittance of the atmosphere are associated with the path and the spectral band. In this study, the MODTRAN computer program [40] was utilized, which is a moderate-resolution atmospheric radiation transfer model developed by LOWTRAN that can provide atmospheric information for different paths and spectral bands, including atmospheric transmittance, background radiation (e.g., rural, urban, marine, and desert), and solar irradiance in different seasons covering ultraviolent, visible, and infrared wave bands.

In the desired wavelength range of λ_l–λ_u, the spectral band is divided into many equally spaced segments $\Delta\lambda = \lambda_{i+1} - \lambda_i$, $i = 1, 2, \dots, n$. When the interval $\Delta\lambda$ is sufficiently small, the atmospheric transmittance $\tau_{\Delta\lambda_i}$ and radiance $L_{\Delta\lambda_i}$ in the interval wavelength of λ_i can be expressed as

$$\tau_{\Delta\lambda_i} = \frac{\tau_{\lambda_i} + \tau_{\lambda_{i+1}}}{2}, \text{ and } L_{\Delta\lambda_i} = \frac{L_{\lambda_i} + L_{\lambda_{i+1}}}{2}. \tag{10}$$

Based on the abovementioned treatment, the atmospheric transmittance and radiance for the detecting distance R have the following expressions:

$$\tau_a(\lambda, R) = \begin{bmatrix} \tau_{R_1,\lambda_1} & \tau_{R_1,\lambda_2} & \cdots & \tau_{R_1,\lambda_n} \\ \tau_{R_2,\lambda_1} & \tau_{R_2,\lambda_2} & \cdots & \tau_{R_2,\lambda_n} \\ \vdots & \vdots & \cdots & \vdots \\ \tau_{R_m,\lambda_1} & \tau_{R_{1m},\lambda_2} & \cdots & \tau_{R_m,\lambda_n} \end{bmatrix}. \tag{11}$$

Similarly, the atmospheric spectral radiation intensity can be also given as

$$I_a(\lambda, R) = \begin{bmatrix} I_{R_1,\lambda_1} & I_{R_1,\lambda_2} & \cdots & I_{R_1,\lambda_n} \\ I_{R_2,\lambda_1} & I_{R_2,\lambda_2} & \cdots & I_{R_2,\lambda_n} \\ \vdots & \vdots & \cdots & \vdots \\ I_{R_m,\lambda_1} & I_{R_{1m},\lambda_2} & \cdots & I_{R_m,\lambda_n} \end{bmatrix}. \tag{12}$$

Based on the self-radiation spectrum and the atmospheric transmittance, the attenuated spectrum can be obtained, and then the radiance can be computed by integrating the attenuated spectrum within the required band.

3.3.3. TIR Smoothing Distribution on the Sensor Aperture

Under detection distances below the MDR, the aircraft surface is partially detected along the detection direction. This means that occlusion occurs in the detecting process. The TIR light rays are emitted from the visible surface through the hot gases and undergo atmospheric attenuation and then arrive at the aperture of the sensor. Usually, the geometric model of an aircraft is complex, and its shell meshes consist of many uniformly arranged grids. However, the pixels of the detector are arranged in an orthogonal array. A common occurrence involves more than one light ray arriving at one pixel. In the pixel, the TIR intensity may be assigned to the center node in the imaging process, which results in an unsmooth TIR image with many bright spots. Therefore, imaging techniques are used to deal with such imaging problems, including the treatment of visible surfaces and mesh clipping.

Visible surface elements that are associated with the LOS direction are required in Equation (7). This is attributed to the fact that the detector only receives the TIR irradiance of partial surfaces. As shown in Figure 2a,b, there are two types of occlusion elements. One is a surface element with radiation directions that have no component in the LOS direction, and the other is a surface element obscured by the other elements. A flag 0 represents the invisible surface elements using the following expression:

$$\text{flag}_{A_i} = \begin{cases} 0, & \vec{s} \cdot \vec{n}_i \leq 0 \\ 0, & (\vec{s} \cdot \vec{n}_i > 0)\&(V_{i,p} \notin \{A_{j,j\neq i}\}), \\ 1, & \text{other} \end{cases} \tag{13}$$

where $\vec{n}_i$ is the outward normal of the target surface element A_i, and $V_{i,p}$ is the pth vertex of the element A_i. $\{A_{j,j\neq i}\}$ represents the set of the surface elements excluding A_i, where $j = 1, 2, \dots, N_{element}$.

The imaging process must calculate the irradiance received by each pixel of the detector, as shown in Figure 2c. In this process, part of the surface element projected onto the pixel needs to be retained for evaluation. In order to produce an accurate image, each separate region should be calculated in each individual pixel. As shown in Figure 2d, mesh clipping can be used for computing the area of the polygon V_4V_1IJKL. This procedure requires the vertices of the two sets (A, B, C, D and V_1, V_2, V_3, V_4) and their candidate intersection points (W_1, I, J, K, L, W_2, W_3, W_4). The desired points should then be selected from these vertices and intersection points. These unordered points must be arranged before forming a closed polygon. A clockwise arrangement is established according to the cosine value of the vertices as shown in Figure 2e.

In Figure 2f, a representative case is shown, in which the pixel element receives a total of TIR radiation intensity from nine surface elements. According to the above image treatment, the irradiance received by each detector pixel can be calculated by the following formula:

$$q_{i,j} = \frac{\sum_k I_k A_{i,j,k} \cos\left(\vec{n}, \vec{s}\right) \tau(\lambda, R)}{R^2}, \tag{14}$$

where q is the radiant energy of the pixel, I is the TIR intensity received by the system which is emitted from the surface element k, and $A_{i,j,k}$ represents the visible area of the kth surface element in the $i \times j$ pixel.

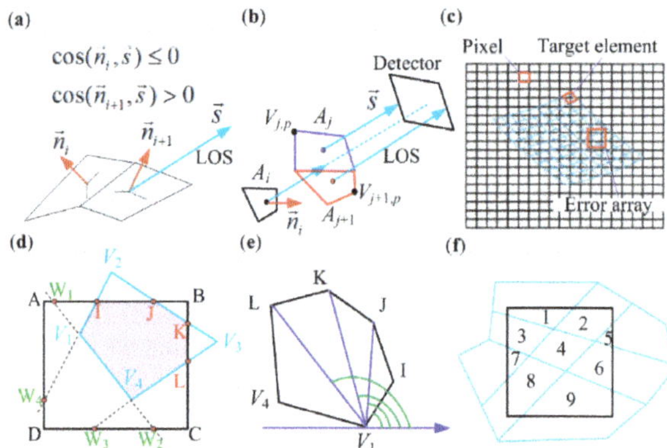

Figure 2. Sketch map of mesh clipping: (**a**) invisible elements caused by the obtuse angle between the line of sight (LOS) and its outer normal vector; (**b**) invisible elements caused by occlusion of other elements; (**c**) pixel array and relation between sensor pixel and target elements imaging; (**d**) distribution of intersection points in mesh clipping; (**e**) vertex sequence in a clockwise arrangement; (**f**) general position relationship between the target element and pixels.

3.4. Computational Flow Chart of MDR

A code was programed in FORTRAN considering above physical models. In these procedures, the fluid computation is decoupled with the radiative computation on the assumption that the gas and surface emissions have little influence on the flow field parameters. The computational flow chart is shown in Figure 3. Firstly, the reacting flow and surface temperature can be obtained using a two-temperature CFD solver based on the known freestream conditions and the structured grid [41]. The radiative properties of gases, including CO_2, NO, and H_2O, are evaluated relying on the HITEMP database. Then, at a specified observation angle, the occlusion effect is considered, and the visible parts of the aircraft are computed. Furthermore, the spectral irradiance is achieved along the LOS direction in a fluid-domain inclusion. An initial detecting distance that is larger than the aircraft's flight altitude is given and used for the computation of the irradiance received by the sensor. Finally, the MDR can be obtained by comparing with the detectivity of the sensor.

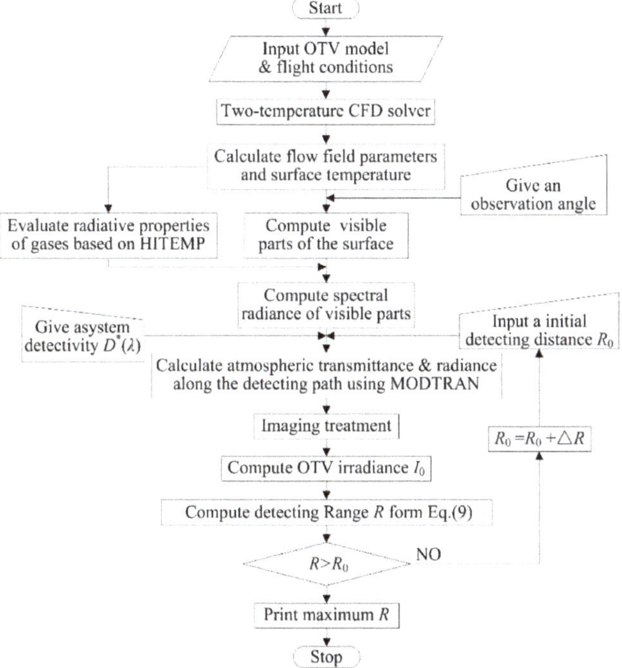

Figure 3. Flow chart of the calculation.

3.5. Validations of Physical Models

At present, to the best of the authors' knowledge, there are few reports on the radiation observation data of hypersonic vehicles. In most cases, it is difficult to obtain the self-radiation intensity of a hypersonic vehicle due to expensive measurement costs, complex test conditions, and strong background noises. Up to now, calculations of the thermo-chemical nonequilibrium flow field and the high-temperature nonequilibrium radiation characteristics of radiating gases are still challenging tasks. Therefore, the physical models are verified separately against reference data in this paper.

3.5.1. Validations of Surface Temperature and Flow Field Parameters

From the two strong TIR radiation sources, accurately predicting the surface temperature and flow field parameters is important. In this section, two available reference data are used for validation studies of the surface temperature and the flow field parameters: (1) double-cone UHTC (Ultra-high temperature ceramics) surface temperature test in the L2K wind tunnel at DLR (German Aerospace Center) Köln in Germany [42], and (2) reference data of the ELECTRE [43] vehicle at 293 s reported by Hao et al. [44]. The detailed computational parameters were given in our previous work [6,41] including the geometry size, material thermal properties, grids, boundary conditions, and so forth. In Figure 4, comparisons between computed results and reference data prove that the current CFD solver has good performance in predictions of the surface temperature and flow field parameters of the hypersonic vehicle. This work can assist in a study of ground-based IR optical observability and imaging for an Earth entry vehicle.

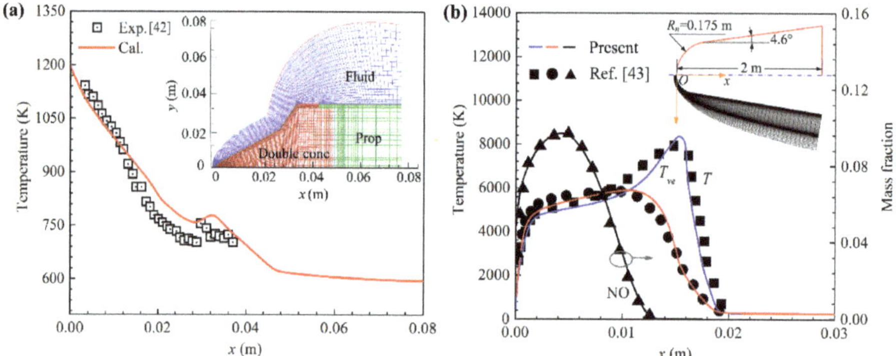

Figure 4. Validation of the surface temperature and main flow field parameters: (**a**) surface temperature along the body at $t = 60$ s for the double-cone experiment in the L2K wind tunnel at DLR (German Aerospace Center) Köln; (**b**) flow field parameters of translational–rotational temperature (T), vibrational–electronic temperature (T_{ve}), and mass fraction of dissociation product NO along the stagnation line for ELECTRE vehicle.

3.5.2. Validations of High-Temperature Optical Radiative Properties

The dual-mode experiment on bow-shock interactions (DEBI) [45,46] was carried out in 2003. In flight measurements, spectrometers mounted in the nose cone of a sounding rocket were used for measuring the forward- and side-looking radiation signatures in the bow-shock layer. Ozawa et al. [46] computed the forward-looking infrared spectrum at 40 km and 3.5 km/s using nonequilibrium radiation distribution (NERD) and the NEQAIR-IR (nonequilibrium air radiation-infrared) program. These data can be used to verify the current optical radiative property computational model.

The DEBI vehicle has a blunt cone with a 0.2032-m-radius nose and a 7.5° half-cone angle. The computational parameters can be seen in Reference [46]. The flow field parameters were computed using the two-temperature CFD solver, which can be treated as the input data in radiation computations. Based on radiative properties of high-temperature gases using the LBL method, the forward-looking infrared spectrum of the shock layer can be obtained, as shown in Figure 5. A comparison of the infrared spectrum between computed and reference data indicates that the current model is in good agreement with the results of NERD and NEQAIR-IR.

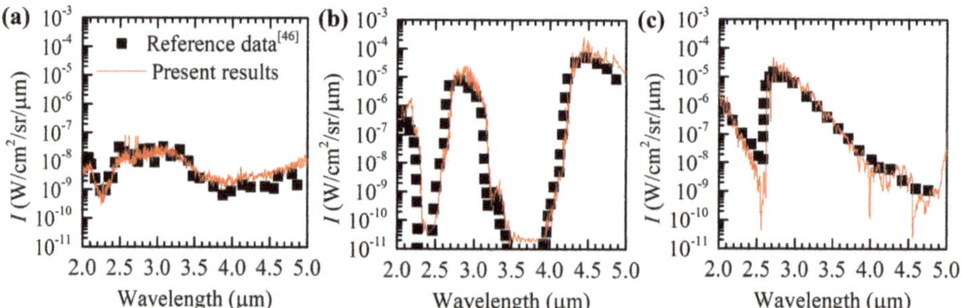

Figure 5. Comparison of spectral radiance between calculated and reference data: (**a**) spectrum of H_2O; (**b**) spectrum of CO_2; (**c**) spectrum of NO.

3.5.3. Validations of Infrared Optical Observability

In this paper, the detection range calculation model was derived from the NEFD model, which is based on the relationship between the total target flux density at the sensor location and the SNR, namely,

NEFD = P_{tar}/SNR. The target flux density P_{tar} was mainly determined by the target's self-radiation intensity and atmospheric transmission. NEFD was determined by the sensor performance and the background radiated noise. The reliability of the NEFD and SNR models was validated against the observation results of a laboratory temperature-controlled blackbody by Richter and Fries [47]. It is demonstrated that the error between the SNR based on the NEFD model and the experimental measurements is less than 5%. Therefore, the reliability of the infrared optical observability module is determined by the target flux density P_{tar}, which depends on the radiation transfer calculation model.

An available reference dataset to verify the current transfer model is the ground-based observation of an Atlas rocket exhaust plume. The observation schematic diagram is shown in Figure 6a. In Reference [48], infrared radiation spectra of the Atlas rocket exhaust plume are numerically presented. Detailed calculation conditions (geometry, boundary, inflow, etc.) of the plume were given in Reference [48] and our prior work [49,50]. In this section, self-radiation of the exhaust plume is studied using our IRSAT (infrared signature analysis tool) code [49], whereby the spectrum can be used to compute the apparent radiation received by the sensor using the model described in Section 3.3. The calculation steps are as follows: (1) the self-radiation spectrum of the plume is obtained without soot at the altitude of H = 40 km by IRSAT, and (2) the apparent radiation spectrum is calculated at the pupil of the sensor using the module in Section 3.3, in which the plume is treated as a point source. A comparison of the apparent radiation spectrum received by the sensor between computed and reference results is shown in Figure 6b. It is indicated that results of the current infrared optical observability model are in good agreement with the reference data.

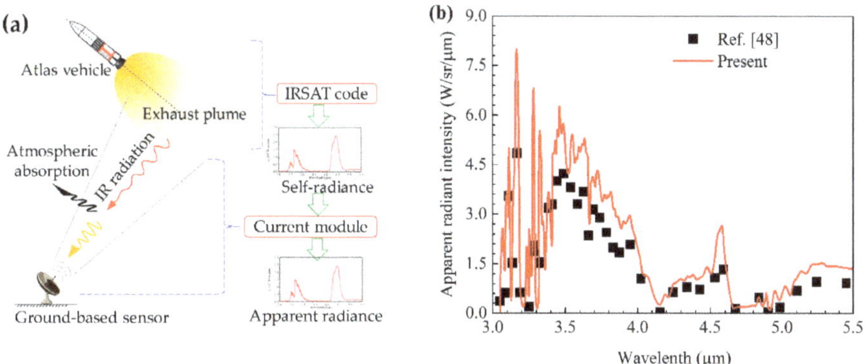

Figure 6. Validation of infrared optical observability: (**a**) observation schematic diagram of the ground-based sensor for Atlas rocket exhaust plume; (**b**) comparison of the spectrum received by the sensor between computed and reference results.

4. Results

4.1. Thermal–Optical Flow Field

In this study, a cube calculation domain was adopted for the OTV. Considering the symmetry of the geometry, one half of the geometric model was used for fluid simulations. All grids were structured, and their distribution is shown in Figure 7. For the conjunction heat transfer calculation, the computational domain was divided into fluid and structure domains. The grids of the two computational domains demonstrated a one-to-one correspondence at the interface. Generating grids in two domains used a total of 102 blocks. The fluid domain consisted of 23 million grids, and the solid domain contained 1.24 million grids with 80,000 shell grids. It was indicated in previous studies that the mesh Reynolds number should be kept below two to guarantee the precision of the heat flux on the surface of a hypersonic aircraft [51]. Therefore, the first wall–normal spacing from the wall was arranged to be approximately 1×10^{-5} m from the wall in this study. In addition, grids near the wall and in the potential shock-layer region were also refined.

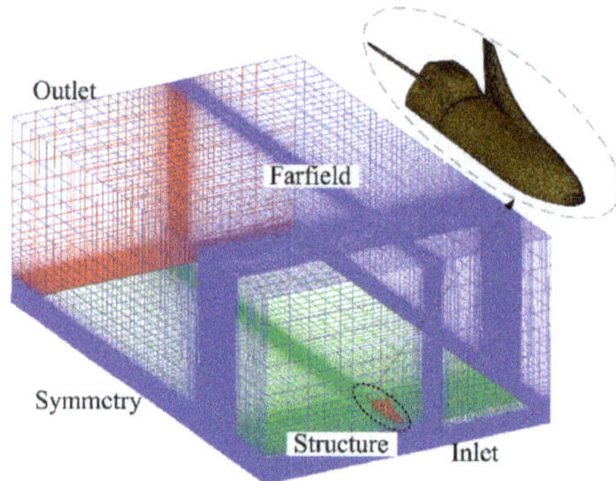

Figure 7. Grid distribution of OTV: fluid computational domain consisting of outlet (red), symmetry plane (green), and far-field and inlet (blue) boundaries. The partially enlarged detail at the top right of the figure shows the structure computational domain.

In the calculations, the inflow and far-field boundaries applied free stream conditions with uniform pressure and temperature. A supersonic outflow boundary was employed. The gas–solid interface was specified for the fluid and structure sides, respectively. The structural materials were assumed to be the stainless steel, whose properties can be seen in Reference [42]. In the structure domain, a radiative transfer wall with an emissivity of 0.85 was specified at the outer surface, and a wall with an initial wall temperature of 300 K was used for the inner surface. According to the OTV's flight regime as shown in Figure 1, seven computing trajectory points were selected for analytical calculations. In this study, it was assumed that the angle of attack (AOA) was zero during the flight and that the flow field reached the steady state at these computing points. The detailed freestream conditions are listed in Table 1.

Table 1. Freestream conditions at computing trajectory points.

Parameter	Values						
h, km	10	20	30	40	50	60	70
p_∞, Pa	26,500	5529	1197	287	79	22	5
T_∞, K	223	216	226	250	271	247	220
u_∞, km/s	0.22	0.38	0.98	1.81	2.84	4.51	6.10

Based on these conditions, the steady reacting flows in these cases were calculated. A machine with 52 central processing unit (CPU) cores was used for parallel computation, taking about 80 h to calculate the flow field for each computational case. The contours of the flow field parameters in the two representative cases (30 km and 70 km) are shown in Figure 8, including the surface temperature, fluid temperature, and species distribution.

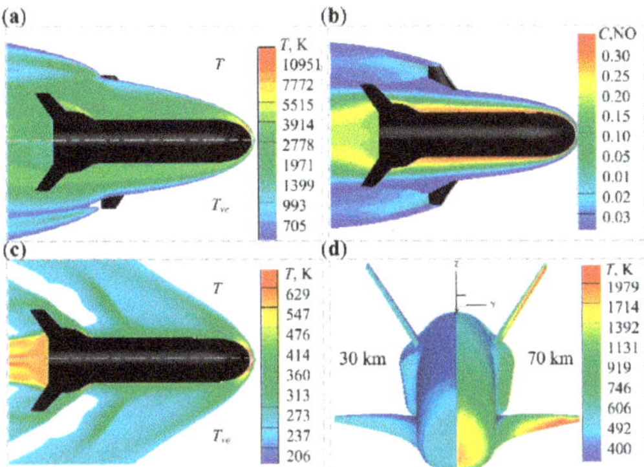

Figure 8. Contours of flow field parameters: (**a**) translational–rotational (upper) and vibrational–electronic (lower) temperatures in the 70-km case; (**b**) mass fraction of NO in the 70-km case; (**c**) translational–rotational (upper) and vibrational–electronic (lower) temperatures in the 30-km case; (**d**) surface temperatures in the 30-km (left) and 70-km (right) cases.

4.2. Self-Emission of OTV

Self-emission is defined as the radiance in the fluid inclusion within an ambient cutoff temperature. It is the radiance of the glowing surface and hot gases occupying a small space before considering atmospheric attenuation. In this study, gas emissions of the four species including NO, CO, CO_2, and H_2O were considered. Profiles of the radiance of two groups of typical detecting angles, described by θ_1 and θ_2, are plotted in Figure 9. It can be seen from these illustrations that the distribution of radiation intensity is associated with spectral bands and detecting angles. For different computing points, the radiance distribution within the same band is similar, but the radiation intensity is quite different. In order to examine the contribution of the gas and the surface to the radiance, the spectrum distribution at the angle of $\theta_1 = 0°$ is shown in Figure 10.

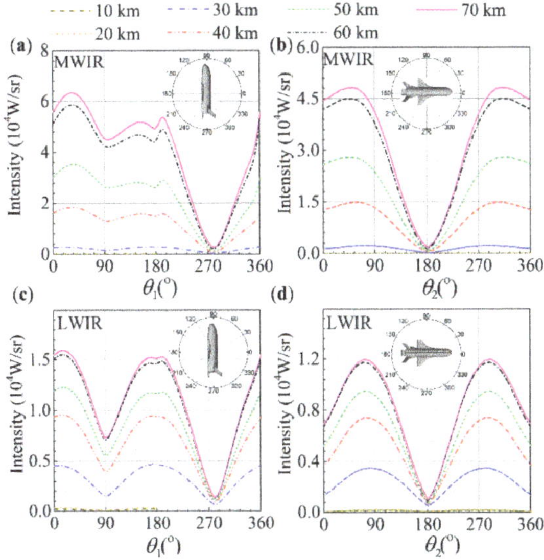

Figure 9. (**a**–**d**) Profiles of radiation intensity in two typical detection surfaces.

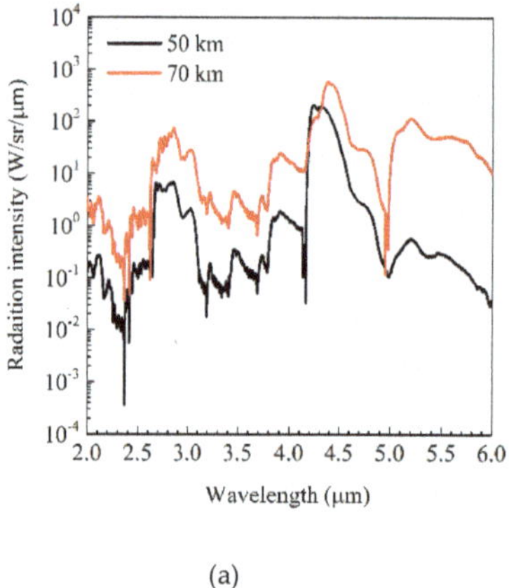 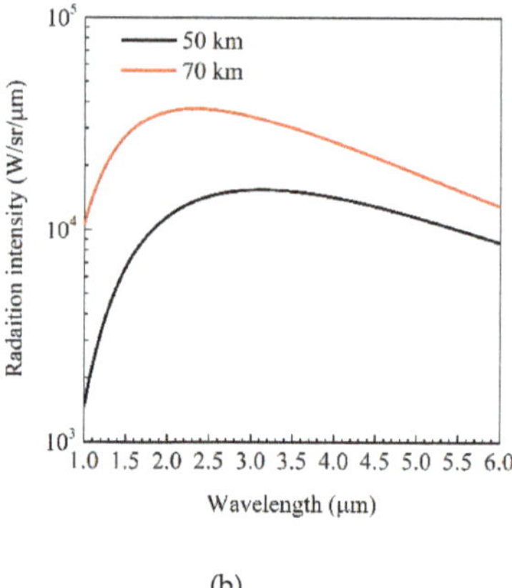

(a)   (b)

Figure 10. Spectrum distributions in the top view ($\theta_1 = 0°$) in the 50-km and 70-km cases: (**a**) gas radiance; (**b**) surface radiance.

As a matter of fact, the detecting angle may be arbitrary during target detection. An angular coordinate system was used to describe the radiance distribution at all possible angle. The observation angle can be described by a pair of the circumferential angle (φ) and the zenith angle (θ). The x-axis is defined as being in the direction toward the nose of the aircraft, while the z-axis is toward the back of the aircraft. φ is the angle between the direction vector and the x-axis within the range of 0°–360°. θ is the angle between the direction vector and the z-axis within the range of 0°–180°.

Figure 11 shows the contours of the radiance for the two representative cases of 30 km and 70 km in 2π space, which shows that the distribution of radiance is strongly dependent on the angle of observation and the spectral band. The peak intensity distributions for different computing points of the two bands are shown in Figure 12.

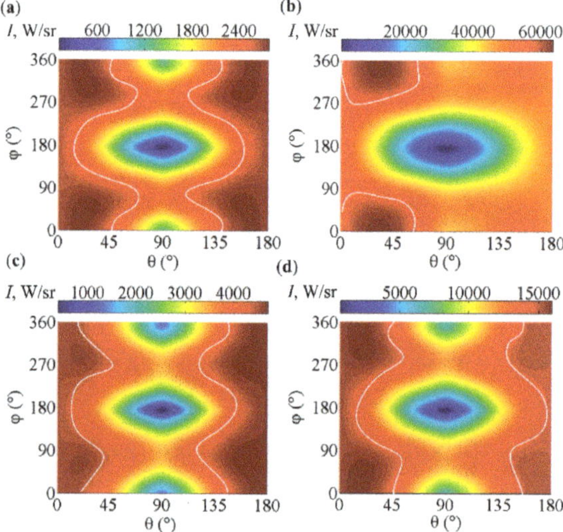

Figure 11. TIR contours: (**a**) medium-wavelength infrared (MWIR) in the 30-km case; (**b**) MWIR in the 70-km case; (**c**) long-wavelength infrared (LWIR) in the 30-km case; (**d**) LWIR in the 70-km case.

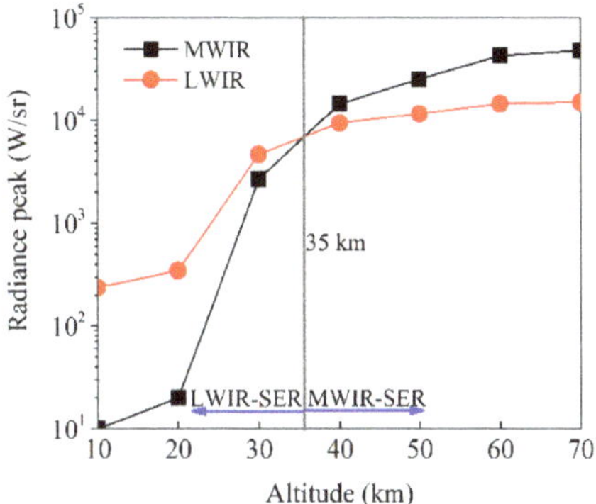

Figure 12. Comparison of maximum TIR radiance for different computing points.

4.3. Detecting Distance of the Ground-Based Sensor

Based on the above radiance, the MDR could be evaluated by considering the atmospheric transmittance. In the calculation, a discrete angle of 10° was used for the angles of θ and φ, and a total of 722 observation angles were considered in 2π space. It should be noted that the occlusion of the horizon was also considered in calculating the detecting distance, as shown in Figure 1.

On the assumption that the NEFD was 10^{-12} W/cm^2, the MDR contours within the MWIR and LWIR bands are shown in Figure 13. It can be seen from the figure that the MDR of the rear-most parts of the aircraft ($\theta = 90°$, $\varphi = 180°$) was the shortest. This was attributed to the low-temperature tail section and parts concealed by the high-temperature gas in the shock layer and partial surfaces. Figure 13c also presents a three-peak structure, which is distinctly different from the other three illustrations. In the 30-km case, the MDR of the MWIR band was greater than that of the LWIR, which was reversed in the 70-km case. This phenomenon was similar to the distribution of the radiance as shown in Figure 11. The MDR profiles of the MWIR and LWIR bands as a function of the altitude are shown in Figure 14.

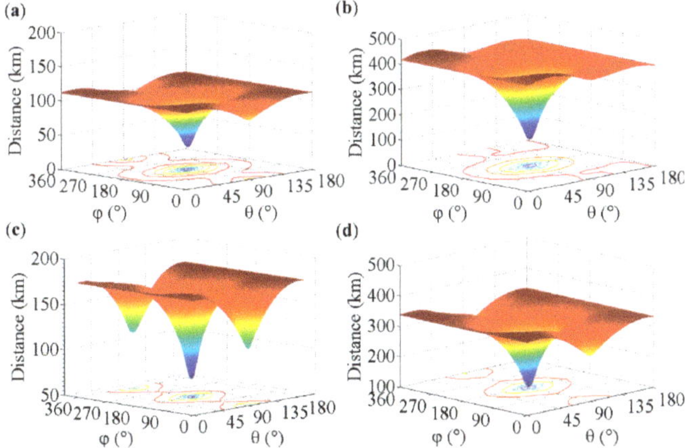

Figure 13. Maximum detector distance distributions: (**a**) MWIR for the 30-km case; (**b**) MWIR for the 70-km case; (**c**) LWIR for the 30-km case; (**d**) LWIR for the 70-km case.

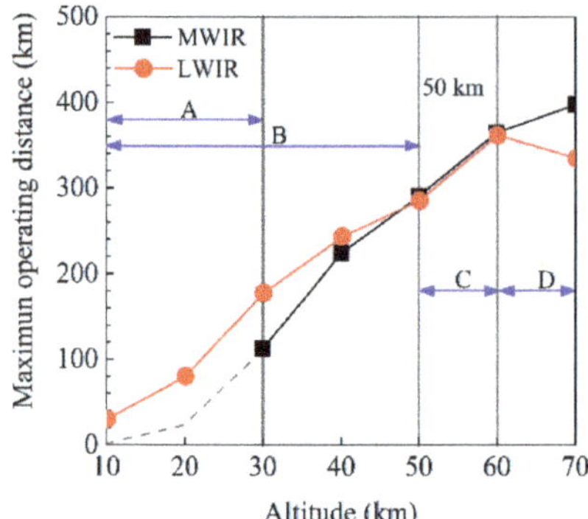

Figure 14. Maximum detection range (MDR) profiles for different test points. There are four typical zones: blind region (A-zone), LWIR strong receiving region (SRR) (B-zone), equivalent zone (C-zone), and MWIR SRR (D-zone).

4.4. Effect of Sensor Detectivity on MDR

To examine the effect of the detectivity on the MDR, a detectivity equivalent of NEFD = 10^{-14} W/cm^2 is employed in this section. Figure 15 shows the profiles of the peak MDR for seven computing points. The MDR profiles for the MWIR and LWIR bands also intersected at the characteristic altitude of 40 km. Comparing these results with Figure 14 shows that the characteristic altitude decreased as the detectivity increased. Above 40 km, the peak MDR did not drop, as shown in Figure 14.

Figure 16 shows the contours of the MDR increment for different detection angles from NEFD = 10^{-12} W/cm^2 to NEFD = 10^{-14} W/cm^2. Profiles of the peak increment of the MDR are shown in Figure 17. It can be seen from this figure that the peak MDR increment increased as the altitude increased. The increments of the two bands were almost identical at altitudes of 30 km and 70 km. In this region, the maximum increment of the MDR in the MWIR band was higher than that in the LWIR band. This indicates that the increase in detectivity was helpful for increasing performance in the MWIR band.

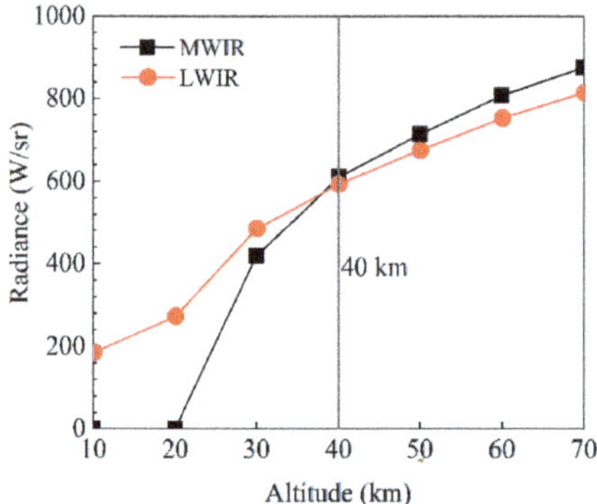

Figure 15. Maximum improve distance for different test points.

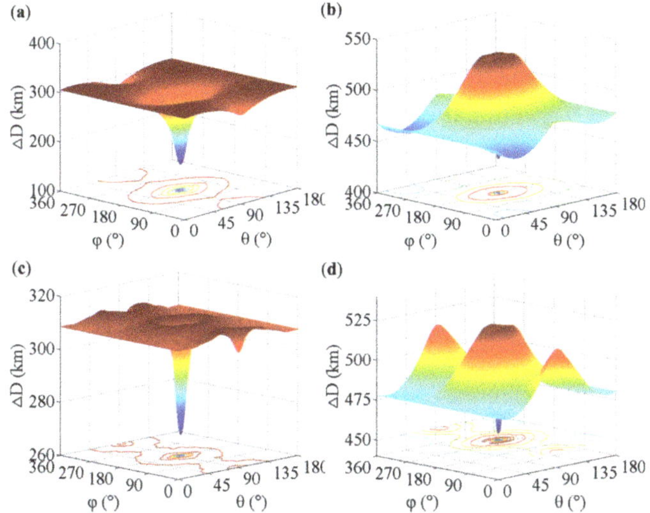

Figure 16. MDR increments: (**a**) MWIR for the 30-km case; (**b**) MWIR for the 70-km case; (**c**) LWIR for the 30-km case; (**d**) LWIR for the 70-km case.

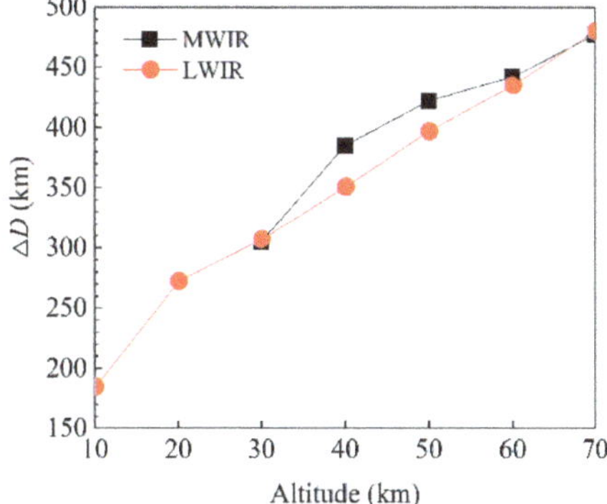

Figure 17. Profiles of MDR maximum increments.

4.5. Infrared Optical Image of the Sensor

While below the MDR, the aircraft's optical signature can be received by the pupil aperture of the detection system, and the TIR image fills the detector's pixels. At greater distances, the TIR intensity is captured by only a few pixels. To examine the distribution of radiant energy in the detector pixels, the imaging characteristics of the OTV are analyzed in this section.

Imaging is associated with the field of view (FOV) and the resolution of the detection system. Usually, the FOV of the scanning telescope has a wide range of 0.01–100 mrad [52–54]. As the FOV and the detection distance increase, the number of pixels receiving the TIR irradiance decreases. This number may even decease to one or a few pixels. In this study, three artificial FOVs were used for analyzing the distribution of TIR images, including $\alpha/2 = 0.01°$, $\alpha/2 = 0.05°$, and $\alpha/2 = 0.1°$. All calculations were simulated on the assumption that the aperture of the system consisted of 100×100 pixels.

To display an enlarged image, partial background regions are removed in Figure 18a. In order to get the images below the same detection distance, an assumed distance of $R = 30$ km was used in both the 30-km and 70-km cases. Figure 18b–f show the upward-view TIR images with $\alpha/2 = 0.05°$. Figure 18b shows the result without the mesh clipping.

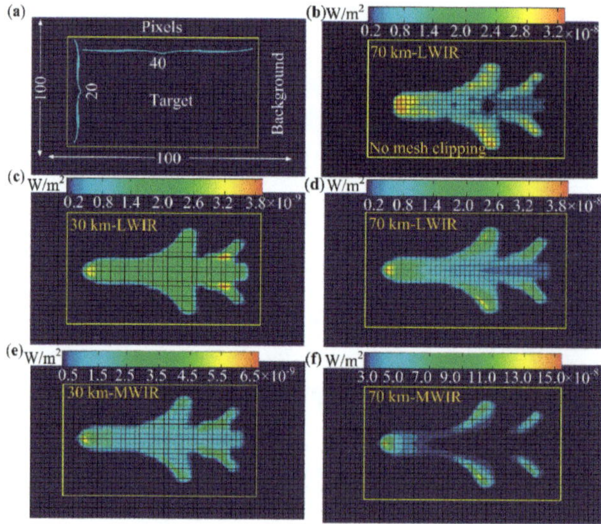

Figure 18. TIR images at an artificial detecting distance $R = 30$ km: (**a**) Pixel arrangement of target and background regions; (**b**) LWIR image without mesh clipping for 70-km case; (**c**) LWIR image for 30-km case; (**d**) LWIR image for 70-km case; (**e**) MWIR image for 30-km case; (**f**) MWIR image for 70-km case.

To analyze the TIR distribution for different observation angles, the computing point of $H = 70$ km was selected. Figure 19 shows the TIR images of $R = 70$ km at $\alpha/2 = 0.01°$ and $\alpha/2 = 0.1°$. Imaging was calculated in the front ($\theta = 90°$, $\varphi = 0°$) and oblique-side ($\theta = 90°$, $\varphi = 135°$) views. The upper left corner of the figure shows plots of the images at $\alpha/2 = 0.1°$, and the right side of the figure presents the ratio of the peak intensity at $\alpha/2 = 0.1°$ to that at $\alpha/2 = 0.01°$. It can be seen from Figure 19 that the image nearly became a point when the FOV increased to $\alpha/2 = 0.1°$.

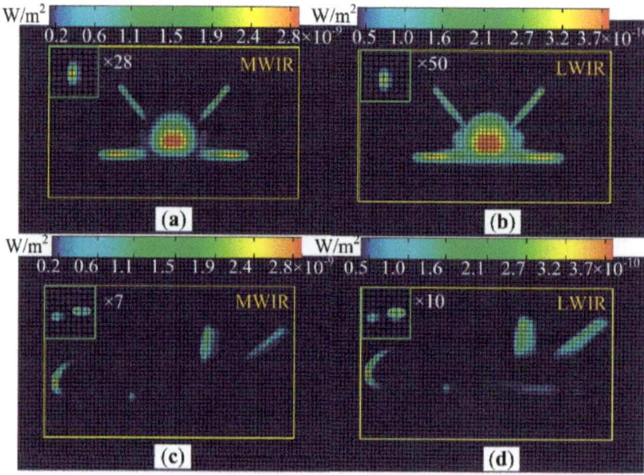

Figure 19. TIR images of two typical observation–angles for the 70-km case: (**a**) MWIR at $\varphi = 0°$, $\theta = 90°$; (**b**) LWIR at $\varphi = 0°$, $\theta = 90°$; (**c**) MWIR at $\varphi = 135°$. $\theta = 90°$; (**d**) LWIR at $\varphi = 135°$, $\theta = 90°$.

Figure 20 shows the peak intensity profiles of the TIR image at $\alpha/2 = 0.01°$ for different trajectory points. All calculations were performed on the assumption that $R = H$. The corresponding images within the MWIR and LWIR bands are illustrated at the top and bottom of the figure, respectively.

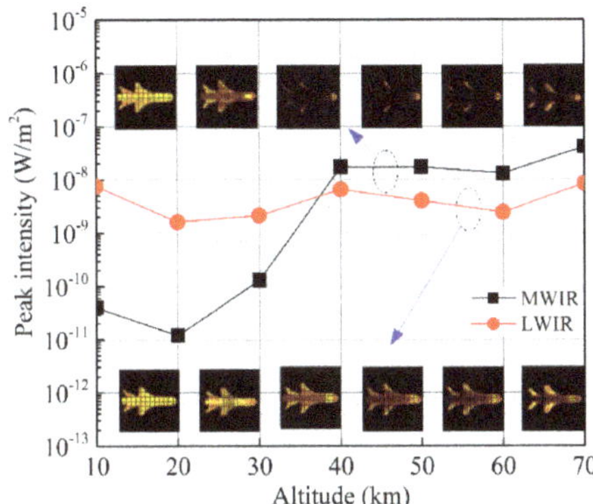

Figure 20. Profiles of peak intensity and TIR images for different computing points.

5. Discussion

The flow field contours of Figure 8 show that the high-temperature region of the surface occurred mainly at the windward surface of the nose and wing leading edges, and the temperature peak reached 2110 K in the 70-km case and 560 K in the 30-km case. In fluid regions, a distinct thermal nonequilibrium effect appeared in the case of 70 km, but these regions, appearing in the shock layer around the nose, were small. The mass fraction of NO generated by air dissociation was as high as 0.35. In the 30-km case, the aerodynamic temperature was drastically reduced in comparison with the 70-km case, and the flow was in thermal equilibrium.

In the top-view observation, the radiation of both the overall wake flows and most parts of the shock layer could be observed. It was demonstrated that the peak intensity radiation of the gases occurred mainly at the 2.7-μm (H_2O), 4.3-μm (CO_2,) and 5.3-μm (NO) bands. The spectrum of the surface radiation was smooth, and its peak intensity could be found around the short-wavelength region. A comparison of the radiation intensity between in Figures 10a and 10b indicates that the gas radiance was at least one order of magnitude lower than that of the surface.

From Figure 11, it can be found that the MWIR radiance was lower than the LWIR for the 30-km case, but this phenomenon was reversed for the 70-km case. These two computational cases were significantly different for the MWIR radiance. There were four high-intensity areas in the case of 30 km and two in the 70-km case. This can be explained by the fact that the peak wavelength of the surface emission (in accordance with Planck's law of gray-body radiation) moved toward the shorter wavelength as the temperature increased. The peak intensity did not occur in the front ($\theta = 90°$, $\varphi = 0°$) or top ($\theta = 0°$, $\varphi = 90°$) view, but in the oblique-side ($\theta = 22.5°$, $\varphi = 67.5°$ or $\varphi = 337.5°$) view. It can be observed that the two profiles intersected at the altitude of 35 km, which was the characteristic altitude H_c that separated the LWIR strong-emission regime (SER) and the MWIR SER.

In Figure 13, several phenomena can clearly be observed. Firstly, the MDR of the LWIR band was larger than that of the MWIR band at altitudes below 50 km (B-zone), which was a strong receiving regime (SRR) in the LWIR band, compared to the results shown in Figure 12, in which the characteristic altitude was shifted back by 15 km. Secondly, the target could not be detected at altitudes below 30 km using the MWIR band, which was a blind region (A-zone). In this figure, the gray dotted line indicates that the MDR was below the flight altitude H. Thirdly, there was an equivalent zone between 50 km and 60 km (C-zone) where the MDR was almost identical in two bands. Lastly, the MDR of the

LWIR band decreased after 60 km in altitude, resulting in the presence of an SRR in the MWIR band. This phenomenon was related to radiance features and atmospheric attenuation.

It can be observed from Figure 16 that the MDR increment had typical characteristics. In the 30-km case, the MDR increments for most angles were approximately distributed on an equivalent flat plane except for a small fluctuation. Figure 16a,c show a heel-shaped distribution due to a low MDR increment in the back view. This indicates that an increase in the detectivity could improve the MDR in low-altitude cases. In the 70-km case, Figure 16b shows a crater-shaped distribution of the MWIR band, which was significantly different from Figure 16a. The values in the center region were larger than those in the marginal regions of the contour. For the LWIR band in Figure 16d, there was a three-peak shape, which reveals that the increase in the detectivity could greatly improve the MDR in the back and front views for high-altitude cases.

From Figure 18, the mesh clipping treatment demonstrated an obvious improvement in the imaging. These images also show that the TIR distributions in the 70-km and 30-km cases were both significantly different, and the peak intensity at 70 km was at least one order of magnitude higher than that at 30 km. It is demonstrated in Figure 19 that a smaller FOV could contribute toward capturing the TIR characteristics, but it required a more sensitive detectivity due to the reduction in TIR intensity for a large FOV. It can be seen from Figure 20 that the peak intensity difference in the LWIR images was within one order of magnitude, whereas the difference varied greatly for the MWIR images. Furthermore, there was an intersection between the two profiles that was attributable to the detection distance and the target's radiance. In addition, TIR features of the nose and flanges became pronounced as the altitude increased. For the LWIR images, the difference in intensity distribution was small above altitudes of 40 km.

6. Conclusions

To examine the detection range and TIR images of an Earth entry vehicle, a complete numerical model was developed by analyzing a ground-based IR detection system and the physical mechanism of the TIR radiation. The proposed model was established considering optical radiative properties, optics propagation, atmospheric attenuation, and TIR arrangements in the pixels. Computer simulations were performed using known parameters for flight conditions and the IR detection system. The simulation results indicated that the radiance was strongly dependent on the observation angle and the spectral band. For the MWIR and LWIR bands, there was a characteristic altitude at which a strong-emission regime was noted. The MDR increased and the characteristic altitude decreased as the detectivity of the detector increased. The improvement in the detectivity could increase the MDR approximately linearly at most observation angles of low altitudes, but the MDR could be greatly improved in high-altitude cases. The TIR images showed that the mesh clipping treatment led to an obvious improvement in the TIR distribution. For the same detection conditions, the difference in the peak intensity for different trajectory points was at least one order of magnitude in scale. In addition, a smaller FOV could contribute toward capturing the TIR characteristics, but it required more sensitive detectivity due to the reduction in TIR intensity. The MWIR TIR features became more pronounced as the altitude increased, and those in the LWIR images were more suitable for detecting the aircraft's configuration.

In further work, a sensitivity study and an uncertainty estimate of the numerical simulation should be carried out. Also, a more refined photodetector model should be used for evaluations of the detectivity of the target, and the effect of weather conditions on infrared optical observability should be considered in future work.

Author Contributions: Conceptualization, Q.N. and X.M.; methodology, Q.N.; software, Q.N.; validation, Q.N., X.M., and Z.H.; writing—original draft preparation, Q.N.; writing—review and editing, Q.N., Z.H., and S.D.; project administration, Q.N.; funding acquisition, S.D.

Funding: This research was funded by the National Nature Science Foundation of China, grant number 51576054.

Acknowledgments: We are grateful to Zhenhua Wang for help with programming.

Conflicts of Interest: The authors declare no conflict of interest.

Nomenclature

A_0	pupil area of the objective lens system, m^2
A_d	pixel area of the detector, m^2
$A_{i,j,k}$	visible area of the kth surface element in the $i \times j$ pixel, m^2
A_t	effective radiation area of the target surface, m^2
c	speed of light, 2.99979×10^8 m/s
C_1, C_2	first and second radiation constants
d	degeneracy factors for state
$D^*(\lambda)$	normalized system detectivity
E_l	energy of the lower state
Δf	frequency bandwidth of the detector circuitry
F_{rot}	term value of rotational state
g	photoconductive gain
G_{vib}	term value of vibrational state
h	Planck constant, $6.6206896 \times 10^{-34}$ J·s
k_B	Boltzmann constant, $1.38064852 \times 10^{-23}$ J·K^{-1}
I	radiation intensity, W/(sr·m^2·µm)
M	number of segments in optical path
N	number density of species
$\mathbf{n}_i$	outward normal of the target surface element A_i
P	spectrum intensity arrived at the detector, W/(sr·µm)
$Q(T)$	partition function
R	distance between the target and the detector, m
q	irradiance received by each detector pixel, W/m^2
s	position
$\mathbf{s}$	optical path vector
$S(T_{ref})$	line intensity under the standard condition
$V_{i,p}$	pth vertex of the element A_i
Greek	
η	wave number, cm^{-1}
Φ	line shape function
ε	emissivity
λ	wavelength, µm
$\tau(\lambda,R)$	atmospheric transmittance with a distance of R
$\tau_0(\lambda)$	spectral transmittance of the optical system
Subscript	
u, l	upper and lower limits of spectral band
tar	target
bg	background
a	atmospheric air
s	surface of aircraft

Abbreviations

ASL	above sea level
AOA	angle of attack
CFD	computational fluid dynamics
FVM	finite volume method
FOV	field of view
HYTHIRM	hypersonic thermodynamic IR measurements
LOS	line-of-sight

LWIR	long-wavelength infrared
MDR	maximum detecting range
non-LTE	local thermodynamic nonequilibrium
NEFD	noise equivalent flux density
OTV	orbital test vehicle
RTE	radiative transfer equation
TIR	thermal infrared
SRC	sample return capsule
SNR	signal-to-noise ratio
SRR	strong receiving region
STS	space transportation system

References

1. Sziroczak, D.; Smith, H. A review of design issues specific to hypersonic flight vehicles. *Prog. Aerosp. Sci.* **2016**, *84*, 1–28. [CrossRef]
2. Grantz, A. X-37B orbital test vehicle and derivatives. In Proceedings of the AIAA SPACE 2011 Conference & Exposition, Long Beach, CA, USA, 27–29 September 2011; p. 7315.
3. Stark, B.; Smith, B.; Chen, Y. Survey of thermal infrared remote sensing for Unmanned Aerial Systems. In Proceedings of the 2014 International Conference on Unmanned Aircraft Systems (ICUAS), Orlando, FL, USA, 27–30 May 2014; pp. 1294–1299.
4. Gong, M.; Guo, R.; He, S.; Wang, W. IR radiation characteristics and operating range research for a quad-rotor unmanned aircraft vehicle. *Appl. Opt.* **2016**, *55*, 8757–8762. [CrossRef]
5. Niu, Q.; He, Z.; Dong, S. Prediction of shock-layer ultraviolet radiation for hypersonic vehicles in near space. *Chin. J. Aeronaut.* **2016**, *29*, 1367–1377. [CrossRef]
6. Niu, Q.; Yuan, Z.; Dong, S.; Tan, H. Assessment of nonequilibrium air-chemistry models on species formation in hypersonic shock layer. *Int. J. Heat Mass Transf.* **2018**, *127*, 703–716. [CrossRef]
7. Bonin, J.; Mundt, C. Full Three-Dimensional Monte Carlo Radiative Transport for Hypersonic Entry Vehicles. *J. Spacecr. Rocket.* **2018**, *56*, 1–9. [CrossRef]
8. Rao, A.G.; Mahulikar, S.P. Effect of atmospheric transmission and radiance on aircraft infared signatures. *J. Aircr.* **2005**, *42*, 1046–1054. [CrossRef]
9. Mahulikar, S.P.; Sonawane, H.R.; Rao, G.A. Infrared signature studies of aerospace vehicles. *Prog. Aerosp. Sci.* **2007**, *43*, 218–245. [CrossRef]
10. Baranwal, N.; Mahulikar, S.P. Aircraft engine's infrared lock-on range due to back pressure penalty from choked convergent nozzle. *Aerosp. Sci. Technol.* **2014**, *39*, 377–383. [CrossRef]
11. Pan, X.; Wang, X.; Wang, R.; Wang, L. Infrared radiation and stealth characteristics prediction for supersonic aircraft with uncertainty. *Infrared Phys. Technol.* **2015**, *73*, 238–250. [CrossRef]
12. Huang, W.; Ji, H. Effect of environmental radiation on the long-wave infrared signature of cruise aircraft. *Aerosp. Sci. Technol.* **2016**, *56*, 125–134. [CrossRef]
13. Beier, K.; Gemperlein, H. Simulation of infrared detection range at fog conditions for enhanced vision systems in civil aviation. *Aerosp. Sci. Technol.* **2004**, *8*, 63–71. [CrossRef]
14. Wang, K.; Dickinson, R.E. Global atmospheric downward longwave radiation at the surface from ground-based observations, satellite retrievals, and reanalyses. *Rev. Geophys.* **2013**, *51*, 150–185. [CrossRef]
15. Horvath, T.J.; Cagle, M.F.; Gibson, D. Remote observations of reentering spacecraft including the space shuttle orbiter. *IEEE Aerosp. Conf.* **2013**. [CrossRef]
16. Spisz, T.S.; Taylor, J.C.; Kennerly, S.W.; Osei-Wusu, K.; Gibson, D.M.; Horvath, T.J.; Zalameda, J.N.; Kerns, R.V.; Shea, E.J.; Mercer, C.D. Processing ground-based near-infrared imagery of space shuttle re-entries. In *Proceedings of Thermosense: Thermal Infrared Applications XXXIV*; Proc. SPIE: Baltimore, MD, USA, 2012; p. 83540G.
17. Horvath, T.J.; Rufer, S.J.; Schuster, D.M.; Mendeck, G.F.; Oliver, A.B.; Schwartz, R.J.; Verstynen, H.A.; Mercer, C.D.; Tack, S.; Ingram, B. Infrared Observations of the Orion Capsule During EFT-1 Hypersonic Reentry. In Proceedings of the AIAA Aviation and Aerospace Forum and Exposition, Washington, DC, USA, 13 June 2016; pp. 1–23.

18. Schuster, D.M.; Horvath, T.J.; Schwartz, R.J. *Remote Imaging of Exploration Flight Test-1 (EFT-1) Entry Heating Risk Reduction*; Report: NASA/TM-2016-219214; NASA Langley Research Center: Hampton, VA, USA, 1 June 2016.

19. Snively, J.B.; Taylor, M.J.; Jenniskens, P.; Winter, M.W.; Kozubal, M.J.; Dantowitz, R.F.; Breitmeyer, J. Near-Infrared Spectroscopy of Hayabusa Sample Return Capsule Reentry. *J. Spacecr. Rocket.* **2014**, *51*, 424–429. [CrossRef]

20. Huang, F.; Wang, Y.; Shen, X.; Li, G.; Yan, S. Analysis of space target detection range based on space-borne fisheye imaging system in deep space background. *Infrared Phys. Technol.* **2012**, *55*, 475–480. [CrossRef]

21. Zhao, Y.; Wu, P.; Sun, W. Calculation of infrared system operating distance by spectral bisection method. *Infrared Phys. Technol.* **2014**, *63*, 198–203. [CrossRef]

22. Ren, K.; Tian, J.; Gu, G.; Chen, Q. Operating distance calculation of ground-based and air-based infrared system based on Lowtran7. *Infrared Phys. Technol.* **2016**, *77*, 414–420. [CrossRef]

23. Suzuki, T.; Fujita, K.; Ando, K.; Sakai, T. Experimental study of graphite ablation in nitrogen flow. *J. Thermophys. Heat Transf.* **2008**, *22*, 382–389. [CrossRef]

24. Lemal, A.; Jacobs, C.; Perrin, M.-Y.; Laux, C.; Tran, P.; Raynaud, E. Prediction of nonequilibrium air plasma radiation behind a shock wave. *J. Thermophys. Heat Transf.* **2015**, *30*, 197–210. [CrossRef]

25. Mikula, D.; Holthaus, M.; Jensen, T.; Kubo, D.; Redgate, M. X-37 Flight Demonstrator system safety program and challenges. In Proceedings of the Space 2000 Conference and Exposition, Long Beach, CA, USA, 19–21 September 2000; p. 5073.

26. Sarma, G. Physico–chemical modelling in hypersonic flow simulation. *Prog. Aerosp. Sci.* **2000**, *36*, 281–349. [CrossRef]

27. Felton, M.; Gurton, K.; Pezzaniti, J.; Chenault, D.; Roth, L. Measured comparison of the crossover periods for mid-and long-wave IR (MWIR and LWIR) polarimetric and conventional thermal imagery. *Opt. Express* **2010**, *18*, 15704–15713. [CrossRef] [PubMed]

28. Roe, P.L. Approximate Riemann solvers, parameter vectors, and difference schemes. *J. Comput. Phys.* **1981**, *43*, 357–372. [CrossRef]

29. Yee, H.C. *On Symmetric and Upwind TVD Schemes*; NASA-TM-88325; NASA Langley Research Center: Washington, DC, USA, 1985.

30. Khodabakhsh, A.; Ramaiah-Badarla, V.; Rutkowski, L.; Johansson, A.C.; Lee, K.F.; Jiang, J.; Mohr, C.; Fermann, M.E.; Foltynowicz, A. Fourier Transform and Vernier Spectroscopy with a Mid-Infrared Optical Frequency Comb. *Proc. Opt. Nanostruct. Adv. Mater. Photovolt.* **2016**, *41*, 2541–2544.

31. Cadiou, E.; Dherbecourt, J.-B.; Raybaut, M.; Gorju, G.; Melkonian, J.-M.; Godard, A.; Pelon, J. Multiple-Species DIAL for H_2O, CO_2, and CH_4 remote sensing in the 1.98–2.30 µm range. In Proceedings of the Laser Applications to Chemical, Security and Environmental Analysis, Orlando, FL, USA, 25–28 June 2018; p. LTu5C.5.

32. Park, C. *Nonequilibrium Hypersonic Aerothermodynamics*; John Wiley & Sons: New York, NY, USA, 1989.

33. Gamache, R.R.; Roller, C.; Lopes, E.; Gordon, I.E.; Rothman, L.S.; Polyansky, O.L.; Zobov, N.F.; Kyuberis, A.A.; Tennyson, J.; Yurchenko, S.N. Total internal partition sums for 166 isotopologues of 51 molecules important in planetary atmospheres: Application to HITRAN2016 and beyond. *J. Quant. Spectrosc. Radiat. Transf.* **2017**, *203*, 70–87. [CrossRef]

34. Rothman, L.; Gordon, I.; Barber, R.; Dothe, H.; Gamache, R.; Goldman, A.; Perevalov, V.; Tashkun, S.; Tennyson, J. HITEMP, the high-temperature molecular spectroscopic database. *J. Quant. Spectrosc. Radiat. Transf.* **2010**, *111*, 2139–2150. [CrossRef]

35. Banwell, C.N.; McCash, E.M. *Fundamentals of Molecular Spectroscopy*; McGraw-Hill: New York, NY, USA, 1994; Volume 851.

36. Sparrow, E.M. *Radiation Heat Transfer*; Routledge: New York, NY, USA, 2018.

37. Olivero, J.J.; Longbothum, R. Empirical fits to the Voigt line width: A brief review. *J. Quant. Spectrosc. Radiat. Transf.* **1977**, *17*, 233–236. [CrossRef]

38. Niu, Q.; Yang, S.; He, Z.; Dong, S. Numerical study of infrared radiation characteristics of a boost-gliding aircraft with reaction control systems. *Infrared Phys. Technol.* **2018**, *92*, 417–428. [CrossRef]

39. Fetter, S.; Sessler, A.M.; Cornwall, J.M.; Dietz, B.; Frankel, S.; Garwin, R.L.; Gottfried, K.; Gronlund, L.; Lewis, G.N.; Postol, T.A. *Countermeasures: A Technical Evaluation of the Operational Effectiveness of the Planned US National Missile Defense System*; Union of Concerned Scientist: Cambridge, MA, USA, 2000.

40. Berk, A.; Bernstein, L.S.; Robertson, D.C. *MODTRAN: A Moderate Resolution Model for LOWTRAN*; Spectral Sciences Inc.: Burlington, MA, USA, 1987.
41. Qinglin, N.; Zhichao, Y.; Biao, C.; Shikui, D. Infrared radiation characteristics of a hypersonic vehicle under time-varying angles of attack. *Chin. J. Aeronaut.* **2019**, *32*, 861–874.
42. Ferrero, P.; D'Ambrosio, D. A numerical method for conjugate heat transfer problems in hypersonic flows. In Proceedings of the 40th Thermophysics Conference, Seattle, DC, USA, 23–26 June 2008; p. 4247.
43. Mallet, M.; Periaux, J.; Rostand, P.; Stoufflet, B. Validation of aerodynamic simulation methods for Hermes spaceplane and future hypersonic vehicles. In Proceedings of the 4th Symposium on Multidisciplinary Analysis and Optimization, Cleveland, OH, USA, 21–23 September 1992.
44. Hao, J.; Wang, J.; Lee, C. Numerical study of hypersonic flows over reentry configurations with different chemical nonequilibrium models. *Acta Astronaut.* **2016**, *126*, 1–10. [CrossRef]
45. Levin, D.A.; Candler, G.V.; Limbaugh, C.C. Multispectral shock-layer radiance from a hypersonic slender body. *J. Thermophys. Heat Transf.* **2000**, *14*, 237–243. [CrossRef]
46. Ozawa, T.; Garrison, M.B.; Levin, D.A. Accurate molecular and soot infrared radiation model for high-temperature flows. *J. Thermophys. Heat Transf.* **2007**, *21*, 19–27. [CrossRef]
47. Richter, R.; Fries, J. Radiometric analysis of infrared sensor performance. *Appl. Opt.* **1988**, *27*, 4771–4776. [CrossRef] [PubMed]
48. Alexeenko, A.; Gimelshein, N.; Levin, D.; Collins, R.; Rao, R.; Candler, G.; Gimelshein, S.; Hong, J.; Schilling, T. Modeling of flow and radiation in the Atlas plume. *J. Thermophys. Heat Transf.* **2002**, *16*, 50–57. [CrossRef]
49. Qinglin, N.; Zhihong, H.; Shikui, D. IR radiation characteristics of rocket exhaust plumes under varying motor operating conditions. *Chin. J. Aeronaut.* **2017**, *30*, 1101–1114.
50. Niu, Q.; Duan, X.; Meng, X.; He, Z.; Dong, S. Numerical analysis of point-source infrared radiation phenomena of rocket exhaust plumes at low and middle altitudes. *Infrared Phys. Technol.* **2019**, *99*, 28–38. [CrossRef]
51. Shao, C.; Nie, L.; Chen, W. Analysis of weakly ionized ablation plasma flows for a hypersonic vehicle. *Aerosp. Sci. Technol.* **2016**, *51*, 151–161. [CrossRef]
52. Coudrain, C.; Bernhardt, S.; Caes, M.; Domel, R.; Ferrec, Y.; Gouyon, R.; Henry, D.; Jacquart, M.; Kattnig, A.; Perrault, P. SIELETERS, an airborne infrared dual-band spectro-imaging system for measurement of scene spectral signatures. *Opt. Express* **2015**, *23*, 16164–16176. [CrossRef]
53. Shibata, Y.; Nagasawa, C.; Abo, M. Development of 1.6 μm DIAL using an OPG/OPA transmitter for measuring atmospheric CO_2 concentration profiles. *Appl. Opt.* **2017**, *56*, 1194–1201. [CrossRef]
54. Meng, L.; Fix, A.; Wirth, M.; Høgstedt, L.; Tidemand-Lichtenberg, P.; Pedersen, C.; Rodrigo, P.J. Upconversion detector for range-resolved DIAL measurement of atmospheric CH_4. *Opt. Express* **2018**, *26*, 3850–3860. [CrossRef]

MDPI

St. Alban-Anlage 66

4052 Basel

Switzerland

Tel. +41 61 683 77 34

Fax +41 61 302 89 18

www.mdpi.com

Remote Sensing Editorial Office

E-mail: remotesensing@mdpi.com

www.mdpi.com/journal/remotesensing